Lecture Notes in Artificial Intelligence 9384

Subseries of Lecture Notes in Computer Science

More information about this series at http://www.springer.com/series/1244

Floriana Esposito · Olivier Pivert
Mohand-Saïd Hacid · Zbigniew W. Raś
Stefano Ferilli (Eds.)

Foundations
of Intelligent Systems

22nd International Symposium, ISMIS 2015
Lyon, France, October 21–23, 2015
Proceedings

 Springer

Editors
Floriana Esposito
Computer Science
University of Bari
Bari
Italy

Olivier Pivert
Enssat
Lannion
France

Mohand-Saïd Hacid
LISI-UFR d'Informatique
Université Claude Bernard Lyon 1
Villeurbanne Cedex
France

Zbigniew W. Raś
University of North Carolina
Charlotte, NC
USA

Stefano Ferilli
Dipartimento di Informatica
Università degli Studi di Bari
Bari
Italy

ISSN 0302-9743   ISSN 1611-3349 (electronic)
Lecture Notes in Artificial Intelligence
ISBN 978-3-319-25251-3   ISBN 978-3-319-25252-0 (eBook)
DOI 10.1007/978-3-319-25252-0

Library of Congress Control Number: 2015950448

LNCS Sublibrary: SL7 – Artificial Intelligence

Springer Cham Heidelberg New York Dordrecht London

Printed on acid-free paper

Springer International Publishing AG Switzerland is part of Springer Science+Business Media
(www.springer.com)

Preface

This volume contains the papers presented at ISMIS 2015: the 22nd International Symposium on Methodologies for Intelligent Systems held during October 21–23, 2015, in Lyon, France. The symposium was organized by members of the LIRIS laboratory at the Claude Bernard University, Lyon 1.

ISMIS is a conference series that started in 1986 and has developed into an established and prestigious conference for exchanging the latest research results in building intelligent systems. The scope of ISMIS represents a wide range of topics on applying artificial intelligence techniques to areas as diverse as decision support, automated deduction, reasoning, knowledge-based systems, machine learning, computer vision, robotics, planning, databases, information retrieval, etc. ISMIS provides a forum and a means for exchanging information for those interested purely in theory, those interested primarily in implementation, and those interested in specific research and industrial applications.

We want to express our special thanks to the Program Committee members and everyone who contributed at any level to the organization of ISMIS 2015. Also, special thanks to our invited speakers Didier Dubois (IRIT-CNRS, University of Toulouse, France), Thomas Lukasiewicz (Oxford University, UK; TU Vienna, Austria), and Marie-Christine Rousset (University of Grenoble, LIG, France). We would like to thank every author who submitted a paper to ISMIS 2015 and finally the team of EasyChair, without whose free software the handling of the submissions and editing of the proceedings could not have been managed so smoothly by a small group of people. Last but not the least, we thank Alfred Hofmann of Springer for his continuous support.

October 2015

Floriana Esposito
Olivier Pivert
Mohand-Saïd Hacid
Zbigniew W. Raś
Stefano Ferilli

Organization

The symposium was organized by members of the LIRIS laboratory at the Claude Bernard University, Lyon 1, France.

Symposium Chairs

Mohand-Saïd Hacid	LIRIS, Claude Bernard University, Lyon, France
Zbigniew W. Raś	University of North Carolina, Charlotte, USA; Warsaw University of Technology, Poland

Program Co-chairs

Floriana Esposito	University of Bari, Italy
Olivier Pivert	IRISA, University of Rennes 1, France

Proceedings Chair

Stefano Ferilli	University of Bari, Italy

Steering Committee

Alexander Felfernig	Graz University of Technology, Austria
Andrzej Skowron	University of Warsaw, Poland
Dominik Ślęzak	Infobright Inc., Canada; University of Warsaw, Poland
Henning Christiansen	Roskilde University, Roskilde, Denmark
Henryk Rybinski	Warsaw University of Technology, Poland
Jaime Carbonell	CMU, USA
Jan Rauch	University of Economics, Prague, Czech Republic
Jiming Liu	Hong Kong Baptist University, Hong Kong, SAR China
Juan Carlos Cubero	University of Granada, Granada, Spain
Li Chen	Hong Kong Baptist University, Hong Kong, SAR China
Lorenza Saitta	University of Piemonte Orientale, Italy
Maria Zemankova	NSF, USA
Marzena Kryszkiewicz	Warsaw University of Technology, Poland
Petr Berka	University of Economics, Prague, Czech Republic
Tapio Elomaa	Tampere University of Technology, Finland
Troels Andreasen	Roskilde University, Roskilde, Denmark
Zbigniew W. Raś	University of North Carolina, Charlotte, USA; Warsaw University of Technology, Poland

Program Committee

Luigia Carlucci Aiello	Sapienza University, Rome, Italy
Bernd Amann	Pierre et Marie Curie University, LIP6, Paris, France
Aijun An	York University, Canada
Annalisa Appice	University of Bari, Italy
Alexandra Balahur	European Commission, Italy
Maria Martín Bautista	University of Granada, Spain
Ladjel Bellatreche	ENSMA, Poitiers, France
Salima Benbernou	University of Paris V, France
Marenglen Biba	University of New York Tirana, Albania
Maria Bielikova	Slovak University of Technology, Slovak Republic
Gloria Bordogna	CNR, Italy
Ivan Bratko	University of Ljubljana, Slovenia
Francois Bry	University of Munich, Germany
Henrik Bulskov	Roskilde University, Denmark
Michelangelo Ceci	University of Bari, Italy
Jianhua Chen	Louisiana State University, USA
Lina Clover	SAS Institute Inc., USA
Luca Console	University of Turin, Italy
Emmanuel Coquery	University of Lyon 1, France
Bruno Cremilleux	University of Caen, France
Alfredo Cuzzocrea	University of Calabria, Italy
Nicola Di Mauro	University of Bari, Italy
Christoph F. Eick	University of Houston, USA
Peter Eklund	University of Wollongong, Australia
Nicola Fanizzi	University of Bari, Italy
Cécile Favre	University of Lyon 2, France
Sébastien Ferré	University of Rennes 1, IRISA, France
Daniel Sanchez Fernandez	University of Granada, Spain
Laura Giordano	University of Piemonte Orientale, Italy
Jacek Grekow	Białystok University of Technology, Poland
Jerzy Grzymala-Busse	University of Kansas, USA
Hakim Hacid	Zayed University, UAE
Allel Hadjali	ENSMA, Poitiers, France
Mirsad Hadzikadic	University of North Carolina, Charlotte, USA
Shoji Hirano	Shimane Medical University, Japan
Lothar Hotz	University of Hamburg, Germany
Nathalie Japkowicz	University of Ottawa, Canada
Matthias Jarke	RWTH Aachen, Germany
Janusz Kacprzyk	Polish Academy of Sciences, Poland
Mieczyslaw Kłopotek	Polish Academy of Sciences, Poland
Bożena Kostek	Gdansk University of Technology, Poland
Patrick Lambrix	Linkoping University, Sweden
Anne Laurent	University of Montpellier 2, LIRMM, France

Additional Reviewers

Ameeta Agrawal
Flavien Balbo
Martin Dimkowski
Massimo Guarascio
Corrado Loglisci

Aleksandra Rashkovska
Ettore Ritacco
Arnaud Soulet
Nouredine Tamani
Daniele Theseider Dupré

Andrea Vanzo
Mario Vento
Antonio Vergari

Invited Talks

The Basic Principles of Information Fusion and Their Instantiations in Various Uncertainty Representation Frameworks

Didier Dubois[1,2], Weiru Liu[2], Henri Prade[1] and Jianbing Ma[3]

[1] IRIT, Université Paul Sabatier, 31062 Toulouse Cedex 9, France
[2] School of Electronics, Electrical Engineering and Computer Science,
Queen's University Belfast, Belfast BT7 1NN, UK
[3] School of Design, Engineering and Computing,
Bournemouth University, Bournemouth, BH12 5BB, UK

Information fusion is a specific aggregation process which aims to extract truthful knowledge out of information coming from various sources. This topic is relevant in many areas: expert opinion fusion in risk analysis, image fusion in computer vision, sensor fusion in robotics, database merging, and so forth. Historically the problem is very old. It lies at the origin of probability theory whose pioneers in the XVIIth century were concerned by merging unreliable testimonies at courts of law. Then, this problem fell into oblivion with the development of statistics in the late XVIIIth century. It was revived in the late XXth century in connection with the widespread use of computers, and the necessity of dealing with large amounts of data coming from different sources, as well as the renewed interest toward processing human-originated information, and the construction of autonomous artifacts that sense their environment and reason with uncertain and inconsistent inputs.

Information fusion is inescapably related to the issue of uncertainty modeling. Indeed, the fact that pieces of information often come from several sources results in conflicts to be solved, as inconsistency threatens in such an environment. The presence of incomplete, unreliable and inconsistent information leads to uncertainty, and the necessity of coping with it, so as make the best of what is available, while discarding the wrong. This is the role of information fusion.

There are many approaches and formats to model information, and several uncertainty theories. The information fusion problem has been discussed in each of these settings almost independently of the other ones. Sometimes, dedicated principles have been stated in order to characterize the specific features of the fusion process in the language of each particular formal setting. Several fusion strategies exist according to the various settings. These strategies share some commonalities but may differ from each other in some aspects due to their specific representation formats (for instance, symbolic vs. numerical).

This work takes an inclusive view of the current available properties from different theories and investigates the common laws that *must be* followed by these fusion strategies. We argue that some properties are mandatory and some are facultative only. The latter can be useful in certain circumstances, or to speed up computation time. It is interesting to notice that although each requested property looks intuitively reasonable on its own, they can be inconsistent when put together. This happens in the problem of merging preferences from several individuals modelled by complete preorderings (Arrow impossibility theorem). However the core mandatory properties of information fusion we propose are often globally consistent. We present general features of what can be called an information item. Such features can be extracted from information items in each representation framework.

The aim of the work is to lay bare the specific nature of the information fusion problem. This general analysis yields a better understanding of what fusion is about and how an optimal fusion strategy (operator) can be designed. In particular, information fusion differs from preference aggregation, whose aim is to find a good compromise between several parties. Noticeably, while the result of information fusion should be consistent with what reliable sources bring about, a good compromise in a multiagent choice problem may turn out to be some proposal no party proposed in the first stand. So while they share some properties and methods, we claim that information fusion and preference aggregation do not obey exactly the same principles.

We also wish to show the deep unity of information fusion methods, beyond the particulars of each representation setting. To this aim, we look at special characteristics of each theory and what becomes of fusion principles, what are the fusion rules in agreement with these principles. We check whether known fusion rules in each theory comply with general postulates of information fusion. We explain how these basic properties can be written in different representation settings ranging from numerical to logic-based representations. These comparisons demonstrate that the proposed core properties truly reflect the nature of fusion in different settings.

We instantiate our principles on various representation settings such as

- The crudest representation of an information item, namely a set of possible values. When such a set basically excludes impossible values, we show that our setting characterizes the method of maximal consistent subsets [5].
- The case of merging propositional belief bases, for which a set of postulates, due to Konieczny and Pino-Perez, exists [4]. It comes down to merging sets of most plausible values.
- the fusion of plausibility rankings of possible values, going from ordinal representations to numerical ones in terms of fuzzy sets representing possibility distributions [3].
- Combination rules for belief functions [6], especially Dempster rule of combination.
- Postulates for merging imprecise probabilities proposed by Peter Walley [7], in the light of our general approach.

Preliminary and partial views of this work were presented in two conferences [1, 2]. A long paper is in preparation.

References

1. Dubois, D., Liu, W., Ma, J., Prade., H.: A principled discussion of information combination rules in different representation settings. In International Conference of Soft Computing and Pattern Recognition (SoCPaR), Dalian, China. Proceedings, pp. 446–451. IEEE (2011)
2. Dubois, D., Liu, W., Ma, J., Prade., H.: Toward a general framework for information fusion. In: Torra, V., Narukawa, Y., Navarro-Arribas, G., Megias, D. (eds.) MDAI 2013. LNCS, vol. 8234, pp. 37–48. Springer, Heidelberg (2013)
3. Dubois, D., Prade, H., Yager, R. R.: Merging fuzzy information. In: Bezdek,J.C., Dubois, D., Prade, H. (eds.) Fuzzy Sets in Approximate Reasoning and Information Systems, The Handbooks of Fuzzy Sets Series, pp. 335–401. Kluwer Academic Publishers, Dordrecht, (1999)
4. Konieczny, S., Pino-Perez, R.: Logic based merging. J. Phil. Log. **40**(2), 239–270 (2011)
5. Rescher, N., Manor, R.: On inference from inconsistent premises. Theor. Decis. **1**, 179–219 (1970)
6. Smets, P.: Analyzing the combination of conflicting belief functions. Inf. Fusion. **8**(4), 387–412 (2007)
7. Walley, P.: The elicitation and aggregation of beliefs. Tech. rep. University of Warwick, (1982)

Uncertainty in the Semantic Web

Thomas Lukasiewicz

Department of Computer Science, University of Oxford, Oxford, UK
thomas.lukasiewicz@cs.ox.ac.uk

During the recent decade, the *Semantic Web* [1–3] has attracted much attention, both from academia and industry, and is commonly regarded as the next step in the evolution of the World Wide Web. It aims at an extension of the current Web by standards and technologies that help machines to understand the information on the Web so that they can support richer discovery, data integration, navigation, and automation of tasks. The main ideas behind it are to add a machine-understandable "meaning" to Web pages, to use ontologies for a precise definition of shared terms in Web resources, to use KR technology for automated reasoning from Web resources, and to apply cooperative agent technology for processing the information of the Web.

The Semantic Web is divided into several hierarchical layers, including the *Ontology layer*, in the form of the *OWL Web Ontology Language* [4, 5]. OWL consists of three increasingly expressive sublanguages, namely, *OWL Lite*, *OWL DL*, and *OWL Full*. OWL Lite and OWL DL are essentially very expressive description logics with an RDF syntax. On top of the Ontology layer, there are sophisticated representation and reasoning capabilities for the *Rules*, *Logic*, and *Proof layers* of the Semantic Web.

A key requirement of the layered architecture of the Semantic Web is in particular to integrate the Rules and the Ontology layer. Here, it is crucial to allow for building rules on top of ontologies, that is, for rule-based systems that use vocabulary from ontology knowledge bases. Another type of combination is to build ontologies on top of rules, where ontological definitions are supplemented by rules or imported from rules. Both types of integration have been realized in recent hybrid integrations of rules and ontologies, called *description logic programs* (or *dl-programs*), which are of the form $KB = (L, P)$, where L is a description logic knowledge base, and P is a finite set of rules involving either queries to L in a loose integration or concepts and roles from L as unary resp. binary predicates in a tight integration.

However, classical ontology languages and description logics as well as formalisms integrating rules and ontologies are less suitable in all those domains where the information to be represented comes along with (*quantitative*) *uncertainty* and/or *vagueness* (or *imprecision*). For this reason, during the recent years, handling uncertainty and vagueness has started to play an important role in research towards the Semantic Web. A recent forum for approaches to uncertainty reasoning in the Semantic Web is the annual *International Workshop on Uncertainty Reasoning for the Semantic Web (URSW)* at the *International Semantic Web Conference (ISWC)*. There has also been a W3C Incubator Group on *Uncertainty Reasoning for the World Wide Web*. The research focuses especially on probabilistic and fuzzy extensions of description logics, ontology languages, and formalisms integrating rules and ontologies. Note that probabilistic formalisms allow to encode ambiguous information, such as "John is a student with the probability 0.7 and a teacher with the probability 0.3" (roughly, John is either a teacher or a student, but more likely a student), while fuzzy approaches allow to encode vague or imprecise information, such as "John is tall with the degree of truth 0.7" (roughly, John is quite tall). Formalisms for dealing with uncertainty and vagueness are especially applied in ontology mapping, data integration, information retrieval, and database querying. For example, some of the most prominent technologies for dealing with uncertainty are probably the ranking algorithms standing behind Web search engines. Other important applications are belief fusion and opinion pooling, recommendation systems, user preference modeling, trust and reputation modeling, and shopping agents. Vagueness and imprecision also abound in multimedia information processing and retrieval, and are an important aspect of natural language interfaces to the Web.

In the invited talk, I give an overview of some own recent extensions of description logics and description logic programs by probabilistic uncertainty and fuzzy vagueness. For a more detailed overview of extensions of description logics for handling uncertainty and vagueness in the Semantic Web, I also refer the reader to the survey [6].

Acknowledgments. This work was supported by the UK EPSRC grant EP/J008346/1 ("PrOQAW: Probabilistic Ontological Query Answering on the Web"), by the EU (FP7/ 2007-2013) ERC grant 246858 ("DIADEM"), and by a Yahoo! Research Fellowship.

References

1. Berners-Lee, T.: Weaving the Web. Harper, San Francisco (1999)
2. Berners-Lee, T., Hendler, J., Lassila, O.: The Semantic Web. Sci. Am. **284**(5), 34–43 (2001)
3. Fensel, D., Wahlster, W., Lieberman, H., Hendler, J. (eds.): Spinning the Semantic Web: Bringing the World Wide Web to Its Full Potential. MIT Press (2002)
4. W3C: OWL Web Ontology Language Overview (2004), W3C Recommendation, 10 February 2004. http://www.w3.org/TR/2004/REC-owl-features-20040210/
5. W3C: OWL 2 Web Ontology Language Document Overview (2012), W3C Recommendation, 11 December 2012. http://www.w3.org/TR/2012/REC-owl2-overview-20121211/http://www.w3.org/TR/2012/REC-owl2-overview-20121211/
6. Lukasiewicz, T., Straccia, U.: Managing uncertainty and vagueness in description logics for the Semantic Web. J. Web Semant. **6**(4) (2008) 291–308

Datalog Revisited for Reasoning
and Answering Queries on Linked Open Data

Marie-Christine Rousset

University of Grenoble, LIG, Grenoble, France
Marie-Christine.Rousset@imag.fr

In this presentation, we will describe a unifying framework for RDF ontologies and databases that we call *deductive RDF triplestores*. It consists in equipping RDF triplestores with Datalog inference rules. This rule language allows to capture in a uniform manner OWL constraints that are useful in practice, such as property transtivity or symmetry, but also *domain-specific* rules with practical relevance for users in many domains of interest.

We will illustrate the expressivity of this framework for modeling Linked Data applications and its genericity for developing inference algorithms. In particular, we will show how it allows to model the problem of data linkage in Linked Data as a reasoning problem on possibly decentralized data. We will also explain how it makes possible to efficiently extract expressive modules from Semantic Web ontologies and databases with formal guarantees, whilst effectively controlling their succinctness.

Experiments conducted on real-world datasets have demonstrated the feasibility of this approach and its usefulness in practice for data linkage, disambiguation and module extraction.

Contents

Machine Learning

Knowledge Representation, Semantic Web

Emotion Recognition, Music Information Retrieval

Network Analysis, Multi-agent Systems

Data Mining Methods

Data Mining with Histograms – A Case Study

Jan Rauch[(⊠)] and Milan Šimůnek

Faculty of Informatics and Statistics, University of Economics,
Prague Nám W. Churchilla 4, 130 67 Prague 3, Czech Republic
{rauch,simunek}@vse.cz

Abstract. Histograms are introduced as interesting patterns for data mining. An application of the procedure CF-Miner mining for various types of histograms is described. Possibilities of using domain knowledge in a process of mining interesting histograms are outlined.

1 Introduction

The goal of this paper is to introduce the procedure CF-Miner mining for interesting histograms. The CF-Miner procedure is one of the GUHA procedures implemented in the LISp-Miner system (http://lispminer.vse.cz). The GUHA method and the LISp-Miner system have been several times published; relevant papers are listed e.g. in [3]. However, there is no English paper describing the CF-Miner procedure, the only relevant publication is a chapter in the book [9] written in Czech. An additional goal is to outline a way how to deal with domain knowledge when mining for interesting histograms.

Main features of the CF-Miner procedure are introduced in Sect. 2. Possibilities of dealing with domain knowledge are outlined in Sect. 3. Conclusions, related research and further work are summarized in Sect. 4.

2 CF-Miner Procedure

2.1 Examples of Histograms CF-Miner Deals with

We use a data matrix *Entry* – a part of a medical data set STULONG (http://euromise.vse.cz/challenge2004/data/). The *Entry* data matrix has 1417 rows corresponding to particular patients and 64 columns corresponding to attributes of patients. We transformed the original column *BMI* (Body Mass Index) to an attribute *BMI* with seven equifrequency categories (i.e. intervals of original values defined such that the numbers of patients in particular intervals are as close as possible): *extra low, very low, lower, avg, higher, very high, extra high.* Frequencies of particular categories are 201, 200, ..., 213, see second row in Table 1 (there are eight missing values). Examples of histograms produced by the CF-Miner are in Fig. 1.

The work described here has been supported by funds of institutional support for long-term conceptual development of science and research at FIS of the University of Economics, Prague.

© Springer International Publishing Switzerland 2015
F. Esposito et al. (Eds.): ISMIS 2015, LNAI 9384, pp. 3–8, 2015.
DOI: 10.1007/978-3-319-25252-0_1

Table 1. Frequencies for the histogram $Hsg(BMI, \chi, Entry, Rel)$

Category	extra low	very low	lower	avg	higher	very high	extra high
number of rows of *Entry*	201	200	198	196	201	200	213
number of rows satisfying χ	12	16	19	23	32	44	70
relative frequency in % for χ	5.97	8.00	9.60	11.73	15.92	22.00	32.86

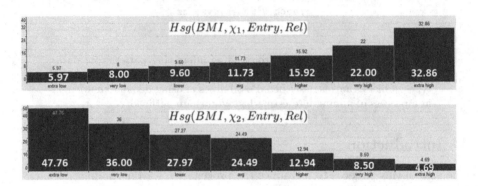

Fig. 1. Histograms $Hsg(BMI, \chi_1, Entry, Rel)$ and $Hsg(BMI, \chi_2, Entry, Rel)$

The histogram $Hsg(BMI, \chi_1, Entry, Rel)$ shows relative frequencies of categories of the attribute BMI concerning rows of data matrix $Entry$ which satisfy the Boolean attribute χ_1. It is $\chi_1 = Subscapular\langle22; 30\rangle \wedge Status(married)$. This means that the histogram $Hsg(BMI, \chi_1, Entry, Rel)$ shows a distribution of relative frequencies of categories of BMI among married patients with a skinfold above musculus subscapularis in the interval $\langle22; 30\rangle$ mm. All relevant frequencies are in Table 1. There are 201 rows belonging to the category *extra low*, 12 of them satisfy χ_1. The relative frequency in % for the category *extra low* is $5.97 = 100 \times \frac{12}{201}$ etc.

The histogram $Hsg(BMI, \chi_2, Entry, Rel)$ is created analogously, it holds $\chi_2 = Diastolic\langle75; 104\rangle \wedge Subscapular(< 16)$. $Diastolic\langle75; 104\rangle$ means that diastolic blood pressure of a patient is in the interval $\langle75; 104\rangle$ mmHg.

2.2   CF-Miner Input and Output

The histogram $Hsg(BMI, \chi_1, Entry, Rel)$ is interesting because the sequence 5.97, 8.00, 9.60, 11.73, 15.92, 22.00, 32.86 of relative frequencies in % is increasing. We can also say that all steps in this histogram are up. Similarly, all steps in the histogram $Hsg(BMI, \chi_2, Entry, Rel)$ are down.

We consider a histogram $\mathcal{H}$ as an expression $\approx A/\chi$ where A is an attribute with categories $a_1, \ldots, a_K$, χ is a Boolean attribute and a symbol $\approx$ defines a criterion of interestingness. A histogram $\approx A/\chi$ is interesting in a data matrix $\mathcal{M}$ if the criterion $\approx$ is satisfied in $\mathcal{M}$. The criterion $\approx$ is called a *CF-quantifier*. It is verified in a CF-Table $CF(A, \chi, \mathcal{M})$ for attribute A, condition χ and a data matrix $\mathcal{M}$, see Table 2. An expression $\mathcal{M}/\chi$ denotes a sub-matrix of $\mathcal{M}$ consisting of all rows of $\mathcal{M}$ satisfying χ.

Table 2. CF-Table $CF(A, \chi, \mathcal{M})$ for attribute A, condition χ and a data matrix $\mathcal{M}$

Frequencies of categories	a_1	$\ldots$	a_K
absolute frequencies in the data matrix $\mathcal{M}/\chi$	n_1	$\ldots$	n_K
absolute frequencies in the data matrix $\mathcal{M}$	m_1	$\ldots$	m_K
relative frequencies between $\mathcal{M}/\chi$ and $\mathcal{M}$	$v_1 = \frac{n_1}{m_1}$	$\ldots$	$v_K = \frac{n_K}{m_K}$

The attribute BMI has seven categories. The criterion of interestingness for the histogram $Hsg(BMI, \chi_1, Entry, Rel)$ is $v_1 < v_2 < \cdots < v_6 < v_7$. Couples $\langle v_1, v_2 \rangle, \ldots, \langle v_6, v_7 \rangle$ are *steps in a histogram of relative frequencies between* $\mathcal{M}/\chi$ and $\mathcal{M}$. We also say that the criterion of interestingness for the histogram $Hsg(BMI, \chi_1, Entry, Rel)$ is that there are six steps up. The criterion of interestingness for the histogram $Hsg(BMI, \chi_2, Entry, Rel)$ is that there are six steps down. There are also CF-quantifiers defined using characteristics of distribution of variables (skewness, asymmetry, variance, ...).

The CF-Miner procedure is based on the GUHA approach [3]. Input of the CF-Miner are an analysed data matrix $\mathcal{M}$ together with parameters defining a set of potentially interesting histograms $\approx A/\chi$. The procedure generates all relevant histograms $\approx A/\chi$ and tests interestingness of each of them. Output of the CF-Miner procedure consists of all interesting histograms $\approx A/\chi$. An example of input of the CF-Miner procedure is in Fig. 2.

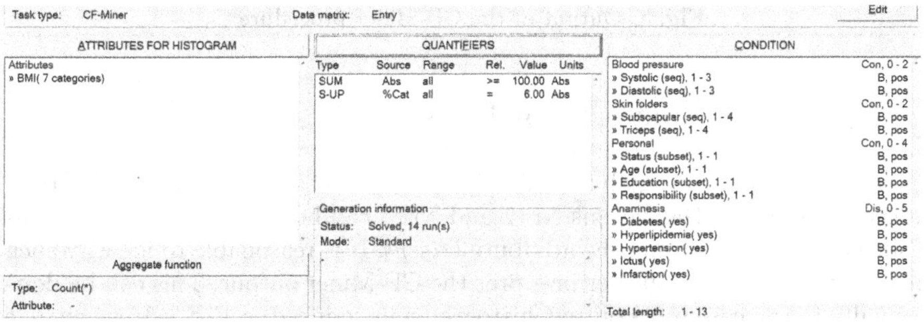

Fig. 2. Input parameters of the CF-Miner procedure

In the column **ATTRIBUTES FOR HISTOGRAM**, there is the attribute BMI which has seven categories. Thus we deal with histograms $\approx BMI/\chi$ verified in CF-table $CF(BMI, \chi, Entry)$ where K $= 7$ (see Table 2). A CF-quantifier $\approx^U_{100,6}$ is defined in the column **QUANTIFIERS**. The first row "SUM Abs All >= 100.00 Abs" means that there are at least 100 rows of the data matrix *Entry* satisfying the condition χ, i.e. $\Sigma_{i=1}^7 n_i \geq 100$. The second row "S-UP %Cat All = 6.00 Abs" means that there are six steps up in the histogram of relative frequencies between *Entry*$/\chi$ and *Entry*. The parameters defining a set of relevant conditions χ are given in the column **CONDITION**. Each condition χ is a conjunction of Boolean characteristics of groups of attributes *Blood pressure*, *Skin folders*,

Personal, and *Anamnesis*. The Boolean characteristics are defined as conjunctions and/or disjunctions of basic Booelan attributes, see e.g. [8]. The Boolean attribute *Subscapular*($\langle 22; 24 \rangle$, $\langle 24; 26 \rangle$, $\langle 26; 28 \rangle$, $\langle 28; 30 \rangle$) is an example of basic Boolean attribute created as a sequence of four categories $\langle 22; 24 \rangle, \ldots, \langle 28; 30 \rangle$) of the attribute *Subscapular*. We write it shortly as *Subscapular*$\langle 22; 30 \rangle$.

The task specified in Fig. 2 was solved in 51 seconds(PC with 2GB RAM and Intel T7200 processor at 2 GHz). 2 013 542 relevant histograms were generated and tested, 28 interesting histograms were found. The five histograms with the highest sum $\Sigma_{i=1}^{7} n_i$ are introduced in Fig. 3. The first histogram is $\approx_{100,6}^{U} BMI/Subscapular\langle 22; 30 \rangle \wedge Status(maried)$ presented in Fig. 1 as the histogram $Hsg(BMI, \chi_1, Entry, Rel)$.

Task type:	CF-Miner			
Task run				
Start 20.3.2015 17:39:46		Total time: 0h 0m 51s		
Number of verifications:	2013542			
Number of hypotheses:	28		Mode: Standard	<u>A</u>dd group De<u>l</u> gr

Actual group of hypotheses:		All hypotheses		
Hypotheses in group:	28	Shown hypotheses: 28	Highlighted: 0	

Nr.	Id	Sum	<u>H</u>ypothesis
1	3	216	**BMI / Subscapular**(*<22;30>*) **& Status**(*maried*)
2	6	208	**BMI / Triceps**(>= *13-14*) **& Status**(*maried*)
3	24	198	**BMI / Diastolic**(*<75;104>*) **& Subscapular**(*<22;30>*)
4	25	171	**BMI / Diastolic**(*<75;104>*) **& Subscapular**(*<22;30>*) **& Status**(*maried*)
5	7	161	**BMI / Triceps**(>= *15-17*)

Fig. 3. Output of the CF-Miner procedure

3   Applying Domain Knowledge

There are groups of mutually similar histograms in the output of the CF-Miner procedure in Fig. 3. Histograms Nr. 2 and Nr. 5 can be understood as a nucleus of such a group concerning the attribute *Triceps*. It is reasonable to identify such groups of histograms to better interpret the CF-Miner output. This can be done manually in the case of 28 output histograms introduced in Fig. 3. If we change the row "SUM Abs All >= 100.00 Abs" in the column QUANTIFIERS in Fig. 2 to "SUM Abs All >= 50.00 Abs", we also change the condition $\Sigma_{i=1}^{7} n_i \geq 100$ concerning CF-table in Table 2 to the condition $\Sigma_{i=1}^{7} n_i \geq 50$ and the CF-quantifier $\approx_{100,6}^{U}$ to the CF-quantifier $\approx_{50,6}^{U}$. This results to 147 output histograms, which can be hardly manually interpreted.

We outline how domain knowledge can be used to interpret resulting set of histograms. We apply principles introduced in [8] to interpret results of data mining with association rules. We use an item *Triceps* $\uparrow\uparrow$ *BMI* of domain knowledge meaning *"if Triceps (i.e. skinfold above musculus triceps) of a patient increases, then his/here BMI increases too"*. We denote it by Ω. We deal with a CF-quantifier $\approx_{B,6}^{U}$ defined by a condition $v_1 < v_2 < \cdots < v_6 < v_7 \wedge \Sigma_{i=1}^{7} n_i \geq B$ related to Table 2. We define a set $CONS(\Omega, \approx_{B,6}^{U})$ of histograms $\approx_{B,6}^{U} BMI/\chi$

which can be considered as consequences of Ω for $\approx^U_{B,6}$. A cooperation with a domain-expert is assumed.

In the first step we define a set $AC(\Omega, \approx^U_{B,6})$ of *atomic consequences* of Ω for $\approx^U_{B,6}$ as a set of all histograms $\approx^U_{B,6} BMI/Triceps(\alpha)$ where $\alpha \subseteq High$. $High = \{9, 10, 11, 12, 13-14, 15-17, 18-35\}$ is a set of categories of the attribute *Triceps* corresponding to higher skinfolds above musculus triceps. The remaining categories are $\{1-4, 5, 6, 7, 8\}$. Histogram $\approx^U_{100,6} BMI/\text{Triceps}(>= 15-17)$ (i.e. Nr. 5 in Fig. 3) is an example of atomic consequence of Ω (it can be written as $\approx^U_{100,6} BMI/\text{Triceps}(\geq 15)$).

Let us emphasize that *Triceps* $\uparrow\uparrow$ *BMI* says nothing about a shape of histograms $\approx^U_{B,6} BMI/Triceps(\alpha)$. However, a role of the domain expert is to guarantee that it is not surprising if such a histogram is increasing when $\alpha \subseteq High$.

In the second step, we create a set $AgC(\Omega, \approx^U_{B,6})$ of *agreed consequences of* Ω for $\approx^U_{B,6}$. A histogram $\approx^U_{B,6} BMI/(Triceps(\alpha) \wedge D_1(\delta_1) \wedge \dots D_u(\delta_u))$ belongs to $AgC(\Omega, \approx^U_{B,6})$ if $\approx^U_{B,6} BMI/Triceps(\alpha)$ is an atomic consequence of Ω, and if it is possible to agree that $\approx^U_{B,6} BMI/(Triceps(\alpha) \wedge D_1(\delta_1) \wedge \dots D_u(\delta_u))$ says nothing new in addition to $\approx^U_{B,6} BMI/Triceps(\alpha)$. This happens if attributes $D_1, \dots D_u$ have no influence on the relation of attributes BMI and *Triceps*. Histogram $\approx^U_{50,6} BMI/ \text{Triceps}(>= 13-14) \wedge \text{Status(married)}$ (i.e. Nr. 2 in Fig. 3) is an example of an agreed consequence of Ω (it can be written as $\approx^U_{50,6} BMI/Triceps(\geq 13) \wedge Status(married)$).

In the third step we define $CONS(\Omega, \approx^U_{B,6}) = AC(\Omega, \approx^U_{B,6}) \cup AgC(\Omega, \approx^U_{B,6})$. We have applied this approach to filtering out histograms – consequences of Triceps $\uparrow\uparrow$ BMI to results of the above introduced runs of the CF-Miner procedure with CF-quantifiers $\approx^U_{100,6}$ and $\approx^U_{50,6}$. We assume that, according to a domain expert, each of attributes used in the column CONDITION in Fig. 2 and distinct from *Triceps* has no influence on the relation of the attributes BMI and *Triceps*.

In the first run, the CF-quantifier $\approx^U_{100,6}$ was used resulting to 28 interesting histograms $\approx^U_{100,6} BMI/\chi$. Among them, there are eight histograms in a form $\approx^U_{100,6} BMI/\chi' \wedge Triceps(\alpha)$ where $\alpha \subseteq High$. Each of them can be considered as a consequence of Triceps $\uparrow\uparrow$ BMI. In addition, there are 20 histograms $\approx^U_{100,6} BMI/\chi$ such that the attribute *Triceps* does not occur in χ.

In the second run, the CF-quantifier $\approx^U_{50,6}$ was used resulting to 147 interesting histograms $\approx^U_{50,6} BMI/\chi$. Among them, there are 68 histograms $\approx^U_{50,6} BMI/\chi' \wedge Triceps(\alpha)$ where $\alpha \subseteq High$, which can be considered as consequences of *Triceps* $\uparrow\uparrow$ *BMI*. There are also 73 histograms $\approx^U_{100,6} BMI/\chi$ such that the attribute *Triceps* does not occur in χ. In addition, there are 6 histograms $\approx^U_{50,6} BMI/\chi' \wedge Triceps(\alpha)$ where $\alpha \not\subseteq High$. These 6 histograms should be further investigated.

Let us note that additional items of domain knowledge can be used to interpret remaining histograms $\approx^U_{100,6} BMI/\chi$ and $\approx^U_{100,6} BMI/\chi$ such that the attribute *Triceps* does not occur in χ. Expressions *Subscapular* $\uparrow\uparrow$ *BMI* and *Diastolic* $\uparrow\uparrow$ *BMI* are examples of such items of domain knowledge.

4   Conclusions

We have described an application of the procedure CF-Miner – one of the procedures implementing the GUHA approach to data mining [3]. There are still few experience with the CF-Miner procedure and additional applications in more areas are necessary.

We have also outlined a way how to use domain knowledge in interpretation of the CF-Miner results; this is inspired by [8]. There is no similar approach to dealing with domain knowledge in mining interesting histograms known to the authors. However, there are various approaches concerning dealing with domain knowledge in data mining, see e.g. [4–7]. Their detailed comparison with the presented approach is left as a further work.

An additional task is to clarify a relation of the CF-Miner procedure to methods for subgroups discovery [1] and contrast patterns discovery [2].

References

1. Atzmueller, M.: Subgroup discovery. WIREs Data Min. Knowl. Discov. **5**, 35–49 (2015)
2. Dong, G., Bailey, J.: Contrast Data Mining: Concepts, Algorithms, and Applications. Chapman and Hall/CRC, Boca Raton (2012)
3. Hájek, P., Holeňa, M., Rauch, J.: The GUHA method and its meaning for data mining. J. Comput. Syst. Sci. **76**, 34–48 (2010)
4. Jaroszewicz, S., Scheffer, T., Simovici, D.A.: Scalable pattern mining with Bayesian networks as background knowledge. Data Min. Knowl. Discov. **18**, 56–100 (2009)
5. Lavrac, N., et al.: The utility of background knowledge in learning medical diagnostic rules. Appl. Artif. Intell. **7**, 273–293 (1993)
6. Mansingh, G., Osei-Bryson, K.-M., Reichgelt, H.: Using ontologies to facilitate post-processing of association rules by domain experts. Inf. Sci. **181**, 419–434 (2011)
7. Phillips, J., Buchanan, B.G.: Ontology guided knowledge discovery in databases. In: Proc. First International Conference on Knowledge Capture, pp. 123–130. ACM, Victoria, British Columbia, Canada (2001)
8. Rauch, J., Šimůnek, M.: Learning association rules from data through domain knowledge and automation. In: Bikakis, A., Fodor, P., Roman, D. (eds.) RuleML 2014. LNCS, vol. 8620, pp. 266–280. Springer, Heidelberg (2014)
9. Rauch, J., Šimůnek M.: Knowledge Discovery in Databases, LISp-Miner and GUHA (in Czech) Economia, Praha (2015)

Discovering Variability Patterns for Change Detection in Complex Phenotype Data

Corrado Loglisci[1](✉), Bachir Balech[2], and Donato Malerba[1]

[1] Dipartimento di Informatica, Università degli Studi di Bari "Aldo Moro",
Bari, Italy
{corrado.loglisci,donato.malerba}@uniba.it
[2] Istituto di Biomembrane and Bioenergetica (IBBE), CNR, Bari, Italy
b.balech@ibbe.cnr.it

Abstract. The phenotype is the result of a genotype expression in a given environment. Genetic and eventually protein mutations and/or environmental changes may affect the biological homeostasis leading to a pathological status of a normal phenotype. Studying the alterations of the phenotypes on a temporal basis becomes thus relevant and even determinant whether considering the biological re-assortment between the involved organisms and the cyclic nature of the pandemic outbreaks. In this paper, we present a computational solution that analyzes phenotype data in order to capture statistically evident changes emerged over time and track their repeatability. The proposed method adopts a model of analysis based on time-windows and relies on two kinds of patterns, *emerging* patterns and *variability* patterns. The first one models the changes in the phenotype detected between time-windows, while the second one models the changes in the phenotype replicated over time-windows. The application to Influenza A virus H1N1 subtype proves the usefulness of our *in silico* approach.

1 Introduction

Phenotype is the set of observable traits of an organism, such as biochemical, physiological and morphological properties, and it is the result of complex factors, especially the influence of environmental factors and random variation on gene expression. Health and disease conditions of an organism have often a direct effect on its phenotype. Therefore, studying changes in phenotypes enable researchers to identify the main events occurring at DNA and protein levels which are responsible of immediate phenotypic responses, such as increased viral virulence or site/s-specific drug resistance. Often, these events are caused by molecular evolution and can occur together, thus they can be interesting for a better comprehension of the disease, but their eventual interaction, in combination with their heterogeneous nature, makes even more complex the investigation in laboratory. The analysis of phenotype data becomes of great importance given the socio-economical impact of the related diseases. In this scenario, the use of technologies able to handle the huge amount of phenotype data, model

© Springer International Publishing Switzerland 2015
F. Esposito et al. (Eds.): ISMIS 2015, LNAI 9384, pp. 9–18, 2015.
DOI: 10.1007/978-3-319-25252-0_2

the complex and heterogeneous aspects of the phenotypes and elicit possible interactions can play a determinant role.

In this paper, we focus on the analysis of phenotypes and propose a data mining method in order to capture statistically evident changes emerged over time and track their repeatability. The proposed method adopts a model of analysis based on time-windows and uses a frequent pattern mining framework as mean for abstracting and summarizing the data. This enables us to search changes as differences between frequent patterns. Since frequency denotes regularity, patterns can provide empirical evidence about real changes. Frequent patterns are discovered from the phenotype data collected by time-windows and thus they reflect co-occurrences of gene expression data, protein sequences and epidemiological impact, which are frequent in specific intervals of time. The changes which can emerge in this setting regard differences between the frequent patterns of two time-windows. In particular, the changes we are interested in correspond to variations of the frequency of the patterns occurred from a time-window to the next time-window. Not all the changes are considered, but only those which are replicated over time. We extend the concept of *Emerging Patterns* in order to depict changes between two time-windows and introduce the notion of *Variability Patterns* in order to characterize changes repeated over time.

2   Basics and Definitions

Most of the methods reported in the literature represent phenotype data by using formalisms based on vectors or attribute-value sets, which model only global descriptive properties of the phenotypes (e.g., presence/absence of mutations). These solutions could be too limiting because they neglect the complex structure of the phenotypes and the inner relationships existing among the biological entities related to the phenotypes. To overcome this drawback, we use the (multi-) relational setting, which has been argued to be the most suitable formalism for representing complex data. In the relational setting, phenotypes and the related biological entities can play different roles in the analysis. We can distinguish them between target objects (TOs) and non-target objects ($NTOs$). The former are the main subjects of the analysis (i.e., phenotypes), while the latter are objects (i.e., DNA level sequences, protein mutations) relevant for the current problem and biologically associated with the former.

Let $\{t_1 \ldots t_n\}$ be a sequence of time-points. At each time-point t_i, a set of instances (TOs) is collected. A *time-window* τ is a sequence of consecutive time-points $\{t_i, \ldots, t_j\}$ ($t_1 \leq t_i, t_j \leq t_n$) which we denote as $[t_i; t_{i+w}]$. The width w of a time-window $\tau = \{t_i, \ldots, t_{i+w}\}$ is the number of time-points in τ, i.e. $w = j - i + 1$. Two time-windows τ and τ' are *consecutive* if $\tau = \{t_i, \ldots, t_{i+w-1}\}$ and $\tau' = \{t_{i+w} \ldots t_{i+2w-1}\}$. Two pairs of consecutive time-windows (τ, τ') and (τ'', τ''') are δ-*separated* if $(j + w) - (i + w) \leq \delta$ ($\delta > 0$, $\delta \geq w$,) with $\tau = \{t_i, \ldots, t_{i+w-1}\}$, $\tau' = \{t_{i+w} \ldots t_{i+2w-1}\}$, $\tau'' = \{t_j \ldots t_{j+w-1}\}$, and $\tau''' = \{t_{j+w} \ldots t_{j+2w-1}\}$). Two pairs of consecutive time-windows (τ, τ') and (τ'', τ''') are *chronologically ordered* if $(j + w) > (i + w)$. We assume that all the time-windows have the same width w and we use the notation τ_{h_k} to refer to a time-window and the notation (τ_{h_1}, τ_{h_2}) to indicate a pair of consecutive time-windows.

Both TOs and $NTOs$ can be represented in Datalog language as sets of ground atoms A ground atom is an n-ary logic predicate symbol applied to n constants. We consider three categories of logic predicates: (1) *key predicate*, which identifies the TOs, (2) *property predicates*, which define the value taken by a property of a TO or of a $NTOs$, and (3) *structural predicates*, which relate TOs with their $NTOs$ or relate the $NTOs$ each other.

The following definitions are crucial for this work:

Definition 1. Relational pattern
A conjunction of atoms $P = p_0(t_0^1), p_1(t_1^1, t_1^2), p_2(t_2^1, t_2^2), \ldots, p_m(t_m^1, t_m^2)$, is a relational pattern if p_0 is the key predicate, p_i, $i = 1, \ldots, m$ is either a structural predicate or a property predicate.

Terms t_i^j are either constants, which correspond to values of property predicates, or variables, which identify TOs or $NTOs$. Moreover, all variables are linked to the variable used in the key predicate [6].

A relational pattern P is characterized by a statistical parameter, namely the *support* (denoted as $sup_{\tau_{h_k}}(P)$), which denotes the relative frequency of P in the time-window τ_{h_k}. It is computed as the number of TOs of τ_{h_k} in which P occurs divided the total number of TOs of τ_{h_k}. When the support exceeds a minimum user-defined threshold $minSUP$, P is said to be *frequent*.

Definition 2. Emerging pattern-EP
Let (τ_{h_1}, τ_{h_2}) be a pair of consecutive time-windows; P be a frequent relational pattern in τ_{h_1} and in τ_{h_2}; $sup_{\tau_{h_1}}(P)$ and $sup_{\tau_{h_2}}(P)$ be the support of the pattern P in τ_{h_1} and in τ_{h_2}. P is an emerging pattern in (τ_{h_1}, τ_{h_2}) iff $\frac{sup_{\tau_{h_1}}(P)}{sup_{\tau_{h_2}}(P)} \geq$

$$minGR \quad \lor \quad \frac{sup_{\tau_2}(P)}{sup_{\tau_1}(P)} \geq minGR$$

where, $minGR$ (> 1) is a user-defined minimum threshold. The ratio $sup_{\tau_{h_1}}(P)/sup_{\tau_{h_2}}(P)$ is denoted with $GR_{\tau_{h_1}, \tau_{h_2}}(P)$ and it is called *growth-rate* of P from τ_{h_1} to τ_{h_2}. When $GR_{\tau_{h_1}, \tau_{h_2}}(P)$ exceeds $minGR$, we have that the support of P decreases from τ_{h_1} to τ_{h_2} by a factor equal to the ratio $sup_{\tau_{h_1}}(P)/sup_{\tau_{h_2}}(P)$, while when $GR_{\tau_{h_2}, \tau_{h_1}}(P)$ exceeds $minGR$, the support of P increases by a factor equal to $sup_{\tau_{h_2}}(P)/sup_{\tau_{h_1}}(P)$.

The concept of emerging patterns is not novel in the literature [2]. In its classical formulation, it refers to the values of support of the same pattern which has been discovered in two different classes of data, while here we extend it to represent the differences between the data collected in two intervals of time, and therefore, we refer to the values of support of the same pattern which has been discovered in two time-windows. In the following, we report an example of EP.

P: phenotype(P), clinical_condition(P,C), dependent_by(C,M), affects(M,N).

with $\tau_{h_1} = [1991;1995]$, $\tau_{h_2} = [1996;2000]$, $sup_{[1991;1995]}(P) = 0.8$ and $sup_{[1996;2000]}(P) = 0.5$. Here, the support of the pattern P decreases, whereby of the growth-rate $GR_{[1991;1995],[1996;2000]}(P)$ is 1.6 (0.8/0.5). By supposing that $minGR=1.5$, the pattern P is considered emerging in ([1991;1995],[1996;2000]).

Definition 3. Variability pattern-VP

Let $T : \langle (\tau_{i_1}, \tau_{i_2}), \dots, (\tau_{m_1}, \tau_{m_2}) \rangle$ be a set of chronologically ordered pairs of time-windows; P be an emerging pattern in all the pairs (τ_{h_1}, τ_{h_2}) with $h = i, \dots, m$; $\langle GR_{\tau_{i_1}, \tau_{i_2}}, \dots, GR_{\tau_{m_1}, \tau_{m_2}} \rangle$ be the values of growth-rate of P in the pairs $\langle (\tau_{i_1}, \tau_{i_2}), \dots, (\tau_{m_1}, \tau_{m_2}) \rangle$ respectively; $\Theta_P : \Re \to \Psi$ be a function which maps $GR_{\tau_{h_1}, \tau_{h_2}}(P)$ into a discrete value $\psi_{\tau_{h_1}, \tau_{h_2}} \in \Psi$ with $h = i, \dots, m$. P is a variability pattern iff:

1. $|T| \geq minREP$
2. (τ_{h_1}, τ_{h_2}) and (τ_{k_1}, τ_{k_2}) are δ-separated for all $h = i, \dots, m - 1$, $k=h+1$ and there is no pair (τ_{l_1}, τ_{l_2}), $h < l$, s.t. (τ_{h_1}, τ_{h_2}) and (τ_{l_1}, τ_{l_2}) are δ-separated
3. $\psi = \psi_{\tau_{i_1}, \tau_{i_2}} = \dots = \psi_{\tau_{m_1}, \tau_{m_2}}$

where $minREP$ is a user-defined threshold.

A VP is a frequent pattern whose support increases (decreases) at least $minREP$ times with an order of magnitude greater than $minGR$. Each change (increase/decrease) occurs within δ time-points and it is characterized by the value ψ. Intuitively, a VP represents a variation of the frequency of the same pattern, which is manifested with a particular regularity. It is quite evident that discovering this kind of information is relevant for studies on epidemics and pandemics. An example of VP is reported here. Consider the following EPs

P: *phenotype(P),clinical_condition(P,C),dependent_by(C,M),affects(M,N)*
  emerging in ([1991;1992],[1993,1994])
P: *phenotype(P),clinical_condition(P,C),dependent_by(C,M),affects(M,N)*
  emerging in ([1996;1997],[1998;1999])
P: *phenotype(P),clinical_condition(P,C),dependent_by(C,M),affects(M,N)*
  emerging in ([1999;2000],[2001;2002])
P: *phenotype(P),clinical_condition(P,C),dependent_by(C,M),affects(M,N)*
  emerging in ([2004;2005],[2006,2007])

Here, $\psi_{[1991;1992],[1993;1994]} = \psi_{[1996;1997],[1998;1999]} = \psi_{[2004;2005],[2006;2007]}$, $\psi_{[1991;1992],[1993;1994]} \neq \psi_{[1999;2000],[2001;2002]}$. By supposing $minREP = 2$ and $\delta = 6$, P is a variability pattern. Indeed, $T : \langle ([1991; 1992], [1993; 1994]),$ $([1996; 1997], [1998, 1999]) \rangle$ meets the conditions (1) and (2) because $|T| = 2$ and (1998-1993)<6; the discrete values of the growth-rate in ([1991;1992],[1993;1994]) and ([1996;1997],[1998;1999]) meet the condition (3). The pair of time-windows ([1999; 2000], [2001; 2002]) is not considered because $\psi_{[1999;2000],[2001;2002]}$ does meet the condition (3), while the pair of time-windows ([2004; 2005],[2006; 2007]) does not meet the condition (3) because (2006-1998)>6.

3   The Algorithm

We propose an algorithm which discovers VPs incrementally as time goes by. It works on the succession $\langle (\tau_{1_1}, \tau_{1_2}), \dots, (\tau_{h_1}, \tau_{h_2}), \dots \rangle$ of pairs of time-windows obtained from $\{t_1, \ldots, t_n\}$. Each time-window τ_{u_v} (except that for the first and last one) is present in two pairs, that is, the pair (τ_{h_1}, τ_{h_2}) where $\tau_{u_v} = \tau_{h_2}$, and the pair $(\tau_{(h+1)_1}, \tau_{(h+1)_2})$ with $\tau_{u_v} = \tau_{(h+1)_1}$. This is done with the intent to capture the changes of support of the patterns from τ_{h_1} to τ_{u_v} and from τ_{u_v} to $\tau_{(h+1)_2}$. For each pair of time-windows (τ_{h_1}, τ_{h_2}), the algorithm performs three steps: (1) Discovery of frequent patterns on the time-windows τ_{h_1} and τ_{h_2} separately; (2) Extraction of EPs by matching the frequent patterns discovered from τ_{h_1} against the frequent patterns discovered from τ_{h_2}. These EPs are stored in a pattern base, which is incrementally updated as the time-windows are processed; (3) Identification of VPs by testing the conditions of Definition 3 on the EPs stored in the base. Note that, when the algorithm processes the pair $(\tau_{(h+1)_1}, \tau_{(h+1)_2})$, it uses the frequent patterns of the time-window τ_{h_2}, which had been discovered when the algorithm had processed the pair (τ_{h_1}, τ_{h_2}). This avoids of performing the step (1) twice on the same time-window. Details on these three steps are reported in the following.

3.1 Relational Frequent Pattern Discovery

Frequent patterns are mined from each time-window by using the method proposed in [5], which enables the discovery of patterns whose support exceeds $minSUP$. It explores level-by-level the lattice of the patterns, from the most general to the more specific ones, starting from the most general pattern (which contains only the key predicate). The lattice is organized according to a generality ordering based on the notion of θ-subsumption [6]. Formally, given two relational patterns $P1$ and $P2$, $P1$ ($P2$) is more general (specific) than $P2$ ($P1$) under θ-subsumption, denoted as $P1 \geqslant_\theta P2$, if and only if $P2$ θ-subsumes $P1$, where $P2$ θ-subsumes $P1$ if and only if a substitution θ exists such that $P2$ $\theta \subseteq P1$. The method adopts a two-stepped procedure: (i) generation of candidate patterns with k atoms (k-th level) by using the frequent patterns with $k-1$ atoms (k-1-th level); (ii) evaluation of the support of the patterns with k atoms.

The monotonicity property of the support value (i.e., a super-set of an non-frequent pattern cannot be frequent) is exploited to avoid the generation of non-frequent relational patterns. In fact, in accordance with the Definition 2, non-frequent patterns are not used for detecting changes and thus we can prune portions of the space containing non-frequent patterns. Thus, given two relational patterns $P1$ and $P2$ with $P1 \geqslant_\theta P2$, if $P1$ is non-frequent in a time-window, then the support of $P2$ is less than the threshold $minSUP$ and it is non-frequent too in the same time-window. Therefore, we do not refine the patterns which are non-frequent.

3.2 Emerging Pattern Extraction

Once the frequent patterns have been discovered from the time-windows τ_{h_1} and τ_{h_2}, they are evaluated in order to check if the growth-rate exceeds the threshold $minGR$. Unfortunately, the monotonicity property does not hold for the growth-rate. In fact, given two frequent patterns $P1$ and $P2$ with $P1 \geqslant_\theta P2$, if $P1$ is not emerging, namely $GR_{\tau_{h_1},\tau_{h_2}}(P1) < minGR$ $(GR_{\tau_{h_2},\tau_{h_1}}(P1) < minGR)$, then the pattern $P2$ may or may not be an EP, namely its growth-rate could exceed the threshold $minGR$. However, we can equally optimize this step by avoiding the evaluation of the refinements of a pattern P discovered from the time-window τ_{h_1} (τ_{h_2}) in the case P is non-frequent in the time-window τ_{h_2} (τ_{h_1}). Note that this operation could exclude EPs with very high values of growth-rate (i.e., the strongest changes), but here we are interested in the changes exhibited by co-occurrences which are statistically evident in both intervals of time.

The EPs extracted on the pairs of time-windows are stored in the pattern base, which hence contains the frequent patterns that satisfy the constraint set by $minGR$ on at least one pair of time-windows. Each EP is associated with two lists, named as $TWlist$ and $GRlist$. $TWlist$ is used to store the pairs of time-windows in which the growth-rate of the pattern exceeds $minGR$, while $GRlist$ is used to store the corresponding values of growth-rate. To distinguish the changes due to the decrease of the support from those due to the increase, we store the values of growth-rate as negative when it decreases.

The base is maintained with two operations, namely insertion of the EPs and update of the lists $TWlist$ and $GRlist$ associated with the EPs. A pattern is inserted if it has not been recognized as emerging in the previous pairs of time-windows, while, if it has been previously inserted, we update the two lists.

3.3   Variability Pattern Identification

The step *(3)* works on the pattern base and filters out the EPs that do not meet the conditions of Definition 3. The function Θ_P implements an equal-width discretization technique. It is applied to two sets of values obtained from the lists $GRlist$ of all the stored EPs, the first set consists of all the positive values of growth-rate, the second one consists of all the negative values. Note that we have not infinite values of growth-rate because all the patterns considered are frequent, i.e., there are no values of support equal to zero. The ranges returned by the discretization technique correspond to the discrete values $\psi_{\tau_h,\tau_{h+1}}$. Thus, we have two sets of ranges Ψ^+ and Ψ^-: Ψ^+ refers to the discrete values obtained from the positive values of growth-rate, while Ψ^- refers to the discrete values obtained from the negative values. We replace the numeric values contained in the lists $GRlist$ with the corresponding ranges in Ψ^+ and Ψ^-. This allows us to obtain two separate sets of discrete values and capture the increases/decreases of the support of the patterns by representing them with a finite number of cases.

This new representation of the growth-rate could suggest to prune the EPs that are more general and conserve the EPs that are more specific when they have the same discrete values. But, this cannot be done because it is not guaranteed that there is equality between the discrete values over all the time-windows.

In this step, the algorithm performs two preliminary operations: (i) removal of the EPs where the lists $TWlist$ and $GRlist$ have length less than the threshold $minREP$; (ii) sorting of the remaining lists $TWlist$ by chronological order. The lists $GRlist$ will be re-arranged accordingly.

The algorithm discovers VPs by working on the EP separately and it can identify more than one VP from a single EP. For each EP, it scans the $TWlist$ once and incrementally builds the set T of each candidate VP. A candidate VP is characterized by one discrete value. During the scan, it evaluates the current pair of time-windows (τ_h, τ_{h+1}) of $TWlist$ and the relative discrete value $\psi_{\tau_h, \tau_{h+1}}$ against with the latest pair of time-windows (τ_k, τ_{k+1}) inserted in the set T of the candidate VP that has the same discrete value: if the pairs of time-windows are δ-separated, then the pair (τ_h, τ_{h+1}) in inserted in the set T of the candidate VP, otherwise it can be considered to start the construction of the set T of a new candidate VP having the same discrete value. Finally, the algorithm filters our the VPs with $|T|$ less than the threshold $minREP$.

In order to clarify how the step *(3)* works, we report an explanatory example. Consider $\Psi^+ = \{\psi', \psi''\}$, $minREP=3$, $\delta=13$ and the lists $TWlist$ and $GRlist$ built as follows:

$TWlist : \langle([1970; 1972], [1973; 1975]), ([1976; 1978], [1979; 1981]), ([1982; 1984], [1985, 1987]),$
$([1988; 1990], [1991; 1993]), ([1994; 1996], [1997; 1999]), ([2010; 2012], [2013; 2015])\rangle$

$GRlist : \langle \qquad\quad \psi', \qquad\qquad\qquad \psi', \qquad\qquad\qquad \psi'',$
$\psi', \qquad\qquad\qquad \psi'', \qquad\qquad\qquad \psi'\rangle$

By scanning the list $TWlist$, we can initialize the set T of a candidate VP' by using the pairs ([1970;1972],[1973;1975]) and ([1976;1978],[1979;1981]) since they are δ-separated (1979-1973$< \delta$) and they have the same discrete value ψ'. The pair ([1982;1984],[1985;1987]) instead refers to a different discrete value (ψ'') and therefore it cannot be inserted into T of VP'. We use it to initialize the set T of a new candidate VP", which thus will include the time-windows referred to ψ''. Subsequently, the pair ([1988;1990],[1991;1993]) is inserted into T of VP' since its distance from the latest pair is less than δ (1991-1979$< \delta$). Then, T of VP" is updated with ([1994;1996],[1997;1999]) since 1997-1985 is less than δ, while the pair ([2010;2012],[2013;2015]) cannot be inserted into T because the distance between 2013 and 1997 is greater than δ. Thus, we use the pair ([2010;2012],[2013;2015]) to initialize the set T of a new candidate VP"'. The set T of VP' cannot be further updated, but, since its size exceeds $minREP$, we consider the candidate VP' as valid variability pattern. Finally, the candidate VP" cannot be considered as valid since its size is less than $minREP$. The candidate VP"' is not even considered since $|T_{\psi'}| < minREP$.

4   A Case Study: Influenza A/H1N1 Virus

We performed an empirical evaluation on the phenotype data concerning the influenza A/H1N1 virus. The flu virus is a common cause of respiratory infection all over the world. The Influenza A virus can infect several species. This virus contains eight segments gene of negative single-stranded RNA ($PB2$, $PB1$, PA, HA, NP, NA, M, and NS) encoding for 11 proteins. The subtype of Influenza A virus is determined by the antigenicity (the capacity to induce an immune response) of the two surface glycoproteins, haemagglutinin (HA) and neuraminidase (NA) [3]. The subtypes circulating in the human populations determining important clinical conditions are H1N1 and H3N2. They cause epidemics and pandemics by antigenic drift and antigenic shift, respectively.

The datasets we use comprise phenotype data describing isolate strains of viruses of three different species, i.e., human, avian and swine. These isolate strains have been registered from 1958 to September 2009, while the datasets have been generated as a view on Influence Research Database hosted at the NIAID BioHealthBase BRC[1] and contain 3221 isolate strains for human, 1119 isolate strains for swine, and 757 isolate strains for avian.

Experiments are performed to study the effect of the thresholds w, δ, $minGR$ and $minREP$ on the discovered variability patterns and emerging patterns. The parameter $minSUP$ is fixed to 0.1. In this case study, the time-points correspond to years, while the number of the ranges produced by the discretization function is fixed to 5. Statistics on the results are collected in Table 1.

In Table 1(a), we have the number of VPs and EPs when tuning w (δ=20, $minGR = 2$, $minREP = 3$). By increasing the width of the time-windows, the overall number of the time-windows decreases, which results in a shorter succession of pairs of time-windows where finding EPs and VPs. This explains the decrease of the number of EPs and VPs for swine and human. We have a different behavior for the phenotypes of avian, where the number of EPs and VPs increases. Indeed, the use of wider time-windows (w=10 and 15 against w=5 and 7) leads to collect greater sets phenotypes having likely higher changeability. In this case, it seems that phenotype change concerns longer periods. In Table 1(b), we have the results when tuning δ (w=5, $minGR = 2$, $minREP = 3$). As expected, higher values of δ allow us to detect a more numerous set of VPs, which comprises both the replications of EPs which are closer and the replications of EPs that are distant. Whilst, when δ is 10, we capture only the VPs that cover at most ten years. The threshold δ does not affect the number of EPs since it operates after the extraction of the EPs. In Table 1(c), we have the results when tuning $minGR$ (w=5, δ=20, $minREP = 3$). We observe that $minGR$ has great effect on the number of VPs and on the number of EPs. Indeed, at high values of $minGR$, the algorithm is required to detect the strongest changes of support of the patterns, which leads to extract only the EPs with the higher values of growth-rate. This explains the decrease of the number of VPs. The threshold $minREP$ has no effect on the EPs since it acts on the VPs only (Table 1(d), w=5, δ=20, $minGR = 2$). As expected, higher values of $minREP$ lead to exclude the EPs that have a low number of replications. This means that the algorithm works on a smaller set of EPs, with the result to have a lower number of VPs. In particular, when $minREP$ is 6, we have no VP that includes EPs repeated six times and distant at most 20 years.

In the following, we report an example of variability pattern discovered by the proposed algorithm from the human dataset with w=10, δ=20, $minGR = 2$, $minREP$=3:   *phenotype(P), epidemiological_condition(P,E), is_a(E,enhanced_Transmission_to_Human), dependent_by(E,M1), is_a(M1,mutation_A199S), mutation_of(M1,T), is_a(T,protein_PB2), dependent_by(E,M2), is_a(M2,mutation_A661T),*

---

mutation_of(M2,T), dependent_by(E,M3), is_a(M3,mutation_K702R), mutation_ of(M3,T).

Here, T : $\langle([1958;1967],[1968;1977]),([1968;1977],[1978;1987]),([1988; 1997],[1998;2009])\rangle$, $\psi=[2;3,5]$. The pattern concerns the epidemiological condition 'enhanced_Transmission_to_Human' with the mutations 'A199S' on the protein 'PB2' and the mutations 'A661T' and 'K702R'. The frequency of this pattern increases three times by a factor included in the range [2;3,5]. This happens between the time-windows [1958;1967] and [1968;1977], [1968;1977] and [1978;1987], [1988;1997] and [1998;2009]. Virologists observed in swine a pattern similar to that illustrated above. Indeed, given that PB2 has an important role in viral replication, transcription and spread, the common pattern between human and swine can explain a possible reassortment event. This could be happened favored by this amino acid mutation allowing viral particles to be exchanged between the two host species.

Table 1. Total number of the variability patterns and emerging patterns discovered on the three species when tuning the width of the time-window w (a), the maximum admissible distance between consecutive pairs of time-windows δ (b), minimum threshold of growth-rate $minGR$ (c) and minimum threshold of repetitions $minREP$ (d). In each cell, we have reported the statistics as number of VPs–number of the EPs.

	w(years)			
	5	7	10	15
swine	11–126	11–126	8–63	2–18
human	20–176	20–176	10–69	2–44
avian	1–4	1–4	2–10	2–10

(a)

	δ(years)			
	10	15	20	25
swine	0–126	6–126	11–126	15–126
human	14–176	18–176	20–176	22–176
avian	0–4	0–4	1–4	1–4

(b)

	$minGR$			
	2	4	6	8
swine	11–126	7–68	0–2	0–0
human	20–176	6–61	0–4	0–0
avian	1–4	0–2	0–2	0–0

(c)

	$minREP$			
	3	4	5	6
swine	11–126	6–126	2–126	0–126
human	20–176	7–176	3–176	0–176
avian	0–6	0–2	0–2	0–2

(d)

5 Related Works and Conclusions

Despite its relevance, the analysis of phenotypes variability is a problem that has attracted attention only recently. In biology, a vast research focused on the evolution of the phenotype in the genome, but very few attempts have been done for analyzing the evolution over short periods of time. Phenotype evolution has been addressed without linking it to genetic information. For instance, in [7], the authors consider the phenotype of patients as temporal clinical manifestations semantically annotated and propose a technique based on the constraint networks to automatically infer phenotype evolution patterns of generic patients. In this work, we mined the relevant proteomic information enclosing many epidemiological and pathogenic variable of Influenza virus to draw their history dynamics over defined time intervals. This comparative proteomics novel approach could emphasize the drivers sub-patterns (key traits) and link them to phenotypes in an epidemiological study framework. It is noteworthy that our approach is complementary to the antigenic trees construction from genetic sequences since these trees use mainly the antigenic phenotypes information.

The use of frequent pattern mining for analyzing dynamic domains is of recent investigation in data mining research, while the repeatability of patterns over time seems novel. In [1], the authors designed a density-based clustering algorithm to detect novelties from complex data in streaming setting. Novelties are captured as patterns whose frequency significantly changes with respect to homogeneous clusters of frequencies computed on previous windows of the data stream. In [4], we investigated the problem of capturing structural changes within heterogeneous and dynamic networks with a new notion of emerging patterns revised to model changes local to the topology of the network.

In this paper we investigated the task of characterizing the temporal variability of complex phenotypes by defining a novel notion of patterns to track the repeatability of such variations. The experimental results highlight the strong influence of the input parameter of minimum growth-rate and minimum number of repetitions on the discovered variability patterns. Although the method has been applied to phenotype data, we plan to explore its viability also in other scenarios of life sciences.

Acknowledgements. The authors would like to acknowledge the support of the European Commission through the project MAESTRA - Learning from Massive, Incompletely annotated, and Structured Data (Grant number ICT-2013-612944).

References

1. Ceci, M., Appice, A., Loglisci, C., Caruso, C., Fumarola, F., Malerba, D.: Novelty detection from evolving complex data streams with time windows. In: Rauch, J., Raś, Z.W., Berka, P., Elomaa, T. (eds.) ISMIS 2009. LNCS, vol. 5722, pp. 563–572. Springer, Heidelberg (2009)
2. Dong, G., Li, J.: Efficient mining of emerging patterns: discovering trends and differences. In: Proceedings of the Fifth ACM SIGKDD International Conference on Knowledge Discovery and Data Mining, pp. 43–52 (1999)
3. Furuse, Y., Suzuki, A., Kamigaki, T., Oshitani, H.: Evolution of the M gene of the influenza A virus in different host species: large-scale sequence analysis. Virol. J. **6**(67), 67–79 (2009)
4. Loglisci, C., Ceci, M., Malerba, D.: Discovering evolution chains in dynamic networks. In: Appice, A., Ceci, M., Loglisci, C., Manco, G., Masciari, E., Ras, Z.W. (eds.) NFMCP 2012. LNCS, vol. 7765, pp. 185–199. Springer, Heidelberg (2013)
5. Loglisci, C., Ceci, M., Malerba, D.: Relational mining for discovering changes in evolving networks. Neurocomputing **150**(Part A), 265–288 (2015)
6. Plotkin, G.D.: A note on inductive generalization. Mach. Intell. **5**, 153–163 (1970)
7. Taboada, M., Alvarez, V., Martnez, D., Pilo, B., Robinson, P., Sobrido, M.: Summarizing phenotype evolution patterns from report cases. J. Med. Syst. **36**(Suppl 1), S25–S36 (2012)

Computation of Approximate Reducts
with Dynamically Adjusted Approximation
Threshold

Andrzej Janusz[1] and Dominik Ślęzak[1,2(✉)]

[1] Faculty of Mathematics, Informatics and Mechanics, University of Warsaw,
ul. Banacha 2, 02-097 Warsaw, Poland
{janusza,slezak}@mimuw.edu.pl
http://www.dominikslezak.org
[2] Infobright Inc., ul. Krzywickiego 34, lok. 219, 02-078 Warsaw, Poland

Abstract. We continue our research on dynamically adjusted approximate reducts (DAAR). We modify DAAR computation algorithm to take into account dependencies between attribute values in data. We discuss a motivation for this improvement and analyze its performance impact. We also revisit a filtering technique utilizing approximate reducts to create a ranking of attributes according to their relevance. As an illustration we study a data set from AAIA'14 Data Mining Competition.

Keywords: Attribute selection · Approximate reducts · Random probes

1 Introduction

Since a proper identification of relevant data aspects plays a vital role in data classification and visualization, attribute selection became a very popular research topic [1]. A vast majority of attribute subset selection methods can be classified either as a wrapper or as a filtering approach [2]. In the first one, attribute subsets are assessed by measuring performance of a predictive model constructed using the attributes from those sets. In the second one, a preselected quality measure is used to create a ranking of attributes and after that, usually, k top ranked attributes are selected, where the parameter k is set by an expert or using an automatic method such as a permutation test [3]. Even though the wrapper approach may allow to obtain better performance for a particular predictive model, the filtering approach is preferable in analysis of large data sets due to its computational efficiency.

Measures utilized in the filtering approach often consider a contribution of an attribute to predictive models, e.g., by taking into account its occurrence in useful decision rules [4]. Furthermore, the optimal value of k for filtering methods

Partially supported by Polish National Science Centre grants DEC-2012/05/B/-ST6/03215 and DEC-2013/09/B/ST6/01568, and by Polish National Centre for Research and Development grants PBS2/B9/20/2013 and O ROB/0010/03/001.

F. Esposito et al. (Eds.): ISMIS 2015, LNAI 9384, pp. 19–28, 2015.
DOI: 10.1007/978-3-319-25252-0_3

is sometimes selected using the wrapper approach, i.e., a ranking of attributes is used to create an ordered list of nested attribute subsets whose quality is later evaluated using a prediction model.

Attribute subset selection is also one of the key research directions within the scope of the rough set theory [5], where a concept of decision reducts was proposed [6]. A number of heuristic algorithms was developed to search for collections of meaningful decision reducts in large high-dimensional data [7]. The notion of a reduct was also investigated in a variety of theoretical frameworks to better combine it with other methodologies [8].

Although the decision reducts attempt to express information about data-based dependencies in a possibly most compact way, in practice they often contain unnecessary attributes. This happens due to a fact that in a finite set of data elements described by a large number of features there are dependencies which are valid only for the available data, but do not hold for new cases. Such dependencies can be called *illusionary* or *random*. Several generalizations of reducts were proposed with this respect, such as approximate decision reducts [9]. Ensembles of diverse approximate decision reducts corresponding to different thresholds proved useful for knowledge representation and data classification [10].

Rough set approach can also be applied to the problem of attribute ranking. Commonly used rough-set-based ranking methods exploit the fact that the informative attributes are more likely to be present in decision reducts. For instance, in a method called rough attribute ranker [11], attributes are ordered based on a frequency of their appearance in approximate decision reducts computed on random subsets of data. This simple method can also be paired with a more complex approach, e.g., Breiman's attribute relevance measure [12].

In this paper, we combine this ranking method with approximate reducts generated using different algorithms. In particular, we utilize the method proposed in [13], which incorporates permutation tests to assess meaningfulness of attributes. Experiments reported in our previous studies show that the resulting attribute subsets – called dynamically adjusted approximate reducts (DAARs) – are more sensitive than other types of reducts over synthetic data sets, when the task is to find truly relevant attribute sets. Hence, it is tempting to verify the usefulness of DAARs also for the attribute ranking purposes.

Our contribution in this paper is twofold. Firstly, we propose an optional modification to the above-mentioned technique, which may make DAARs more intuitive. It generates random probes in a way that does not interfere with dependencies between a currently investigated attribute and attributes added by the algorithm in earlier steps. Secondly, we extend our previous experimental framework onto the task of attribute ranking and we compare several rough-set-based methods on a real-life data set using, as an evaluation reference, rankings produced out of top solutions submitted to one of data mining competitions that we conducted in 2014 [14].

The paper is organized as follows: In Sect. 2, we recall the basics of approximate reducts and random probes. In Sect. 3, we explain the DAAR method and discuss its above-mentioned modification. In Sect. 4, we present a case study related to attribute ranking in real-life data, where DAARs are compared to approximate decision reducts derived in a more traditional way. In Sect. 5, we conclude the paper with some directions for further research.

2   Preliminary Notions and Notations

Let us recall the basic notions from the rough set theory. We assume that a data set is represented by a decision system understood as a tuple $\mathbb{S}_d = (U, A \cup \{d\})$, where U is a set of objects, A is a set of their attributes and $d \notin A$ is called a decision or class attribute. The task in the attribute selection is to find a subset of A which compactly represents relevant information about objects in U and allows an accurate prediction of values of d for objects outside U.

Typically, in the rough set theory selecting compact yet informative sets of attributes is conducted by computing so called reducts [6]. For a subset of attributes $B \subseteq A \cup \{d\}$, we say that objects $u_1, u_2 \in U$ satisfy an indiscernibility relation IND_B iff they have equal attribute values for every $a \in B$, i.e., $(u_1, u_2) \in IND_B \Leftrightarrow \forall_{a \in B} a(u_1) = a(u_2)$. Otherwise u_1 and u_2 are regarded as discernible. An equivalence class of IND_B containing u is usually denoted by $[u]_B$. One of possible formulations of the notion of a decision reduct is as follows:

Definition 1. *Let* $\mathbb{S}_d = (U, A \cup \{d\})$ *be a decision system. A subset of attributes* $DR \subseteq A$ *is called a decision reduct iff the following conditions are met:*

1. *For each* $u \in U$, *if its indiscernibility class relative to* A *is a subset of some decision class, then its indiscernibility class relative to* DR *should also be a subset of that decision class, i.e.,* $\forall_{u \in U} [u]_A \subseteq [u]_d \Rightarrow [u]_{DR} \subseteq [u]_d$.
2. *There is no proper subset* $DR' \subsetneq DR$ *for which the first condition holds.*

Some of decision reducts may contain attributes which are not truly related to d, even if they are crucial to keep indiscernibility classes of some of objects fully within their corresponding decision classes. This situation happens simply by a chance. It can be explained either by random disturbances in data or by the fact that even a randomly generated binary attribute would discern approximately a half of pairs of objects. Such attributes are useless for the predictive data analysis. It is not easy to distinguish between truly relevant and irrelevant attributes, especially when some of them are *partially* correlated with the decision, but this dependence is lost when they are considered in a context of some other attributes. These problems were addressed by several extensions of the decision reduct concept. One of those is the notion of approximate reduct [9]:

Definition 2. *Let* $\mathbb{S}_d = (U, A \cup \{d\})$ *be a decision system and let* $\phi_d : 2^A \to \mathbb{R}$ *be an attribute subset quality measure which is nondecreasing with regard to inclusion. A subset* $AR \subseteq A$ *is called an* (ϕ_d, ε)-*approximate decision reduct for an approximation threshold* $\varepsilon \in [0, 1)$ *iff the following conditions are met:*

1. $\phi_d(AR) \geq (1 - \varepsilon)\phi_d(A)$.
2. *There is no proper subset* $AR' \subsetneq AR$ *for which the first condition holds.*

One of possible versions of ϕ_d is to the cardinality of so called positive region, i.e., a set of objects $u \in U$ for which $[u]_{AR} \subseteq [u]_d$ holds [6]. As another example, among possibilities for ϕ_d reported in [9], we considered the *Gini Gain* that can be treated as an expected decrease in the impurity of decisions after a division of data into indiscernibility classes defined by an attribute subset.

Regardless of the formula for ϕ_d, it is not an easy task to choose ε that corresponds to the actual level of noise or uncertainty in data. Multiple costly tests have to be conducted on additional on validation data to tune the value of ε. To overcome this issue, we proposed the dynamically adjusted approximate reduct (DAAR) heuristic [13]. It is a modification of the greedy algorithm for searching for decision reducts which uses the notion of random probes to measure randomness of the attributes. Each attribute in $\mathbb{S}_d$ can be associated with a random variable that could have generated the values of this attribute within U. In practice, we may not have explicit knowledge about distributions of such random variables but we can assume their existence even though we can reason about them only through their valuations in data.

Definition 3. *Let* $\mathbb{S}_d = (U, A \cup \{d\})$ *be a decision system. A random probe for an attribute* $a \in A$ *is an artificial attribute* $\hat{a}$, *such that the corresponding random variable has an identical distribution as values of a but is independent from the random variables that correspond to the attribute a and the decision d.*

To obtain a random probe for a given attribute, we can permute its values for objects in U. Another solution is to use the bootstrapping technique, i.e., sampling values with repetitions. An advantage of that former method is that it does not change the distribution of values. It guarantees that the distributions of the selected attribute and all corresponding random probes are the same.

Let us recall the pseudo code of DAAR heuristic in Algorithm 1, slightly modified comparing to [13]. Originally, quality of all attributes was measured using ϕ_d in every iteration. Now, only a randomly chosen subset of attributes is assessed. By doing so we make it possible to generate many different approximate decision reducts for a single decision system. Also, by setting the parameter defining the size of the attribute subsets to a value of order $O(\sqrt{|A|})$, we guarantee a linear time complexity of the algorithm with regard to $|A|$.

Ability to compute many diverse decision reducts is important for constructing robust classification and knowledge representation models. It is also useful when applying the attribute ranking method described in Sect. 1. The first way to generate a bigger ensemble of reducts is to derive all reducts of a decision system and then to consider a subset of the reducts that fulfill some prerequisites. However, this method is never used for high-dimensional decision systems due to its high computational cost. The second way is to rely on a heuristic approach which typically involves Monte Carlo generation of candidate attribute subsets. From this perspective, Algorithm 1 can be indeed treated as a modification to the greedy heuristic for computation of approximate decision reducts, allowing a Monte-Carlo-style search through the attribute space [7].

3   Utilization of Random Probes in DAAR Heuristic

A random probe should have no influence on values of the decision attribute and as such, it can be used to exemplify fake dependencies. If we have a sufficient number of random probes for attribute a, then we can estimate a probability that the observed dependency between a and d is in fact illusionary, i.e., we can estimate the probability that a is *irrelevant*. This can be done in a few different ways. In [13], the probability of no relation between a and d was estimated based on a distribution of scores obtained by the generated random probes. Quality of a and the corresponding probes was assessed using a predefined measure. The scores of the random probes were compared with the score of a. The ratio between the number of probes which received a higher score than a and the number of all probes was taken as the estimation of irrelevance of a.

In such a way we can estimate the probability that an attribute with a given distribution could have obtained as high evaluation score as a only by a chance. Thus, we silently *assume* that a is in fact irrelevant and our test aims at disapproving this assumption. If our estimation is low, then we can safely say that a is *not irrelevant*. However, if the estimation is high, then we cannot claim that a is surely irrelevant. Having that in mind, we will use the term *'probability of being irrelevant'* as a kind of a mental shortcut.

Comparing to [13], we would like to suggest an optional way to estimate the probability $P\Big(\phi_d\big(B \cup \{a\}\big) \leq \phi_d\big(B \cup \{\hat{a}\}\big)\Big)$. It is due to the question whether the process of random probe generation can be allowed to disturb potential inter-attribute dependencies in data. At each iteration of the main loop of Algorithm 1 the assessment of attribute relevance is made in a context of already selected attributes. If we use the straightforward method for generating random probes, then it is likely that some of its values will be inconsistent with the domain-specific data dependencies. For example, let us consider a decision system describing vehicles and imagine that in the first iteration we selected an attribute giving a number of wheels in a vehicle and in the second iteration we want to test relevance of an attribute which gives a make of the vehicle. Some combination of those two attributes' values cannot exist in data, e.g. Toyota does not produce motorcycles and there are no cars constructed by Yamaha. However, if we computed random probes for the second attribute by randomly permuting its values, we would often end up with conflicting values and use such unrealistic probes in the relevance estimation.

Consequently, we propose Algorithm 2 where, instead of always generating random probes using global permutations, they can be constructed by locally permuting or bootstrapping values of the original attribute within indiscernibility classes defined by the previously selected attributes. This method guarantees that values of random probes will never be conflicting with the values of real attributes and thus, in some situations it may have better chances for obtaining a reliable estimation of the attribute relevance. However, since this procedure does not have to fit to all types of data, the global permutation method can still be chosen as in [13], using an additional parameter *globalPerms*. In both cases,

---

Algorithm 1. A dynamically adjusted approximate reduct calculation

---

Input: a decision system $\mathbb{S}_d = (U, A \cup \{d\})$; quality measure $\phi_d : 2^A \to \mathbb{R}$;
acceptable probability of adding irrelevant attribute $p_{probe} \in [0, 1)$;
attribute sample size $mTry$;

Output: an approximate decision reduct AR of $\mathbb{S}_d$

1 **begin**
2     $AR := \emptyset$; $\phi_{max} := -\infty$;
3     $stopFlag := FALSE$;
4   **while** $stopFlag = FALSE$ **do**
5   randomly select subset $A' \subseteq A \setminus AR$ consisting of $mTry$ attributes;
6   **foreach** $a \in A' \setminus AR$ **do**
7           $AR' := AR \cup \{a\}$;
8   **if** $\phi_d(AR') > \phi_{max}$ **then**
9              $\phi_{max} := \phi_d(AR')$;
10              $a_{best} := a$;
11   **end**
12   **end**
13   **if** $P\Big(\phi_d(AR \cup \{a_{best}\}) \leq \phi_d(AR \cup \{\hat{a}_{best}\})\Big) \leq p_{probe}$ **then**
14           $AR := AR \cup \{a_{best}\}$;
15   **end**
16   **else**
17           $stopFlag := TRUE$;
18   **end**
19   **end**
20   **foreach** $a \in AR$ **do**
21        $AR' := AR \setminus \{a\}$;
22   **if** $\phi_d(AR') \geq \phi_d(AR)$ **then**
23           $AR := AR'$;
24   **end**
25   **end**
26   **return** AR;
27 **end**

---

the Laplace estimation is computed in the line 19 by simply adding 1 and 2 to the nominator and the denominator, respectively. Comparing to our previous studies, the Laplace estimator seems to better fit our purposes [15].

4 Case Study

We conducted experiments on a data set from *AAIA'14 Data Mining Competition* [14], organized at Knowledge Pit platform[1]. This competition task was an example of a large scale, real-life attribute subset selection problem. The data set describes a total of $100,000$ incident reports ($50,000$ in the training set) issued by firefighters after a variety of fire&rescue actions which took place

---

[1] https://knowledgepit.fedcsis.org/contest/view.php?id=83.

Algorithm 2. Estimation of $P\Big(\phi_d\big(B \cup \{a\}\big) \le \phi_d\big(B \cup \{\hat{a}\}\big)\Big)$

Input: a decision system $\mathbb{S}_d = \big(U, A \cup \{d\}\big)$; quality measure $\phi_d : 2^A \to \mathbb{R}$; attribute subset $B \subseteq A$; attribute $a \in A$; positive integer $nProbes$; boolean value $globalPerms$;

Output: an estimation of $P\Big(\phi_d\big(B \cup \{a\}\big) \le \phi_d\big(B \cup \{\hat{a}\}\big)\Big)$

1 **begin**
2 $\quad \phi_a := \phi_d(B \cup \{a\})$;
3 $\quad higherScoreCount = 0$;
4 $\quad$ **if** $globalPerms == FALSE$ **then**
5 $\quad\quad$ compute the indiscernibility relation IND_B;
6 $\quad\quad$ **foreach** $i \in 1 : nProbes$ **do**
7 $\quad\quad\quad$ generate a random probe $\hat{a}_i$ either by permuting or bootstrapping values of a, separately for each indiscernibility class of IND_B;
8 $\quad\quad\quad$ **if** $\phi_a \le \phi_d\big(B \cup \{\hat{a}_i\}\big)$ **then**
9 $\quad\quad\quad\quad |\;\; higherScoreCount = higherScoreCount + 1$;
10 $\quad\quad\quad$ **end**
11 $\quad\quad$ **end**
12 $\quad$ **end**
13 $\quad$ **else**
14 $\quad\quad$ generate a random probe $\hat{a}_i$ either by permuting or bootstrapping values of a;
15 $\quad\quad$ **if** $\phi_a \le \phi_d\big(B \cup \{\hat{a}_i\}\big)$ **then**
16 $\quad\quad\quad |\;\; higherScoreCount = higherScoreCount + 1$;
17 $\quad\quad$ **end**
18 $\quad$ **end**
19 $\quad$ **return** $\frac{higherScoreCount+1}{nProbes+2}$;
20 **end**

---

within the city of Warsaw and its surroundings in years 1992–2011. Each report is represented by values of 11,852 symbolic, mostly sparse attributes. In the competition, participants were asked to indicate sets of attributes that allow to accurately identify major incidents, i.e., those which involved injuries. The competition data has three binary decision attributes indicating injuries among firefighters, children and all civilians. All decision attributes are highly imbalanced – proportions of the positive classes were ≈ 0.004, ≈ 0.007 and ≈ 0.059, respectively.

For simplicity, we aggregated the three original decision attributes into a new binary attribute which indicates whether there were any injuries at all. For this new attribute we computed 40 sets of approximate decision reducts with different settings of the approximation threshold, ranging from $\varepsilon = 0.0$ (standard decision reducts) to 0.95. Each of the sets consisted of 100 reducts. All the computations were done using the greedy reduct computation heuristic implemented in *RoughSets* package for R System[2] [16]. We also computed two

---

[2] Version 1.2.2 of the package was used in all experiments described in this paper.

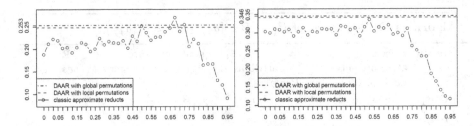

Fig. 1. Comparison of attribute subsets defined by approximate decision reducts with fixed approximation thresholds and those computed using DAAR heuristic. Vertical axes show values of F_1-*score*. On the left plot are the results of comparison with top 3 solutions. The right plot shows the comparison to solutions with a score 0.95 or higher.

additional sets of 100 approximate decision reducts using DAAR heuristic, for both settings of *globalPerms* in Algorithm 2. In all cases, the computations were performed only on the training part of data.

Parameters of algorithms were set to their defaults, apart from the attribute sample size (*mTry* in Algorithm 1) which was set to 300. Both reduct computation methods used the same quality measure ϕ_d, i.e., the *Gini Gain* that was mentioned in Sect. 2. For each set of approximate reducts we computed rankings of attributes using the simplest method mentioned in Sect. 1. We also constructed a single set of attributes for each set of approximate reducts, by taking the union of all attributes that occurred at least twice.

Table 1 shows basic characteristics of selected groups of reducts. As expected, with the increasing value of approximation threshold decreases the average size of approximate reducts. Interestingly, there is also a statistically significant difference in the average sizes of reducts computed using the DAAR heuristic. The reducts for which random probes were generated using global permutations were on average shorter by 1.12 attribute. This fact is also reflected by the actual approximation thresholds computed for DAAR reducts. Their value is different for every reduct but their average values are significantly lower for the DAAR reduct with probes generated using the local permutations.

In order to compare quality of the obtained reduct sets, we decided to take advantage of the fact that AAIA'14 Data Mining Competition attracted a large number of participants from all over the world, who altogether sent us over 1,300 solutions. Quality of quite many submitted solutions was high and the best performing one exceeded the baseline by over 5.5 % (the AUC value of the best solution was ≈ 0.9623). Consequently, as the reference attribute set we took a union of the attributes occurring in the best solutions in the competition – the attributes occurring in the top three solutions, as well as the attributes occurring in all solutions with a score of 0.95 or higher (top 23 solutions).

We computed $F_1 scores$ between the attribute sets corresponding to different approximate reduct sets and the two reference sets. Figure 1 shows the results for different approximate threshold values. Additionally, these plots show the score achieved by the attributes from the approximate reducts computed using DAAR heuristic (the horizontal lines). It is clearly visible that both methods

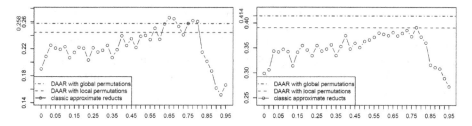

Fig. 2. Comparison of Spearman's correlations between attribute rankings defined by approximate reducts with fixed thresholds and those computed using DAAR heuristic. On the left plot are the results of comparison with top 3 solutions. The right plot shows the comparison to solutions with a score 0.95 or higher.

Table 1. Basic characteristics of selected groups of approximate reducts.

	Approximate reducts:				
Approx. threshold:	0.00	0.15	0.30	0.45	0.55
Avg. size ± stdev.:	35.08 ± 3.17	17.88 ± 1.74	14.96 ± 1.46	12.93 ± 1.44	11.48 ± 1.32
	approximate reducts:			DAAR reducts:	
Approx. threshold:	0.65	0.75	0.90	0.61 ± 0.09	0.55 ± 0.11
Avg. size ± stdev.:	9.34 ± 1.06	5.99 ± 1.38	1.77 ± 0.62	10.38 ± 2.21	11.50 ± 2.40

for generation of random probes allowed to automatically identify nearly optimal stop moments during computation of approximate reducts. For the second reference set the results obtained using DAAR heuristic were even better than the best results achieved by the traditional method of computation of approximate reducts. This phenomenon can be interpreted as an argument supporting a claim that for real-life data sets there is no such a thing as a globally optimal approximation threshold. We also checked the Spearman's correlations between attribute rankings created using different sets of reducts and the reference solutions. These results are shown in Fig. 2. They confirm that DAAR heuristic can be successfully applied to compute meaningful attribute rankings.

5   Conclusions

The presented case study confirms efficiency of the methods that use random probes in the computation of approximate decision reducts. We showed that this technique makes it possible to compute approximate decision reducts with approximation thresholds adjusted to a given data set. Nevertheless, the presented framework still requires far more tests to verify its sensitivity with regard to parameters that guide computations of random probes. In particular, the parameter p_{probe} that sets the acceptable probability of including unnecessary attribute into a reduct may have a significant role in the process of finding useful attribute sets. This way, we will be also able to proceed with more thorough comparisons with other approaches to the attribute ranking.

We are also interested in new heuristics for computation of reducts. We believe that certain modifications of the attribute quality criteria can lead to finding reducts which are more suitable for some types of classifiers. From this perspective, it is important to investigate whether procedures for random probe generation can remain transparent with respect to specific functions ϕ_d, or whether they should be additionally adjusted to their particular families.

References

1. Guyon, I., Gunn, S., Nikravesh, M., Zadeh, L.A. (eds.): Feature Extraction: Foundations and Applications. STUDFUZZ. Springer, Heidelberg (2006)
2. Kohavi, R., John, G.H.: Wrappers for feature subset selection. Artif. Intell. **97**, 273–324 (1997)
3. Kruczyk, M., Baltzer, N., Mieczkowski, J., Dramiński, M., Koronacki, J., Komorowski, J.: Random reducts: a monte carlo rough set-based method for feature selection in large datasets. Fundamenta Informaticae **127**(1–4), 273–288 (2013)
4. Błaszczyński, J., Słowiński, R., Susmaga, R.: Rule-based estimation of attribute relevance. In: Yao, J.T., Ramanna, S., Wang, G., Suraj, Z. (eds.) RSKT 2011. LNCS, vol. 6954, pp. 36–44. Springer, Heidelberg (2011)
5. Pawlak, Z.: Rough sets: present state and the future. Found. Comput. Decis. Sci. **18**(3–4), 157–166 (1993)
6. Świniarski, R.W., Skowron, A.: Rough set methods in feature selection and recognition. Pattern Recogn. Lett. **24**(6), 833–849 (2003)
7. Janusz, A., Ślęzak, D.: Rough set methods for attribute clustering and selection. Appl. Artif. Intell. **28**(3), 220–242 (2014)
8. Yao, Y.: The two sides of the theory of rough sets. Knowl.-Based Syst. **80**, 67–77 (2015)
9. Ślęzak, D.: Normalized decision functions and measures for inconsistent decision tables analysis. Fundamenta Informaticae **44**(3), 291–319 (2000)
10. Wróblewski, J.: Ensembles of classifiers based on approximate reducts. Fundamenta Informaticae **47**(3–4), 351–360 (2001)
11. Janusz, A., Stawicki, S.: Applications of approximate reducts to the feature selection problem. In: Yao, J.T., Ramanna, S., Wang, G., Suraj, Z. (eds.) RSKT 2011. LNCS, vol. 6954, pp. 45–50. Springer, Heidelberg (2011)
12. Breiman, L.: Random forests. Mach. Learn. **45**(1), 5–32 (2001)
13. Janusz, A., Ślęzak, D.: Random probes in computation and assessment of approximate reducts. In: Kryszkiewicz, M., Cornelis, C., Ciucci, D., Medina-Moreno, J., Motoda, H., Raś, Z.W. (eds.) RSEISP 2014. LNCS, vol. 8537, pp. 53–64. Springer, Heidelberg (2014)
14. Janusz, A., Krasuski, A., Stawicki, S., Rosiak, M., Ślęzak, D., Nguyen, H.S.: Key risk factors for polish state fire service: a data mining competition at knowledge pit. In: Ganzha, M., Maciaszek, L.A., Paprzycki, M. (eds.) Proceedings of FedCSIS 2014, pp. 345–354 (2014)
15. Jaynes, E.T.: Probability Theory: The Logic of Science. Cambridge University Press, Cambridge (2003)
16. Riza, L.S., Janusz, A., Bergmeir, C., Cornelis, C., Herrera, F., Ślęzak, D., Benitez, J.M.: Implementing algorithms of rough set theory and fuzzy rough set theory in the r package 'roughsets'. Inf. Sci. **287**, 68–89 (2014)

Databases, Information Retrieval, Recommender Systems

A New Formalism for Evidential Databases

Fatma Ezzahra Bousnina[1(✉)], Mohamed Anis Bach Tobji[2],
Mouna Chebbah[1], Ludovic Liétard[3], and Boutheina Ben Yaghlane[4]

[1] LARODEC/ISG-University of Tunis, Tunis, Tunisia
{fatmaezzahra.bousnina,chabbeh.mouna}@gmail.com
[2] LARODEC/ESEN-University of Manouba, Manouba, Tunisia
anis.bach@isg.rnu.tn
[3] IRISA/ENSSAT-University of Rennes 1, Rennes, France
ludovic.lietard@univ-rennes1.fr
[4] LARODEC/IHEC-University of Carthage, Tunis, Tunisia
boutheina.yaghlane@ihec.rnu.tn

Abstract. This paper is about modeling and querying evidential databases. This kind of databases copes with imperfect data which are modeled via the evidence theory. Existing works on such data deal only with the compact form of the database. In this article, we propose a new formalism for modeling and querying evidential databases based on the possible worlds form. This work is a first step toward the definition of a strong representation system.

Keywords: Evidential databases · Dempster-Shafer theory · Strong representation system · Possible worlds

1 Introduction

Many theories are used to represent imperfect data. The most famous one is the probability theory. Other theories have been introduced to offer other models like possibility theory [15] and evidence theory [13]. In order to store and manage imperfect information, specified database models, based on these theories were introduced [2,4,5].

In this paper, we consider the general framework of evidential databases where attributes' values are defined using evidence theory. This model was introduced in [2,11]. We define a new representation model of evidential databases based on *possible worlds*, contrarily to existing evidential database formalisms which are based on the *compact form*, such in [2,6,10,11]. Such a new representation opens the way to the definition of a *strong representation system* for evidential databases. A database model is a strong representation system if the result of any query processed on the *compact form* of the database is equivalent to the result of the same query over the set of its *possible worlds*.

The Table 1 presents an example of evidential database that stores medical diagnoses. This example is used throughout this article. This relation contains two attributes *Disease* and *Blood Type*. Being an evidential database, the value

© Springer International Publishing Switzerland 2015
F. Esposito et al. (Eds.): ISMIS 2015, LNAI 9384, pp. 31–40, 2015.
DOI: 10.1007/978-3-319-25252-0_4

Table 1. A medical evidential table

ID	Disease	BloodType
1	Flu 0.7	A
	{Asthma, Anemia} 0.3	
2	Anemia	B 0.5
		{B,O} 0.5

of each attribute is imperfect. It is expressed through an evidence theory distribution (called a *basic belief assignment*) as shown in Sect. 2.

For this database example, world $W_1 = (1, flu, A), (2, Anemia, B)$ is a possible world among six possible ones. The aim of this article is to generate all possible worlds for one evidential database, and to quantify the amount of belief for each one.

This paper is organized as follows: in Sect. 2, we review the basic notions of the evidence theory. In Sect. 3, we briefly present the definition of evidential databases. Section 4 is the main contribution of this paper. We detail our approach to generate possible worlds from an evidential database and we define our new representation formalism. We focus in the last part of Sect. 4 on querying the new representation model. In Sect. 5, conclusion and future works are drawn.

2   Evidence Theory

Evidence theory also called *the theory of belief functions or the Dempster-Shafer theory*, was introduced by Dempster [7] and mathematically formalized by Shafer [13].

The theory of belief functions models and combines imperfect information. It provides an explicit representation of uncertainty, imprecison and inconsistency. In the theory of belief functions, a set $\Theta = \{\theta_1, \theta_2, ..., \theta_n\}$ is a finite, non empty and exhaustive set of n elementary and mutually exclusive hypotheses related to a given problem. The set Θ is called the *frame of discernment* or *universe of discourse*.

The *power set* $2^\Theta = \{\varnothing, \theta_1, \theta_2, ..., \theta_n, \{\theta_1, \theta_2\}, .., \{\theta_1, \theta_2, ..., \theta_n\}\}$ is the set of all subsets of Θ. A *basic belief assignement* (*bba*), noted m, is a mapping from 2^Θ to the interval $[0, 1]$ that assesses a degree of belief to some elements of the power set. A bba also called *mass function* is defined such that:

$$\sum_{x \subseteq \Theta} m(x) = 1 \qquad (1)$$

The amount $m(x)$ is the *mass* that represents our belief on the truth of the hypothesis x. The mass $m(x)$ is the degree of faith on x that is not distributed on its subsets. When $m(x) > 0$, x is called *focal element*. We denote F the set of all focal elements.

The couple $\{F, m\}$ is called *body of evidence*. The union of all focal elements is called *core* and is defined as follows:

$$\varphi = \bigcup_{x \in F: m(x) > 0} x \tag{2}$$

The mass $m(x)$ represents the part of belief degree that is committed exactly to x. To compute all the belief committed to x, we must take into account the masses also committed to all subsets of x. It is computed as follows:

$$bel(x) = \sum_{y, x \subseteq \Theta: y \subseteq x} m(y) \tag{3}$$

The *plausibility function* noted pl is the *maximal* amount of belief on x. It is computed as follows:

$$pl(x) = \sum_{y, x \subseteq \Theta: x \cap y \neq \varnothing} m(y) \tag{4}$$

2.1 Combination Rules

Combination rules merge mass functions provided by distinct and independent sources. Many combination rules are proposed in the framework of the theory of belief functions. Smets [14] introduced the *conjunctive rule of combination* for two mass functions m_1 and m_2 defined on the same frame of discernment Θ as follows:

$$m_1 \bigcirc m_2(z) = \sum_{x, y \subseteq \theta: x \cap y = z} m_1(x).m_2(y) \tag{5}$$

The conjunctive rule of combination merges bbas m_1 and m_2 provided by different sources; focal elements of the combined bba are the intersection of those of m_1 and m_2. Smets [14] proposed also the disjunctive rule of combination based on the union of focal elements. Combining bbas m_1 and m_2 with the disjunctive rule leads to a combined bba which focal elements are the union of focal elements of m_1 and m_2. Combining two bbas m_1 and m_2 defined on the same frame of discernment Θ is defined as follows:

$$m_1 \bigcirc m_2(z) = \sum_{x, y \subseteq \theta: x \cup y = z} m_1(x).m_2(y). \tag{6}$$

The disjunctive and conjunctive rules are generalized to combine more than two mass functions.

A mass function that is defined on a given frame of discernment can be also defined on compatible frame as we will show in the next section.

2.2 Vacuous Extension

In some cases, we need to combine two *bbas* m_1 and m_2 which are not defined on the same frame of discernment. However, all combination rules require that *bbas* have the same frame of discernment. The *vacuous extension of belief functions* [13] is a tool that defines bbas on a compatible frame of discernment.

It consists in extending the frames of discernment Θ_1 and Θ_2, corresponding to the mass functions m_1 and m_2, to the joint frame of discernment Θ defined as:

$$\Theta = \Theta_1 \times \Theta_2$$

Each focal element is extended to its cylindrical extension ($A \times \Theta_2$ is the cylindrical extension of $A \subseteq \Theta_1$)

The extended mass function of any evidential value of the extended focal element A, denoted by m, is defined as follows:

$$m^{\Theta_1 \uparrow \Theta_1 \times \Theta_2}(A) = \begin{cases} m_1(B) & \text{where } A = B \times \Theta_2,\, B \subseteq \Theta_1 \\ 0 & \text{otherwise} \end{cases} \tag{7}$$

3   Evidential Databases

Evidential databases (EDB), also called *D-S databases* are used to store both perfect and imperfect data unlike classic databases that are used to store only certain and precise data. Imperfect values in evidential databases are expressed via evidence theory presented above.

An *EDB* has N objects and D attributes. The value of an attribute c for an object l is a bba. It is called *evidential value* and is denoted by V_{lc}.

$$V_{lc} : 2^{\Theta_l} \rightarrow [0, 1] \tag{8}$$

$$\text{with } m_{lc}(\varnothing) = 0 \text{ and } \sum_{x \subseteq \Theta_l} m_l(x) = 1 \tag{9}$$

The set of focal elements relative to the bba V_{lc} is F_{lc} such that:

$$F_{lc} = \{x \subseteq \Theta / m_{lc}(x) > 0\}$$

Example 1. Table 1 is an evidential database that stores various types of data. When focal element is a singleton and its mass is equal to one, the bba is called *certain*. We are in the case of *a perfect information* such for the disease of the second patient. When focal elements are singletons, we are in the case of *probabilistic information* and the bba is called *bayesian*. When focal elements are nested, the bba is called *consonant* and we are the case of *possibilistic information*. When the unique focal element is Θ whose mass is equal to 1, we are in the case of the *total ignorance*. When none of the previous cases is present, we are in the case of *a generic evidential information* such for the disease of the first patient.

Data are stored to be further queried. Some models have been proposed for this purpose. Bell et al. [2] and Choenni et al. [6] and Lee in [10, 11] defined models of querying an evidential database using relational operators like selection, projection and join. Their models are based on a compact form of the database. To show if these models are strong representation systems, the defined operations should give the same result if applied on the possible worlds form.

4   Evidential Database as Possible Worlds

4.1   Definition of Possible Worlds

An evidential database EDB, as defined in Sect. 3 on its *compact form*, is a set of N objects (tuples) and D attributes (see Table 1). The *non-compact form* of EDB is a finite set of *imprecise possible worlds* such that $EDB = \{IW_1, IW_2, ..., IW_k\}$. Each imprecise possible world includes N objects, where each object contains one focal element per attribute. The Table 2 is an example of an imprecise possible world induced from Table 1.

Table 2. Example of an imprecise possible world of Table 1

ID	Disease	BloodType
1	Flu	A
2	Anemia	(B,O)

Possible worlds are qualified by imprecise because they include imprecise values such as $\{B, O\}$ which is the value of the attribute *BloodType* for the second object in Table 2.

Note that the number of all imprecise possible worlds, k, is induced from sizes of sets F_{lc} as follows:

$$k = \prod_{l=1}^{N} \prod_{c=1}^{D} |F_{lc}| \tag{10}$$

For example, Table 3 includes the four imprecise possible worlds from Table 1. The set of imprecise possible worlds defines the non-compact form of an EDB.

Table 3. Imperfect worlds of EDB

IW_1	IW_2	IW_3	IW_4
$(1, Flu, A)$	$(1, Flu, A)$	$(1, \{Asthma, Anemia\}, A)$	$(1, \{Asthma, Anemia\}, A)$
$(2, Anemia, B)$	$(2, Anemia, \{B,O\})$	$(2, Anemia, B)$	$(2, Anemia, \{B,O\})$

A classical database is the relational union of its objects. Analogically, an evidential database is the union of its N evidential objects [1].

Definition 1 (Evidential Object). *An evidential object is a basic belief assignement issued from the combination of all evidential values of that object. Its frame of discernment is the joint frame of all attributes' domains denoted by Θ. The mass funtion m_l^Θ related to the evidential object of line l is the conjunctive combination of all evidential values m_{lc} extended to the joint frame. An evidential object is defined as follows:*

$$m_l^\Theta = \bigcirc_{c \in [1,D]} m_{lc}^{\Theta_c \uparrow \Theta} \ with \ \Theta = \prod_{c \in [1,D]} \Theta_c \tag{11}$$

Table 4. Extension of bbas from the evidential Table 1

ID	$\Theta_{DE} \uparrow \Theta$	$\Theta_{BT} \uparrow \Theta$
1	$Flu \times \Theta_{BT}$	$A \times \Theta_{DE}$
	$\{Asthma, Anemia\} \times \Theta_{BT}$	
2	$Anemia \times \Theta_{BT}$	$B \times \Theta_{DE}$
		$\{B, O\} \times \Theta_{DE}$

Table 5. Combination of the extended bbas using the conjunctive rule of combination

$\bigcirc\!\!\!\!\cap$
$(1, Flu, A)0.7 = 1 \times 0.7 \times 1$
$(1, \{Asthma, Anemia\}, A)0.3 = 1 \times 0.3 \times 1$
$(2, Anemia, B)0.5 = 1 \times 1 \times 0.5$
$(2, Anemia, \{B, O\})0.5 = 1 \times 1 \times 0.5$

Example 2. To obtain evidential objects of Table 1: First, their evidential values are extended to the joint frame of discernment as detailed in Table 4. Then, the extended evidential values are combined using the conjunctive rule as shown in Table 5. Note that Θ_{DE} is the frame of discernement relative to attribute *Disease* and that Θ_{BT} is the frame of discernment relative to attribute *BloodType*.

Formally, an evidential database is the disjunctive combination of its evidential objects; hence the following definition of the non-compact form of an evidential database:

Definition 2 (Non-compact Evidential Database). *A non-compact evidential database is presented through a bba issued from the disjunctive combination of all its evidential objects. That bba is defined as follows:*

$$m^\Theta = \bigcirc\!\!\!\!\cup_{l \in [1,N]} m_l^\Theta \tag{12}$$

Example 3. Table 6 is the disjunctive combination of all evidential objects detailed in Table 5. The combined bba is defined on a frame of discernment $\Theta_{IW} = \{IW_1, IW_2, IW_3, IW_4\}$ such that each IW_i is an imprecise possible world.

Table 6. Combination of the evidential objects of *EDB* using the disjunctive rule of combination

$\bigcirc\!\!\!\!\cup$			
IW_1	IW_2	IW_3	IW_4
$(1, Flu, A)$	$(1, Flu, A)$	$(1, \{Asthma, Anemia\}, A)$	$(1, \{Asthma, Anemia\}, A)$
$(2, Anemia, B)$	$(2, Anemia, \{B, O\})$	$(2, Anemia, B)$	$(2, Anemia, \{B, O\})$
0.35	0.35	0.15	0.15

Each imprecise possible world IW_i can be expanded into different precise states called possible worlds W_j, as shown in Table 7. A same world W_j can be a precise state of several imprecise possible worlds. For example, the possible world W_1 is derived from imprecise worlds IW_1 and IW_2.

Each possible world is a candidate to represent the evidential relation. The degree of belief on a possible world to be candidate is given from the belief of its imprecise possible worlds.

Table 7. Possible worlds of EDB

IW_1			
W_1			
$(1, Flu, A)$			
$(2, Anemia, B)$			

IW_2		
W_1		W_2
$(1, Flu, A)$		$(1, Flu, A)$
$(2, Anemia, B)$		$(2, Anemia, O)$

IW_3	
W_3	W_4
$(1, Asthma, A)$	$(1, Anemia, A)$
$(2, Anemia, B)$	$(2, Anemia, B)$

IW_4			
W_3	W_4	W_5	W_6
$(1, Asthma, A)$	$(1, Anemia, A)$	$(1, Asthma, A)$	$(1, Anemia, A)$
$(2, Anemia, B)$	$(2, Anemia, B)$	$(2, Anemia, O)$	$(2, Anemia, O)$

Example 4. We illustrate here the basic belief assignment of possible worlds related to Table 1:

$$m^{\Theta_{IW}}(IW_1) = m^{\Theta_W}(\{W_1\}) = 0.35$$
$$m^{\Theta_{IW}}(IW_2) = m^{\Theta_W}(\{W_1; W_2\}) = 0.35$$
$$m^{\Theta_{IW}}(IW_3) = m^{\Theta_W}(\{W_3; W_4\}) = 0.15$$
$$m^{\Theta_{IW}}(IW_4) = m^{\Theta_W}(\{W_3; W_4; W_5; W_6\}) = 0.15$$

Note that $m^{\Theta_{IW}}$ is defined on Θ_{IW} and that m^{Θ_W} is defined on Θ_W which is the set of all possible worlds obtained from Θ_{IW}.

4.2   Querying Possible Worlds

Let Q be the query processed on each possible world W_j. Querying each possible world W_j (noted $Q(W_j)$) gives a possible answer R_u:

$$R_u = Q(W_j)$$

Definition 3 (Mass Function on Result Sets). *Let be the mass function* $m^{\Theta w}$, *defined over possible worlds of an evidential database* EDB, *where* $F^{\Theta w}$ *is the set of focal elements. The mass function* $m^{\Theta R}$ *is derived* $m^{\Theta w}$, *and is defined on the result sets* R_u *returned by a query* Q:

$$m^{\Theta R}(\{R_u/R_u = Q(W_j), W_j \in F_j, F_j \in F^{\Theta w}\}) = m^{\Theta w}(F_j/F_j \in F^{\Theta w}) \quad (13)$$

Example 5. We consider the following query and its results on each possible world of Table 1.

$Q : SELECT * FROM EDB WHERE (Disease = \{Anemia\})$

$Q(W_1) = \{(2, Anemia, B)\} = R_1$
$Q(W_2) = \{(2, Anemia, O)\} = R_2$
$Q(W_3) = \{(2, Anemia, B)\} = R_1$
$Q(W_4) = \{(1, Anemia, A); (2, Anemia, B)\} = R_3$
$Q(W_5) = \{(2, Anemia, O)\} = R_2$
$Q(W_6) = \{(1, Anemia, A); (2, Anemia, O)\} = R_4$

From Examples 5 and 3 we deduce that:

$m^{\Theta R}(\{R_1\}) = 0.35$
$m^{\Theta R}(\{R_1; R_2\}) = 0.35$
$m^{\Theta R}(\{R_1; R_3\}) = 0.15$
$m^{\Theta R}(\{R_1; R_2; R_3; R_4\}) = 0.15$

Now, we can assign to each tuple t belonging to the union of sets R_u an interval limited by a belief and a plausibility measures, called the confidence level. The belief and the plausibility of a tuple t quantify the amount of belief and plausibility that t is a part of the query result. These measures are derived from the belief and the plausibility of results' sets that returned t.

Definition 4 (Confidence Level). *For each tuple* t *belonging in the set of responses* R_u, *a confidence level* CL *is calculated such that:*

$$t.CL = [bel(S), pl(S)] \text{ where } S = \{R_u/t \in R_u\} \quad (14)$$

Tuples, that satisfy a query Q, belong to the union of results' sets R_u. Assume they compose the set $R = \bigcup R_u$. Each tuple $t \in R$ is a potential result of the query Q applied on EDB. To quantify satisfaction belief and plausibility of each one, we use Definition 4.

Example 6. The union of results' sets, denoted by $R = \bigcup R_u$, include following tuples:
$R = R_1 \cup R_2 \cup R_3 \cup R_4 = \{(1, Anemia, A); (2, Anemia, B); (2, Anemia, O)\}$
We compute the confidence level of each tuple in R:

- $t_1 = (1, Anemia, A)$ is the answer appearing in R_3 and R_4. Thus, $t_1.CL = [bel(S), pl(S)]$; with $S = \{R_3; R_4\}$. Consequently, $t_1.CL = [0, 0.3]$.

- $t_2 = (2, Anemia, B)$ is the answer appearing in R_1 and R_3. Thus, $t_2.CL = [bel(S), pl(S)]$; with $S = \{R_1; R_3\}$. Consequently, $ID_2.CL = [0.5, 1]$.
- $t_3 = (2, Anemia, O)$ is the answer appearing in R_2 and R_4. Thus, $t_2.CL = [bel(S), pl(S)]$; with $S = \{R_2; R_4\}$. Consequently, $ID_2.CL = [0, 0.5]$.

5 Conclusion

In this paper, we proposed a new approach for modeling and querying evidential databases based on possible worlds.

At first, we used Dempster-Shafer theory tools, mainly the conjunctive and disjunctive rules to quantify the uncertainty about possible worlds generated from a compact evidential database. Uncertainty quantifies the degree of belief on each possible world to be a candidate to the evidential database. This result is very important in the literature since it impacts on all previous works on querying and analyzing evidential data. At the best of our knowledge, all contributions about querying [2,6], analyzing [8] and mining [1,3,9,12] evidential data are based on the compact form. Models cited above are not validated in comparison to the possible worlds form. In other words, it has not been proved that they constitute strong representation systems. Then, we formalized queries of evidential databases by the way of possible worlds form. The definition we introduced was about a generic relational query Q. For short, the example we gave is about selection operator only.

As future works, all relational evidential operators may be detailed. In fact, for seek of simplicity, we focused on the selection operator. However, all relational evidential operators namely the projection, join, union and intersect may also be detailed in a future wok. In addition, previous works on evidential databases introducing new relational operators (selection, projection, join, union and intersection); new analysis operators (such as skyline operator) and new data mining methods should be reviewed and validated with regard to the possible worlds form. Finally, implementation of relational operators on possible worlds is a matter of a future work. Scalability is the main problem to solve, because the number of possible worlds increases dramatically when sizes and numbers of focal elements increase.

References

1. Tobji, M.A.B., Yaghlane, B.B., Mellouli, K.: A new algorithm for mining frequent itemsets from evidential databases. In: Proceedings of the 12th International Conference on Information Processing and Management of Uncertainty in Knowledge-Based Systems, Málaga, Spain, pp. 1535–1542 (2008)
2. Bell, D.A., Guan, J.W., Lee, S.K.: Generalized union and project operations for pooling uncertain and imprecise information. Data Knowl. Eng. **18**, 89–117 (1996)
3. Hariz, S.B., Elouedi, Z.: New dynamic clustering approaches within belief function framework. Intell. Data Anal. **18**(3), 409–428 (2014)
4. Bosc, P., Pivert, O.: About projection-selection-join queries addressed to possibilistic relational databases. IEEE Trans. Fuzzy Syst. **13**(1), 124–139 (2005)

5. Cavallo, R., Pittarelli, M.: The theory of probabilistic databases. In: Proceedings of the Thirteenth VLDB Conference, Brighton, pp. 71–81 (1987)

6. Choenni, S., Blok, H.E., Leertouwer, E.: Handling uncertainty and ignorance in databases: a rule to combine dependent data. In: Li Lee, M., Tan, K.-L., Wuwongse, V. (eds.) DASFAA 2006. LNCS, vol. 3882, pp. 310–324. Springer, Heidelberg (2006)

7. Dempster, A.P.: Upper and lower probabilities induced by a multiple valued mapping. Ann. Math. Stat. **38**(2), 325–339 (1967)

8. Elmi, S., Benouaret, K., Hadjali, A., Tobji, M.A.B., Yaghlane, B.B.: Computing skyline from evidential data. In: Straccia, U., Calì, A. (eds.) SUM 2014. LNCS, vol. 8720, pp. 148–161. Springer, Heidelberg (2014)

9. Hewawasam, K.K.R.G.K., Premaratne, K., Shyu, M.-L.: Rule mining and classification in a situation assessment application: a belief theoretic approach for handling data imperfections. IEEE Trans. Syst. Man Cybern. **37**(6), 1446–1459 (2007)

10. Lee, S.K.: An extended relational database model for uncertain and imprecise information. In: Proceedings of the 18th Conference on Very Large Data Bases, Canada, pp. 211–220 (1992)

11. Lee, S.K.: Imprecise and uncertain information in databases: an evidential approach. In: Proceedings of the 8th International Conference on Data Engineering, pp. 614–621 (1992)

12. Samet, A., Lefèvre, E., Yahia, S.B.: Mining frequent itemsets in evidential database. In: Huynh, V.N., Denoeux, T., Tran, D.H., Le, A.C., Pham, B.S. (eds.) KSE 2013, Part II. AISC, vol. 245, pp. 377–388. Springer, Heidelberg (2014)

13. Shafer, G.: A Mathematical Theory of Evidence. Princeton University Press, Princeton (1976)

14. Smets, P.: Belief functions: the disjunctive rule of combination and the generalized bayesian theorem. Int. J. Approximate Reasoning **9**, 1–35 (1993)

15. Zadeh, L.A.: Fuzzy sets as a basis for a theory of possibility. Fuzzy Sets Syst. **1**, 3–28 (1978)

Ubiquitous City Information Platform Powered by Fuzzy Based DSSs to Meet Multi Criteria Customer Satisfaction: A Feasible Implementation

Alberto Faro[✉] and Daniela Giordano

Department of Electrical, Electronics and Computer Engineering, University of Catania,
viale A. Doria 6, 95125 Catania, Italy
albfaro@gmail.com

Abstract. Aim of the paper is to illustrate a methodology to implement an ubiquitous city platform called Wi-City provided with centralized and mobile Decision Support Systems (DSSs) that take advantage from all the data of city interest including location, social data and data sensed by monitoring devices. The paper proposes to extend the existing Wi-City DSSs based on location intelligence with an advanced version based on multi criteria customer satisfaction expressed by the users grouped by age where the weights of the criteria are provided by the users instead of expert decision makers, and the rating of the aspects involved in the criteria depends on the evaluation expressed by all the service customers. Including such advanced DSSs in Wi-City makes the platform ready to provide information to users of *intelligent cities* where recommendations should depend on location and collective intelligence.

Keywords: Intelligent systems · Decision support systems · Recommender systems · Computing with words · Mobile information systems

1 Introduction

Almost all the proposals of smart and intelligent cities, e.g., [1, 2], point out that the basic infrastructure of the such cities is the ubiquitous information platform able to provide the mobile users with: (a) mobility and logistics suggestions, e.g., nearest services depending on the user position and on traffic/environment sensed data, (b) timely alerts about critical mobility conditions, (c) continuous monitoring services for people with specific disease, and (d) recommendations taking into account all the databases at urban scale, personal information (e.g., age, health status, preferences, and current task), and social data. A platform, called Wi-City, that satisfies the above requirements has been proposed by the authors in [3], where decision support systems are under test to meet the mentioned requirements by means of simple rule based systems that may be implemented on both the user mobiles and the main servers.

In [4] we illustrated how the computing with words paradigm [5] could be an effective way to achieve feasible mobile implementations of DSS instead of other available techniques such as Bayesian logic or decision trees. Indeed, as explained in the

© Springer International Publishing Switzerland 2015
F. Esposito et al. (Eds.): ISMIS 2015, LNAI 9384, pp. 41–51, 2015.
DOI: 10.1007/978-3-319-25252-0_5

concluding remarks, this allows the DSS to recommend the best course of actions that meets the user expectations by taking into account many conflicting rules and with a guaranteed customer satisfaction. Also, in the previous papers, we have demonstrated how such DSS may works in simple user scenarios.

Aim of the paper is to propose a DSS, that may reside on either the main server or the user mobiles, that works as simply as possible to identify the best course of actions that maximize multi criteria customer satisfaction where the weights of the criteria and their rating are not given by a limited number of expert decision makers, but depend on the user and on the evaluation expressed by all the service customers.

Including such advanced DSSs in Wi-City makes the platform ready to provide information to users of intelligent cities where suggestions depend on both location and collective intelligence [2]. Section 2 illustrates the methodology to implement feasible DSSs that take into account basic personal data and location contexts. Section 3 proposes the methodology to provide mobile users with recommendations that take into account different types of customer satisfaction derived from social data and user expectations. Section 4 shows how such DSS works in practice to help mobile using users in scenarios where decisions derive from a multi-criteria decision-making (MCDM) process. Concluding remarks briefly point out that the recommendations derived from the proposed methodology outperform the ones obtained from the available DSSs little equipped with AI tools, and envisage DSSs that improve the effectiveness of its recommendations using a smart city ontology, e.g. [6].

2   DSS Based on Location Intelligence Rules

To point out the basic features of an intelligent system, such a DSS, that supports mobile users in smart city scenario, let us consider the internal structure of the mentioned ubiquitous information platform shown in Fig. 1 called Wi-City. It consists of a central information system controlling three main subsystems: (1) the sub-system that monitors the current traffic parameters to identify origin-destination travel times and traffic congestion, as well as the environment conditions that influence user decisions (e.g., rain and temperature) monitored by networked measuring devices; (2) the sub-system that interconnects all the public and private data bases of city interest; (3) the sub-system that aims at collecting data constrained by privacy conditions and at supporting the user-system interactions through mobile user devices.

Currently, all the relevant data are stored in relational tables managed by a MySQL server, but a mapping of such data to RDF format is under test to evaluate how a data ontology, queried through a Sesame server, may facilitate data integration, thus improving the effectiveness of the information platform. All the sensed data, as well as the administrative and business data are taken into account by a fuzzy logic based DSS resident at the main server. Also, simple local DSSs resident on the user mobiles have been implemented to recommend to the mobile users the best course of actions when the personal data play an important role.

The responses to the user queries are sent from the main server to the user's mobiles through JQMobile scripts so that the results of the query may be displayed with the same

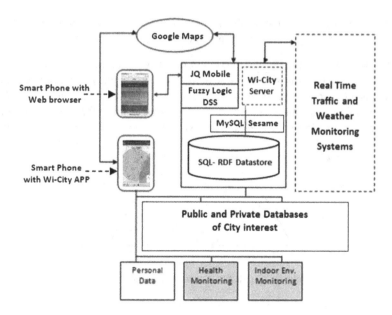

Fig. 1. The internal structure of Wi-City points out that it is an ubiquitous city information platform provided with DSSs at both central site and mobile devices. The data of user interest are displayed on the user mobiles using Google Maps API and JQMobile scripts.

format independently on the mobile brand. The information are geo-localized using Google-Maps API that makes very easy for the users to understand the DSS suggestions by a familiar GIS. The local DSS implemented on the user devices have been developed using the Flash Builder framework [7] to collect data and control the relevant parameters that deal with user health status and indoor conditions.

The rules currently managed by the DSSs aim at finding, upon user request, the nearest services depending on the user age and weather conditions expressed by simple fuzzy rules such as the following ones:

- *the higher the age, the closer the service*
  the antecedent and consequent fuzzy sets are shown side by side in Fig. 2 so that it is easy to compute the evidence e_A of the expression "the higher the age" from the user age and how close should be the service passing from e_A to dmax in the fuzzy set representing "the closer the service".
- *the higher the external temperature T_{ext} differs from 20 °C, the closer the service*
  the antecedent fuzzy set is shown in Fig. 3 left, whereas the consequent is given in Fig. 2 right. It is easy to compute the evidence e_T of the expression "the higher the external temperature T_{ext} differs from 20 °C" from the current temperature and how close should be the service passing from e_T to d_{max} in the fuzzy set representing "the closer the service".
- *the higher the rain the closer the service*
  the antecedent fuzzy set is shown in Fig. 3 right, whereas the consequent is given in Fig. 2 right. It is easy to compute the evidence e_R of the expression "the higher the

rain" from the current rain intensity and how close should be the service passing from e_R to d_{max} in the fuzzy set representing "the closer the service".

In the paper, for simplicity, the maximum distance is obtained by a defuzzification operation that returns the maximum value of the α-cut, as shown in Fig. 2. Consequently, if a person is 38 years old, the external temperature is 15 °C and the rain is 2 mm/h, the suggested maximum distance, according to fuzzy logic is the 350 m, i.e., the minimum among 420, 400 and 350 m (see Figs. 2 and 3).

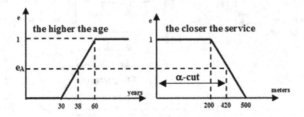

Fig. 2. Maximum acceptable walking distance according to the rule *the higher the age the closer the service*.

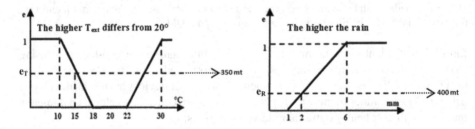

Fig. 3. The maximum acceptable walking distances according to the rules "the higher the external temperature T_{ext} differs from 20 °C, the closer the service" and "the higher the rain, the closer the service" are obtained by passing from e_T and e_R to the distance in the fuzzy set "the closer the service" shown in Fig. 2, thus obtaining 350 m and 400 m respectively.

Similar rules may be used to find the maximum time, i.e., tmax, within which the desired service should be reached using the car. The above fuzzy rules allow the DSS to extend the "near me" information provided by the service locators available on the web since in Wi-City the maximum acceptable distance of the recommended services depends on the user age and on the weather conditions. Also, other rules are taken into account to further filter the services, e.g., the recommended services should be open, should have vacancy, and might depend on the user task and health status. In the next section, we propose the fuzzy rules that allow the DSS to suggest the services that best meet the customer satisfaction to better satisfy the user preferences.

3   DSS Based on Location Intelligence and Customer Satisfaction

In the advanced version of Wi-City, the DSSs will be able to suggest the services that meet the customer satisfaction among those services that are located within a reasonable distance around the user location, or that are reachable in a suitable time interval.

Recommending the services that have a high customer satisfaction aims at better satisfying the user preferences. This hypothesis is valid if higher levels of customer satisfaction are positively associated with higher levels of customer trust, loyalty, and repurchase intention as suggested in [8]. Indeed, in such case, as pointed out in [9], we may assume that a customer will reuse/repurchase the same or similar services used/ purchased in the past.

However, the above hypothesis is not valid in general, but depends on the business sector. For example, it is valid in the retail sector, but not in the mobile phone market where product features may influence repurchase more than customer satisfaction [10], as well as in the hotel industry where not only customer satisfaction but also the hotel design features may affect loyalty and repurchase [11]. This implies that psychographic/ behavioral aspects should be taken into account for better targeting the customers and their preferences [12].

Therefore, a multi-criterial approach, to take into account psychological, emotional, behavioral and service quality aspects, may be necessary in some cases to clearly identify the customer preferences. On the basis of these considerations, in the following we illustrate a methodology that allows the DSS to recommend: (1) services that have a high average customer satisfaction if they belong to sectors featured by a linear relation between repurchase intent and customer satisfaction and the satisfaction is few differentiated across age groups, e.g., parking areas and gas stations [13]; (2) services that received a high average score from the customers of a given age group if such services belong to sectors where demographic segmentation positively affect the customer satisfaction, e.g., hospitals and pharmacies [14]; (3) services that highly meet multi-criteria customer satisfaction and whose weights should be assigned by the user if they belong to sectors where service quality and emotional/behavioral aspects play an important role, e.g., parking garages, restaurants and hotels [11].

To simplify the implementation, we assume that: (a) the scores are normally distributed with a mean value λ and a standard deviation σ, and (b) the fuzzy set representing a *high score* is the one shown in Fig. 4.

Under these assumptions, in [15] we have demonstrated, using the computing with words paradigm, that the fuzzy set *high average score* is obtained by moving to the left of about 2σ the lower corner of the oblique side of the fuzzy set high score. Such fuzzy set allows the DSS to find the evidence $e_{h-\lambda\sigma-s}$ that a service with a mean customer satisfaction λ has a high average score.

After determined such fuzzy set, the DSS could recommend the services whose λ corresponds to a high customer satisfaction with a high evidence, e.g., $e_{h-\lambda\sigma-s} \geq 0,7$. Since this implies a risk of about 50 % that the service will not meet the customer expectations, the DSS may recommend the services that received a high score from 90 % of the customers with high evidence. To check such constraint, the DSS should verify, as pointed out in [15], that the mean customer satisfaction λ is greater than the one

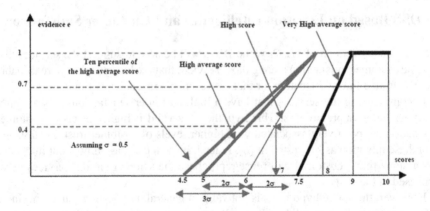

Fig. 4. Fuzzy set *high score*. The fuzzy set *high average score* is obtained by displacing of about 2σ the fuzzy set *high score*. The fuzzy set related to the ten percentile of the high average score is obtained by displacing of about σ the fuzzy set *high average score*.

corresponding to $e_{vh-\lambda\sigma-s} = 0.7$, i.e., in fuzzy terms, that the service has a very high (vh) average score with a high evidence, e.g., from Fig. 4 we derive that if a service has a mean customer satisfaction equal to 8, then 90 % of its customers expressed a score of 7 with evidence of 0.4 (service not to recommend), whereas if $\lambda = 8.5$, then 90 % of customers expressed a score of 7.5 with evidence of 0.7 (service to recommend).

To better satisfy the user expectations, we may group the scores by age, e.g., young, middle-age and elderly people so that the DSS may suggest to the users not only the services that are located (reachable) at a reasonable distance (time), but also the ones that better meet the user expectations, e.g., the nearest restaurants that highly satisfy the young people. Let a_j the interval $[(j-1) \times 10, j \times 10]$, a_N the highest interval, $e_{vh-\lambda\sigma-s}[a_j]$ the evidence that the scores expressed about a service by persons in the age range a_j with mean customer satisfaction $\lambda(a_j)$ and standard deviation $\sigma(a_j)$ is very high, and said $e_y(a_j)$ the evidence that a person in the age range a_j is young, the evidence that the customer satisfaction of young persons is very high is:

$$e_{y-vh-s} = \sum_{a1}^{aN} e_y\left(a_j\right) e_{vh-\lambda\sigma-s}\left[a_j\right] / \sum_{a1}^{aN} e_y\left(a_j\right) \tag{1}$$

After computing e_{y-vh-s}, the DSS should recommend the services whose e_{y-vh-s} is high, e.g., $e_{y-vh-s} \geq 0,7$. Modifying the previous formula to identify the service featured by a very high customer satisfaction of middle aged and elderly people is straightforward. However, even more often, the customer satisfaction is based on several aspects, e.g., price, food quality, and cleaning for the restaurants. In such cases, the customers should weight each aspect, and the DSS should follow a multiple criteria based customer satisfaction procedure. This may be achieved by fuzzifying the Technique for Order of Preference by Similarity to Ideal Solution (TOPSIS), e.g. the one described in [16], as follows:

- the user gives weights $w(i)$ to the aspects, e.g., $w_{price} = 7$, $w_{food\_quality} = 9$, $w_{cleaning} = 8$
- the DSS identifies the best service as the one that has a customer satisfaction closest as possible to the positive ideal solution and farthest from the most negative ideal one, i.e., the service that maximizes the following expression:

$$\max_j NS_j / \left(PS_j + NS_j \right) \tag{2}$$

where $NS_j = \{ (\Sigma_i \, [w(i) \, e_{y-vh-s}(i,j) - \min_j(w(i) \, e_{y-vh-s}(i,j))]^2 \}^{1/2}$ is the distance of service j from the negative ideal service, i.e., the one with minimum score $\min_j(w(i) \, e_{y-vh-s}(i,$ j)) for every aspect i, and $PS_j = \{ (\Sigma_i \, [w(i) \, e_{y-vh-s}(i,j) - \max_j(w(i) \, e_{y-vh-s}(i,j))]^2 \}^{1/2}$ is the distance of service j from the positive ideal service, i.e., the one with maximum score $\max_j(w(i) \, e_{y-vh-s}(i,j))$ for every aspect i

Alternatively, it is possible to use the classical rule table that indicates the customer satisfaction expressed by three terms: Low, Medium and High (L,M and H) in correspondence to each aspect configuration with the caution of deducing the customer satisfaction evidence by the fuzzy sets Low-Medium (LM), Medium-High (MH), and VH (Very High) instead of the fuzzy sets L, M and H. In fact, generalizing the results illustrated in [15], this assures that 90 % of times the customer will find that the recommended service meet the highly satisfactory features and don't present the unwanted ones. Under this assumption, the Sugeno defazzification rule indicates that the DSS should recommend the services with the highest values of the expression:

$$E = \left[X_{LM} \left(B_{LM} + \alpha_{y-LM-s} \right) e_{y-LM-s} + X_{MH} \left(B_{MH} + \alpha_{y-MH-s} \right) e_{y-MH-s} + X_{VH} \left(B_{VH} + \alpha_{y-VH-s} \right) e_{y-VH-s} \right] / 2A \tag{3}$$

where (as illustrated in Fig. 5) we have:

- B_{LM} B_{MH} and B_{VH} are the length of the major bases of the trapezoids of the fuzzy sets LM, MH and VH respectively.
- e_{y-LM-s} e_{y-MH-s} and e_{y-VH-s} are the evidences that the overall customer satisfaction for the service expressed by 90 % of young people is Low, Medium and High respectively. Such values should be obtained by computing for each service the consequence of each rule using the formula (1) where vh are substituted by LM, MH and VH and then by putting e_{y-LM-s} e_{y-MH-s} and e_{y-VH-s} equal to the maximum evidences of the rules having as consequence L, M and H respectively.
- α_{y-LM-s} α_{y-MH-s} and α_{y-VH-s} are the α-cuts of the fuzzy sets LM, MH and VH made at $e = e_{y-LM-s}$, e_{y-MH-s} and e_{y-VH-s} respectively; such cuts are the minor bases of the trapezoids obtained by cutting the fuzzy sets LM, MH and VH.
- X_{LM} X_{MH} X_{VH} are the x-coordinates of the midpoints of the cuts α_{y-LM-s} α_{y-MH-s} and α_{y-VH-s} respectively.
- $(B_{LM} + \alpha_{y-LM-s}) \, e_{y-LM-s}/2$ $(B_{MH} + \alpha_{y-MH-s}) \, e_{y-MH-s}/2$ and $(B_{VH} + \alpha_{y-VH-s}) \, e_{y-VH-s}/2$ are the areas of the trapezoids obtained by the cuts of the fuzzy sets LM, MH and VH made at the evidences $e = e_{y-LM-s}$ $e = e_{y-MH-s}$ and $e = e_{y-VH-s}$

- $A = [(B_{LM} + \alpha_{y-LM-s}) e_{y-LM-s}/2 + (B_{MH} + \alpha_{y-MH-s}) e_{y-MH-s}/2 + (B_{VH} + \alpha_{y-VH-s}) e_{y-VH-s}] /2]$ is the sum of the mentioned areas.

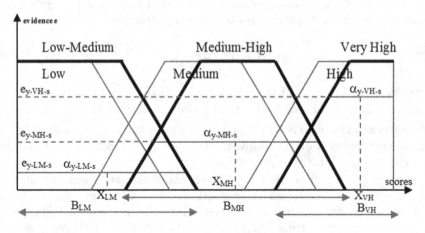

Fig. 5. Graphical representation of the fuzzy sets and variables needed to compute the customer satisfaction using the rule table.

4   An Example

In the current implementation, the *services near me* are displayed on the user mobiles within a circle of radius dmax (or tmax) around the user location computed by the DSS using the mentioned location based fuzzy rules. Figure 6 shows the pharmacies recommended to the users depending on the user age and traffic/weather conditions.

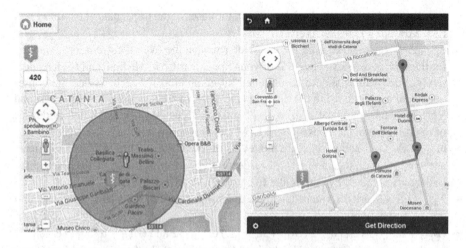

Fig. 6. Pharmacies recommended to the mobile users depending on the user status and traffic/weather conditions (left) and suggested path to the pharmacy chosen by the user.

The advanced version of DSS proposed in the paper, currently under test, filters, upon user request, the services contained in the mentioned circle around the user using the type of customer satisfaction rules chosen by the user. If the users are interested in the third type of customer satisfaction, they should weight each aspect, or alternatively should give a fuzzy rating, i.e., L, M or H, to each aspect configuration. To find the right services, the DSS will apply formula (2) or (3) respectively.

For example, in case the DSS should apply formula (3), the values of the evidences e_{y-LM-s} e_{y-MH-s} and e_{y-VH-s} will be computed using a rule table as the one illustrated in Fig. 7 left dealing with the case in which the user is searching a parking garage. Each aspect is rated by three terms, i.e., Poor, Medium, Good.

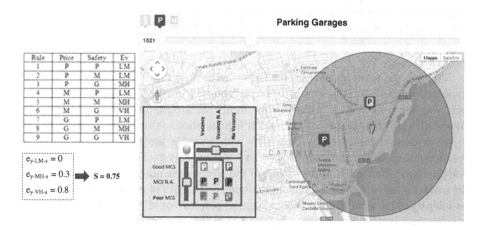

Fig. 7. How a parking garage score S is computed using formula (3) on the left, and how the most recommended garages are signaled to the user on the right, where MCS and NA indicate Mean Customer Satisfaction and Not Available information respectively.

Assuming that a garage has the following aspect evidences: Price: $e_P = 0$, $e_M = 0.2$, $e_G = 0.85$, Safety: $e_P = 0$, $e_M = 0.3$, $e_G = 0.8$, it will be recommended or not by the DSS depending on the following procedure: (a) compute the evidence of each rule as the minimum of the evidences of the involved aspects, (b) assign the maximum evidence of the rules 1,2,4,7 to e_{y-LM-s}, the maximum evidence of the rules 3, 5, 8 to e_{y-MH-s}, and the maximum evidence of the rules 6, 9 to e_{y-VH-s} (see Fig. 7 left), (c) use the fuzzy sets of Fig. 5 to obtain the parking garage score S depending on the user weights, and d) suggest the parking garage if $S \geq 0.7$. Since in our case we have that $S = 0.75$, the parking garage will be recommended to the user as shown in Fig. 7 right.

5   Concluding Remarks

The paper has illustrated how a fuzzy logic based DSS may provide the mobile users of a smart/intelligent city with relevant information that may address suitably the course of their actions. The proposed methodology allows the DSS to take into account many, even conflicting, criteria by a very limited computational effort and to identify solutions that have a practical value for the users.

The proposed approach outperforms the available navigational information systems, either the general navigators such as TomTom and Google Maps or the specialized ones, e.g., [17], since they offer paths to the interest points and relevant services without filtering the solutions to meet suitably the user preferences.

The fuzzification proposed in the paper of the TOPSIS algorithm allows the DSS to support the user decision making process with a simple mathematical apparatus that is able to derive from both crisp and statistical data and conflicting rules the services that have a suitable customer satisfaction. For this reason, neither crisp decision trees or probabilistic models, e.g., Bayesian logic, have been used. In fact, the former approach would recommend the city services according to a priority list that not necessarily meets at the best all the requisites, whereas the latter will recommend the services using probabilistic criteria that don't assure that the suggested service will meet *almost always* the user expectations.

Although the paper recognizes the importance of all the MCDM approaches, we privileged TOPSIS for "its simplicity in perception and usage" [18]. Also, the fuzzy logic was used not only to express with words the aspects involved in TOPSIS as suggested in [19] but also for a fuzzy market segmentation so that TOPSIS may be based on the evaluation expressed by customers with a profile similar to the current user. Decision tables with approximation schema were currently excluded since it is difficult to approximate the user profile with one of the profiles emerging from such tables, but this subject is for further study. The shortcoming of TOPSIS of not being able to handle the interdependence among the criteria, e.g., as pointed out in [20], may be overcome by using the rule table based approach discussed in Sect. 4, where the evaluation is based on the combination of all the aspects involved in the decision.

Future works deal with the user evaluation of the proposed methodology and the development of an ontological version of Wi-City to improve the interoperability of the data stores of city interest and the DSS effectiveness.

References

1. Chourabi, H., et al.: Understanding smart cities: an integrative framework. In: 45th Hawaii International Conference on System Sciences (2012)
2. Berthon, B., et al.: Building and managing an intelligent City (2011). http://www.accentureduckcreekpolicy.com/SiteCollectionDocuments/PDF/Accenture-Building-Managing-Intelligent-City.pdf
3. Costanzo, A., Faro A., Giordano, D.: Wi-City: living, deciding and planning using mobiles in intelligent cities. In: 3rd International Conference on Pervasive Embedded Computing and Communication Systems, PECCS, Barcelona, Spain (2013)
4. Costanzo, A., Faro, A.: Real time decision support systems for mobile users in intelligent cities. In: IEEE Proceedings of the 6th International Conference on Application of Information and Communication Technologies (AICT), Tbilisi, Georgia (2012)
5. Wang, P.P.: Computing With Words. Wiley, New York (2001)

6. Zhai, J., Jiang, J., Yu, Y., Li, J.: Ontology-based integrated information platform for digital city. In: IEEE Proceedings of Wireless Communication, Networking and Mobile Computing, WiCOM 2008 (2008)
7. Corlan, M.: Adobe Flash Platform Tooling: Flash Builder (2009)
8. Yi, Y., La, S.: What influences the relationship between customer satisfaction and repurchase intention? Psychol. Mark. **21**, 351–373 (2004)
9. Chinho, L., Watcharee, L.: Factors affecting online repurchase intention. Ind. Manage. Data Syst. **114**(4), 597–611 (2014)
10. Haverila, M.: Mobile phone feature preferences, customer satisfaction and repurchase intent among male users. Australas. Mark. J. (AMJ) **19**(4), 238–246 (2011)
11. Carev, D.: Guest satisfaction and guest loyalty study for the hotel industry. Dissertation Collection, Rochester Institute of Technology (2008)
12. Kotler, P., Armstrong, G.: Principles of Marketing. Prentice Hall, Englewood Cliffs (2013)
13. CSD: Which Fueling Stations Are Customers' Favorites?. Conv. Dec. Store, July 2015
14. Lee, S., Smith, S., Kim, E.: Factors affecting customer satisfaction with pharmacy franchises. J. Mark. Thought **1**(3), 59–67 (2014)
15. Faro, A., Giordano, D., Spampinato, C.: Evaluation of the traffic parameters in a metropolitan area by fusing visual perceptions and CNN processing of webcam images. IEEE Trans. Neural Netw. **19**(6), 1108–1129 (2008)
16. Wang, Y.-J., Lee, H.S.: Generalizing TOPSIS for fuzzy multiple-criteria group decision-making. Comput. Math. Appl. **53**(11), 1762–1772 (2007)
17. Letchner J., et al.: Trip router with individualized preferences (trip): incorporating personalization into route planning. In: Eighteenth Conference on Innovative Applications of Artificial Intelligence, IAAI 2006, July 2006
18. Aruldos, M., Lakshmi, T.M., Venkatesan, V.P.: A survey on multi criteria decision making methods and its applications. Am. J. Inf. Syst. **1**(1), 31–43 (2013)
19. Hsu, T.K., Tsai, Y.F., Wu, H.H.: The preference analysis for tourist choice of destination: a case study of Taiwan. Tourism Manage. **30**(2), 288–297 (2009)
20. Li, G., Vu, H.Q., Law, R., Rong, J.: Discovering the hotel selection preferences of Hong Kong inbound travelers using the Choquet Integral. Tourism Manage. **36**, 321–330 (2013)

A Framework Supporting the Analysis of Process Logs Stored in Either Relational or NoSQL DBMSs

Bettina Fazzinga[2], Sergio Flesca[1], Filippo Furfaro[1], Elio Masciari[2],
Luigi Pontieri[2(✉)], and Chiara Pulice[1]

[1] DIMES, University of Calabria, Rende, Italy
{flesca,furfaro,cpulice}@dimes.unical.it
[2] ICAR-CNR, Rende, Italy
{fazzinga,masciari,pontieri}@icar.cnr.it

Abstract. The issue of devising efficient and effective solutions for supporting the analysis of process logs has recently received great attention from the research community, as effectively accomplishing any business process management task requires understanding the behavior of the processes. In this paper, we propose a new framework supporting the analysis of process logs, exhibiting two main features: a flexible data model (enabling an exhaustive representation of the facets of the business processes that are typically of interest for the analysis) and a graphical query language, providing a user-friendly tool for easily expressing both selection and aggregate queries over the business processes and the activities they are composed of. The framework can be easily and efficiently implemented by leveraging either "traditional" relational DBMSs or "innovative" NoSQL DBMSs, such as *Neo4J*.

Keywords: Business process logs · Querying

1 Introduction

The increasing diffusion of powerful storage, computing and collaboration technologies has caused a deep change in the way business processes take place. This forced many organizations to deal with an unprecedented amount of process data (in particular, of execution logs), which have become a common kind of organizational asset [11], representing a rich source of knowledge on the behavior of business processes.

From a business company point of view, these data need to be used intelligently in order to compete with other organizations in terms of efficiency, prompt reaction to the market and quality of service. In this respect, the challenge is shifting from the efficient storage of large process data volumes to a new intriguing goal: *How gather actionable information from (big) process data?* Answering this question is rather complex and implies detecting and analyzing typical execution patterns, and evaluating how well a process is aligned to business objectives. Obviously, as process logs are essentially composed of time-related events,

F. Esposito et al. (Eds.): ISMIS 2015, LNAI 9384, pp. 52–58, 2015.
DOI: 10.1007/978-3-319-25252-0_6

one could think of adopting data analysis paradigms defined for non process-aware logs, such as web access logs, search query logs, transaction logs, and several other kinds of event-centered repositories [3,10]. However, these paradigms are not suitable for analyzing process logs, as they disregard the information that some events happened during the execution of the same process instance, which are useful for understanding process logs.

Traditionally, analyzing process data has been the main objective of *Business Process Analysis* approaches, which can be roughly grouped in two main categories: *process discovery methods*, allowing for automatically extracting some kind of process model [9] from log data; and *deductive techniques*, allowing for formulating and evaluating queries [2,4,6,8] against log contents. In this work, we specifically focus on the second flavor of analysis, which can help test hypotheses on the behavior of a business process, and retrieve or aggregate process instances that follow a pattern of interest (involving, e.g., inefficient or forbidden execution flows), and may deserve further investigation/diagnoses.

Unfortunately, none of the currently available log querying tools represents an optimal solution, ensuring satisfactory levels for easiness of use, flexibility/expressiveness, and efficiency of query evaluation. Most of current solutions, indeed, simply reuse general-purpose data and query models, often paying little attention to usability issues. An interesting exception is the work in [2], where a graph-based querying framework is defined for analyzing the instances of an *a-priori known* BPEL process model. Essentially, each process instance is viewed as a nested set of DAGs, while the queries are still based on DAGs specifying execution patterns of interest, based on the usage of typical control-flow constructs and of transitive edges (resp., nodes) for modeling arbitrarily-long execution paths (resp., recursive deepening into a composite activity). Ad-hoc query optimization methods were also defined in [2], exploiting the knowledge on the structure of the process that generated the traces in order to prevent the evaluation of redundant sub-queries. However, such optimization mechanisms cannot be exploited in real scenarios where *no a-priori model* is available for the process (e.g., the case of many CSCW/ERP systems).

We here propose a framework supporting the analysis of business process logs that, taking into account the very nature of business processes, allows (even non-expert) users to query the logs intuitively and efficiently. Our proposal bases on a general data model for a process log, viewed as a set of process instances (*p-instances*). In turn, any p-instance consists of activity instances (*a-instances*), each referring to a process activity (i.e., a well-defined task, performed for the p-instance).

In order to ease the analysis tasks for non expert users, we propose a novel expressive graph-based query language, that allows the analyst to query process data at different levels of detail. A query is based on the notion of *Precedence Graph Pattern* (*PGP*), that allows selection conditions to be expressed on both the structure of a p-instance (both in terms of involved activities/executors and of order relationships among a-instances), and on properties of its a-instances (e.g., starting/ending time, executor, etc.). Besides returning the p-instances that match a pattern, the language allows aggregate statistics to be computed.

The proposed framework is apt to be implemented using different physical data management infrastructures, based either on the robust "traditional" technology of relational DBMSs, or on the "innovative" technology of NoSQL DBMSs (like, e.g., *Neo4J*).

2 Storing Log Data

Retrieving actionable information from log data requires to represent information about both process instances and their associated activity instances. Specifically, for each process instance the analyst may be interested in knowing aggregate information, such as the overall elapsed time. Moreover, for each activity executed in a process instance the analyst may be interested in knowing detailed information about its execution, such as its starting and ending times, the executor that performed the activity and the activity execution context (e.g., the place where the activity is performed). We next introduce the logical model of log data adopted in this work, which can be used as a basis for storing (and efficiently querying) log data in both RDBMSs and NoSQL databases.

Process instances and activity executions are logically represented in two separate tables: P-INSTANCES and A-INSTANCES. Each tuple in P-INSTANCES (called *p-instance*) stores high level information about an *overall* process instance, and has the form $\langle id, type, n_A, n_e, n_c, n_s, t_{tot} \rangle$, where:

- id is a process instance identifier;
- $type$ is a process instance category. This attribute is intended to allow analysts to define a taxonomy (such as *Urgent* or *Not Urgent*) over process instances.
- n_A is the number of distinct activities pertaining the p-instance;
- n_e is the number of different executors involved in the p-instance;
- n_c is the number of different contexts where the process activities took place;
- n_s is the number of a-instances performed within the p-instance;
- t_{tot} is the overall duration of the process instance.

Each tuple in A-INSTANCES (referred as *a-instance*) describes an activity instance (i.e. an activity execution), and takes the form $\langle id, p, A, e, c, \tau_s, \tau_d \rangle$, where:

- id is an a-instance identifier;
- p is a p-instance identifier;
- A is an activity;
- e is an executor;
- c is an execution context, i.e. this attribute can model either a physical device devoted to the activity execution (e.g. a specific server within the organization) or a place where the activity was executed (e.g. a dock);
- τ_s, τ_d are the starting time and the duration of the activity execution.

Basically, an a-instance $\langle id, p, A, e, c, \tau_s, \tau_d \rangle$ describes a "step" of the p-instance p, as it says that this p-instance contains an execution of the activity A, performed by e within the context c starting at time τ_s, and ending at time $\tau_s + \tau_d$. In the following, given a log-tuple t, we will say that t has been "*performed*" (or, equivalently, "*executed*") referring to the execution of the process step represented by t.

3   Querying Log Data

In order to extract actionable information from log data an analyst should be able to pose queries at different levels of detail. In this section, we first introduce a new query language allowing users to pose queries over log data easily and effectively. We then briefly discuss how such queries can be translated into the native languages of both RDBMSs and NoSQL databases, and their potentiality to support OLAP analyses.

3.1   Log Activity Queries

We first introduce the notion of *Activity Query Graph* (AQG). An AQG is a pair $\langle V, E \rangle$ where V is a set of nodes and E is a set of edges. A node $v \in V$ represents an activity instance. Moreover, $v.A$, $v.e$, $v.c$, $v.\tau_s$ and $v.\tau_d$ represent the activity type, executor, context, starting time and duration of the activity instance v, respectively. Edges in E can be either *solid* or *dashed*. A *solid* edge is a pair $\langle v_i, v_j \rangle$, and it specifies that v_i and v_j represent activities *sequentially* executed one after the other. Analogously, a *dashed* edge is a pair $[v_i, v_j]$, and expresses the condition that the activity instance v_j is executed after v_i, but not necessarily sequentially after v_i. Moreover, we point out that all the nodes in V refer to a-instances belonging to the same p-instance.

We can now introduce the notion of *precedence graph pattern - PGP*. A *PGP* π is a triple $\langle AQG, p, \varphi_\pi \rangle$ where AQG is an Activity Query Graph, p is a variable denoting the p-instance including the a-instances onto which the nodes of AQG are mapped and φ_π is a Boolean formula whose terms are of the form $\phi \theta k$, where θ is a comparison operator, k is a constant, and ϕ is linear combination of terms of the form $x.y$, with $x \in V_{AQG} \cup \{p\}$ and either (i) $y \in \{A, e, c, \tau_s, \tau_d\}$, in the case that $x \neq p$, or (ii) $y \in \{type, n_A, n_e, n_c, n_s, t_{tot}\}$, in the case that $x = p$.

We can now define *Log Activity Queries* exploited in our framework.

Definition 1 (Log Activity Query). *A log activity query is a pair $\langle \mathcal{F}, \pi \rangle$, where π is a PGP and $\mathcal{F}$ is either:*

1. *a linear combination of terms of the form $x.y$, with $x \in V_{AQG} \cup \{p\}$ and either (i) $y \in \{A, e, c, \tau_s, \tau_d\}$, in the case that $x \neq p$, or (ii) $y \in \{id, type, n_A, n_e, n_c, n_s, t_{tot}\}$, in the case that $x = p$;*
2. *COUNT*
3. *an aggregation function $\gamma(\psi)$, where $\gamma \in \{MIN, MAX, AVG\}$ and ψ is a linear combination of terms of the form $x.y$, with $x \in V_{AQG} \cup \{p\}$ and either (i) $y \in \{A, e, c, \tau_s, \tau_d\}$, in the case that $x \neq p$, or (ii) $y \in \{type, n_A, n_e, n_c, n_s, t_{tot}\}$, in the case that $x = p$.*

We refer to these kind of queries as log activity queries as they represent a general form of queries that can be instantiated as: *(i) Process queries* if at least a term of the form $p.y$ appears in $\mathcal{F}$ or φ_π, or *(ii) Instance Queries*, otherwise.

Log activity queries can be expressed graphically using our querying interface.

Example 1. Consider two log activity queries Q_1 and Q_2 defined as follows. Q_1 asks for the identifiers of the p-instances containing three activity executions of the same activity $A1$, performed by three different executors and one sequentially after the other, and Q_2 counts the p-instances whose identifiers are returned by Q_1. Query Q_1 can be expressed as a pair $\langle p.id, \pi \rangle$, such that $\pi = \langle AQG, p, \varphi_\pi \rangle$, where: *(i)* $AQG = \langle \{v_1, v_2, v_3\}, \{\langle v_1, v_2 \rangle \langle v_2, v_3 \rangle\} \rangle$, and *(ii)* $\varphi_\pi = v_1.A = A1 \wedge v_2.A = A1 \wedge v_2.A = A1 \wedge v_1.e \neq v_2.e \wedge v_2.e \neq v_3.e \wedge v_1.e \neq v_3.e$. Query Q_2 can be written as $\langle COUNT, \pi \rangle$.

As regard the semantics of log activity queries, it can be easily given in terms of homomorphisms from nodes of the AQG to a-instances.

3.2 Query Rewriting

Queries in the language presented before can be easily translated into the query languages of both relational and NoSQL databases (such as Neo4J), once log data are suitably stored in such databases. Due to space constraint, we only illustrate, in Table 1, a possible translation of the query Q_1 in Example 1 into both SQL and Cypher.

The SQL version of Q_1 is quite straightforward, apart from a technical detail: the usage of an additional table (named here PREC) to store a triple $\langle a_2, a_1, p \rangle$ for each pair of activities a_1, a_2 such that a_2 was performed sequentially after a_1, both within p. This avoids producing an SQL query with negation, whenever it comes to check whether two activities are executed one sequentially after the other.

The translation of Q_1 into Cypher bases on the assumption that log data are represented through the following node types: *Process Instance* and *Activity Instance* (representing p-instances and a-instances, respectively), *Executor Context*, and *Activity*. Each Activity Instance node is also assumed to be linked to a Process Instance node, an Executor node, a Context node, and an Activity

Table 1. Translation of Q_1 in SQL (left) and Cypher (right).

SELECT p.idProcess	MATCH
FROM P-INSTANCES p, A-INSTANCES n0, PREC p01,	(n0.ActivityInstance) - [:INVOLVE] → (p.ProcessInstance),
A-INSTANCES n1, PREC p12, A-INSTANCES n2	(n1:ActivityInstance) - [:INVOLVE] → (p),
WHERE p.idProcess = n0.idProcess	(n2:ActivityInstance) - [:INVOLVE] → (p),
AND p.idProcess = n1.idProcess	(n0) ← [:EXECUTE] (e1.Executor), (n0) - [:NEXT] (n1),
AND p.idProcess = n2.idProcess	(n1) ← [:EXECUTE] (e2.Executor), (n1) - [:NEXT] (n2),
AND n0.IdInstance = p01.idPreviousInstance	(n2) ← [:EXECUTE] (e3.Executor),
AND n1.IdInstance = p01.idInstance	(n0) - [:RECORD] (a0:Activity),
AND n1.IdInstance = p12.idPreviousInstance	(n1) - [:RECORD] (a1:Activity),
AND n2.IdInstance = p12.idInstance	(n2) - [:RECORD] (a2:Activity)
AND n0.executor <> n1.executor	WHERE e1.name <> e2.name AND e1.name <> e3.name
AND n0.executor <> n2.executor	AND e2.name <> e3.name AND a0.type = 'A1'
AND n1.executor <> n2.executor	AND a1.type = 'A1' AND a2.type = 'A1'
AND n0.activity='A1' AND n1.activity='A1'	RETURN p
AND n2.activity='A1'	

node (through the relationships INVOLVE, EXECUTE, HAVE, and RECORD, respectively), as well as to an arbitrary number of Activity Instance nodes (via a relationship NEXT), representing a-instances executed sequentially after the current one.

3.3   Multidimensional Analysis of Log Data

Log Activity Queries can be used to support multidimensional analyses of log data. Indeed, rather than providing only multidimensional analysis features on classical dimensions such as activities, executors and contexts, we allow log data to be analyzed according to an additional *process structure* dimension, associated with each p-instance. The values of this dimension are defined in terms of different behavioral patterns, each specified in terms of a PGP.

Beside allowing users to pose queries at a variable level of granularity for the dimensions of activities, executors and contexts in a classical way, we support two specific drill-down and roll-up operations for the *process structure* dimension.

Given a Log Activity Query $Q = \langle \mathcal{F}, \pi \rangle$, a *drill-down* operation over the dimension of process structure correspond to replacing a dashed edge $[v_i, v_j]$ in π with a *PGP* π'. Specifically when replacing $[v_i, v_j]$ with π' in π, we add an edge from v_i to each of the nodes in π' having no ingoing edge and an edge from each of the nodes in π' having no outgoing edge to v_j.

Conversely, a *roll-up* operation over the dimension of process structure corresponds to the replacement of a subpattern of π with a dashed edge. This capability distinguishes our approach from previous work on process-oriented datacubes [1,4–7], where aggregation hierarchies for process instances are usually defined on non-structural properties of the process (e.g., process type).

To the best of our knowledge, none of the state-of-the-art approaches allows to flexibly and dynamically partition p-instances based on their structural properties. Anyway, in order to speed up the computation of drill-down and roll-up operations over the dimension of process structure, one could think of materializing a specific hierarchy of log activity queries, which is expected to be reused frequently in subsequent analyses. This issue, however, goes beyond the scope of the paper, and it is left for future work.

References

1. Casati, F., Castellanos, M., Dayal, U., Salazar, N.: A generic solution for warehousing business process data. In: Proceedings of VLDB, pp. 1128–1137 (2007)
2. Deutch, D., Milo, T.: Type inference and type checking for queries over execution traces. VLDB J. **21**(1), 51–68 (2012)
3. Gao, X.: Towards the next generation intelligent BPM – in the era of big data. In: Daniel, F., Wang, J., Weber, B. (eds.) BPM 2013. LNCS, vol. 8094, pp. 4–9. Springer, Heidelberg (2013)
4. Grigori, D., Casati, F., Castellanos, M., Dayal, U., Sayal, M., Shan, M.-C.: Business process intelligence. Comput. Ind. **53**(3), 321–343 (2004)

5. Ribeiro, J.T.S., Weijters, A.J.M.M.: Event cube: another perspective on business processes. In: Meersman, R., Dillon, T., Herrero, P., Kumar, A., Reichert, M., Qing, L., Ooi, B.-C., Damiani, E., Schmidt, D.C., White, J., Hauswirth, M., Hitzler, P., Mohania, M. (eds.) OTM 2011, Part I. LNCS, vol. 7044, pp. 274–283. Springer, Heidelberg (2011)

6. Schiefer, J., List, B., Bruckner, R.M.: Process data store: a real-time data store for monitoring business processes. In: Mařík, V., Štěpánková, O., Retschitzegger, W. (eds.) DEXA 2003. LNCS, vol. 2736, pp. 760–770. Springer, Heidelberg (2003)

7. van der Aalst, W.M.P.: Process cubes: slicing, dicing, rolling up and drilling down event data for process mining. In: Song, M., Wynn, M.T., Liu, J. (eds.) AP-BPM 2013. LNBIP, vol. 159, pp. 1–22. Springer, Heidelberg (2013)

8. van der Aalst, W.M.P., de Beer, H.T., van Dongen, B.F.: Process mining and verification of properties: an approach based on temporal logic. In: Meersman, R., Tari, Z. (eds.) OTM 2005. LNCS, vol. 3760, pp. 130–147. Springer, Heidelberg (2005)

9. van der Aalst, W.M.P., van Dongen, B.F., Herbst, J., Maruster, L., Schimm, G., (Ton) Weijters, A.J.M.M.: Workflow mining: a survey of issues and approaches. Data & Knowl. Eng. 47(2), 237–267 (2003)

10. van der Aalst, W.M.P.: A decade of business process management conferences: personal reflections on a developing discipline. In: Barros, A., Gal, A., Kindler, E. (eds.) BPM 2012. LNCS, vol. 7481, pp. 1–16. Springer, Heidelberg (2012)

11. Wang, S., Lv, C., Wen, L., Wang, J.: Managing massive business process models and instances with process space. In: Business Process Management Demos, p. 91 (2014)

An Approximate Proximity Graph Incremental Construction for Large Image Collections Indexing

Frédéric Rayar[1]([✉]), Sabine Barrat[1], Fatma Bouali[2], and Gilles Venturini[1]

[1] Laboratoire d'Informatique, Université François Rabelais de Tours, Tours, France
{frederic.rayar,sabine.barrat,gilles.venturini}@univ-tours.fr
[2] Université Lille 2 - Droit et Santé, IUT, Lille, France
fatma.bouali@univ-lille2.fr

Abstract. This paper addresses the problem of the incremental construction of an indexing structure, namely a proximity graph, for large image collections. To this purpose, a local update strategy is examined. Considering an existing graph G and a new node q, how only a relevant sub-graph of G can be updated following the insertion of q? For a given proximity graph, we study the most recent algorithm of the literature and highlight its limitations. Then, a method that leverages an edge-based neighbourhood local update strategy to yield an approximate graph is proposed. Using real-world and synthetic data, the proposed algorithm is tested to assess the accuracy of the approximate graphs. The scalability is verified with large image collections, up to one million images.

Keywords: Image indexing · Large image collections · Incremental · Proximity graphs

1 Introduction

Dealing with large amount of data has become a great challenge. Advances in technology allow to collect data from almost everywhere, everything and everyone in a continuous way. This permanent flow of data can be nearly infinite and occurs in various fields. A perfect illustration of this phenomenon is the exponential growth of images. Thousands of photos are added each minute on online platforms such as Flickr, Instagram or Facebook.

One challenge, along with the storage of this huge amount of images, is the exploration of these image collections. In order to extract relevant information from these images, one needs to have a relevant representation to observe their global topology and search local information. Proximity graphs [6] have the property of extracting the structure of the data they represent. Each piece of data is represented by a vertex, and two vertices are linked by an edge if they are close enough to be considered as neighbours. Such graphs fit perfectly for purposes such as clustering and outlier detection, but also for indexing and retrieval tasks.

© Springer International Publishing Switzerland 2015
F. Esposito et al. (Eds.): ISMIS 2015, LNAI 9384, pp. 59–68, 2015.
DOI: 10.1007/978-3-319-25252-0_7

Unfortunately, most of these proximity graphs have been studied as static structures, *i.e.* one considers the whole studied collection, and builds a proximity graph for further studies. However, as previously mentioned, as images keep flooding in a continuous way, it becomes mandatory to handle graph structure in a dynamic way: images and their links can be either added, removed or edited if needed. Thus, it becomes essential to have incremental algorithms to build and update proximity graphs, in order to keep organizing images in a dynamic, yet consistent way.

Since one goal is to handle large image collections, one must take into account the following constraints: *(i)* large matrices, namely the adjacency matrix and the distance matrix, must not be stored, *(ii)* parallelism should be leveraged when feasible and *(iii)* heuristics should be considered and assessed to cut the time complexity of the algorithms.

In this paper, we perform a study of the most recent incremental relative neighbourhood graph (RNG) construction algorithm proposed in the literature, to the best of our knowledge, and highlight some drawbacks of this method. Then, an *edge-based neighbourhood* local update strategy is proposed to incrementally build an approximate RNG. Experiments, on both synthetic and real-world datasets, are presented to assess the quality of the proposed strategy and its scalability to handle large image collections.

The rest of the paper is organized as follows: Sect. 2 introduces the relative neighbourhood graph. Related works on proximity graphs incremental construction and their limits are studied. In Sect. 3, the edge-based neighbourhood local update strategy is defined and the related insertion algorithm is given. Experiments on several datasets to assess the proposed algorithm are presented in Sect. 4. Finally, we conclude the paper in Sect. 5.

2   Relative Neighbourhood Graph

Proximity graphs [6] are weighted graphs with no loops. They aim at extracting the structure of a data point set, where each point is represented by a node. They associate an edge between two points if they are close enough to be considered as neighbours. The notable proximity graphs include k-nearest neighbour graph, relative neighbourhood graph, Gabriel graph and Delaunay graph.

In the present paper, we will focus our attention on the RNG. Indeed, it is the smallest connected proximity graph that embeds local information about vertices neighbourhood. The connectivity property guarantees that each image can be reachable during a content-based exploration.

2.1   Definition

The relative neighbourhood graph has been introduced in the work of Toussaint [5]. The construction of this graph is based on the notion of *relatively close* neighbours, that defines two points as relative neighbours if they are at least as close to each other as they are to any other points. From this definition, we can define $RNG = (V, E)$ as the graph built from the points of D where distinct points p and q of D are connected by an edge $\overline{pq}$ if and only if they are relative neighbours. Thus,

$$E(RNG) = \{\overline{pq} \mid p, q \in D, p \neq q, \delta(p,q) \leq max(\delta(p,r), \delta(q,r)), \forall r \in D\backslash\{p,q\}\}.$$

where $\delta : D \times D \to \mathbb{R}$ is a distance function. An illustration of the *relative neighbourhood* of two point $p, q \in \mathbb{R}^2$ is given in Fig. 1.

Fig. 1. Relative neighbourhood (grey area) of two points $p, q \in \mathbb{R}^2$. If no other point lays in this neighbourhood, then p and q are relative neighbours.

The main drawback of the RNG is its construction. The classical and brute-force construction has a complexity of $O(n^3)$, where $n = |D|$ is the number of considered data point. A few works in the literature address this complexity for 2D and 3D points. Their key idea is to build a supergraph of the RNG (*e.g.* the Delaunay graph), and adopt a strategy to eliminate some edges to yield the RNG. Thus, one can find in the literature [2] algorithms for 2D and 3D points, whose complexity are $O(nlog(n))$ and $O(n^{\frac{23}{12}}log(n))$, respectively.

2.2 Incremental Construction

To the best of our knowledge, only few works have been done in the literature regarding the incremental construction of the RNG. Scuturici et al. [4] explain that they insert the new vertex in the existing RNG graph by verifying the relative neighbourhood criteria specified in Sect. 2. The authors state that the graph is locally updated, but no details are given. They experimented up to 10,000 images and evaluated their work using classification performance metrics, namely precision and recall.

In [1], Hacid et al. propose an algorithm to perform local update of a RNG following the insertion of a new vertex. This algorithm is leveraged to incrementally build the RNG. Given a set of vertices V, the incremental construction of the RNG proposed by Hacid et al. consists in *(i)* randomly selecting 2 vertices of V and creating an edge between them and *(ii)* iteratively inserting the other vertices by locally updating the RNG. The insertion algorithm (Algorithm 1) is detailed below.

Let RNG be the relative neighbourhood graph built from the vertices of V, q be a new vertex to be inserted, and $\epsilon \in \mathbb{R}^+$. First the nearest vertex nn of q is sought in V (line 1). The farthest relative neighbour fn of nn is retrieved in the graph RNG (line 2). A hypersphere SR centred around q is then computed as its neighbourhood. All vertices that lay in that hypersphere are retrieved (lines 6–11). The radius of this hypersphere corresponds to the sum of the distances between q and nn, and the one between nn and fn. Note that this hypersphere radius can be magnified thanks to the parameter ϵ (line 3). The neighbourhood relationships of the hypersphere SR are updated (line 12) with the classical brute-force algorithm.

Algorithm 1. Hacid et al.'s insertion algorithm

Input: $RNG = (V, E)$, q, ϵ
Output: $RNG' = (V', E')$
 1: $nn = nearest\_vertex(q, V)$
 2: $fn = farthest\_relative\_neighbour(nn, RNG)$
 3: $sr = (\delta(q, nn) + \delta(nn, fn)) * (1 + \epsilon)$
 4: $V' = V \cup \{q\}$
 5: $E' = E$
 6: $SR = \emptyset$
 7: **for** each $p \in V'$ **do**
 8:     **if** $\delta(p, q) \leq sr$ **then**
 9:         $SR = SR \cup \{p\}$
10:     **end if**
11: **end for**
12: $E' = Update(SR)$
13: **return** $RNG' = (V', E')$

The complexity of this insertion algorithm is $O(2n + n'^3)$, where $n = |E|$ and $n' = |SR|$. The $2n$ term corresponds to the search of the nearest neighbour and the search of vertices that lay in the hypersphere. The second term is the time for updating the neighbourhood relations between the points within the hypersphere with the classical RNG algorithm. The authors state that the incrementally built RNG corresponds exactly to the RNG built with a brute-force algorithm, using a recall measure and graph correspondence.

We have noticed several drawbacks regarding this insertion algorithm. First, the choice of the parameter ϵ, used by the authors to expand the neighbourhood of q. It is empirically set at $\epsilon = 0.1$ in [1]. However, no proof that relative neighbours of the newly inserted point must lay in this magnified hypersphere is given. This could be the cause of losing relative neighbours as illustrated in Fig. 2 (left). Second, due to the spherical definition of the neighbourhood SR, the update step may create false edges. Indeed, the classical RNG algorithm is performed only considering the vertices laying in SR. Figure 2 (right) illustrates such a erroneous edge creation. Thus, the insertion algorithm described above might not incrementally yield the exact RNG, as stated by the authors, due to the loss of edges or the inclusion of bad ones. This has been observed and reported in Sect. 4. Third, an assumption is done stating that $n' << n$, *i.e.* the number of vertices in the hypersphere is way less than the number of previously added points. It may not be the case, for instance, if a set of dense points laying in the same part of the space is considered. This has been experimentally observed for a few datasets. Thus the term n'^3 in the complexity might be an issue.

In the next section, we define an edge-based neighbourhood and propose a strategy that leverages it to locally update an RNG, following the insertion of a new vertex. This strategy is then used to incrementally yield an approximate RNG. Thus, we aim at reducing the time complexity of the insertion algorithm while yielding a refined approximation of the exact RNG.

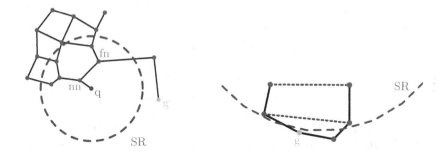

Fig. 2. (Left) Vertex g is a relative neighbour of vertex q. Algorithm 1 fails to retrieve this relationship because g does not lay in the (dashed blue) neighbourhood SR of q, due to ϵ value. **(Right)** False (dotted blue) edges are created. Indeed, as g does not lay in the (dashed blue) neighbourhood SR, it was not considered during the creation of the (dotted blue) edges (Color figure online).

3 Edge-Based Neighbourhood for Local Update

3.1 Definition

Let us consider a simple weighted graph $G = (V, E)$, and a vertex $q \in V$. We introduce the notion of *neighbourhood order*, and recursively define the lth-*order vertex neighbours* (Eq. 1) and the lth-*order edge neighbours* (Eq. 2) of a vertex q. Such sets are illustrated in Fig. 3.

$$\begin{cases} N^1(q) = \{p \in V \mid \overline{pq} \in E\} \\ N^l(q) = \{p \in V \mid \overline{pr} \in E, r \in N^{l-1}(q)\}, \text{ for } l > 1 \end{cases} \quad (1)$$

$$\begin{cases} N_e^1(q) = \{\overline{pq} \in E \mid p \in N^1(q)\} \\ N_e^l(q) = \{\overline{pr} \in E \mid p \in N^l(q) \text{ and } r \in N^{l-1}(q)\} \end{cases} \quad (2)$$

Thus, for a given order L, one can define an edge-based neighbourhood of a vertex q given by:

$$N_e^L(q) = \bigcup_{i=1}^{L} N_e^i(q)$$

3.2 Algorithm

We propose an insertion algorithm (Algorithm 2) that relies on our edge-based neighbourhood N_e^L and a local update strategy.

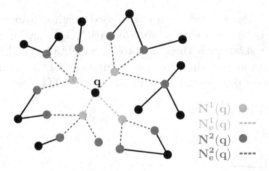

Fig. 3. First and second order *vertex neighbours* (in lightgrey and grey respectively) and *edge neighbours* (in dotted lightgrey and grey respectively) of the vertex q.

The main steps of Algorithm 2 are as follows. The first steps (lines 1–9) are the same as in Algorithm 1. As explained in the previous section, the hypersphere SR centred around q and its content are computed. Then, we retrieve the relative neighbours of q in SR (lines 12–16). For each vertex p in SR, the pair (p, q) is considered. For each vertex r in SR, we check if r lays in the relative neighbourhood of the pair (p, q). If no vertices lay in this relative neighbourhood, then p and q are relative neighbours, and the edge $\overline{pq}$ is created. This step is carried out in $O(n'^2)$, where $n' = |SR|$. STEP_1 gathers all the edges that belong to the edge-based neighbourhood $N_e^L(q)$ of q, given an order L. This is performed with a recursive algorithm. First, we initialise an empty set of edges A. For each relative neighbour q' of q in RNG, we recursively compute the $(L-1)^{th}$-order edge neighbours of q' and store them in A. At the end of this step, the set A contains the list of edges that belong to $N_e^L(q) \backslash N_e^1(q)$. Finally, the effective update is made in STEP_2: for each edge e in A, we check if e has to be removed due to the apparition of q, *i.e.*, if q lays in the relative neighbourhood of the two endpoints of the considered edge. The overall complexity of the proposed insertion algorithm is $O(2n + n'^2 + deg^L)$, where deg is the average degree of the graph RNG.

Thus, we propose here an algorithm that reduces the time complexity of the local update strategy. Moreover, the edge-based neighbourhood allows to verify more edges that may be concerned by the apparition of a new vertex. The trade-off between computation time and accuracy that can be achieved will be studied in the experiments.

4   Experiments

4.1   Experimental Setup

The Algorithms 1 and 2 presented in this paper were implemented in C++. The classical $O(n^3)$ RNG algorithm was also implemented for reference, to assess the graph accuracy. For a fair comparison, the algorithm described by Hacid et al. was implemented under the same constraints as our algorithm (only the

---
Algorithm 2. Edge-based neighbourhood local update strategy

Input: $RNG = (V, E)$, q, ϵ, L
Output: $RNG' = (V', E')$
1: $nn = nearest\_vertex(q, V)$
2: $fn = farthest\_relative\_neighbour(nn, RNG)$
3: $sr = (\delta(q, nn) + \delta(nn, fn)) * (1 + \epsilon)$
4: $SR = \emptyset$
5: **for** each $p \in V$ **do**
6:    **if** $\delta(p, q) \leq sr$ **then**
7:       $SR = SR \cup \{p\}$
8:    **end if**
9: **end for**
10: $V' = V \cup \{q\}$
11: $E' = E$
12: **for** each $p \in SR$ **do**
13:    **if** $are\_relative\_neighbours(p, q)$ **then**
14:       $E' = E' \cup \overline{pq}$
15:    **end if**
16: **end for**
17: **Step_1**: $A = compute\_edge\_neighbourhood(q, L, RNG)$
18: **Step_2**: $update\_edges\_in\_neighbourhood(q, A, E')$
19: **return** $RNG' = (V', E')$

---

local update strategy differs). ϵ value was set to 0.1 as in [1]. In order to speed up some operations (*e.g.* the nearest neighbours search), they were parallelised using OpenMP[1]. In the present experiments, the whole dataset was loaded in memory, and then each piece of data was inserted one by one. The graph was stored as an adjacency list. For runtime experiments, we used an Intel Xeon CPU W3520 (quadcore) at 2.66 Ghz, with 8 Go of RAM.

4.2 Datasets

Five datasets were selected (available on the online UCI machine learning repository[2]). They are either artificial or real world multidimensional datasets. Table 1 summarizes the specifications of the datasets. The three first can be considered as *small* datasets, *i.e.* their distance matrices can be stored in the memory. They were used mainly to assess the validity of our algorithm and the accuracy of the resulting graphs. The two last, which are large image collections up to one million images, were used to verify the scalability of the algorithm. For these datasets, the exact computation of the RNG is not tractable (*n.t.*) in reasonable time with the $O(n^3)$ algorithm. Therefore, a CPU/GPU RNG construction method [3], which can handle up to 300.000 entries, was used to generate the exact graph for the Corel68k dataset. Regarding the MIRFLICKR-1M[3]

---
[1] http://www.openmp.org/.

[2] http://archive.ics.uci.edu/ml.

[3] http://press.liacs.nl/mirflickr.

(MF-1M) image collection, its RNG is not tractable at all, thus the number of edges does not appear in Table 1.

Table 1. Datasets used for experiments. The number of vertices, their dimension and the number of edges in the exact RNG are given.

D	Type	| V |	d	| E(RNG) |
Iris	real world	150	4	195
WDBC	real world	569	30	712
Breiman	artificial	5000	40	17,837
Corel68 k	real world	68,040	57	190,410
MF-1M	real world	1,000,000	150	*n.t.*

All the five datasets share one common property: their attributes are numerical, hence the euclidean distance was used for data comparison. Note that this work can be applied to the data described by categorical features with an appropriate distance function.

4.3 Accuracy Evaluation

First, we evaluate the accuracy of the proposed algorithm and the sensitivity with regards to the parameter L. The exact RNG was computed for the four first datasets. The number of edges of these graphs are used as ground truth and the graph correspondence is computed to evaluate the approximate graphs. Table 2 gives the number of erroneously added edges and removed edges. Since exact RNG could not be produced in reasonable time, this experiment is not reported for the largest dataset, namely *MF-1M*.

One interesting observation in this experiment is that the main difference between the graph produced incrementally and the exact graph is often the addition of wrong edges. Actually, it is not the addition of wrong edges, but rather the fact that some edges are not invalidated after an insertion due to the proposed edge-based neighbourhood. Thus, our algorithm leads to create a few number of false similarities between data, which may not be critical in some applications (*e.g.* similar images retrieval or user recommendation systems).

We notice that Hacid et al.'s algorithm does not always incrementally yield the exact RNG as stated in their paper. Furthermore, our proposed algorithm performs at least as well as, if not better, than Hacid et al.'s algorithm in terms of accuracy, considering low edge-based neighbourhood order ($L = 4$).

As expected, the number of the wrongly added or removed edges in the approximate graphs decreases as the edge-based neighbourhood order increases. Indeed, as more edges are checked, less erroneous edges are left, thus improving the accuracy of the approximate RNG. It is possible to build such a graph with less than 1 % of wrongly added or removed edges considering low order of edge-based neighbourhood (such as $L = 4$).

Table 2. Number of wrongly added edges and removed edges in the RNGs computed by Algorithms 1 and 2. The symbol == means that the approximate graph corresponds exactly to the exact graph.

	| E(RNG) |	Algorithm 1	Algorithm 2		
			$L = 2$	$L = 3$	$L = 4$
Iris	195	+10/− 2	+8/− 1	==	==
WDBC	712	+2/− 1	+10/− 0	+3/− 0	==
Breiman	17837	+0/− 0	+1161/− 0	+299/− 0	+26/− 0
Corel68 k	190410	+20363/− 11	+9089/− 356	+2165/− 388	+637/− 397

4.4 Computation Time Evaluation

Table 3 presents the computation time of Algorithms 1 and 2. Algorithm 2 results are given over the edge-based neighbourhood order L.

We present the computation times for the small *Breiman* dataset in order to highlight the fact that the computation is high for Algorithm 1. Indeed, this dataset illustrates the incremental worst case scenario. Due to the topology of this peculiar dataset, at almost every iteration, the hypersphere SR contains all the vertices previously added. Thus, this dataset is a perfect counterexample of the assumption $n' << n$, made by Hacid et al., as mentioned in Sect. 3.

For the *Corel68k* dataset, the computation time of Algorithm 1 is five days while our algorithm yields the graph in less than half an hour with $L = 4$. The achieved speed-up ratio is slightly less than 273. For the million images collection, the proposed algorithm succeeds in computing an approximate RNG while Hacid et al.'s algorithm is not tractable in reasonable time. Algorithm 2 computation time might seem quite high, yet this time can be reduced by considering an hybrid approach. First, computing the exact RNG of a subset of images with the embarrassingly parallel $O(n^3)$ algorithm. This subset may contain as much images as possible, provided that the memory can handle its distances matrix. Second, processing the rest of the images using our incremental approach. Indeed, experiments have shown that one can expect an average insertion time of less than 500 ms. This is promising for almost real time updates.

Table 3. Comparison of the computation times of Algorithms 1 and 2. Computation times are given in seconds.

	Algorithm 1	Algorithm 2		
		$L = 2$	$L = 3$	$L = 4$
Breiman	7692	16	25	178
Corel68 k	122 h	889	1371	1604
MF-1M	>> 250 h	145 h	151 h	181 h

5   Conclusion

In this paper, the problem of the incremental construction of a proximity graph for large image collection indexing is addressed.

Considering constraints such as memory availability, continuous images incoming, and fast computation, we have leveraged a local update strategy to update only a relevant sub-graph of the existing graph, following the insertion of a new image. This strategy allows to incrementally yield an approximate RNG. The proposed algorithm outperforms the existing work in terms of computation time while yielding a refined approximation. Synthetic datasets have assessed the performance of the algorithm. Moreover, scalability is tested on real-world image collections.

Immediate future work will be to implement the hybrid approach in order to reduce computation time for larger image collections. Then, we plan to leverage this incremental proximity graph construction for clustering and large image collection visualisation.

References

1. Hacid, H., Yoshida, T.: Incremental neighborhood graphs construction for multidimensional databases indexing. In: Canadian Conference on AI (2007)
2. Jaromczyk, J.W., Toussaint, G.T.: Relative neighborhood graphs and their relatives. Proc. IEEE **80**, 1502–1517 (1992)
3. Liu, T., Bouali, F., Venturini, G.: EXOD: a tool for building and exploring a large graph of open datasets. Comput. Graph. **39**, 117–130 (2014)
4. Scuturici, M., Scuturici, V.-M., Clech, J., Zighed, D.A.: Navigation dans une base d'images à l'aide de graphes topologiques. In: Inforsid (2004)
5. Toussaint, G.T.: The relative neighbourhood graph of a finite planar set. Pattern Recogn. **12**, 261–268 (1980)
6. Toussaint, G.T.: Some unsolved problems on proximity graphs (1991)

Experimenting Analogical Reasoning in Recommendation

Nicolas Hug[(✉)], Henri Prade, and Gilles Richard

Institut de Recherche en Informatique de Toulouse, Université Paul Sabatier,
31062 Toulouse Cedex 09, France
{nicolas.hug,henri.prade,gilles.richard}@irit.fr

Abstract. Recommender systems aim at providing suggestions of interest for end-users. Two main types of approach underlie existing recommender systems: content-based methods and collaborative filtering.
In this paper, encouraged by good results obtained in classification by
analogical proportion-based techniques, we investigate the possibility of
using analogy as the main underlying principle for implementing a prediction algorithm of the collaborative filtering type. The quality of a
recommender system can be estimated along diverse dimensions. The
accuracy to predict user's rating for unseen items is clearly an important
matter. Still other dimensions like *coverage* and *surprise* are also of great
interest. In this paper, we describe our implementation and we compare
the proposed approach with well-known recommender systems.

1 Introduction

In a world of information overload, automatic filtering tools are essential to
extract relevant information from basic noise. In the field of e-commerce, recommender systems play the role of search engines when surfing the entire web:
they filter available items to provide relevant suggestions to customers.

Besides, analogical reasoning is widely acknowledged as an important feature
of human intelligence [4,9]. It is a powerful way for establishing parallels between
apparently non related objects, and then guessing relations, properties, or values,
on the basis of the observed similarities and dissimilarities. As a consequence, we
may consider an analogy-based system as a suitable candidate for providing a
user with relevant, and possibly surprising, recommendations. Indeed, providing
the user with both accurate and surprising recommendation has become a key
challenge. This is the option that we investigate in this paper.

Our paper is organized as follows. In the next section, we provide a brief
survey of recommender system technologies. We also review diverse dimensions
along which such a system can be evaluated. In Sect. 3, we provide the necessary
background about analogical proportions and how they can be the basis of an
inference process. In Sect. 4, we investigate how analogical proportions may provide a clean underlying framework to design a recommender system. In Sect. 5,
we report the results obtained on a benchmark and we compare them to those
of well-known existing approaches. Finally, we conclude and provide lines for
future research in Sect. 6.

© Springer International Publishing Switzerland 2015
F. Esposito et al. (Eds.): ISMIS 2015, LNAI 9384, pp. 69–78, 2015.
DOI: 10.1007/978-3-319-25252-0_8

2   Background on Recommender Systems

The aim of a recommendation system is to provide users with lists of relevant personalized items. Let us now formalize the problem.

2.1   Problem Formalization

Let U be a set of users and I a set of items. For some pairs $(u, i) \in U \times I$, a rating r_{ui} is supposed to have been given by u to express if he/she likes or not the item i. It is quite common that $r_{ui} \in [1, 5]$, 5 meaning a strong preference for item i, 1 meaning a strong rejection, and 3 meaning indifference, or just an average note. Let us denote by R the set of ratings recorded in the system. It is well known that, in real systems, the size of R is very small with regard to the potential number of ratings which is $|U| \times |I|$, as a lot of ratings are missing. In the following, U_i denotes the set of users that have rated item i, and I_u is the set of items that user u has rated.

To recommend items to users, a recommender system will proceed as follows:

1. Using a prediction algorithm A, estimate the unknown ratings r_{ui} (i.e. $r_{ui} \notin R$). This estimation $A(u, i)$ is usually denoted $\hat{r}_{ui}$.
2. Using a recommendation strategy S and in the light of the previously estimated ratings, recommend items to users. For instance, a basic yet common strategy is to suggest to user u the items $i \notin I_u$ with the highest $\hat{r}_{ui}$.

The two main prediction techniques are commonly referred to as *content-based* and *collaborative filtering*, that we both briefly review below.

2.2   Content-Based Techniques

Content-based algorithms use the metadata of users and items to estimate a rating. Metadata are external information that can be collected. Typically:

– for users: gender, age, occupation, location (zip code), etc.
– for items: it depends on the type of items, but in the case of movies, it could be their genre, main actors, film director, etc.

Based on these metadata, a content-based system will try to find items that are similar to the ones for which the target user has already expressed a preference (for instance by giving a high rating). This implies the need for a similarity measure between items. A lot of options are available for such metrics. They will not be discussed here as our method is of a collaborative nature.

Indeed, a well-known drawback of content-based techniques is their tendency to recommend only items that users may already know, and therefore the recommendations lack in novelty, surprise and diversity. In the following, we only use collaborative filtering techniques, so we do not consider the use of any metadata.

2.3 Collaborative Filtering Techniques

By collaborative filtering, we mean here algorithms that only rely on the set of known ratings R to make a prediction: to predict $r_{ui} \notin R$, the algorithm A will output $\hat{r}_{ui}$ based on R or on a carefully chosen subset. The main difference between collaborative and content-based method is that in the former, metadata of items and users are not used, and in the latter the only ratings we may take into account are that of the target user. A popular collaborative filtering technique is neighbourhood-based, that we describe here in its simplest form.

To estimate the rating of a user u for an item i, we select $N_i^k(u)$, the set of k users that are most similar to u and that have rated i. Here again, there is a need for a similarity measure (between users, and based on their respective ratings). The estimation of r_{ui} is computed as follows:

$$\hat{r}_{ui} = \underset{v \in N_i^k(u)}{\mathrm{aggr}} \ r_{vi},$$

where the aggregate function aggr is usually a mean weighted by the similarity between u and v. A more sophisticated prediction, popularized by [2] is as follows:

$$\hat{r}_{ui} = b_{ui} + \underset{v \in N_i^k(u)}{\mathrm{aggr}} \ (r_{vi} - b_{vi}),$$

where b_{ui} is a baseline (or bias) related to user u and item i. It is supposed to model how u tends to give higher (or lower) ratings than the average of ratings μ, as well as how i tends to be rated higher or lower than μ. As it uses the neighbourhood of users to output a prediction, this technique tends to model local relationships in the data.

Note that it is perfectly possible to proceed in an item-based way. Indeed, rather than looking for users similar to u, one may look for items similar to i, which leads to formulas dual of the above ones.

2.4 Recommender System Evaluation

Providing an accurate measure of the overall quality of a recommender system is not a simple task and diverse viewpoints have to be considered (see [14] for an extensive survey).

Accuracy. The performance of the algorithm A is usually evaluated in terms of accuracy, which measures how close the rating prediction $\hat{r}_{ui}$ is to the true rating value r_{ui}, for every possible prediction. To evaluate the accuracy of a prediction algorithm, one usually follows the classical machine learning framework: the set of ratings R is divided into two disjoint sets R_{train} and R_{test}, and the algorithm A has to predict ratings in R_{test} based on the ones belonging to R_{train}.

The Root Mean Squared Error (RMSE) is a very common indicator of how accurate an algorithm is, and is calculated as follows:

$$\mathrm{RMSE}(A) = \sqrt{\frac{1}{|R_{test}|} \sum_{r_{ui} \in R_{test}} (\hat{r}_{ui} - r_{ui})^2}$$

To better reflect the user-system interaction, other precision-oriented metrics are generally used in order to provide a more informed view.

Precision and Recall. Precision and recall help measuring the ability of a system to provide relevant recommendations. In the following, we denote by I_S the set of items that the strategy S will suggest to the users using the predictions coming from A. For ratings in the interval $[1,5]$, a simple strategy could be for example to recommend an item i to user u if the estimation rating $\hat{r}_{ui}$ is greater than 4.

$$I_S = \{i \in I | \exists u \in U, \hat{r}_{ui} \geq 4\}.$$

Let I_{relev} be the set of items that are actually relevant to users (i.e. the set of items that would have been recommended to users if all the predictions made by A were exact). The precision of the system is defined as the fraction of recommended items that are relevant to the users:

$$\text{Precision} = \frac{|I_S \cap I_{relev}|}{|I_S|},$$

and the recall as the fraction of recommended items that are relevant to the users out of all possible relevant items:

$$\text{Recall} = \frac{|I_S \cap I_{relev}|}{|I_{relev}|},$$

If accurate predictions are crucial, it is widely agreed that it is insufficient for deploying an effective recommendation engine. Indeed, still other dimensions are worth estimating in order to get a complete picture of the performance of a system. [5,6,8]. For instance, one may naturally expect from a recommender system not only to be accurate, but also to be surprising, and to be able to recommend a large number of items. When evaluating the recommendation strategy, one must keep in mind that its performance is closely related to that of the algorithm A, as the recommendation strategy S is based on the predictions provided by A.

Coverage. Coverage, in its simplest form, is used to measure the ability of a system to recommend a large amount of items: it is quite easy indeed to create a recommender system that would only recommend very popular items. Such a recommender system would drop to zero added value. Coverage can be defined as the proportion of recommended items out of all existing items:

$$\text{Coverage} = \frac{|I_S|}{|I|}.$$

Surprise. Users expect a recommender system to be surprising: recommending an extremely popular item is not really helpful. Following the works of [6], surprise of a recommendation can be evaluated with the help of the pointwise mutual information (PMI). The PMI between two items i and j is defined as follows:

$$PMI(i,j) = -\log_2 \frac{p(i,j)}{p(i)p(j)} / \log_2 p(i,j),$$

where $p(i)$ and $p(j)$ represent the probabilities for the items to be rated by any user, and $p(i,j)$ is the probability for i and j to be rated together : $p(i) = \frac{|U_i|}{|U|}$ and $p(i,j) = \frac{|U_i \cup U_j|}{|U|}$. PMI values fluctuate between the interval $[-1,1]$, -1 meaning that i and j are never rated together and 1 meaning that they are always rated together. To estimate the surprise of recommending an item i to a user u we have two choices:

– either to take the maximum of the PMI values for i and all other items rated by u, with $Surp^{max}(u,i) = \max_{j \in I_u} PMI(i,j)$

– or to take the mean of these PMI values with $Surp^{avg}(u,i) = \frac{\sum_{j \in I_u} PMI(i,j)}{|I_u|}$

Then the overall capacity of a recommender to surprise its users is the mean of the surprise values for all predictions.

3 Background on Analogical Proportions

An analogical proportion "a is to b as c is to d" states analogical relations between the pairs (a,b) and (c,d), as well as between the pairs (a,c) and (b,d). It is only rather recently that formal definitions have been proposed for analogical proportions, in different settings [7,10,16]. In this section, we provide a brief account of a formal view of analogical relations that underlie their use in the proposed algorithm. For more details, see [11–13].

Formal Framework. It has been agreed, since Aristotle time, that an analogical proportion T, as a quaternary relation, satisfies the three following characteristic properties:

1. $T(a,b,a,b)$ (reflexivity)
2. $T(a,b,c,d) \implies T(c,d,a,b)$ (symmetry)
3. $T(a,b,c,d) \implies T(a,c,b,d)$ (central permutation)

There are various models of analogical proportions, depending on the target domain. When the underlying domain is fixed, $T(a,b,c,d)$ is simply denoted $a:b::c:d$. Standard examples are:

– Domain $\mathbb{R}$: $a:b::c:d$ iff $a/b = c/d$ iff $ad = bc$ (geometric proportion)
– Domain $\mathbb{R}$: $a:b::c:d$ iff $a-b = c-d$ iff $a+d = b+c$ (arithmetic proportion)
– Domain $\mathbb{R}^n$: $\boldsymbol{a}:\boldsymbol{b}::\boldsymbol{c}:\boldsymbol{d}$ iff $\boldsymbol{a}-\boldsymbol{b} = \boldsymbol{c}-\boldsymbol{d}$. This is just the extension of arithmetic proportion to real vectors. In that case, the 4 vectors $\boldsymbol{a},\boldsymbol{b},\boldsymbol{c},\boldsymbol{d}$ build up a parallelogram.
– Domain $\mathbb{B} = \{0,1\}$: $a:b::c:d$ iff $(a \wedge d \equiv b \wedge c) \wedge (a \vee d \equiv b \vee c)$
– Domain $\mathbb{B}^n$: $\boldsymbol{a}:\boldsymbol{b}::\boldsymbol{c}:\boldsymbol{d}$ iff $\forall i \in [1,n], \quad a_i : b_i :: c_i : d_i$.

Other definitions are available when dealing with matriceces, formal concepts or lattices, etc. (see [7,16] for other options). In this paper, we are interested in evaluating analogical proportions between ratings, which might be Boolean (*like/dislike*), or in our case study, integer-valued (using the scale $\{1,2,3,4,5\}$).

Equation Solving. Starting from such an analogical proportion, the equation solving problem amounts to finding a fourth element x to make the incompletely stated proportion $a : b :: c : x$ to hold. As expected, the solution of this problem depends on the domain on which the analogy is defined. For instance, in the case of extended arithmetic proportions, the solution always exists and is unique: $x = b - a + c$. In terms of geometry, this simply tells us that given 3 points, we can always find a fourth one to build a parallelogram.

The existence of a unique solution is not always granted: for instance in the Boolean setting, the solution may not exist [10].

Analogical Inference. In this perspective, analogical reasoning can be viewed as a way to infer new plausible information, starting from observed analogical proportions. The *analogical jump* is an unsound inference principle postulating that, given 4 vectors a, b, c, d such that the proportion holds on some components, then it should also hold on the remaining ones. This can be stated as (where $a = (a_1, a_2, \cdots a_n)$, and $J \subset [1, n]$):

$$\frac{\forall j \in J, a_j : b_j :: c_j : d_j}{\forall i \in [1, n] \setminus J, a_i : b_i :: c_i : d_i} \quad (analogical\ inference)$$

This principle leads to a prediction rule in the following context:

- 4 vectors a, b, c, d are given where d is partially known: only the components of d with indexes in J are known.
- Using analogical inference, we can predict the missing components of d by solving (w.r.t d_i) the set of equations (in the case they are solvable):

$$\forall i \in [1, n] \setminus J, \quad a_i : b_i :: c_i : d_i.$$

In the case where the items are such that their last component is a label, applying this principle to a new element d whose label is unknown leads to predict a candidate label for d.

4 Analogical Recommendation

The main idea is that if an analogical proportion stands between four users a, b, c, d, meaning that for each item j that they have commonly rated, the analogical proportion $r_{aj} : r_{bj} :: r_{cj} : r_{dj}$ holds, then it should also hold for an item i that a, b, c have rated but d has not (i.e. r_{di} is the missing component). This leads us to estimate r_{di} as the solution $x = \hat{r}_{di}$ of the following analogical equation:

$$r_{ai} : r_{bi} :: r_{ci} : x.$$

Given a pair (u, i) such that $r_{ui} \notin R$ (i.e. there is no available rating from user u for item i), the main procedure is as follows:

1. find the set of 3-tuples of users a, b, c such that an analogical proportion stands between a, b, c, and u and such that the equation $r_{ai} : r_{bi} :: r_{ci} : x$ is solvable.

2. solve the equation $r_{ai} : r_{bi} :: r_{ci} : x$ and consider the solution x as a candidate rating for r_{ui}.
3. set $\hat{r}_{ui}$ as an aggregate of all candidate ratings.

The first step simply states that users a, b, c and u, considered as vectors of ratings, constitute a parallelogram. This condition is a bit strong and we may want to relax it by allowing some deformation of this parallelogram. This can be done by choosing another condition, such as $||(a - b) - (c - d)|| \le \lambda$ where λ is a suitable threshold and $||.||$ denotes the Euclidean norm.

Another modification to the algorithm would be to only search for the users a, b, and c in a subset of U. One may consider the set of the k-nearest neighbours of d, using the assumption that neighbours are the most relevant users to estimate a recommendation for d.

Obviously, analogical proportion may be applied in an item-based way rather than in a user-based way, as in the case of standard techniques. Both views will be considered in the experimentation.

Implementation. In our implementation, we consider the basic definition of analogy using the arithmetic definition in $\mathbb{R}^n$: $a : b :: c : d \iff a - b = c - d$. Given 4 users, they are represented as real vectors of dimension m, where m is the number of item they have rated in common. This dimension can change with the 4-tuples of users that we consider. Then, the threshold λ has to be a function of this dimension m, as the range of values that $||.||$ may take depends on it. Using cross-validation, we have found that $\lambda = \frac{3}{2} \cdot \sqrt{m}$ showed the best results and acts as a kind of normalization factor. Our pseudo-code is described in Algorithm 1.

5   Experiments and Results

Our algorithm *Analogy* ($k = 20$) is compared to the neighbourhood-based algorithms described in Sect. 2.3, referred to as *kNN* for the basic model and *Bsln-kNN* for the extended model using a baseline predictor (with $k = 40$). For each of the algorithms, we have estimated the metrics described in Sect. 2.4. The recommendation strategy S is to recommend i to u if $\hat{r}_{ui} \ge 4$.

Dataset. We have tested and compared our algorithm on the Movielens-100K dataset[1], composed of 100,000 ratings from 1000 users on 1700 movies. Each rating belongs to the interval $[1, 5]$.

Evaluation Protocol. In order to obtain meaningful measures, we have run a five-folds cross-validation procedure: for each of the five steps, the set of ratings R is split into two disjoint sets R_{train} and R_{test}, R_{train} containing five times more ratings than R_{test}. The reported measures are averaged over the five steps.

Results and Comments. Table 1 shows the performances of the algorithms applied in a user-based way. Similar experiments have been led in a movie-based setting, exhibiting very similar results, slightly worse for RMSE (about 5‰ higher).

---

[1] http://grouplens.org/datasets/movielens/.

Algorithm 1. Analogy

Input: A set of known ratings R, a user u, an item i such that $r_{ui} \notin R$.
Output: $\hat{r}_{ui}$, an estimation of r_{ui}
Init:
$C = \varnothing$ // list of candidate ratings
for all users a, b, c such that

1. $r_{ai} \in R, r_{bi} \in R, r_{ci} \in R$
2. $r_{ai} - r_{bi} = r_{ci} - x$ is solvable // i.e. the solution $x \in [1, 5]$
3. $||(a - b) - (c - d)|| \leq \lambda$ // Analogy almost stands between a, b, c, d considered as real vectors

do
  $x \leftarrow r_{ci} - r_{ai} + r_{bi}$
  $C \leftarrow C \cup \{x\}$ // add x as a candidate rating
end for
$\hat{r}_{ui} = \underset{x \in C}{\text{aggr}} \, x$

Table 1. Performance of algorithms

	RMSE	Prec	Rec	Cov	$Surp^{max}$	$Surp^{avg}$	Time
Analogy	.898	89.1	43.3	31.2	0.433	0.199	2 h
kNN	.894	89.1	44.1	27.8	0.432	0.198	10 s
Bsln-kNN	.865	88.4	44.0	44.7	0.431	0.199	10 s

As expected, the Bsln-kNN algorithm is more accurate than the basic collaborative filtering method (KNN). The two classical collaborative algorithms perform better than the new proposed analogy-based method in terms of RMSE. Still, there seems to be some room for improvement for the analogical approach, with the help of a careful analysis of the behaviour of the algorithm.

As for performances other than RMSE, we see that the figures obtained by the three algorithms are quite close. Regarding surprise, which is a delicate notion to grasp, one may also wonder if the used measure is fully appropriate.

As usual, analogy-based algorithms suffer from their inherent cubic complexity. In the case of recommender systems where millions of users/items are involved, this is also a serious issue.

6   Conclusion and Future Research

This is clearly a preliminary study of an analogical approach to prediction in recommendation, a topic that has never been addressed before. First results are not better than the ones obtained with standard approaches. Even if they do not look that far, the difference is still significant enough to have a genuine impact on the users' experience. However, it is interesting to observe that approaches based on quite different ideas may lead to comparable results.

Besides, it should be recognized that the prediction part of the recommendation problem, although somewhat similar to a classification problem (for which analogical proportion-based classifiers are successful), presents major differences, since grades on items (playing here the role of descriptive features) may be both quite redundant and somewhat incomplete for providing a meaningful profile of a user. This may explain that the application of an analogical proportion-based approach is less straightforward in recommendation than in classification.

Analogical proportion-based methods have also been used recently for predicting missing Boolean values in databases [3]. The recommendation problem can be also viewed as a problem of missing values, but here the proportion of unknown data is very high, and data are not Boolean. This again suggests that the recommendation task is more difficult.

There are quite a number of issues to further explore, such as understanding on what types of situation an analogical proportion-based approach would perform better, and when another view is preferable. Another basic issue is the fact that users likely use the rating scale $\{1, 2, 3, 4, 5\}$ in an ordinal way rather than in an absolute manner, since the meaning of a rating may change with users. This calls for the use of analogical proportion between ordinal data [1].

Lastly, the ideas of the exploitation of the creative power of analogy for (i) proposing items never considered by a user, but having some noticeable common features with items (s)he likes [15], (ii) of the explanation power of analogy for suggesting to the user why an item may be of interest for him, are still entirely to explore.

References

1. Barbot, N., Miclet, L.: La proportion analogique dans les groupes. applications à l'apprentissage et à la génération. In: Proceedings Conference Francophone sur l'Apprentissage Artificiel (CAP), Hammamet, Tunisia (2009)
2. Bell, R.M., Koren, Y.: Lessons from the netflix prize challenge. SIGKDD Explor. Newsl. **9**(2), 75–79 (2007)
3. Correa Beltran, W., Jaudoin, H., Pivert, O.: Estimating null values in relational databases using analogical proportions. In: Laurent, A., Strauss, O., Bouchon-Meunier, B., Yager, R.R. (eds.) IPMU 2014, Part III. CCIS, vol. 444, pp. 110–119. Springer, Heidelberg (2014)
4. Gentner, D., Holyoak, K.J., Kokinov, B.N.: The Analogical Mind: Perspectives from Cognitive Science. Cognitive Science, and Philosophy. MIT Press, Cambridge (2001)
5. Herlocker, J.L., Konstan, J.A., Terveen, L.G., Riedl, J.T.: Evaluating collaborative filtering recommender systems. ACM Trans. Inf. Syst. **22**(1), 5–53 (2004)
6. Kaminskas, M., Bridge, D.: Measuring surprise in recommender systems. In: Adamopoulos, P., et al. (ed.) Proceedings of the Workshop on Recommender Systems Evaluation: Dimensions and Design (Workshop Programme of the 8th ACM Conference on Recommender Systems) (2014)
7. Lepage, Y.: De l'analogie rendant compte de la commutation en linguistique. Habilit. à Diriger des Recher., Univ. J. Fourier, Grenoble (2003)

8. McNee, S.M., Riedl, J., Konstan, J.A.: Being accurate is not enough: how accuracy metrics have hurt recommender systems. In: Olson, G.M., Jeffries, R., (eds.) Extended Abstracts Proceedings of the 2006 Conference on Human Factors in Computing Systems (CHI 2006), Montréal, Québec, Canada, 22–27 April, pp. 1097–1101 (2006)

9. Melis, E., Veloso, M.: Analogy in problem solving. In: Handbook of Practical Reasoning: Computational and Theoretical Aspects. Oxford University Press (1998)

10. Miclet, L., Prade, H.: Handling analogical proportions in classical logic and fuzzy logics settings. In: Sossai, C., Chemello, G. (eds.) ECSQARU 2009. LNCS, vol. 5590, pp. 638–650. Springer, Heidelberg (2009)

11. Prade, H., Richard, G.: Analogical proportions and multiple-valued logics. In: van der Gaag, L.C. (ed.) ECSQARU 2013. LNCS, vol. 7958, pp. 497–509. Springer, Heidelberg (2013)

12. Prade, H., Richard, G.: From analogical proportion to logical proportions. Log. Univers. 7(4), 441–505 (2013)

13. Prade, H., Richard, G.: Homogenous and heterogeneous logical proportions. IfCoLog J. Log. Appl. 1(1), 1–51 (2014)

14. Ricci, F., Rokach, L., Shapira, B., Kantor, P.B.: Recommender Systems Handbook. Springer, Cambridge (2011)

15. Sakaguchi, T., Akaho, Y., Okada, K., Date, T., Takagi, T., Kamimaeda, N., Miyahara, M., Tsunoda, T.: Recommendation system with multi-dimensional and parallel-case four-term analogy. In: Proceedings of IEEE International Conference on Systems, Man, and Cybernetics (SMC 2011), pp. 3137–3143 (2011)

16. Yvon, F., Stroppa, N.: Formal models of analogical proportions. Technical report D008, Ecole Nationale Supérieure des Télécommunications, Paris (2006)

Personalized Meta-Action Mining for NPS Improvement

Jieyan Kuang[1], Zbigniew W. Raś[1,2($\boxtimes$)], and Albert Daniel[1]

[1] KDD Laboratory, College of Computing and Informatics,
University of North Carolina, Charlotte, NC 28223, USA
{jkuang1,ras}@uncc.edu

[2] Institute of Computer Science, Warsaw University of Technology,
00-665 Warsaw, Poland

Abstract. The paper presents one of the main modules of HAMIS recommender system built for 34 business companies (clients) involved in heavy equipment repair in the US and Canada. This module is responsible for meta-actions discovery from a large collection of comments, written as text, collected from customers about their satisfaction with services provided by each client. Meta-actions, when executed, trigger action rules discovered from customers data which are in a table format. We specifically focus on the process of mining meta-actions, which consists of four representative and characteristic steps involving sentiment analysis and text summarization. Arranging these four steps in proposed order distinguishes our work from others and better serves our purpose. Compared to procedures presented in other works, each step in our procedure is adapted accordingly with respect to our own observations and knowledge of the domain. Results obtained from the experiments prove the high effectiveness of the proposed approach for mining meta-actions.

1 Introduction

Improving companies' NPS (Net Promoter System) has become one of the hottest topics nowadays since NPS is the most popular measurement for evaluating the performance of a company's growth engine. Generally speaking, NPS systems categorizes customers into three groups: *Promoter, Passive* and *Detractor*, which describe customers' satisfaction, loyalty and likelihood to recommend the company in a descending order [14].

The dataset we have contains over 42,000 records that are collected from 34 clients dealing with similar businesses crossing US and Canada. Each record represents answers to a questionnaire sent to a randomly selected customer. The questionnaire asks customers' personal information, general information about the service, but more importantly, customers' feeling on the service, such as *"was the job completed correctly"* and *"are you satisfied with the job"*. To answer such questions, customers assign scores ranging from 0 to 10 (higher the score is, more satisfied the customer is), and detailed comments in text format are also recorded if customers left any. Based on the numerical values given by customers,

F. Esposito et al. (Eds.): ISMIS 2015, LNAI 9384, pp. 79–87, 2015.
DOI: 10.1007/978-3-319-25252-0_9

the average score of each customer can be computed and used to determine their NPS status: 9–10 is promoter, 7–8 is passive and 0–6 is detractor. Additionally, the NPS rating of each client can be calculated as the percentage difference between customers that are promoter and customers that are detractor.

In our dataset, NPS rating of individual client ranges from 0.503 to 0.86. Our ultimate goal is to provide proper suggestions for improving NPS rating of every client, in other words, improving customer satisfaction. To achieve this goal, we intend to build a hierarchically structured recommender system driven by action rules and meta-actions. We already built a hierarchical dendrogram by applying agglomerative clustering algorithm to the semantic distance matrix covering 34 clients [6]. Based on the dendrogram, we also proposed a strategy called Hierarchically Agglomerative Method for Improving NPS (HAMIS) to extend every client's dataset by merging it with other clients' datasets which are relatively close in the dendrogram, have better NPS, and classifiers extracted from them have higher precision and recall [7]. After a new maximally enlarged dataset is assigned to a client by HAMIS, action rules and meta-actions have to be extracted from it.

In this paper, we mainly focus on mining meta-actions. Many strategies have been designed for discovering action rules, but the area of mining meta-actions is still blossoming. The concept of action rules and meta-actions will be recalled in the next section. After that, the process of mining meta-actions will be explained thoroughly via four steps: *(1)* Identifying opinion sentences and their orientation with localization; *(2)* Summarizing each opinion sentence using discovered dependency templates; *(3)* Opinion summarizations based on identified feature words; *(4)* Generating meta-actions with regard to given suggestions. To test the proposed method for extracting meta-actions, experiments with a sample dataset are made. Evaluation results prove its high accuracy and effectiveness.

2   Action Rules, Meta-Actions and the Process of Generating Them

The concept of an action rule was firstly proposed by Ras and Wieczorkowska in [12] and investigated further in [3,16]. It is defined as a term $[(\omega) \wedge (\alpha \rightarrow \beta)] \Rightarrow (\phi \rightarrow \psi)$, where $(\omega \wedge \alpha) \Rightarrow \phi$ and $(\omega \wedge \beta) \Rightarrow \psi$ are classification rules, ω is a conjunction of stable attribute values, $(\alpha \rightarrow \beta)$ shows changes of flexible attribute values, and $(\phi \rightarrow \psi)$ shows desired effect of the action. Now we give an example assuming that a is stable attribute, b is flexible attribute and d is decision attribute. Terms (a, a_2), $(b, b_1 \rightarrow b_2)$, $(d, d_1 \rightarrow d_2)$ are examples of atomic actions. Expression $r = [(a, a_2) \wedge (b, b_1 \rightarrow b_2)] \Rightarrow (d, d_1 \rightarrow d_2)$ is an example of an action rule saying that if value a_2 of a in a given object remains unchanged and its value of b will change from b_1 to b_2, then its value of d is expected to transition from d_1 to d_2.

Meta-actions are the actions that need to be executed in order to trigger corresponding atomic actions. The concept of *meta-action* was initially proposed in [17]. But unlike the traditional understanding of meta-actions in [16], in our domain one atomic action can be invoked by more than one meta-action. So a set of meta-actions will trigger an action rule which consists of atomic actions covered by these meta-actions. Also some action rules can be invoked by more than one set of meta-actions. By selecting a proper set of meta-actions we could benefit in triggering larger number of action rules.

As meta-actions are the actual tools to trigger action rules and ulteriorly improve NPS ratings, the process of discovering them is what we need and have accomplished in this paper. Triggers aiming at different action rules should be extracted from respectively relevant comments left by customers in our domain. Let's assume that action rule r mentioned earlier is our target, and two classification rules r_1 and r_2 have been used to construct r [13], so the clues for generating meta-actions are in the comments stored in records satisfying the description $(a, a_2) \wedge (b, b_1) \wedge (d, d_1)$ or $(a, a_2) \wedge (b, b_2) \wedge (d, d_2)$.

To generate meta-actions from a determined set of comments, four steps mentioned earlier are designed to accomplish this task. The whole process involves not only the sentiment analysis and text summarization, but also generation of appropriate suggestions as meta-actions, which is more characteristic for our purpose. Before going into details of each step, adjusting the order of steps 1–3 is another uniqueness of our approach. This adjustment is made because of two reasons: *(a)* Data is uncleaned. Unlike reviews on particular products or experience mentioned in other research, some comments in our domain are useless due to lack of opinion orientation; *(b)* Former steps can benefit latter steps. After comparing the methods of extracting aspects (features) from different formatted comments, it turns out that dealing with reviews in short segments is more efficient without compromising their effectiveness than dealing with them in long segments. Therefore, ordering the first three steps in a way we proposed accelerates the process by eliminating useless information and makes preparing the data easier to handle for the next step. Most crucially, the accuracy will get improved.

2.1 Identification of Opinion Sentences and the Orientation with Localization

To identify an opinion sentence which expresses customers' sentiment, the presence of opinion words is considered as a standard sign. Initially adjectives are usually used as the main opinion words, like Hu and Liu have used only adjectives in [4]. Two sets of opinion words expressing positive and negative feelings are generated. Although these sets of opinion words are still growing continually, using them as the only references to detect opinion words and their orientation is not sufficient due to its generality. In some local scenarios, the lists of opinion words can be expanded more broadly by considering some neutral words with implicit polarity. For example, a comment *"the charge was too high"* can not be associated with given lists because the adjective *"high"* is not recognized as a positive nor negative word. However it definitely presents a useful message reflecting customers' negative opinion about the price, so the word *"high"* can be treated as a negative opinion word in this case. Similarly, other special neutral words or phrases can be added as opinion words if they reflect oriented meanings under certain circumstances without confusion. Such addition strongly relies on designers' knowledge about the domain.

Hence, based on our own experience, some neutral adjectives and verbs are added into our library of opinion words with clarified orientations. Four types of words that could have orientations are used to filter the appearance of opinion words and they are: *verb, adjective, adverb* and *noun*. As long as a word in a sentence tagged as any one of the four types exists in an extended list of opinion words, this sentence is an opinion sentence and the orientation of a tagged opinion word depends on its ascription to which list. The orientation of a sentence is determined by following the basic principles summarized in [9], when there is only one opinion word. Otherwise, the orientation of a sentence is a collection of the orientations of all subsentences associating with corresponding opinion words.

2.2 Summarization of Opinion Sentence Based on Dependency Relationships

With opinion sentences identified, shortening them into segments is an important procedure. Relevant research like [15,18] constructs feature-opinion pairs with grammatical rules describing the relationships between features and opinion words. Without pre-identified features in opinion sentences, extracting summaries from every sentence by following certain grammatical relations associated with opinion words solely is also applicable and sufficient for two reasons. Firstly, unlike other relevant works, there is no need of Part-of-Speech (POS) [10] tagging, as the grammatical structure of a sentence is the only factor that we depend on. Secondly, the grammatical relations of the expected portion most closely connecting to the opinion words in a sentence can be summarized based on the knowledge of linguistics and used to extract a short but meaningful segment from a complete sentence.

The foundation of this step is based on the grammatical relations defined by Stanford Typed Dependencies Manual [2] and generated by Stanford Parser. A dependency relationship describes a grammatical relation between a governor word and a dependent word in a sentence and it is represented as $d(G, D)$, where d is one type of dependency among approximately 50 defined dependencies in [2], G and D are the governor and dependent respectively. With the comprehensive representation of dependencies, the nearest necessary relations associated with opinion words can be identified. Moreover, the types of dependencies that could link to opinion words straightforwardly rely on the tags of opinion words. Table 1 demonstrates all the discovered dependency templates for four types of opinion words respectively.

In Table 1, for each type of opinion words, all the other words linked with opinion words directly or indirectly are labeled as D_* regarding the dependency type. In one template, there could be more than one dependency, and the segment result is the combination of all involved words in a sequence as shown in the third column. When there are two or more templates involved with one opinion word,

Table 1. Dependency templates for extracting sentence segments

Type of opinion words	Dependency template	Segment result
Noun	$\mathrm{nsubj}(op, D_{nsubj})$	$D_{nsubj} + op$
	$\mathrm{prep}(op, D_{prep}) + \mathrm{pobj}(D_{prep}, D_{pobj})$	$op + D_{prep} + D_{pobj}$
	$\mathrm{dobj}(D_{dobj}, op)$	$D_{dobj} + op$
Adjective	$\mathrm{nsubj}(op, D_{nsubj})$	$D_{nsubj} + op$
	$\mathrm{amod}(op, D_{amod}) + \mathrm{vmod}(D_{amod},$ $D_{vmod}) + \mathrm{dobj}(D_{vmod}, D_{dobj})$	$op + D_{amod} + D_{vmod} +$ D_{dobj}
	$\mathrm{xcomp}(op, D_{xcomp}) + \mathrm{dobj}(D_{xcomp},$ $D_{dobj})$	$op + D_{xcomp} + D_{dobj}$
	$\mathrm{prep}(op, D_{prep}) + \mathrm{pobj}(D_{prep}, D_{pobj})$	$op + D_{prep} + D_{pobj}$
	$\mathrm{pcomp}(op, D_{pcomp}) + \mathrm{dobj}(D_{pcomp},$ $D_{dobj})$	$op + D_{pcomp} + D_{dobj}$
	$\mathrm{vmod}(op, D_{vmod}) + \mathrm{dobj}(D_{vmod}, D_{dobj})$	$op + D_{vmod} + D_{dobj}$
Adverb	$\mathrm{advmod}(op, D_{advmod})$	$D_{advmod} + op$
Verb	$\mathrm{dobj}(op, D)$	$op + D$
	$\mathrm{prep}(op, D_{prep}) + \mathrm{pobj}(D_{prep}, D_{pobj})$	$op + D_{prep} + D_{pobj}$
	$\mathrm{xcomp}(op, D_{xcomp}) + \mathrm{dobj}(D_{xcomp},$ $D_{dobj})$	$op + D_{xcomp} + D_{dobj}$
	$\mathrm{advcl}(op, D_{advcl}) + \mathrm{dobj}(D_{advcl}, D_{dobj})$	$op + D_{advcl} + D_{dobj}$

[a] *op* denotes opinion word.

words from all detected templates will be combined sequentially as the words appear in the sentence. For example, if only dependency *nsubj* is discovered in a sentence associated with a *noun* opinion word, then the final segment is "D_{nsubj} *op*" as the first row in Table 1 shows. If additional dependencies *prep* and *pobj* are discovered and they are involved with the same *noun* opinion word in a sentence, then the final segment result becomes "D_{nsubj} *op* D_{prep} D_{pobj}" by combining the results from both templates and ordering the words according to their locations in the sentence. During the process of exploring the dependencies in a sentence, it is necessary to detect the existence of a negation relation linked to a opinion word, if there is, then the opinion word *op* will be changed to *not op* before being used in the final segment.

2.3 Opinion Summarization Based on Identified Feature Words

Identifying feature words from opinion summarizations is a simpler case now, because there is at most one feature in each segment with one opinion word, sometimes there is no valid feature existing in invalid segments. As the supervised pattern mining method - label sequential rule mining in [8,9] is proposed to handle reviews formated similarly as our segments, the similar idea is borrowed but broadened with our own observations to fulfill this step.

In the training dataset, the column is used to mark sequence of words in each segment. For example, the longest segment contains 5 words, then there

Table 2. Feature classes and relevant feature words

Feature class	Feature words
Service	service, job, work, part, done, completed
Communication	communication, communicate, contact, reply, hear back
Staff	staff, dealer, mechanic, manager, guy, attitude, knowledgeable
Invoice/Price	invoice, price, charge, amount

are five columns in training dataset and each of them has assigned name "*word#*" to indicate the position of values in segments. In each row, every word in one segment is put in its corresponding column from the beginning, along with their POS tags. Last but not the least, every feature word in each segment will be identified and replaced with label [feature] manually, so a segment like "pleased with attitude" will be represented as "pleased_VB with_IN [feature]_NN" in our training dataset, where tags VB, IN and NN denote for verb, preposition or conjunction and noun respectively. Then association rules are mined from the training set with assistance of WEKA, and only the ones with label [feature] at the right hand side are kept and transformed into patterns. Following the example given above, the association rule generated for it is: word1=pleased_VB ==> word3=[feature]_NN and the pattern transformed from it becomes: <pleased_VB> <>< [feature]_NN>. Inspired by summarizing quite a lot of valid association rules and patterns retrieved from them, we cannot help thinking that the tags actually help to generalize the recognition of features, especially there are limited kinds of tags appearing in our segments. For instance, <excellent_JJ> <[feature]_NN> and <hard_JJ><[feature]_NN> (JJ denotes adjective) are two patterns which form a more general one <JJ><[feature]_NN> indicating that the noun appearing right after the adjective could be the feature under certain possibility. With regards to such observations, we are more inclined to build general patterns with only tags to predict features, which remarkably decreases the number of useful patterns and increases the efficiency.

In many sentiment analysis works, it is necessary to generate a final review summary for all discovered information about features and opinions, and also rank them by their appearances in the reviews. Besides that, more attention in our strategy is put on avoiding the redundancy of features words. To remove the redundant features words, feature classes are defined with a list of seed words or provided phrases based on our knowledge about the domain. Table 2 shows the examples of four representative feature classes and their feature words in our domain. Thus a segment will be clustered into a feature class if its feature word belongs to that class. As learning process continues, larger set of feature words for each class can be retrieved to enlarge its coverage.

2.4   Generation of Personalized Meta-Actions

The positive opinions indicate the satisfying behaviors that should remain, so the meta-actions for them are called *keeping* actions. Negative opinions show the undesirable behaviors that should be fixed, so their solutions are referred as *fixing* actions. Sometimes it is not hard to create *keeping* actions, since the positive segments can be used directly and they are explicit enough for users to understand and adopt. However, for negative segments, reversing them literally or removing the negation is not always right. To provide the most suitable fixing actions, consulting with company members who have expertise in this field is necessary and useful. In our case, a list of *fixing* actions to commonly discovered problems is collected and labeled with feature classes and subclasses. By subclasses, we mean the more specific aspects that could be designated by segments. For example, staff's attitude and staff's expertise are subclasses in class *Staff*. To map a segment to its meta-actions, we check if its opinion word is a synonym or antonym to the subclasses clarified in the list. If yes, then the meta-action for this segment is found; otherwise, the meta-action is not successfully matched.

3   Experiments

To implement our system for mining meta-actions, several existing tools from other projects are used. Stanford NLP part-of-speech tagger and lexicalized parser are used for generating POS tags [10] and identifying the dependency relations [2]. The lists containing positive and negative words respectively from Liu [8] are applied to detect opinion words and their polarities, ulteriorly the orientation of the segments. WordNet [11] is used to find the set of synonyms or antonyms. The system is built on JAVA and the sample used to test our approach contains 116 sentences which are manually labeled with relevant information including all expected results for each step, such as opinion sentence (or not) and opinion words orientation for the first step, and so on.

Table 3. Experiment results of major steps

	Precision	Recall	F-score
Opinion sentence identifier	0.833	0.696	0.758
Opinion sentence summarizer	0.883	0.8	0.839
Feature words identifier	0.81	0.71	0.757
Feature aggregator	0.78	0.733	0.753
Meta-action generator	0.78	0.75	0.764

After the sample data is processed with the proposed procedures, precision, recall and F-score are computed and shown in Table 3. Firstly, although there is no other comparable results for *Opinion Sentence Identifier*, its performance is very satisfying, its F-score is over 0.75 and the precision is over 0.8. Secondly, if comparing the performance of *Opinion Sentence Summarizer* and *Feature Words Identifier* in our work to the average results of feature-opinion pair mining and feature mining using approaches from [4, 18] respectively, our approach achieves much better results in all three measurements. The accuracy of *Feature Aggregator* is very optimistic. For *Meta-Action Generator*, there are 30 *fixing* actions provided, and the performance of mapping them to specific segments is very acceptable, and its F-score is 0.764. Thus, the experiments confirm our expectation in the proposed method.

4   Conclusion

Generally speaking, the typical procedure of feature-based sentiment analysis in [1,4,5] proceeds without opinion sentence summarization. Later, although [15,18] have completed some work on opinion summarization by mining feature-opinion pairs, our approach accomplishes opinion summarization by following the discovered templates of dependency relations involving expanded opinion words solely, and features in the summarized opinion segments are recognized by applying tag-dominated patterns transformed from association rules. Compared with other relevant work, our *Sentence Summarizer* and *Feature Words Identifier* achieve higher accuracy in the experiments, which proves the effectiveness of the atypically ordered and accordingly adjusted procedures. Besides adapting the traditional sentiment analysis into our project, designing the unique procedure - generation of meta-actions - resolves the demands for providing proper solutions to exposed problems, and the experiments also demonstrate its very positive effect. Moreover, we believe this process can be applied to other areas for solving the discovered problems with their personalizations.

References

1. Blair-Goldensohn, S., Hannan, K., McDonald, R., Neylon, T., Reis, G.A., Reynar, J.: Building a sentiment summarizer for local service reviews. In: WWW Workshop on NLP in the Information Explosion Era, vol. 14 (2008)
2. De Marneffe, M.C., Manning, C.D.: Stanford typed dependencies manual (2008). http://nlp.stanford.edu/software/dependenciesmanual.pdf
3. He, Z., Xu, X., Deng, S., Ma, R.: Mining action rules from scratch. Expert Syst. Appl. **29**(3), 691–699 (2005)
4. Hu, M., Liu, B.: Mining and summarizing customer reviews. In: Proceedings of the Tenth ACM SIGKDD International Conference on Knowledge Discovery and Data Mining, pp. 168–177, ACM (2004)
5. Hu, M., Liu, B.: Mining opinion features in customer reviews. AAAI **4**(4), 755–760 (2004)
6. Kuang, J., Daniel, A., Johnston, J., Raś, Z.W.: Hierarchically structured recommender system for improving NPS of a company. In: Cornelis, C., Kryszkiewicz, M., Ślęzak, D., Ruiz, E.M., Bello, R., Shang, L. (eds.) RSCTC 2014. LNCS, vol. 8536, pp. 347–357. Springer, Heidelberg (2014)
7. Kuang, J., Raś, Z.W., Daniel, A.: Hierarchical agglomerative method for improving NPS. In: Kryszkiewicz, M., Bandyopadhyay, S., Rybinski, H., Pal, S.K. (eds.) PReMI 2015. LNCS, vol. 9124, pp. 54–64. Springer, Heidelberg (2015)

8. Liu, B., Hu, M., Cheng, J.: Opinion observer: analyzing and comparing opinions on the web. In: Proceedings of the 14th International Conference on World Wide Web, pp. 342–351. ACM (2005)

9. Liu, B.: Sentiment analysis and subjectivity. Handb. Nat. Lang. Process. **2**, 627–666 (2010)

10. Marcus, M.P., Santorini, B., Marcinkiewicz, M.A.: Building a large annotated corpus of english: the Penn treebank. Comput. Linguist. **19**(2), 313–330 (1993)

11. Miller, G.A., Beckwith, R., Fellbaum, C., Gross, D., Miller, K.J.: Introduction to WordNet: an on-line lexical database*. Int. J. Lexicogr. **3**(4), 235–244 (1990)

12. Raś, Z.W., Wieczorkowska, A.A.: Action-rules: how to increase profit of a company. In: Zighed, D.A., Komorowski, J., Żytkow, J.M. (eds.) PKDD 2000. LNCS (LNAI), vol. 1910, pp. 587–592. Springer, Heidelberg (2000)

13. Ras, Z., Dardzinska, A.: From data to classification rules and actions. Int. J. Intell. Syst. **26**(6), 572–590 (2011). In the Special Issue on Rough Sets. Theory and Applications, Wiley

14. Reichheld, F.F.: The one number you need to grow. In: Harvard Business Review, pp. 1–8, December 2003

15. Somprasertsri, G., Lalitrojwong, P.: Mining feature-opinion in online customer reviews for opinion summarization. J. UCS **16**(6), 938–955 (2010)

16. Tzacheva, A., Ras, Z.W.: Association action rules and action paths triggered by meta-actions. In: Proceedings of 2010 IEEE Conference on Granular Computing, pp. 772–776. IEEE Computer Society, Silicon Valley, CA (2010)

17. Wang, K., Jiang, Y., Tuzhilin, A.: Mining actionable patterns by role models. In: Proceedings of the 22nd International Conference on Data Engineering, pp. 16–25. IEEE Computer Society (2006)

18. Zhuang, L, Jing, F., Zhu,, X.Y.: Movie review mining and summarization. In: Proceedings of the 15th ACM International Conference on Information and Knowledge Management, pp. 43–50. ACM (2006)

On the Qualitative Calibration of Bipolar Queries

Jalel Akaichi[1], Ludovic Liétard[2]([✉]), Daniel Rocacher[2], and Olfa Slama[2]

[1] BESTMOD Laboratory, ISGT, 41, Rue de la Liberté,
2000 Le Bardo, Tunis, Tunisia
jalel.akaichi@isg.rnu.tn
[2] IRISA/ENSSAT, 6, Rue de Kérampont, 22305 Lannion Cedex, France
ludovic.lietard@univ-rennes1.fr,
{daniel.rocacher,olfa.slama}@enssat.fr

Abstract. This article considers the bipolar approach to define database queries expressing users' preferences (flexible queries). An algebraic framework for the definition of flexible queries of relational databases using fuzzy bipolar conditions of type and-if-possible and or-else has been considered. This paper defines some qualitative calibrations of such queries to specify a minimal quality of answers and to reduce their number. Different operators (extended α-cuts) are defined and studied in this article. They can apply on the set of answers to express a qualitative calibrations of bipolar fuzzy queries. Some properties of these extended α-cuts are pointed out and some of their applications for query evaluation are shown.

Keywords: Fuzzy set · Flexible querying · Bipolarity · Qualitative calibration · Extended α-cut

1 Introduction

Flexible querying aims at allowing users to express preferences in their queries. In this context, several preference operators and languages relying on various logical frameworks have been proposed, such as Preference SQL [1], SQLf language [2], winnow operator [3], among many other works. We consider in this paper fuzzy set theory [4][1] as a general model to express flexibility through fuzzy conditions[2]. In this line, fuzzy bipolar conditions [5–11] are particular conditions to express users' preferences (where satisfaction is not necessary the negation of rejection and vice-versa). An algebraic framework for the definition of flexible queries of

---

[1] A fuzzy set A defined on an universe U is a set whose membership is gradual and belongs to $[0, 1]$. Membership degrees are given by a membership function μ_A ($\forall x \in U$, the more $\mu_A(x)$ is high, the more x belongs to A, when $\mu_A(x) = 1$, x fully belongs to A, when $\mu_A(x) = 0$, x does not belong to A).

[2] A fuzzy condition is defined by a fuzzy set. The membership degree expresses the preference (the more it is high for a given element, the more the element is preferred).

© Springer International Publishing Switzerland 2015
F. Esposito et al. (Eds.): ISMIS 2015, LNAI 9384, pp. 88–97, 2015.
DOI: 10.1007/978-3-319-25252-0_10

relational databases using fuzzy bipolar conditions of type and-if-possible and or-else has been proposed [12–15] and the SQL language has been extended to this context [16,17].

This paper is about the integration of some qualitative calibrations into bipolar fuzzy queries [18]. Such a calibration aims at specifying a minimal quality of answers and it reduces the size of the set of answers. Section 2 is a recall of fuzzy bipolar queries and introduces their calibrations. Different operators (extended α-cuts) which can be the basis to define the calibrations are introduced in Sect. 3, along with some of their properties. The use of these calibration operators in fuzzy bipolar querying and some applications of their properties for query evaluation are introduced in Sect. 4. As an example, the use of some of these properties leads to obtain a Boolean query from the fuzzy query when a calibration (qualitative threshold) is specified. The processing of the Boolean query is expected to be quicker than that of the flexible query and time can be saved.

2 Bipolarity in Flexible Querying of Relational Databases

We recall in this section the approach considered to define bipolar conditions, its related logical operators, and the definition of fuzzy bipolar relations and the Bipolar SQLf language.

2.1 Fuzzy Bipolar Conditions

A bipolar condition is made of two poles: a negative pole and a positive pole. The negation of the former expresses a constraint that every accepted elements must satisfy. The latter expresses a wish that distinguishes among accepted elements those which are optimal. We consider it is not possible to aggregate these two poles because they are not commensurable (other approaches to define bipolar conditions are based on an aggregation of these two poles, as the one proposed by De Tré, Zadrozny and Kacprzyk [9–11]). In this framework, we consider two kinds of bipolar conditions: (i) conjunctive bipolar conditions of type and-if-possible, and (ii) disjunctive bipolar conditions of type or-else.

In this context, a bipolar condition is made of two conditions defined on the same universe:

(i) a constraint c, which describes the set of acceptable elements,
(ii) a wish w which defines the set of desired or wished elements.

The negation of c is the set of rejected elements since it describes non-acceptable elements. Since it is not coherent to wish a rejected element, the following property of coherence holds: $w \subseteq c$. In addition, condition c is mandatory because an element which does not satisfy c is rejected; $\neg c$ is then considered as the negative pole of the bipolar condition. Condition w is optional because its non-satisfaction does not automatically mean the rejection. But condition w describes the best elements and w is then considered as the positive pole of the bipolar condition.

If c and w are boolean conditions, the satisfaction with respect to (c, w) is an ordered pair from $\{0, 1\}^2$. When querying a database with such a condition, tuples satisfying the constraint and the wish are returned in priority to the user. They are the top answers. Tuples satisfying only the constraint are delivered but are ranked after the top answers (if there are no top answers, they are considered as the best answers it is possible to provide to the user).

If c and w are fuzzy conditions (defined on the universe U), the property of coherence becomes: $\forall u \in U, \mu_w(u) \leq \mu_c(u)$[3]. The satisfaction is a pair of degrees where $\mu_w(u)$ expresses the optimality while $\mu_c(u)$ expresses the non-rejection. When dealing with such conditions, two different attitudes can be considered. The first one is to assess that the non-rejection is the most important pole and the fuzzy bipolar condition is an *and-if-possible* condition (to satisfy c *and if possible* to satisfy w). The idea is then "to be not rejected and if possible to be optimal". The satisfaction with respect to such a fuzzy bipolar condition is denoted $(\mu_c(u), \mu_w(u))$ (due to the property of coherence, we have $\mu_c(u) \geq \mu_w(u)$). The second attitude is the opposite one, it considers that the optimality is the most important pole and the fuzzy bipolar condition is an *or-else* condition (to satisfy w *or else* to satisfy c). The idea is then "to be optimal or else to be not rejected". The satisfaction with respect to such a fuzzy bipolar condition is denoted $[\mu_w(u), \mu_c(u)]$ (due to the property of coherence, we have $\mu_w(u) \leq \mu_c(u)$). The satisfactions with respect of both types of conditions can be sorted using the lexicographical order and the minimum and the maximum on the lexicographical order are respectively the conjunction and disjunction of fuzzy bipolar conditions [14]. The negation of an and-if-possible condition is an or-else condition (and vice-versa) and an algebraic framework can be defined [14] to handle these two types of fuzzy bipolar conditions[4]. These fuzzy bipolar conditions generalize fuzzy conditions, a fuzzy condition F can be rewritten "F *and if possible F*" or "*F or else F*".

In this algebraic framework, the satisfaction with respect to a fuzzy bipolar condition can be rewritten using the same syntax: ⟨a,b⟩ to denote either [a,b] or (a,b) (where $a, b \in [0, 1]$):

- ⟨a,b⟩ with a < b represents the satisfaction with respect to a fuzzy bipolar condition of type or-else,
- ⟨a,b⟩ with a = b can be interpreted as a satisfaction with respect to a fuzzy condition,
- ⟨a,b⟩ with a > b represents the satisfaction with respect to a fuzzy bipolar condition of type and-if-possible.

The satisfaction of an element x to a fuzzy bipolar condition C is denoted $\langle \mu_1^C(x), \mu_2^C(x) \rangle$. If C is a fuzzy bipolar condition of type or-else we have: $\mu_1^C(x) \leq$

---

[3] The inclusion of two fuzzy sets A and B from the same universe U is defined by: $A \subseteq B \Leftrightarrow \forall x \in U, \mu_A(x) \leq \mu_B(x)$.

[4] Both types of conditions are different because it is possible to show [14] that the same values of satisfaction for w and c does not lead to a same ordering whether it is an *and-if-possible* condition or an *or-else* condition.

$\mu_2^C(x)$; if it is of type and-if-possible, we have: $\mu_1^C(x) \geq \mu_2^C(x)$; if it is a fuzzy condition, we have: $\mu_1^C(x) = \mu_2^C(x)$.

Couples of satisfaction to a fuzzy bipolar conditions C (which can be the satisfaction with respect to a query Q involving fuzzy bipolar conditions) are ranked using the lexicographical order to state that tuple t_1 is preferred to tuple t_2. More precisely, t_1 is preferred to t_2 if and only if:

$$(\mu_1^C(t_1) > \mu_1^C(t_2)) \text{ or } ((\mu_1^C(t_1) = \mu_1^C(t_2)) \wedge (\mu_2^C(t_1) > \mu_2^C(t_2))),$$

which is denoted $\langle \mu_1^C(t_1), \mu_2^C(t_1) \rangle \succ \langle \mu_1^C(t_2), \mu_2^C(t_2) \rangle$. We denote $\langle \mu_1^C(t_1), \mu_2^C(t_1) \rangle \succeq \langle \mu_1^C(t_2), \mu_2^C(t_2) \rangle$ when $\langle \mu_1^C(t_1), \mu_2^C(t_1) \rangle \succ \langle \mu_1^C(t_2), \mu_2^C(t_2) \rangle$ or $\langle \mu_1^C(t_1), \mu_2^C(t_1) \rangle = \langle \mu_1^C(t_2), \mu_2^C(t_2) \rangle$.

Based on the lexicographical order, the conjunction (resp. disjunction) of bipolar fuzzy conditions can be defined by the $lmin$ (resp. $lmax$) operator. They are respectively defined as follows (where $\mu, \mu', \eta, \eta' \in [0,1]$ and $\langle \mu, \eta \rangle$ and $\langle \mu', \eta' \rangle$ are two satisfactions with respect to fuzzy bipolar conditions):

$$lmin(\langle \mu, \eta \rangle, \langle \mu', \eta' \rangle) = \langle \mu, \eta \rangle$$
$$if \; \mu < \mu' \vee (\mu = \mu' \wedge \eta < \eta'),$$
$$= \langle \mu', \eta' \rangle \text{ otherwise.}$$

$$lmax(\langle \mu, \eta \rangle, \langle \mu', \eta' \rangle) = \langle \mu, \eta \rangle$$
$$if \; \mu > \mu' \vee (\mu = \mu' \wedge \eta > \eta'),$$
$$= \langle \mu', \eta' \rangle \text{ otherwise.}$$

The $lmin$ (resp. $lmax$) operator is commutative, associative, idempotent and monotonic. The pair of grades $\langle 1,1 \rangle$ is the neutral (resp. absorbing) element of the operator $lmin$ (resp. $lmax$) and the pair $\langle 0,0 \rangle$ is the absorbing (resp. neutral) element of the operator $lmin$ (resp. $lmax$). In case of fuzzy bipolar conditions being fuzzy conditions, the $lmin$ operator (resp. $lmax$ operator) reverts to the t-norm min (resp. t-conorm max).

2.2 The Bipolar SQLf Language

The Bipolar SQLf language [16,17] allows the expression of simple and complex queries with bipolar conditions (based on projections, restrictions, nesting, partitioning with bipolar quantified propositions, divisions involving bipolar relations). Its basic statement is a combination of a bipolar projection and a bipolar selection. It is expressed as:

select $[(t_1, t_2)_\geq | first(t_1)]$ * **from** $relations$ **where** E;

The fuzzy bipolar condition E can refer either to an and-if-possible or-else fuzzy bipolar condition. It can also refer to an aggregation using the conjunction (**and**), the disjunction (**and**) and the negation (**not**) of several fuzzy bipolar conditions.

The qualitative calibration parameter $[(t_1, t_2)_\geq | first(t_1)]$ is optional. If no calibration parameter is set in the query, the answer to this query is made of the tuples t from R attached with the pair of grades $\langle \mu_1^E(t), \mu_2^E(t) \rangle$ (and a so-called fuzzy bipolar relation is obtained). A tuple t is then denoted $\langle \mu_1^E(t), \mu_2^E(t) \rangle / t$ and any tuple t such that $\langle \mu_1^E(t), \mu_2^E(t) \rangle = \langle 0, 0 \rangle$ does not belong to the fuzzy bipolar relation.

The calibration parameter helps to reduce the size of the set of answers to the query:

- The parameter $(t_1, t_2)_{\geq}$ is a couple of degrees representing a minimal level of satisfaction (qualitative calibration). In this case, the query returns the answers having a couple of satisfactions above (or equal to) level (t_1, t_2) in the lexicographical order.
- The parameter $first(t_1)$ retrieves the elements such that their first degree (from the couple of degrees presenting their satistaction to the query) is above or equal to t_1. It delivers a fuzzy set of these elements, the membership degree being the first element from the couple (the degree involved in the calibration).

Example 1. A user is looking for a train ticket to go to Paris. He would like an *early arrival* (with a *short travel time*) or a *cheap price* with a *good seat*. This requirement is a disjunction of two conditions, and for the first one, the most important is to have an *early arrival*. The first condition is denoted A and is an and-if-possible fuzzy bipolar condition: $A = $ *an early arrival and if possible (early arrival $\wedge$ a short travel time)*. The second condition is denoted B and is an or-else fuzzy bipolar condition: $B = $ *(a cheap price $\wedge$ a good seat) or else a cheap price*. It is an or-else condition because the requirement *a cheap price $\wedge$ a good seat* is more important than *a cheap price*.

The fuzzy conditions *earlyArrival*, *shortTravelTime* and *cheapPrice* are given by Fig. 1. The seats are divided into five different categories, denoted $C1$, $C2$, $C3$, $C4$ and $C5$. The quality of a seat, according to the user, is given by the fuzzy set: $goodSeat = 1/C1 + 0.8/C2 + 0.6/C3 + 0.4/C4 + 0.2/C5$.

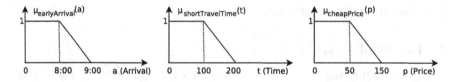

Fig. 1. Fuzzy conditions *earlyArrival*, *shortTravelTime*, *cheapPrice*.

Let R be the relation describing ticket offers to go to Paris (see Table 1), the query is then $R : (A \vee B)$. The query is written in Bipolar SQLf language as:

select * **from** R
where *an early arrival and if possible (early arrival $\wedge$ a short travel time)*
or *(a cheap price $\wedge$ a good seat) or else a cheap price*.

The satisfactions of each ticket to the fuzzy bipolar conditions A, B (and their disjunction $A \vee B$) is given by Table 2. The disjunction operator is evaluated with the operator *lexmax*, it retrieves the maximum according to the

Table 1. Example of extension of the relation R.

#Ticket	Arrival	Travel time	Price	Seat
#1	8:12	140	40	C1
#2	8:24	160	80	C2
#3	8:18	140	100	C3
#4	8:18	170	30	C5
#5	8:30	120	60	C3

Table 2. Satisfactions to the fuzzy bipolar conditions.

#Ticket	A	B	Satisfaction to $R : (A \vee B)$
#1	$\langle 0.8, 0.6 \rangle$	$\langle 1, 1 \rangle$	$\langle 1, 1 \rangle$
#2	$\langle 0.6, 0.4 \rangle$	$\langle 0.7, 0.7 \rangle$	$\langle 0.7, 0.7 \rangle$
#3	$\langle 0.7, 0.6 \rangle$	$\langle 0.5, 0.5 \rangle$	$\langle 0.7, 0.6 \rangle$
#4	$\langle 0.7, 0.3 \rangle$	$\langle 0.2, 1 \rangle$	$\langle 0.7, 0.3 \rangle$
#5	$\langle 0.5, 0.5 \rangle$	$\langle 0.6, 0.9 \rangle$	$\langle 0.6, 0.9 \rangle$

lexicographical order. The satisfactions to the query are given by the last column of Table 2. The answers can be ranked in decreasing order of satisfaction: ticket 1 with satisfaction $\langle 1, 1 \rangle$, ticket 2 with satisfaction $\langle 0.7, 0.7 \rangle$, ticket 3 with satisfaction $\langle 0.7, 0.6 \rangle$, ticket 4 with satisfaction $\langle 0.7, 0.3 \rangle$ and ticket 5 with satisfaction $\langle 0.6, 0.9 \rangle$. Some final results depending on the calibration parameter are given by Table 3.

Table 3. Calibrations.

Calibration	Result
$(0.7, 0.4)_{\geq}$	$\{\#1, \#2, \#3\}$
$first(0.7)$	$1/\#1 + 0.7/\#2 + 0.7/\#3 + 0.7\#4$

3   Extended α-Cuts for Fuzzy Bipolar Queries

In this section, we extend the definition of an α-cut of a fuzzy set[5] to the bipolar context. In Sect. 3.1, different definitions for an extended α-cut are proposed and some of their properties are introduced in Sect. 3.2.

3.1   Definitions

Let $B = \{\langle \mu_B^1(t), \mu_B^2(t) \rangle | t \in U \}$ be a set of elements from an universe U with their bipolar evaluation (a fuzzy bipolar set). Such a set can be the result of a

---

[5] The α-cut of a fuzzy set F defined on the universe U is the regular set F_α made of its elements having a membership degree greater than (or equal to) the level α:
$F_\alpha = \{t \in U | \mu_F(t) \geq \alpha\}$.

(bipolar) flexible querying. We define two extended α-cuts which correspond to the qualitative calibrations in Bipolar SQLf queries (see Sect. 2.2):

1. $B_{\geq(\alpha_1,\alpha_2)} = \{t \in U | \langle \mu_B^1(t), \mu_B^2(t) \rangle \succeq \langle \alpha_1, \alpha_2 \rangle\}$ where $\succeq$ is the lexicographical order. This α-cut retrieves the elements such that their position in the lexicographical order is above (or equal to) $\langle \alpha_1, \alpha_2 \rangle$.
2. $B_{first(\alpha)} = \{\mu_B^1(t) \setminus t | t \in U, \mu_B^1(t) \geq \alpha\}$. This α-cut delivers a *fuzzy set* made of elements such that their first satisfaction degree is greater then (or equal to) a given threshold.

Other α-cuts, which seem to be less useful from a end-user point of view, can be defined:

3. $B_{(\alpha_1,\alpha_2)} = \{t \in U | (\mu_B^1(t) \geq \alpha_1) \wedge (\mu_B^2(t) \geq \alpha_2)\}$. Two different thresholds respectively apply to the two degrees from the couple of satisfactions.
4. $B_{1(\alpha)} = \{\mu_B^2(t) \setminus t | t \in U, \mu_B^1(t) \geq \alpha\}$. This α-cut delivers a *fuzzy set* made of elements such that their first satisfaction degree is greater then (or equal to) a given threshold.
5. $B_{2(\alpha)} = \{\mu_B^1(t) \setminus t | t \in U, \mu_B^2(t) \geq \alpha\}$. This α-cut delivers a *fuzzy set* made of elements such that their second satisfaction degree is greater then (or equal to) a given threshold.

3.2 Properties

These extended α-cuts share interesting properties (their proofs are omitted due to a lack of space but they do not bring any difficulties). In the following, B and A are two sets with bipolar evaluations for their elements (two fuzzy bipolar sets) defined on the same universe and $\alpha, \alpha', \beta, \beta'$ four values in $[0, 1]$. The $\cap$ and $\cup$ operators have been extended using *lexmin* and *lexmax* operators respectively (we recall that the support of a fuzzy set F defined on the universe U is the regular set $support(F)$ made of its elements having a membership degree strictly greater than 0; $support(F) = \{t \in U | \mu_F(t) > 0\}$).

1. $B_{(\alpha,\beta)} = (B_{1(\alpha)})_\beta$.
2. $B_{(\alpha,\beta)} = (B_{2(\beta)})_\alpha$.
3. $B_{(\alpha,\beta)} \subseteq B_{\geq(\alpha,\beta)}$.
4. $B_{(\alpha,0)} = support(B_{1(\alpha)})$.
5. $B_{(0,\beta)} = support(B_{2(\beta)})$.
6. $B_{(\alpha,0)} = B_{\geq(\alpha,0)}$.
7. $B_{(\alpha,0)} = support(first(\alpha))$.
8. $\alpha < \alpha' \Rightarrow B_{1(\alpha')} \subseteq B_{1(\alpha)}$.
9. $\beta < \beta' \Rightarrow B_{2(\beta')} \subseteq B_{2(\beta)}$.
10. $\alpha < \alpha' \Rightarrow B_{first(\alpha')} \subseteq B_{first(\alpha)}$.

11. $(B \cup A)_{\geq(\alpha,\beta)} = B_{\geq(\alpha,\beta)} \cup A_{\geq(\alpha,\beta)}$.
12. $(B \cap A)_{\geq(\alpha,\beta)} = B_{\geq(\alpha,\beta)} \cap A_{\geq(\alpha,\beta)}$.
13. $(B \cup A)_{first(\alpha)} = B_{first(\alpha)} \cup A_{first(\alpha)}$.
14. $(B \cap A)_{first(\alpha)} = B_{first(\alpha)} \cap A_{first(\alpha)}$.
15. $(B \cup A)_{(\alpha,\beta)} \subseteq B_{(\alpha,\beta)} \cup A_{(\alpha,\beta)}$.
16. $(B \cap A)_{(\alpha,\beta)} \supseteq B_{(\alpha,\beta)} \cap A_{(\alpha,\beta)}$.
17. $(B \cup A)_{1(\alpha)} \subseteq B_{1(\alpha)} \cup A_{1(\alpha)}$.
18. $(B \cap A)_{1(\alpha)} \supseteq B_{1(\alpha)} \cap A_{1(\alpha)}$.
19. $(B \cup A)_{2(\beta)} \subseteq B_{2(\beta)} \cup A_{2(\beta)}$.
20. $(B \cap A)_{2(\beta)} \supseteq B_{2(\beta)} \cap A_{2(\beta)}$.

21. $\alpha < \alpha', \beta < \beta' \Rightarrow B_{(\alpha',\beta')} \subseteq B_{(\alpha,\beta)}$.

22. If F is a fuzzy set defined on the universe U and B its representation as a fuzzy bipolar condition (i.e., $\forall t \in U, \mu_1^B(t) = \mu_2^B(t) = \mu_F(t)$), we have: $F_\alpha = B_{\geq(\alpha,\alpha)} = B_{(\alpha,\alpha)} = B_{(\alpha,0)} = B_{(0,\alpha)}$.

4 On the Use of These Extended α-Cuts

The qualitative calibration of a fuzzy bipolar query is nothing but an extended α-cut of the set of its answers. If AQ is the set of bipolar answers to the query, the calibration parameter $(t_1, t_2)_\geq$ in Q leads to determine the extended α-cut $AQ_{\geq (t_1,t_2)}$ as result to deliver to the user. Similarly, the use of the calibration parameter $first(t_1)$ in Q leads to determine the extended α-cut $AQ_{first(t_1)}$. The other types of extended α-cuts defined in Sect. 3.1 can also be used to calibrate the result of a fuzzy bipolar query. However, to show their utility for calibration needs and in-depth study which is not introduced in his paper (due to a lack of space).

In case of a qualitative calibration with parameter $(t_1, t_2)_\geq$ or $first(t_1)$, the properties of the extended α-cuts can be used to improve the query processing and the query processing time.

As an example, in some particular cases and when the calibration parameter is $(t_1, t_2)_\geq$, it is possible to obtain, from the fuzzy bipolar query, an SQL query which delivers the set of answers (or a superset). It is very appealing because the evaluation of the SQL query is expected to be quicker than that of the fuzzy bipolar query since no particular computations of degrees are needed (and the calibration has not to be performed). A simple example to illustrate the idea is provided hereinafter.

Example 2. We consider the query Q of Example 1 with a calibration with parameter $(0.7, 0.4)_\geq$. The query is then $R : (A \lor B)_{\geq(0.7,0.4)}$ where $A = an$ *early arrival and if possible (early arrival $\land$ a short travel time)* and $B = (a$ *cheap price $\land$ a good seat) or else a cheap price.*

Thanks to property 11, the query can be rewritten $R : (A_{\geq(0.7,0.4)} \lor B_{\geq(0.7,0.4)})$. An SQL query is then obtained because the two conditions $A_{\geq(0.7,0.4)}$ and $B_{\geq(0.7,0.4)}$ are Boolean. Condition $A_{\geq(0.7,0.4)}$ can be rewritten:

$$(\mu_{earlyArrival}(t.Arrival) > 0.7) \lor$$
$$(\mu_{earlyArrival}(t.Arrival) = 0.7 \land$$
$$\min(0.7, \mu_{shortTravelTime}(t.TravelTime)) \geq 0.4)$$

which gives:

$$(t.Arrival < 8 : 18) \lor ((t.Arrival = 8 : 18) \land (t.TravelTime \leq 160)).$$

Similar computations for $B_{\geq(0.7,0.4)}$ give:

$$(Price \leq 80) \land (Seat \in \{C1, C2\}).$$

It means that query Q with calibration $(0.7, 0.4)_\geq$ can be rewritten by the following Boolean query:

$$R : (((t.Arrival < 8 : 18) \lor ((t.Arrival = 8 : 18) \land (t.TravelTime \leq 160))) \lor$$
$$((Price \leq 80) \land (Seat \in \{C1, C2\}))).$$

The processing of this Boolean query is expected to be quicker than that of query Q and time can be saved. The Boolean query delivers $\{\#1, \#2, \#3\}$ and it can be checked in Table 3 that it corresponds to calibration $(0.7, 0.4)_\geq$. •

If needed, the satisfaction with respect to the fuzzy bipolar query can be computed on this set of answers delivered by the SQL query (instead of the whole database).

Similarly, in some particular cases and when the calibration parameter is $first(t_1)$, we expect to obtain an SQLf query (a fuzzy query [2] and not a Boolean query) from the fuzzy bipolar query in order to obtain the set of answers (or a superset). Here again, it is appealing because the evaluation of the SQLf query is expected to be quicker than that of the fuzzy bipolar query (furthermore, in some cases, this SQLf query can also be translated into a Boolean SQL query).

5   Conclusion

We have considered the bipolar approach to define database queries expressing users' preferences (flexible queries). An algebraic framework for the definition of flexible queries of relational databases using fuzzy bipolar conditions of type and-if-possible and or-else has been considered.

We have defined different extended α-cuts and have shown their interest to express a qualitative calibrations of bipolar fuzzy queries. Such a calibration aims at specifying a minimal quality of answers and it also reduces their number. Some properties of these extended α-cuts and their use for fuzzy bipolar query evaluation have been pointed out.

This article is a first attempt to define the concept of qualitative calibration in fuzzy bipolar queries. In the future, we aim at completing this work by studying other properties and to show other interesting applications. We would like to focus on the use of these properties to improve the query evaluation process (both from a theoritical and implementation aspects).

References

1. Chomicki, J.: Querying with intrinsic preferences. In: Jensen, C.S., Jeffery, K., Pokorný, J., Šaltenis, S., Bertino, E., Böhm, K., Jarke, M. (eds.) EDBT 2002. LNCS, vol. 2287, pp. 34–51. Springer, Heidelberg (2002)
2. Bosc, P., Pivert, O.: Sqlf: a relational database langage for fuzzy querying. IEEE Trans. Fuzzy Syst. **3**(1), 1–17 (1995)
3. Kiesling, W.: Foundation of preferences in database systems. In: Proceedings of the 28th VLDB Conference, Hong Kong (2002)
4. Zadeh, L.: Fuzzy sets. Inf. Control **8**(3), 338–353 (1965)
5. Bordogna, G., Pasi, G.: A fuzzy query language with a linguistic hierarchical aggregator. In: Proceedings of the ACM SAC 1994, pp. 184–187. ACM, USA (1994)
6. Dubois, D., Prade, H.: Bipolarity in fexible querying. In: Andreasen, T., Motro, A., Christiansen, H., Larsen, H.L. (eds.) FQAS 2002. LNCS (LNAI), vol. 2522. Springer, Heidelberg (2002)
7. Dubois, D., Prade, H.: Handling bipolar queries in fuzzy information processing. In: Handbook of Research on Fuzzy Information Processing in Databases, pp. 97–114. IGI Global (2008)

8. Dubois, D., Prade, H.: An overview of the asymmetric bipolar representation of positive and negative information in possibility theory. Fuzzy Sets Syst. **160**(10), 1355–1366 (2009)

9. De Tré, G., Zadrozny, S., Bronselaer, A.: Handling bipolarity in elementary queries to possibilistic databases. IEEE Trans. Fuzzy Syst. **18**(3), 599–612 (2010)

10. De Tré, G., Zadrożny, S., Matthé, T., Kacprzyk, J., Bronselaer, A.: Dealing with positive and negative query criteria in fuzzy database querying. In: Andreasen, T., Yager, R.R., Bulskov, H., Christiansen, H., Larsen, H.L. (eds.) FQAS 2009. LNCS, vol. 5822, pp. 593–604. Springer, Heidelberg (2009)

11. Zadrożny, S., Kacprzyk, J.: Bipolar queries: a way to enhance the flexibility of database queries. In: Ras, Z.W., Dardzinska, A. (eds.) Advances in Data Management. SCI, vol. 223, pp. 49–66. Springer, Heidelberg (2009)

12. Liétard, L., Rocacher, D.: On the definition of extended norms and co-norms to aggregate fuzzy bipolar conditions. In: IFSA/EUSFLAT, pp. 513–518 (2009)

13. Liétard, L., Rocacher, D., Bosc, P.: On the extension of SQL to fuzzy bipolar conditions. In: 28th North American Information Processing Society Conference (2009)

14. Liétard, L., Rocacher, D., Tamani, N.: A relational algebra for generalized fuzzy bipolar conditions. In: Pivert, O., Zadrozny, S. (eds.) Flexible Approaches in Data, Information and Knowledge Management. SCI, vol. 497, pp. 45–69. Springer, Heidelberg (2013)

15. Liétard, L., Tamani, N., Rocacher, D.: Fuzzy bipolar conditions of type "or else". In: FUZZ-IEEE, pp. 2546–2551 (2011)

16. Tamani, N., Liétard, L., Rocacher, D.: Bipolar SQLf: a flexible querying language for relational databases. In: Christiansen, H., De Tré, G., Yazici, A., Zadrozny, S., Andreasen, T., Larsen, H.L. (eds.) FQAS 2011. LNCS, vol. 7022, pp. 472–484. Springer, Heidelberg (2011)

17. Tamani, N.: Interrogation personnalisée des systêmes d'information dédiés au transport : une approche bipolaire floue. Ph.D. thesis. Université de Rennes 1 (2012)

18. Slama, O.: About Alpha-Cuts in Bipolar and Multipolar Queries in Relational Databases. Master Degree thesis. University of Tunis (2014)

Machine Learning

A Scalable Boosting Learner Using Adaptive Sampling

Jianhua Chen[1]($\boxtimes$), Seth Burleigh[2], Neeharika Chennupati[1],
and Bharath K. Gudapati[1]

[1] Division of Computer Science and Engineering, School of Electrical Engineering and
Computer Science, Louisiana State University, Baton Rouge, LA 70803-4020, USA
`jianhua@csc.lsu.edu, nchenn2@lsu.edu, bharathgudapati@gmail.com`
[2] Department of Physics and Astronomy, Louisiana State University, Baton Rouge,
LA 70803-4020, USA
`wburle@gmail.com`

Abstract. Sampling is an important technique for parameter estimation and hypothesis testing widely used in statistical analysis, machine learning and knowledge discovery. Sampling is particularly useful in data mining when the training data set is huge. In this paper, we present a new sampling-based method for learning by Boosting. We show how to utilize the adaptive sampling method in [2] for estimating classifier accuracy in building an efficient ensemble learning method by Boosting. We provide a preliminary theoretical analysis of the proposed sampling-based boosting method. Empirical studies with 4 datasets from UC Irvine ML database show that our method typically uses much smaller sample size (and is thus much more efficient) while maintaining competitive prediction accuracy compared with Watanabe's sampling-based Boosting learner Madaboost.

Keywords: Adaptive sampling · Sample size · Boosting · Adaboost · Madaboost

1 Introduction

An important problem in data mining and machine learning is to develop methods capable of learning from huge data sets. This issue of scalability is fundamental to the success of practical applications of machine learning and knowledge discovery. Among various approaches to scalability is the often-used method of random sampling. With random sampling, classifiers can be learned based on a subset of samples drawn randomly from the entire instance space. Moreover the accuracies of learned classifiers can also be estimated from a random sample set. Random sampling could also be used to estimate values of features for describing data. Clearly, a well-designed random sampling scheme combined with a machine learning algorithm could greatly improve the efficiency and scalability of learning while keeping competitive accuracies of the learned classifiers.

© Springer International Publishing Switzerland 2015
F. Esposito et al. (Eds.): ISMIS 2015, LNAI 9384, pp. 101–111, 2015.
DOI: 10.1007/978-3-319-25252-0_11

Random sampling has its root in statistics and it is widely used in various science and engineering areas. A key issue in designing a sampling scheme is to determine *sample size*, the number of sampled instances sufficient to assure the estimation accuracy and confidence. In the context of sampling for learning a good classifier or estimating classifier performance, the sample size refers to the number of training/testing data points enough to guarantee with high confidence that the estimation of classifier performance (in terms of prediction accuracy, for example) is sufficiently accurate. Clearly, it is desirable to keep the sample size small subject to the constraint of estimation accuracy and confidence. This would save not only computation time, but also the cost of generating the extra random samples when such costs are significant.

How to determine the sufficient sample size? Well-known theoretical bounds such as the Chernoff bound and Hoeffding bound are commonly used in the theoretical computer science research for computing the sufficient sample size. An *adaptive* sampling algorithm, in contrast to *batch* sampling method, draws random samples one by one in an online fashion. It decides whether it has seen enough samples dependent on some measures related to the samples seen so far. This adaptive nature of sequential sampling method is attractive from both computational and practical perspectives. Moreover as recent studies [4,16] pointed out, there are situations in which using Chernoff/Hoeffding bounds in "static" (non-adaptive) sampling would require a sample size that is unnecessarily large.

Statisticians have investigated adaptive sampling procedures for estimating parameters under the heading of *sequential estimation* (see, [9] for a comprehensive treatment). In recent works [6,7,15,16], Watanabe et. al. proposed techniques for adaptive sampling and applied the methods to Boosting, an ensemble learning method.

In [4], an adaptive, multi-stage sampling framework has been proposed. The framework is more general than the adaptive sampling methods in [6,7,15,16]. Closely related to [4], a new adaptive sequential sampling method was developed in [2] that can handle cases of controlling absolute error and relative error.

In this paper we present a new adaptive sampling-based method for learning by Boosting. The key component of the new method is the criterion for stopping the sampling, which is an application of the adaptive sequential sampling method in [2] to the Boosting task. We present a preliminary theoretical analysis of the proposed boosting method. In addition we report the results of empirical studies that compare the performance of our method with Madaboost by Watanabe. The results indicate that our method is much more efficient while maintaining competitive classification accuracy. A preliminary version of this work was presented in [3], but the current version has substantially more theoretical and empirical results. Due to space limits, we skip the proofs of the theorems which would be included in the expanded version in the future.

2   Background

2.1   Boosting and Adaboost

Boosting is an ensemble learning method which constructs a strong classifier from a sequence of weak classifiers. Boosting proceeds by constructing a sequence of classifiers h_1, h_2, ... , h_T in an iterative fashion such that the combination of these classifiers will produce a strong classifier with high classification accuracy. The well-known Adaboost method [8] proceeds as follows. The algorithm starts with a data set D which could be obtained through sampling from a larger data collection. But once D is obtained, the algorithm works with this **fixed** dataset. A probability distribution $\mathscr{D}(1)$ over the data set D is also given, which is typically assumed to be uniform. Then at each boosting round $1 \leq t \leq T$, Adaboost will generate a classifier/hypothesis h_t which has minimal classification error over D according to the probability distribution $\mathscr{D}(t)$. A positive weight α_t is assigned to h_t which is proportional to its (weighted) classification accuracy on D. Moreover the distribution $\mathscr{D}(t)$ is updated to generate $\mathscr{D}(t+1)$ such that the data points misclassified by h_t would have their weights (probabilities) increased while the weights on other data points decreased. The final ensemble classifier f_T classifies a new data point x by weighted majority vote of h_1, h_2, ... , h_T. In the learning task, each instance x is from the **instance space** $\mathscr{X}$. And we deal with a binary classification problem so that the two classes are labeled as $+1$ and -1 respectively.

2.2   Madaboost

One variation of Adaboost is the Madaboost approach proposed by Domingo and Watanabe et al. [6]. The contention is that when the data set D is very large, we may not want to explicitly handle the distributions $\mathscr{D}_t$ and use all data points in D for selecting h_t. Instead, we can use sampling on D (according to distribution $\mathscr{D}_t$) to get a subset of training data S_t from D and construct h_t from S_t. In [6], a stopping condition that adaptively determines the sample size for S_t was proposed. Moreover theoretical analysis and experimental results were presented in [6] showing that Madaboost can generate classifiers with comparable accuracies and better efficiency.

The heart of Madaboost (when being used in a **filtering** mode) consists of two procedures: the FiltEx algorithm for sampling an instance x according to the distribution $\mathscr{D}_t$, and the $ADSS$ procedure that calls the FiltEx procedure repeatedly and stops the sampling when the stopping criterion is satisfied so that a subset S_t of D is obtained and a hypothesis h_t is constructed for each boosting round t. These two procedures actually implement the important modifications of $Madaboost$ over $Adaboost$.

At each boosting round, the termination condition in $ADSS$ of the Madaboost algorithm is specified (in [6]) as follows: On seeing a random sample S of size n, compute $a_n = (1 + 1/\varepsilon)\sqrt{(2\ln\tau - \ln\ln\tau + 1)/n}$ where $\tau = Nn(n+1)/(2\delta\sqrt{\pi})$ and for each hypothesis (classifier) $h \in H_{DS}$ (H_{DS} is the set of all candidate hypotheses of the form *decision stump* and $N = |H_{DS}|$), define $U(h, S) = \frac{|\{x \in S: \ h \ predicts \ correctly \ on \ x\}|}{n} - 1/2$, the *utility of h* after seeing S. If there is a hypothesis $h \in H_{DS}$ such that $U(h, S) \geq a_n$, then terminate the current boost round and return h as the best classifier for this round.

A probabilistic analysis of the Madaboost algorithm leads to the following theorem in [6]. Here "advantage" of a hypothesis h is the difference between the true prediction accuracy of h and $1/2$.

Theorem 1. *Let $\mathscr{D}$ be the filtering distribution at certain boosting round. Let h_* be the best hypothesis in H_{DS} with respect to $\mathscr{D}$ with advantage $\gamma_* = accu_{\mathscr{D}}(h_*) - 1/2$ and assume $\gamma_* > 0$. Under the choice of $\varepsilon = 1/2$ for the ADSS algorithm for any $0 < \delta < 1$, ADSS will produce, with probability of at least $1 - \delta$, a hypothesis h such that the advantage of h is at least $\gamma_*/2$.*

3   The New Adaptive Sampling Method

It is quite obvious that the adaptive boosting method Madaboost is closely related to the problem of adaptively estimating the probability of correct classification by a specific hypothesis/classifier. So it is natural to apply adaptive sampling techniques to develop adaptive boosting method for learning. In this section we briefly review the adaptive sampling method [2] relevant to boosting.

3.1   Estimating the Parameter for a Bernoulli Variable

Estimating the probability $p = \Pr(A)$ of a random event A from observational data is the same as estimating the parameter $p = \Pr\{X = 1\} = 1 - \Pr\{X = 0\}$ for a Bernoulli variable X from i.i.d. samples $X_1, X_2, \cdots$ of X. An estimator for p can be taken as the relative frequency $\widehat{p}_{\mathbf{n}} = \frac{\sum_{i=1}^{n} X_i}{\mathbf{n}}$, where $\mathbf{n}$ is the sample number at the termination of experiment. Estimating the probability of correct classification for a hypothesis/classifier h from training samples $\{(x_1, c(x_1)), ..., (x_m, c(x_m))...\}$ where each x_j is drawn drawn according to a distribution $\mathscr{D}$ amounts to estimating the parameter P_h for the Bernoulli variable V_h defined by h and $\mathscr{D}$. Here $V_h = 1$ if $h(x) = c(x)$, and $V_h = 0$ otherwise, where each object x is drawn from the instance space D according to the distribution $\mathscr{D}$, and $c(x)$ is the true classification. The probability $Pr\{V_h = 1\} = P_h$ is the expected value of V_h, which is also called the *accuracy* of h. The *true error* of h (w.r.t. distribution $\mathscr{D}$) denoted by $error_{\mathscr{D}}(h)$, is defined as $1 - P_h$. The estimated value for P_h after seeing a set S of samples is given by $P_{h,S} = \frac{|\{x: \ x \in S \ \wedge \ h(x) = c(x)\}|}{|S|}$.

For controlling absolute error in estimating a Bernoulli variable, the following problem is relevant:

Problem A: Controlling Absolute Error

We want to design an adaptive sampling scheme (stopping rule) such that for any *a priori* margin of absolute error $\varepsilon \in (0.1)$, and confidence parameter $\delta \in (0, 1)$ and any $\boldsymbol{p} \in (0, 1)$ we have

$$\Pr\{|\widehat{p}_n - p| \geq \varepsilon\} \leq \delta$$

when sampling is terminated after seeing n samples.

Here ε is called the *accuracy* level and $1 - \delta$ is called the *confidence* level, and $\Pr\{|\widehat{p}_n - p| < \varepsilon\}$ is called the *coverage probability*.

The frequently-used Chernoff-Hoeffding bound asserts that, for any $\varepsilon \in (0, 1)$ $\delta \in (0, 1)$, the coverage probability $\Pr\{|\widehat{p}_n - p| < \varepsilon\}$ is greater than $1 - \delta$ for any $p \in (0, 1)$ provided that $n > \frac{\ln \frac{2}{\delta}}{2\varepsilon^2}$. The Chernoff-Hoeffding bound significantly improves upon the Bernoulli bound which states that $\Pr\{|\widehat{p}_n - p| < \varepsilon\} > 1 - \delta$ if $n > \frac{1}{4\varepsilon^2\delta}$. Clearly, the coverage probability $\Pr\{|\widehat{p}_n - p| < \varepsilon\}$ tends to 1 as n tends to infinity. The objective of adaptive sampling is to use even fewer number of samples than **batch sampling** with the *pre-calculated* sample bound $n > \frac{\ln \frac{2}{\delta}}{2\varepsilon^2}$.

3.2   Controlling Absolute Error

The method in [2] for adaptive sampling with respect to absolute error (Problem A) is summarized below.

The function $\mathscr{U}(z, \theta)$ is defined below which will be useful for specifying the sampling method.

$$\mathscr{U}(z, \theta) = \begin{cases} z \ln \frac{\theta}{z} + (1 - z) \ln \frac{1-\theta}{1-z} & z \in (0, \ 1), \quad \theta \in (0, \ 1) \\ \ln(1 - \theta) & z = 0, \ \theta \in (0, \ 1) \\ \ln(\theta) & z = 1, \ \theta \in (0, \ 1) \\ -\infty & z \in [0, \ 1], \ \theta \notin (0, \ 1) \end{cases} \tag{1}$$

Let $W(x) = \frac{1}{2} - |\frac{1}{2} - x|$ for $x \in [0, \ 1]$. Clearly $W(x) = x$ for $x \in [0, \ \frac{1}{2}]$ and $W(x) = 1 - x$ for $x \in (\frac{1}{2}, \ 1]$.

Let $0 < \varepsilon < 1, 0 < \delta < 1$. The sampling scheme proceeds as follows.

Algorithm A.
$n \leftarrow 0, X \leftarrow 0$
$\widehat{p_n} = 0$
Repeat
    Draw a random sample Y
    $X \leftarrow X + Y, n \leftarrow n + 1$
    $\widehat{p_n} \leftarrow \frac{X}{n}$
Until $n > 30$ *and* $n \geq \frac{\ln\frac{\delta}{2}}{\mathscr{U}(W(\widehat{p_n}), \ W(\widehat{p_n})+\varepsilon)}$
Output $\widehat{p}_n, n$.

Some remarks are due here about Algorithm A. Note that the algorithm requires that at least 30 samples must be seen before testing the stopping condition $n \geq \frac{\ln\frac{\delta}{2}}{\mathscr{U}(W(\widehat{p_n}), \ W(\widehat{p_n})+\varepsilon)}$. This is apparently intended to avoid pre-mature termination of the sampling and to assure that the central limit theorem condition is satisfied when the algorithm terminates.

Properties of the Sampling Method. Without the assumption of the *central limit theorem (CLT)*, the following theorem has been established [2]:

Theorem 2. *Assume that the true probability p to be estimated satisfies $p \leq \frac{1}{2} - 2\varepsilon$. Then with probability of at least $1 - \frac{\delta}{2}$, Algorithm A will stop at step $n \leq n_1 = \lceil \frac{Log\frac{\delta}{2}}{\mathcal{U}(p+\varepsilon, p+2\varepsilon)} \rceil$ and produce $\widehat{p_n}$ which satisfies $\widehat{p_n} \leq p + \varepsilon$. Similarly, if $p \geq \frac{1}{2} + 2\varepsilon$, with probability of at least $1 - \frac{\delta}{2}$, the sampling algorithm will stop at step $n \leq n_1$ and produce $\widehat{p_n}$ which satisfies $\widehat{p_n} \geq p - \varepsilon$.*

If we assume that the central limit theorem is satisfied by the random variable $\widehat{p_n}$ where n is the sample size when Algorithm A is terminated, then the following holds true.

Theorem 3. *Assuming CLT being satisfied by $\widehat{p_n}$, we have*

$$\Pr\{|\widehat{p_n} - p| < \varepsilon\} > 1 - \delta$$

where n is the sample size when Algorithm A is terminated.

The above theorem could be interpreted as the *asymptotic consistency* of Algorithm A, whereas the Theorem 2 shows a partial result of the *exact consistency* of the algorithm.

4   The New Boosting Method by Adaptive Sampling

In this section we describe our new method for boosting by adaptive sampling. The key difference between our new boosting method and Madaboost is the stopping condition when selecting a hypothesis in the *ADSS* algorithm. Our stopping criterion is obtained as an adaptation of Algorithm A in [2] for estimating a Bernoulli variable. This criterion also leads to a new way of solving the utility function estimation problem posed in [16].

4.1   Estimating a Utility Function

Remember that at each round of Boosting, we are trying to select the best hypothesis h^* with the *maximal* accuracy P_{h^*} (assuming $P_{h^*} > 1/2$). When we use sampling to estimate the accuracy P_h of each hypothesis h, the key issue is to determine a "reasonable" sample size for the estimation such that the size is sufficient to guarantee with high confidence that the selected hypothesis based on the sample is "close" enough to the best hypothesis h^*. There are various ways to define "closeness" between two hypotheses. One could define it by the absolute difference between their P_h values. But the most important issue in Boosting is that at least the "weak" hypothesis selected at each round should have accuracy above 1/2. So one very modest requirement of "closeness" between the selected hypothesis h and the best one h^* is that if $P_{h^*} > 1/2$, then $P_h > 1/2$. Focusing on each individual hypothesis h, this requirement could be formulated as follows (per [6,16]). We introduce the utility function U_h for each hypothesis h and define $U_h = P_h - 1/2$. Here the utility function could take negative value in case $P_h < 1/2$. We want to estimate the utility function values close with high confidence. Recalling that P_h is estimated by $P_{h,S} = \frac{|\{x:\ x\in S\ \wedge\ h(x)=c(x)\}|}{|S|}$ with a sample S, the sampling problem is then boiled down to the problem:

Problem B - Adaptive Sampling for Estimating a Utility Function:
We want to design an adaptive sampling scheme such that when sampling is stopped with $n = |S|$,

$$\Pr\{|U_{h,S} - U_h| \geq \varepsilon|U_h|\} \leq \delta.$$

Here $U_{h,S} = P_{h,S} - 1/2$.

Note that $U_{h,S} - U_h = P_{h,S} - 1/2 - (P_h - 1/2) = P_{h,S} - P_h$ for the boosting problem. So we are trying to select a stopping rule on sample size $|S|$ such that

$$\Pr\{|P_{h,S} - P_h| \geq \varepsilon|P_h - 1/2|\} \leq \delta.$$

The Algorithm A for controlling absolute error can be adapted for the general form of problem B in which the utility function U_h does NOT have to be defined as $P_h - 1/2$. In fact the following Algorithm B is applicable for any utility function of the form $U_h = P_h - c$ where $0 < c < 1$. The most important adaptation is to replace the *fixed* ε in Algorithm A by $\frac{\varepsilon}{(1+\varepsilon)} \times U_n$, where U_n is the estimation of U_h after seeing n samples. This gives rise to the following Algorithm B:

Algorithm B.
$n \leftarrow 0,\ X \leftarrow 0$
Repeat
   Draw a random sample Y
   /* Y is a random sample of the Bernoulli variable */
   $X \leftarrow X + Y,\ n \leftarrow n + 1$
   $\widehat{p_n} \leftarrow \frac{X}{n}$
   Compute the utility function U_n from $\widehat{p_n}$
      /* Note U_n is a general function of $\widehat{p_n}$ */
Until $n > 30$ *and* $n \geq \dfrac{\ln\frac{\delta}{2}}{\mathcal{U}(W(\widehat{p_n}),\ W(\widehat{p_n})+\varepsilon')}$
      Here $\varepsilon' = \frac{\varepsilon}{(1+\varepsilon)} \times |U_n|$
Output $\widehat{p_n}, U_n$, and n.

Note that Algorithm B can be applied to a particular hypothesis $h \in H_{DS}$ to estimate P_h. In this case, the $\widehat{p_n} = P_{h,S}$ and $U_n = U(h, S)$ where $n = |S|$.

Property of Algorithm B.

Theorem 4. *Assuming* CLT *is satisfied by* $\widehat{p_n}$, *where* $\mathbf{n}$ *is the number of samples seen at the termination of Algorithm B, we have*

$$\Pr\{|U_{h,S} - U_h| \geq \varepsilon|U_h|\} \leq \delta.$$

4.2   The New Hypothesis Selection Procedure for Boosting

Now we can define the new Boosting by adaptive sampling algorithm. The algorithm is very similar to Madaboost except that the following Algorithm C replaces *ADSS*. In other words, the termination criterion in our method differs from that in *ADSS*.

Algorithm C.
$S \leftarrow \{\}, n \leftarrow 0$
Repeat
    Use FiltEx to draw a random training example x
    (according to distribution $\mathscr{D}_t$ at round t)
    $S \leftarrow S \cup \{x\}, n \leftarrow n + 1$
    Compute $P_{h,S}$ and $U_{h,S}$ for each $h \in H_{DS}$
Until $n > 30 \wedge \exists h \in H_{DS}$ such that
$$U_{h,S} > 0 \text{ and } n \geq \frac{\ln \frac{\delta}{2}}{\mathscr{U}(W(P_{h,S}), \hat{W}(P_{h,S}) + \varepsilon'))}$$
$$\text{Here } \varepsilon' = \frac{\varepsilon}{(1+\varepsilon)} \times |U_{h,S}|$$
Output h, $P_{h,S}$ and n.

Property of Algorithm C.

Theorem 5. *Assume CLT is satisfied when Algorithm C terminates. In addition we assume the decision-stump hypothesis space H_{DS} is complete in the sense that for each hypothesis $h \in H_{DS}$, H_{DS} also contains the complement hypothesis h' (which has the same test condition at the internal node of the tree, but the opposite labeling for the two leaf nodes). Under these assumptions, with probability of at least $1 - \delta$, Algorithm C will output a hypothesis h such that $P_h > 1/2$, i.e., h has accuracy above $1/2$.*

Note that the above theorem provides a rigorous assurance that the new boosting by adaptive sampling method will produce a good classifier with high probability.

5   Experimental Results

We have conducted empirical studies on the above Algorithm C using several benchmark datasets from the UC Irvine ML database. For the sake of comparison, we also implemented the Madaboost algorithm with the more efficient version of the termination condition [16] with $a_n = (1 + \frac{1}{\varepsilon})\sqrt{\frac{\ln \frac{n(n+1)}{\delta}}{2n}}$. As in [6], decision stumps are used as base classifiers for boosting. Each decision stump has one internal node and two leaf nodes with binary (0 or 1) class labels.

Table 1 shows a summary description of the four data sets from the UC Irvine Machine Learning Data Repository that were used in our experiments. In all the experiments we performed a 4-fold cross validation. There is nothing particular about the use of 4-fold cross validation, versus, for example 5-fold or 10-fold. We in fact tried various k-fold cross validations with different values for k, and the results are similar. Here H_{DS} is the set of all decision-stump hypotheses that were searched by the boosting algorithm. For numerical valued attributes, a discretization is performed that constructs 6 equal-sized intervals each corresponding to a new binary attribute. Here the data set name "German" refers to

Table 1. Data sets used in the experimental studies

Data set	Data size	Number of attributes	$|H_{DS}|$
Adult	32561	14	274
KR-KP	319600	36	146
Iono	35100	34	390
German	100000	20	172

Table 2. Accuracy and computation time comparison of the new method vs. Watanabe's

Data set	δ	Accu-new	Accu-Wata	$|S|$-new	$|S|$-Wata	Time-new	Time-Wata
Adult	0.5	0.815	0.821	468	8847	0.065	1.136
	0.4	0.815	0.820	516	8849	0.070	1.181
	0.3	0.819	0.818	833	9504	0.113	1.293
	0.2	0.812	0.820	934	8692	0.126	1.159
	0.1	0.815	0.821	1345	10103	0.180	1.345
KR-KP	0.5	0.925	0.937	119	2666	0.0009	0.177
	0.4	0.905	0.926	163	2860	0.012	0.194
	0.3	0.912	0.933	231	2682	0.016	0.178
	0.2	0.923	0.934	297	2800	0.020	0.186
	0.1	0.929	0.935	352	2977	0.025	0.201
Iono	0.5	0.908	0.897	81	2745	0.013	0.423
	0.4	0.899	0.905	131	2882	0.021	0.433
	0.3	0.896	0.897	167	2983	0.0268	0.464
	0.2	0.903	0.899	272	2989	0.0508	0.460
	0.1	0.897	0.899	344	3249	0.055	0.504
German	0.5	0.738	0.744	481	12698	0.036	0.905
	0.4	0.742	0.742	670	13590	0.050	0.966
	0.3	0.742	0.744	849	13649	0.064	0.981
	0.2	0.740	0.737	1259	13320	0.094	0.955
	0.1	0.754	0.753	1280	13977	0.094	0.995

the "Statlog German Credit" data set, "KR-KP" refers to the chess end game "King-Rook vs. King-Pawn" data set, and "Iono" refers to the "Ionosphere" data set. To test the scalability of our algorithm on large data sets, some of the original data sets from the Irvine repository were duplicated by a factor of 100 to make a larger data set. This approach was also taken by some of the previous researchers, in particular, the authors of *Madaboost* also used these expanded data sets in their experiments [6].

Table 2 shows the experimental results that compare the prediction accuracy and sample size as well as execution time of our algorithm vs. the one in [16]. In all the experiments, we set the parameter $\epsilon = 1/2$, and we varied the value of δ from 0.1 to 0.5 with 0.1 increment. Our implementation is done using C++ on a computer with Intel Petinium 2.8 Ghz processor running Linux. The execution time is recorded in seconds. Here $|S|$ denote the sample size used by the boosting method for estimating the prediction accuracy of a decision stump. In the table heading, "Accu-new" and "Accu-Wata" represent the prediction accuracies (through 4-fold cross validation) of the proposed algorithm and the Madaboost algorithm in [16] respectively. The columns "$|S|$-new"("$|S|$-Wata") and "Time-new"("Time-Wata") indicate the number of samples used and the computation time in seconds for the new method and Watanabe method respectively.

From the results shown in Table 2, it is evident that our method achieved competitive prediction accuracy while using much smaller sample size (less than 10 percent of the data size used in [6]) and thus being much more efficient. The much-improved computational efficiency of our method makes it an excellent candidate for scalable learning with huge data sets.

6   Conclusions and Future Work

We present in this paper a method for learning by Boosting using adaptive sampling. The new boosting method utilizes the sampling method in [2] for estimating classifier accuracy to build an efficient boosting algorithm. We provide a preliminary theoretical analysis of the proposed sampling-based boosting method that offers a rigorous assurance for the validity of the approach. Empirical studies with 4 datasets from UC Irvine ML database show that our method typically uses much smaller sample size (and is thus much more efficient) while maintaining competitive prediction accuracy compared with Watanabe's sampling-based Boosting learner Madaboost [6]. Future works would include more experiments with even larger data sets and more thorough theoretical analysis on the boosting properties of the algorithm. In addition, it would be desirable to compare the performance of our sampling-based boosting method with the recently proposed "Empirical Bernstein Stopping" approach [14] applied to boosting.

References

1. Chernoff, H.: A measure of asymptotic efficiency for tests of a hypothesis based on the sum of observations. Ann. Math. Statist. **23**, 493–507 (1952)
2. Chen, J., Chen, X.: A new method for adaptive sequential sampling for learning and parameter estimation. In: Proceedings of International Symposium on Methodologies for Intelligent Systems, pp. 220–229, Warsaw, Poland, June 2011
3. Chen, J.: Scalable ensemble learning by adaptive sampling. In: Proceedings of International Conference on Machine Learning and Applications (ICMLA), pp. 622–625, December 2012
4. Chen, X.: A New Framework of Multistage Estimation. ArXiv:0809.1241v20 [Math.ST]

5. Chen, X.: A new framework of multistage parametric inference. In: Proceeding of SPIE Conference, Orlando, Frioda, U.S.A., April 2010

6. Domingo, C., Watanabe, O.: Scaling up a boosting-based learner via adaptive sampling. In: Terano, T., Liu, H., Chen, A.L.P. (eds.) PAKDD 2000. LNCS, vol. 1805, pp. 317–328. Springer, Heidelberg (2000)

7. Domingo, C., Watanabe, O.: Adaptive sampling methods for scaling up knowledge discovery algorithms. In: Proceeings of 2nd International Conference on discovery Science, Japan, December 1999

8. Freund, Y., Schapire, R.E.: A decision-theoretic generalization of on-line learning and an application to boosting. J. Comput. Syst. Sci. **55**(1), 119–139 (1997)

9. Ghosh, M., Mukhopadhyay, N., Sen, P.K.: Sequential Estimation. Wiley, New York (1997)

10. Hoeffding, W.: Probability inequalities for sums of bounded variables. J. Am. Stat. Assoc. **58**, 13–29 (1963)

11. Lipton, R., Naughton, J., Schneider, D.A., Seshadri, S.: Efficient sampling strategies for relational database operations. Theoret. Comput. Sci. **116**, 195–226 (1993)

12. Lipton, R., Naughton, J.: Query size estimation by adaptive sampling. J. Comput. Syst. Sci. **51**, 18–25 (1995)

13. Lynch, J.F.: Analysis and application of adaptive sampling. J. Comput. Syst. Sci. **66**, 2–19 (2003)

14. Minh, V., Szepesvari, C., Audibert, J.-Y.: Emoirical Bernstein stopping. In: Proceedings of the 25th International Conference on Machine Learning (ICML 2008), Helsinki, Finland, pp. 672–679 (2008)

15. Watanabe, O.: Simple sampling techniques for discovery sciences. IEICE Trans. Inf. Sys. **ED83–D**(1), 19–26 (2000)

16. Watanabe, O.: Sequential sampling techniques for algorithmic learning theory. Theo. Comput. Sci. **348**, 3–14 (2005)

WPI: Markov Logic Network-Based Statistical Predicate Invention

Stefano Ferilli[1,2]([✉]) and Giuseppe Fatiguso[1]

[1] Dipartimento di Informatica, Università di Bari, Bari, Italy
stefano.ferilli@uniba.it
[2] Centro Interdipartimentale per la Logica e sue Applicazioni,
Università di Bari, Bari, Italy

Abstract. Predicate Invention aims at discovering new emerging concepts in a logic theory. Since there is usually a combinatorial explosion of candidate concepts to be invented, only those that are really relevant should be selected, which cannot be done manually due to the huge number of candidates. While purely logical automatic approaches may be too rigid, statistical solutions provide more flexibility in assigning a degree of relevance to the various candidates in order to select the best ones. This paper proposes a new Statistical Relational Learning approach to Predicate Invention. It was implemented and tested on a traditional problem, yielding interesting results.

1 Introduction

In symbolic Machine Learning, Predicate Invention (PI) deals with the problem of discovering new emerging concepts in the available knowledge. Associated to inductive learning, in particular, PI would allow to obtain a compression of the learned theory or to catch exceptions in the learned clauses. Two fundamental problems in PI are how to handle the combinatorial explosion of candidate concepts to be invented, and how to select only those that are really relevant. Indeed, a huge number of possible concepts might be defined with the given knowledge, most of which are just casual or otherwise irrelevant aggregations of features. So, proper filtering must be carried out in order to keep the most significant and relevant concepts only. Due to the large number of candidates, manual selection by domain experts is infeasible, and automatic techniques are needed to assign a degree of relevance to the various candidates and select the best ones.

Inductive Logic Programming (ILP) is the branch of Machine Learning interested in dealing with logic-based (especially first-order) representation formalisms. However, using a purely logical approach, ILP techniques are sometimes too rigid to deal with noisy data. Statistical Predicate Invention (SPI) aims at discovering new concepts from relational data by means of Statistical Relational Learning (SRL) approaches. This paper proposes *Weighted Predicate Invention* (WPI), a SPI approach that focuses on the discovery of tacit relations inside the theories learned by a traditional ILP system with the objective of making the theory more

© Springer International Publishing Switzerland 2015
F. Esposito et al. (Eds.): ISMIS 2015, LNAI 9384, pp. 112–121, 2015.
DOI: 10.1007/978-3-319-25252-0_12

compact, improving its readability and possibly making it a better representation of the domain of interest. After discussing related works in the next section, we recall some preliminaries in Sect. 3 and describe WPI in Sect. 4. Experimental results on a standard ILP dataset are reported in Sect. 5. Finally, Sect. 6 concludes the paper and proposes future work.

2 Related Works

Historically, the problem of PI originated in the ILP field, where typical approaches have been based on the analysis of the structure of clauses. In the classical vision of PI in ILP, predicates are invented by using different techniques based on the analysis of the structure of clauses. The goal is to obtain a compression of a first-order theory, or to improve the performance of inductive learning tasks, or to catch exceptions in learned clauses [7]. The problem has been explored by analyzing first-order formulas, trying to apply restructuring operators, such as inter-construction (that looks for commonalities among clauses), as in [7,16], or intra-construction (that looks for differences among clauses), as in [7,9]. Many of these works introduced PI in the context of learning systems such as Golem and Cigol [7]. Some results in this field also came from the work of Pazzani [14], in which the search for a predicate to invent is on a second-order instantiation of the logic theory. Other approaches to PI rely on the idea of finding tacit concepts that are useful to catch irregularities in the patterns obtained for a given induction problem [8,15].

Another stream of research that can be considered as related to PI was carried out in the pure SRL setting. Specifically, the proposal in [2] works in statistical learning. It consists of the search for hidden variables in a Bayesian network, by looking for structural patterns, and applies a subsequent development using a clustering method to group observed variables and find hidden ones for each analyzed group. A task in SRL that is very similar to the task of PI in ILP is known as Hidden Variables Discovery. The approaches to PI proposed in this stream are more relevant to the field of Hidden Variable Discovery, which is pure SRL, than they are to the original ILP concept of PI. Recently some works better merged together the ideas of PI with those coming from Hidden Variable Discovery. In [6] attention is moved to the problem of SPI in a context of MLNs using a multiple clustering approach to group both relations and constants in order to improve Structure Learning techniques for Markov Logic. Another approach [1] invents predicates in a statistical context and then exploits the invented predicate in the FOIL learning system. In [12] a particular version k-means is used to cluster separately constants and relations. [11] applied a number of techniques to cluster multi-relational data.

Only in the last few years there have been some developments in the field of SPI. Some of the proposed approaches use some version of relational clustering to aggregate concepts and relations into new ones [5]. E.g., [4] aims at avoiding the limitations of the pure ILP approach (that is prone to overfitting on noisy data, which causes the generation of useless predicates) using a statistical bottom-up approach to invent predicates only in a statistical-relational domain.

3   Preliminaries

Our approach to SPI uses Datalog (a function-free fragment of Prolog) as a representation language. The basic elements in this setting are atoms, i.e. claims to which a truth value can be assigned. An *atom* is a predicate applied to its arguments. The number of arguments required by a predicate is called its *arity*. The arguments of a predicate must be *terms* (i.e., variables or constants). A *literal* is either an atom (positive literal) or a negation of an atom (negative literal). A *clause* is a disjunction of literals; if such a disjunction involves at most one positive literal then it is called a Horn clause, its meaning being that, if all atoms corresponding to negative literals (called the *body* of the clause) are true, then the positive literal (called the *head* of the clause) must be true. A clause may be represented as a set of literals. An atom, a literal or a clause is called *ground* if its terms are all constants.

A clause made up of just the head is a *fact*, while a clause made up of a head and a non-empty body is a *rule*. A rule is linked if, given any two literals in it, one can find a chain of literals in the rule such that any two adjacent literals in the chain share at least one term. We adopt the Object Identity (OI) assumption, by which terms denoted by different symbols in a clause must refer to distinct objects [13]. So, a variable in a clause cannot be associated to another variable or constant in the same clause, nor can two variables in the same clause be associated to the same term.

For instance, $p(X, a, b, Y)$ is an atom (and a positive literal), while $\neg p(X, a, b, Y)$ is a (negative) literal, where p is the predicate, the arity of p is n, and X, a, b, Y are terms (specifically, X, Y are variables and a, b are constants). $\{p(t_1)\}$ and $\{p(t_1), \neg q(t_1, t_2), \neg r(t_2)\}$ are (Horn) clauses. The former is a fact, and the latter is a (linked) rule, with head $p(t_1)$ and body $\{q(t_1, t_2), r(t_2)\}$. Under OI, in a clause $\{p(X), \neg q(X, Y)\}$ it is implicitly assumed that $X \neq Y$.

The vocabulary on which we build our representations is a triple $\mathcal{L} = \langle P, C, V \rangle$ where P is a set of predicate symbols, C is a set of constant symbols and V is a set of variable symbols. So, $C \cup V$ is the set of terms in our vocabulary. In the following, predicates and constants will be denoted by lowercase letters, and variables by uppercase letters.

A Markov Logic Network (MLN) consists of a set of first-order formulas associated to weights that represent their relative strength. Given a finite set of constants C, the set of weighted first-order formulas can be used as a template for constructing a Markov Random Field (MRF) by grounding the formulas by the constants in all possible ways. The result is a graph in which nodes represent ground atoms, and maximal cliques are groundings of first-order formulas, also called 'features' in the MRF. The joint probability for atoms is given by:

$$P(X = x) = \frac{1}{Z} exp \left(\sum_{i=1}^{|F|} w_i n_i(x) \right) \qquad (1)$$

where Z is a normalization constant, $|F|$ is the number of first order formulas in the set, $n_i(x)$ is the number of true groundings of formula F_i given the set of constants C, and w_i is the weight associated to formula F_i. In the Markov Logic framework two types of parameters learning can be distinguished. Structure Learning consists in inducing first-order formulas given a set of ground atoms as evidence and a query predicate, by maximizing the joint probability on the MRF based on the evidence. Weight Learning consists in estimating the weight of each first-order formula given a first-order theory and the facts in the evidence, by maximizing the likelihood of a relational database using formula (1). The weight of a formula captures the importance of the formula itself, i.e. its inclination to be true in all possible worlds defined by the set of constants observed in the evidence.

4   Weighted Predicate Invention

As already pointed out, PI aims at generating new symbols that define latent concepts in a bunch of relational data or in a logic theory. Such new symbols are defined in terms of the previously available symbols. Approaches to PI can be bottom-up (starting from ground atoms) or top-down (starting from a first-order theory). We propose a top-down approach to SPI based on the use of Markov Logic Networks, called *Weighted Predicate Invention (WPI)*. WPI is motivated by the observation that some concepts that are latent in a logic theory may be useful in accomplishing some tasks related to the given problem, but not all such concepts are really significant. While this intuition is not novel [7], it is one of the major problems that PI has encountered so far. Our approach distinguishes significant concepts, to be invented, from insignificant ones by their weight, obtained by considering all possible worlds according to evidence data. If the weights, as in MLNs, express the likelihood that a certain world, described by the rules in the theory, is verified given the data, and the addition of the invented predicate (i.e., of the rule defining it) to the theory increases the likelihood expressed by those weights, then one may draw the conclusion that the latter is a better description of the world, although formally equivalent to the former.

PI can be carried out on two different kinds of sources: on a background knowledge made up of facts, or on a general theory made up of rules. In a Machine Learning setting, both are available: the theory is the model obtained as a result of running some learning system on the facts that describe the observations and the examples. In some sense, such a model has already performed a kind of selection of relevant information in the background knowledge (in this case, information that may characterize a set of classes and/or discriminate them from other classes). So, working on the theory should make the predicate invention problem somehow easier while still accounting also for the background knowledge.

The basic idea underlying WPI is to analyze an inductively learned theory consisting of a set of first-order rules under OI, with the main objective of producing a compact version of the theory itself in which new interesting concepts emerge. It finds common patterns between the rules' bodies working in different steps. First a bipartite graph is created, whose nodes are distinguished between predicates and rules appearing in the theory. Then, a pattern is searched in the graph and checked for matching with subsets of literals in the body of rules. Specifically, the pattern consists of predicate nodes that may be involved in the definition of the predicate to be invented. This problem is complex due to the indeterminacy that characterizes the first-order logic setting. Indeed, it may not have a solution at all. If a matching is found, a candidate rule to be invented is built, and checked for validation before definitely including it in the existing first-order theory. The validation process is based on the use of Weight Learning in the MLN framework to assign a weight to the rules in the first-order theory. We produce two MLNs, one that exploits the invented predicate and one that does not exploit it. We consider the new rule as validated if weights in the latter are non-decreasing with respect to the former. In this case we can add the invented rule into the first-order theory.

Call R the set of first-order rules in the theory. Given a rule $r_i \in R$, let us denote by b_i the set of literals in the body of r_i, and by c_i a subset of b_i. Also, call P the set of all predicates in bodies of rules in the theory, where a literal l is represented as a pair $l = (p, V)$ with p its predicate and V the list of its arguments. We define a bipartite graph $G = \langle R \cup P, E \rangle$ where $E \subseteq R \times P$ is the set of edges such that $\{r, p\} \in E$ iff $\exists l = (p, V) \in R$. We call *upper nodes* the elements of R and *lower nodes* the elements of P. So, every rule is connected to all the predicate symbols appearing in its body.

To find commonalities in the bodies of clauses we consider all possible pairs $I = (\pi, \rho)$, where $\pi \subseteq P$ is a set of lower nodes (made up of at least two elements) that are connected to the same upper-node and $\rho \subseteq R$ is the set of rules in the theory that include π. Among all possible such pairs, we look for one that maximizes (with respect to set inclusion) π. The model can be applied iteratively, so we just look for maximization pairs, and in the next iteration we can invent (if any) the others. Predicates appearing in a maximal I will be used to form a candidate pattern to define a predicate to be invented.

Given a pattern, we aim at finding a subset of literals in the theory rules that matches it. Clearly, only the rules in ρ are involved in this operation. For each rule $r_i \in \rho$, consider the subset of its literals that are relevant to the pattern, $c_i = \{l \in r_i | l = (p, V), p \in \pi\}$. Note that c_i may not be linked. For each predicate p_j in the pattern, call n_j the minimum number of occurrences across the c_i's. We try to invent predicates defined by n_j occurrences of predicate p_j, for $j = 1, \ldots, |\pi|$. The underlying rationale is that this should ensure to have more chances to find that set of literals in the rules of the theory.

Now, for each c_i, consider the set Γ_i of all of its subsets that include exactly n_j occurrences of each predicate p_j in π. We call each element of a Γ_i a *configuration*, and look for a configuration that is present in all Γ_i's (modulo variable renaming). More formally, we look for a

$$\overline{\gamma} \text{ s.t. } \forall i = 1, \ldots, |\rho| : \exists j_i \in \{1, \ldots, |\Gamma_i|\} \text{ s.t. } \overline{\gamma} \equiv \gamma_{ij_i} \in \Gamma_i$$

The existence of a solution is not guaranteed. In case it does not exist, we proceed by removing one occurrence of a predicate and trying again, until subsets made up of two literals are tried (inventing a predicate defined by a single literal would be nonsense, since it would just be a synonym).

The selected configuration $\overline{\gamma}$ becomes the body of the *invented rule* $\overline{r}$ that defines the predicate to be invented. If i is the name of the invented predicate, then the head of $\overline{r}$ is obtained by applying i to a list of arguments corresponding to all the variables in $\overline{\gamma}$. After adding this rule to the theory, for each rule $r_i \in \rho$ the configuration $\overline{\gamma_i} \in \Gamma_i$ equivalent to $\overline{\gamma}$ can in principle be removed and replaced by the corresponding instance of predicate i.

Example 1. Consider a theory made up of three rules, $R = \{r_1, r_2, r_3\}$, where:

$$r_1 : q(X) :- \quad a(X), b(Y), b(W), c(X, Y), d(Y, W).$$
$$r_2 : q(X) :- a(X), b(W), c(X, Y), c(Y, W), g(X), h(Z, Y).$$
$$r_3 : q(X) :- \quad a(X), f(Z, Y), h(X, Y).$$

The following predicates are available in each rule:

$$r_1 \rightarrow \quad \{a/1, b/1, c/2, d/2\}$$
$$r_2 \rightarrow \{a/1, b/1, c/2, g/1, h/2\}$$
$$r_3 \rightarrow \quad \{a/1, f/2, h/2\}$$

After building the bipartite graph, the maximal intersection of lower-nodes is $I = (\pi, \rho)$ where $\pi = \{a/1, b/1, c/2\}$ and $\rho = \{r_1, r_2\}$. The minimum number of literals for all predicates in the pattern $\{a/1, b/1, c/2\}$ is one, thus the best subset of literals to match is $\{a(\cdot), b(\cdot), c(\cdot, \cdot)\}$. So, we have $\Gamma_1 = \{\gamma_{11}, \gamma_{12}\}$ and $\Gamma_2 = \{\gamma_{21}, \gamma_{22}\}$ where:

$$\gamma_{11} = \{a(X), b(Y), c(X, Y)\}, \gamma_{12} = \{a(X), b(W), c(X, Y)\}$$

$$\gamma_{21} = \{a(X), b(W), c(X, Y)\}, \gamma_{22} = \{a(X), b(W), c(Y, Z)\}$$

The best configuration is $\gamma_{12} \equiv \gamma_{21}$. The candidate invented rule would be

$$\overline{r} : i(X, Y, W) :- a(X), b(W), c(X, Y).$$

yielding:

$$\overline{r} : i(X, Y, W) :- \quad a(X), b(W), c(X, Y).$$
$$r_1 : q(X) :- \quad b(Y), d(Y, W), i(X, Y, W).$$
$$r_2 : q(X) :- c(Y, W), g(X), h(Z, Y), i(X, Y, W).$$
$$r_3 : q(X) :- \quad a(X), f(Z, Y), h(X, Y).$$

To prevent the invention of useless predicates, a validation step must be run that determines whether the invented predicate is actually relevant. The idea is that the introduction of the invented rule in the original theory must not decrease the relevance of the existing rules. We propose to use the weights learned by the weight learning functionality of the MLN framework as an estimator of the relevance of a rule in the context defined by the given theory and the set of evidence facts in the background knowledge. So, we build two MLNs. The former simply adds the invented rule to the initial theory. The latter also applies the invented rule to the existing rules, replacing the subset of literals in each rule, that match the invented rule body, with the head of the invented rule properly instantiated. In the former, the invented rule is disjoint from the rest of the graph, because the invented predicate is not present in the other rules. This causes the weights of the other rules not to be affected by the presence of the new one. In the latter, the body of some rules in the original theory has changed so that the invented predicate is used, and thus the invented rule is no more disjoint in the graph. Based on this, we expect a variation of the rule weights in the two cases. Comparing the two weights, we consider the invented predicate as relevant if the weight in the latter template is greater than the weight in the former.

We run Discriminative Weight Learning on both templates, obtaining two sets of weighted first-order rules with respect to the evidence facts and the query predicate for the problem defined by theory. Call $w'_0, w'_1, \ldots, w'_k$ the weights of rules in the former MLN, and $w''_0, w''_1, \ldots, w''_k$ the weights of rules in the latter MLN. Then, we pairwise compare the weights of rules, and specifically the weight of the invented rule and the weights of the rules in ρ involved in the predicate invention process. The invented rule is considered as validated if no weight after the application of the invented predicate is less than it was before:

$$\forall i = 0, \ldots, k : w'_i \leq w''_i$$

Otherwise, if the introduction of the invented rule in the initial theory causes a decrease in the relative importance of any rule, then the invented rule is not added to the theory.

Given the new theory, WPI can be run again on it in order to invent, if possible, further predicates that define implicit concepts. Iterating this procedure yields a wider theory restructuring, resulting in progressively more compact versions of the theory. Every invented predicate subsumes the body of rules involved in its invention.

5   Experiments

The WPI approach was implemented to test its effectiveness, both as regards predicate invention and as regards theory restructuring. To manage MLNs and Discriminative Weight Learning, WPI relies on Tuffy [10], an implementation of the MLN framework based on a Relational DBMS to scale up learning. Then, an experiment was devised. To simulate a totally automatic setting, the theories for the experiment were learned using InTheLEx [3]. InTheLEx was chosen because it embeds multistrategy operators that provide deduction and abstraction capabilities. So, InTheLEx would be actually able to exploit the invented rule when in operation. Also, introducing the predicate invention feature would extend and enhance its functionality, because it would allow the system to automatically obtain information for deduction or abstraction that otherwise should be manually provided by human experts.

A toy dataset was exploited in the experiment, for a twofold reason. First, because of the computational complexity of the SPI approach. Second, because it allows an easier insight into the results. Specifically, we considered the classical *Train Problem*, well-known in ILP, in its extended version given by Muggleton. It includes 20 examples of Eastbound or Westbound trains, with the goal to predict Eastbound ones. Due to the small size of this dataset we applied a Leave-One-Out Cross-Validation technique to avoid the problem of overfitting. This resulted in a total of 20 folds, each using one example for testing purposes and the remaining 19 for training. For each fold, InTheLEx was run on the training set to inductively learn a theory[1]. Of course, each fold resulted in a different theory, which in turn allowed different possibilities to predicate invention. So, we applied WPI to each learned theory and compared the resulting restructured theory with the one originally provided by InTheLEx. Experimental results confirmed that different theories were learned in the different folds, yielding different invented predicates and, thus, different restructured theories.

The quantitative comparison focused on the number of new (invented) concepts, the number of rules in the theories before and after restructuring, and the number of literals per rule in these theories. A statistical analysis of the outcomes shows that the number of rules in the theories significantly increases on average after invention and restructuring. However, the number of invented predicates/rules shows some variability in the different folds, including folds in which no predicate could be invented at all. Applying the invented predicates to the initial theory should result in a compression of the theory itself, because the definition of the invented predicate is removed from the other rules and replaced by a single literal that is an instance of the invented predicate. The significant compression obtained in the experiment can be appreciated from the number of literals per rule. So, overall, as expected, the number of rules increases on average, but their size decreases on average. Specifically, 4.25 new concepts were invented on average in each of the 20 folds, which more than doubled the size of the theories on average. However, the average number of literals per rule dropped from 18.41 to 5.30 on average, which (also considering the increase in number of rules) yields an average compression factor of 71.21 %.

From a qualitative point of view, we looked for possible interesting new concepts emerging from the inductively learned theories. An interesting result is, for example, the invention in many folds of the following predicate:

i(Train, Car) :- car(Car), has_car(Train, Car).

corresponding to the concept that any railway car in the train is somehow connected to the locomotive: $\{car(Car), has\_car(Train, Car)\}$. This concept

---

[1] In a normal procedure, the learned theory is to be tested on the remaining example; however, for the purposes of this paper we are not interested in the predictive capabilities of the theories, so we skip this step.

catches a general intuition about trains, and factorizing it makes the theory more understandable for two reasons (both connected to the fact that a single literal is used to convey a meaning that previously required two literals): first, because the rules describing the trains become more compact and thus more readable; second, because the invented predicate highlights a concept that is meaningful to the reader and that was previously split into several literals that were possibly scattered throughout the description.

6   Conclusions and Future Works

Predicate Invention is the branch of symbolic Machine Learning aimed at discovering new emerging concepts in the available knowledge. Two fundamental problems in Predicate Invention are how to handle the combinatorial explosion of candidate concepts to be invented, and how to select only those that are really relevant. Due to the huge number of candidates, there is a need for automatic techniques to assign a degree of relevance to the various candidates and select the best ones. Purely logical approaches may be too rigid for this purpose, while statistical solutions may provide the required flexibility. This paper proposed a new Statistical Relational Learning approach to Predicate Invention. It was implemented and tested on a traditional problem in Inductive Logic Programming, yielding interesting results in terms of invented predicates and consequent theory restructuring.

Future work will include: optimizing the implementation; devising more informative heuristics for selecting and choosing candidate definitions for the predicate to be invented; running more extensive experiments, also on real-world datasets.

Acknowledgments. This work was partially funded by the Italian PON 2007–2013 project PON02_00563_3489339 'Puglia@Service'.

References

1. Craven, M., Slattery, S.: Relational learning with statistical predicate invention: better models for hypertext. Mach. Learn. **43**(1/2), 97 (2001)
2. Elidan, G., Friedman, N.: Learning hidden variable networks: the information bottleneck approach. J. Mach. Learn. Res. **6**, 81–127 (2005)
3. Esposito, F., Semeraro, G., Fanizzi, N., Ferilli, S.: Multistrategy theory revision: induction and abduction in inthelex. Mach. Learn. J. **38**(1/2), 133–156 (2000)
4. Kemp, C., Tenenbaum, J.B., Griffiths, T.L., Yamada, T., Ueda, N.: Learning systems of concepts with an infinite relational model. In: AAAI, pp. 381–388. AAAI Press (2006)
5. Kok, S., Domingos, P.: Toward statistical predicate invention. In: SRL2006: Open Problems in Statistical Relational Learning (2006)
6. Kok, S., Domingos, P.: Statistical predicate invention. In: Ghahramani, Z. (ed.) ICML of ACM International Conference Proceeding Series, vol. 227, pp. 433–440. ACM (2007)

7. Kramer, S.: Predicate invention: a comprehensive view. Technical report TR-95-32, Oesterreichisches Forschungsinstitut fuer Artificial Intelligence, Wien, Austria (1995)
8. Muggleton, S.: Predicate invention and utilization. J. Exp. Theor. Artif. Intell. **6**(1), 121–130 (1994)
9. Muggleton, S., Buntine, W.: Machine invention of first-order predicates by inverting resolution. In: MLC-88 (1988)
10. Niu, F., Ré, C., Doan, A., Shavlik, J.W.: Tuffy: scaling up statistical inference in markov logic networks using an RDBMS (2011). CoRR, abs/1104.3216
11. Perlich, C., Provost, F.: Aggregation-based feature invention and relational concept classes. In: Domingos, P., Faloutsos, C., Senator, T., Kargupta, H., Getoor, L. (eds.) Proceedings of the Ninth ACM SIGKDD International Conference on Knowledge Discovery and Data Mining (KDD-03), pp. 167–176. ACM Press, New York (2003)
12. Popescul, A., Ungar, L.H.: Cluster-based concept invention for statistical relational learning. In: Kim, W., Kohavi, R., Gehrke, J., DuMouchel, W. (eds.) KDD, pp. 665–670. ACM (2004)
13. Semeraro, G., Esposito, F., Malerba, D., Fanizzi, N., Ferilli, S.: A logic framework for the incremental inductive synthesis of datalog theories. In: Fuchs, N.E. (ed.) LOPSTR 1997. LNCS, vol. 1463, pp. 300–321. Springer, Heidelberg (1998)
14. Silverstein, G., Pazzani, M.: Relational cliches: constraining constructive induction during relational learning. In: Proceedings of the Sixth International Workshop on Machine Learning. Kaufmann, Los Altos (1989)
15. Srinivasan, A., Muggleton, S., Bain, M.: Distinguishing exceptions from noise in non monotonic learning. In: Rouveirol, C. (ed.) Proceedings of the ECAI-92 Workshop on Logical Approaches to Machine Learning (1992)
16. Wogulis, J., Langley, P.: Improving efficiency by learning intermediate concepts. In: Proceedings of the Eleventh International Joint Conference on Artificial Intelligence, pp. 657–662. Morgan Kaufmann, Detroit (1989)

Learning Bayesian Random Cutset Forests

Nicola Di Mauro[1]([✉]), Antonio Vergari[1], and Teresa M.A. Basile[2]

[1] Department of Computer Science, University of Bari "Aldo Moro",
Via E. Orabona 4, 70125 Bari, Italy
{nicola.dimauro,antonio.vergari,teresamaria.basile}@uniba.it
[2] Department of Physics, University of Bari "Aldo Moro", Via G. Amendola 173,
70126 Bari, Italy

Abstract. In the Probabilistic Graphical Model (PGM) community there is an interest around tractable models, i.e., those that can guarantee exact inference even at the price of expressiveness. Structure learning algorithms are interesting tools to automatically infer both these architectures and their parameters from data. Even if the resulting models are efficient at inference time, learning them can be very slow in practice. Here we focus on Cutset Networks (CNets), a recently introduced tractable PGM representing weighted probabilistic model trees with tree-structured models as leaves. CNets have been shown to be easy to learn, and yet fairly accurate. We propose a learning algorithm that aims to improve their average test log-likelihood while preserving efficiency during learning by adopting a random forest approach. We combine more CNets, learned in a generative Bayesian framework, into a generative mixture model. A thorough empirical comparison on real word datasets, against the original learning algorithms extended to our ensembling approach, proves the validity of our approach.

1 Introduction

A key task in *Probabilistic Graphical Models* (PGMs) [11] is *inference*, i.e., the possibility to answer queries about the probability of observing some states of the random variables whose joint distribution is compactly represented in the model. *Exact* inference is known to be a hard task in general, but even approximate inference routines are unfeasible sometimes [22]. In order to gain efficiency, and preserving exactness, one has to renounce expressiveness, that is coping with the possibility of not capturing all the independencies among random variables. The growing data availability demands PGMs able to guarantee tractable inference leading to *tractable* PGMs, an accepted trade-off in the AI community.

Works on learning tractable models structure from data date back to *tree-structured models*, like those inferred by the classical Chow-Liu algorithm [4]. Recently, more accurate tractable PGMs have been proposed such as extensions of tree-structured models by composing them in mixtures [16] or by introducing latent variables [3], Arithmetic Circuits (ACs) capturing the expressiveness of more complex Bayesian and Markov Networks [13,14], or by compacting latent interactions in a deep architecture as done by Sum-Product Networks

© Springer International Publishing Switzerland 2015
F. Esposito et al. (Eds.): ISMIS 2015, LNAI 9384, pp. 122–132, 2015.
DOI: 10.1007/978-3-319-25252-0_13

(SPNs) [18]. As the expressiveness of these models increases, the complexity of learning their structure increases as well, being, in general, the problem formulated as a search in the structure space guided by complex statistical independence tests.

With the objective of making structure learning efficient, *Cutset Networks* (CNets) have been recently introduced in [19] as tractable PGMs embedding Pearl's conditioning algorithm [17]. They are weighted probabilistic model trees in the form of OR-trees having tree-structured models as leaves, and non-negative weights on inner edges emanating from inner OR nodes, representing conditioning on the values of the random variables associated to those nodes. Structure learning in [19] corresponds to a greedy top-down search process in the OR trees space that leverages decision tree learning heuristics to determine the random variable to condition on at each step. The corresponding proposed algorithm, CNet, recursively partitions the data instances into subsets by selecting heuristically the best variable that maximizes an approximation of a reformulation of the information gain based on the joint entropy. Once this variable has been found, the algorithm creates a corresponding inner OR node and proceeds until no more splits can be done, in which case a tree learned with the Chow-Liu algorithm [4] is introduced as a leaf node to approximate the distribution on the current data partition. The cheap computation of an entropy based heuristic makes learning efficient, however the resulting models are *far* from being accurate as density estimators compared to other PGM structure learners [16,21]. Introducing mixtures of CNets learned via the Expectation Maximization algorithm (EM) lead to really competitive results. Such results shed light on the trade-off between the simplicity of such models and their accuracy: to effectively make them accurate one cannot ignore direct likelihood maximization and shall recur to ensemble techniques, making, in the end, learning more expensive if not complex.

In [7] a more principled way to learn CNets has been proposed. Structure learning is reformulated in a Bayesian framework as a likelihood-principled search guided by the Bayesian Information Criterion (BIC). A tractable model score, obtained by exploiting the decomposability of the likelihood of such models, enables the feasibility of finding the best feature for an OR split by directly evaluating part of the model on a portion of the data. Regularization is achieved both by the introduction of the BIC score and by putting informative Dirichlet priors on the probability parameters. The introduced algorithm, dCSN, proved to be more accurate than the original one. As expected, mixtures of models learned in this way, ensembled via *bagging* [10], outperformed both CNet and dCSN, again at the expense of learning time.

Here we extend the work in [7], by learning ensembles of CNets as density estimators by adopting a *random forest* approach [2]. Our main objective is to greatly reduce the evaluation time to choose the best variable to split on in dCSN by considering only a fraction of all candidates variables. Nevertheless, such state space reduction could potentially lead to less than optimal models, likelihood-wise, in our framework. We devised a fair and thorough set of experiments to test our proposed approach along the two dimensions of accuracy and time.

We evaluated our mixture models on 18 real world datasets against the original algorithm as proposed in [19] extended to our ensemble framework, proving that not only learning times are reduced, but also the model accuracy increased.

2   Cutset Networks

Let $\mathcal{D} = \{\xi_1, \ldots, \xi_M\}$ be a set of M i.i.d. instances over $\mathbf{X} = \{X_1, \ldots, X_n\}$ discrete random variables (features), whose domains are the sets $\text{Val}(X_i) = \{x_i^j\}_{j=1}^{k_i}, i = 1, \ldots, n$. We denote as $\xi_m[X_i]$ the value assumed by an instance ξ_m in correspondence to a particular variable X_i .

A *Cutset Network* (CNet) is a pair $\langle \mathcal{G}, \gamma \rangle$, where $\mathcal{G} = \mathcal{O} \cup \{\mathcal{T}_1, \ldots, \mathcal{T}_L\}$ is the graphical structure, composed by a rooted OR tree, $\mathcal{O}$, and by leaf trees $\mathcal{T}_l$; and $\gamma = \boldsymbol{w} \cup \{\boldsymbol{\theta}_1, \ldots, \boldsymbol{\theta}_L\}$ is the parameter set containing the OR tree weights $\boldsymbol{w}$ and the leaf tree parameters $\boldsymbol{\theta}_l$. The *scope* of a CNet $\mathcal{G}$ (resp. a leaf tree $\mathcal{T}_l$), denoted as $\text{scope}(\mathcal{G})$ (resp. $\text{scope}(\mathcal{T}_l)$), is the set of random variables that appear in it. Each node in the OR tree is labeled by a variable X_i, and each edge emanating from it represents the conditioning of X_i by a value $x_i^j \in Val(X_i)$, weighted by the conditional probability $w_{i,j}$ of conditioning the variable X_i to the value x_i^j.

A CNet can be thought of a model tree associating to each instance a weighted probabilistic leaf model. From now on, for the sake of simplicity, we will refer to the leaf trees as CLtrees as they may be learned by the classical Chow-Liu algorithm [4] as the trees best approximating a probability distribution from data in the terms of the Kullback-Leibler divergence, as done in [19].

In [7] a recursive definition of a CNet has been proposed along with the resulting log-likelihood and BIC score decomposition, leading to the principled dCSN algorithm for learning the structure of CNets, that we briefly review here.

Definition 1 (Cutset network [7]). *Let $\mathbf{X}$ be a set of discrete variables, a Cutset Network is defined as follows:*

1. *a CLtree, with scope $\mathbf{X}$, is a CNet;*
2. *given $X_i \in \mathbf{X}$ a variable with $|Val(X_i)| = k$, graphically conditioned in an OR node, a weighted disjunction of k CNets $\mathcal{G}_i$ with same scope $\mathbf{X}_{\backslash i}$ is a CNet, where all weights $w_{i,j}$, $j = 1, \ldots, k$, sum up to one, and $\mathbf{X}_{\backslash i}$ denotes the set $\mathbf{X}$ minus the variable X_i.*

Following this definition, the computation of the log-likelihood of a CNet can be decomposed as follows [7]. Given a CNet $\langle \mathcal{G}, \gamma \rangle$ over variables $\mathbf{X}$ and a set of instances $\mathcal{D}$, its log-likelihood $\ell_{\mathcal{D}}(\langle \mathcal{G}, \gamma \rangle)$ can be computed as:

$$\ell_{\mathcal{D}}(\langle \mathcal{G}, \gamma \rangle) = \begin{cases} \sum_{\xi \in \mathcal{D}} \sum_{i=1}^n \log P(\xi[X_i]|\xi[\text{Pa}_i]), & \text{if } \mathcal{G} = \{\mathcal{T}\} \\ \sum_{j=1}^k M_j \log w_{i,j} + \ell_{\mathcal{D}_j}(\langle \mathcal{G}_j, \gamma_{\mathcal{G}_j} \rangle), & \text{otherwise,} \end{cases} \quad (1)$$

where the first equation refers to the case of a CNet composed by a single CLtree, while the second one specifies the case of an OR tree rooted on the variable X_i, with $|Val(X_i)| = k$, where, for each $j = 1, \ldots, k$, $\mathcal{G}_j$ is the CNet involved in the disjunction with parameters $\gamma_{\mathcal{G}_j}$, $\mathcal{D}_j = \{\xi \in \mathcal{D} : \xi[X_i] = x_i^j\}$ is the slice of $\mathcal{D}$ after conditioning on X_i, and $M_j = |\mathcal{D}_j|$ its cardinality. $\ell_{\mathcal{D}_j}(\langle \mathcal{G}_j, \gamma_{\mathcal{G}_j} \rangle)$ denotes the log-likelihood of the sub-CNet $\mathcal{G}_j$ on $\mathcal{D}_j$.

The dCSN algorithm, proposed in [7], exploits a different approach from that in [19], avoiding decision tree heuristics and instead choosing the best variable by directly maximizing the data log-likelihood. By exploiting the recursive nature of Definition 1, a CNet is grown top-down allowing further expansion, i.e., substituting a CLtree with an OR node only if it improves the log-likelihood (it is clear that maximizing the second term in Eq. 1 results in maximizing the global score). In detail, dCSN starts with a single CLtree, for variables $\mathbf{X}$, learned from $\mathcal{D}$ and it checks whether there is a decomposition, i.e. an OR node applied on as many CLtrees as the values of the best variable X_i, providing a better log-likelihood than that scored by the initial tree. If a such decomposition exists, than the decomposition process is recursively applied to the sub-slices $\mathcal{D}_i$, testing each leaf for a possible substitution.

In order to penalize complex structures and thus avoiding overfitting, a Bayesian Information Criterion (BIC) [8] has been adopted. The BIC score of a CNet $\langle \mathcal{G}, \gamma \rangle$ on data $\mathcal{D}$ is defined as $\text{score}_{\text{BIC}}(\langle \mathcal{G}, \gamma \rangle) = \log P_{\mathcal{D}}(\langle \mathcal{G}, \gamma \rangle) - \frac{\log M}{2} \text{Dim}(\mathcal{G})$, where $\text{Dim}(\mathcal{G})$ is the model dimension, set to the number of OR nodes appearing in $\mathcal{G}$, and M is the size of the dataset $\mathcal{D}$. Given $\mathcal{G}$ and $\mathcal{G}'$ be two CNets, where $\mathcal{G}'$ has been obtained from $\mathcal{G}$ substituting a leaf tree by adding a new sub-CNet rooted in an OR node, then: $\text{score}_{\text{BIC}}(\langle \mathcal{G}', \gamma' \rangle) - \text{score}_{\text{BIC}}(\langle \mathcal{G}, \gamma \rangle) = \ell_{\mathcal{D}}(\langle \mathcal{G}', \gamma' \rangle) - \ell_{\mathcal{D}}(\langle \mathcal{G}, \gamma \rangle) - \frac{\log M}{2}$. Hence, $\mathcal{G}'$ is accepted when $\ell_{\mathcal{D}}(\langle \mathcal{G}', \gamma' \rangle) - \ell_{\mathcal{D}}(\langle \mathcal{G}, \gamma \rangle) > \frac{\log M}{2}$, i.e., a leaf node may be decomposed when the improvement on the global loglikelihood is greater than $\frac{\log M}{2}$.

Proposition 1 (CNet BIC score decomposition [7]). *Given a CNet $\langle \mathcal{G}, \gamma \rangle$, over variables $\mathbf{X}$ and instances $\mathcal{D}$, made up of $\{T_l\}_{l=1}^{L}$ CLtrees, a decomposition of a tree T_l, having scope $\mathbf{X}_l \subset \mathbf{X}$, with parameters $\boldsymbol{\theta}_l$, with a sub-CNet $\mathcal{G}_i$ rooted in a OR node associated to the variable $X_i \in \mathbf{X}_l$ with parameters γ_i, leading to a new CNet $\langle \mathcal{G}', \gamma' \rangle$, is accepted iff $\ell_{\mathcal{D}_l}(\langle \mathcal{G}_i, \gamma_i \rangle) - \ell_{\mathcal{D}_l}(\langle T_l, \boldsymbol{\theta}_l \rangle) > \frac{\log M}{2}$, where $M = |\mathcal{D}|$, and D_l is the slice of D containing only instances associated to T_l.*

Again, due to the decomposability of the likelihood score, instead of recomputing it on the complete dataset $\mathcal{D}$, we can evaluate only the local improvements.

A Bayesian approach to learn a CLtrees T with parameters $\boldsymbol{\theta}$ from data $\mathcal{D}$ has been adopted, by exploiting as scoring function $P(\boldsymbol{\theta}|\mathcal{D}) \approx P(\mathcal{D}|\boldsymbol{\theta})P(\boldsymbol{\theta})$. Considering Dirichlet priors, and indicating with Pa_i the parent variable of X_i, the regularized model parameter estimates are:

$$\theta_{x_i|\text{Pa}_i} \approx E_{P(\theta_{x_i|\text{Pa}_i}|\mathcal{D},T)}[\theta_{x_i|\text{Pa}_i}] = \frac{M_{x_i,\text{Pa}_i} + \alpha_{x_i|\text{Pa}_i}}{M_{\text{Pa}_i} + \alpha_{\text{Pa}_i}},$$

where $M_{\mathbf{z}}$ is the number of entries in a dataset $\mathcal{D}_{\mathbf{z}}$ having the set of variables $\mathbf{Z}$ instatiated to $\mathbf{z}$. As pointed out in [8], a different Dirichlet prior for each distribution of X_i given a particular value of its parents may be used, leading to choosing the regularized parameter estimates as:

$$\hat{\theta}_{X_i|\mathrm{Pa}_i} = \frac{M \cdot P(\mathrm{Pa}_i)P(X_i|\mathrm{Pa}_i)}{M \cdot P(\mathrm{Pa}_i) + \alpha_{X_i|\mathrm{Pa}_i}} + \frac{\alpha_{X_i|\mathrm{Pa}_i}\theta^0(X_i|\mathrm{Pa}_i)}{M \cdot P(\mathrm{Pa}_i) + \alpha_{X_i|\mathrm{Pa}_i}},$$

where $\theta^0(X_i|\mathrm{Pa}_i)$ is the prior estimate of $P(X_i|\mathrm{Pa}_i)$ and $\alpha_{X_i|\mathrm{Pa}_i}$ is the confidence associated with that prior. A reasonable choice uses the marginal probability of X_i in the data as the prior probability. Thus, $\theta^0(X_i|\mathrm{Pa}_i)$ has been set to $P_D(X_i)$, and with fixed $\alpha_{x_i|\mathrm{Pa}_i} = \alpha$, then $\hat{\theta}_{X_i|\mathrm{Pa}_i}$ has been set to $\frac{M_{x_i,\mathrm{Pa}_i}+\alpha P_D(X_i)}{M_{\mathrm{Pa}_i}+\alpha}$.

The dCSN algorithm starts by learning a single CLTree on the whole dataset $\mathcal{D}$, and then calls a *decomposition* procedure on this tree. Two input parameters, δ and σ, act as regularizers, halting the decomposition process by requiring a minimum number of instances, resp. of features, in a slice to split it. The aim of dCSN is to attempt to extend the model by replacing one of the CLtree leaf nodes with a new CNet on the same variables. In particular, the decomposition prodecure checks for each variable X_i on the slice $\mathcal{D}$, whether the OR decomposition associated to that variable (a new CNet) has a better score than that of the input CLtree. If a better decomposition is found, it then recursively tries to decompose the sub-CLtrees of the newly introduced CNet. It is clear that the evaluation of all possible decompositions, i.e. of all possible X_i, even if the likelihood score is decomposable, extends learning time.

3   Random Forests of CNets

The joint probability distribution defined by a mixture K of probabilistic models can be formulated as $P(\mathbf{X}) = \sum_{i=1}^{K} \mu_i P(\mathbf{X} : \mathcal{Q}_i)$, where $\{\mathcal{Q}_i\}_{i=1}^{K}$ are the mixture *components* and $\{\mu_i\}_{i=1}^{k}$ are their *responsibilities*, i.e. the positive weights summing to one representing the contribution of each component to the final model. In our context, the components would be CNets, i.e. $\mathcal{Q}_i = \langle \mathcal{G}_i, \gamma_i \rangle$; instead, for MT [16], a highly accurate algorithm learning mixtures of tree structured models, they would be CLTrees.

The most common algorithmic choice when learning the structure of a mixture of PGMs is the Expectation Maximization algorithm EM [6]. In a nutshell, by considering the mixture responsibilities as the probabilities of observing as many values of a latent variable, the optimization carried out by EM involves the iteration of two steps: the computation of the expected mixture components from the current distribution, followed by the maximization of the likelihood as a function of the responsibilities given the previously computed components [16,19]. To shorten unfeasible learning times, since learning the components structure from zero is quite expensive, in practice, only the responsibilities are updated, having the components structures being learned only once, at the first iteration.

In [7] we explored another direction in mixture structure learning, ensembling techniques exploiting bagging [10]. Each component is learned once on bootstrapped sample $\mathcal{D}_{B_i}$ of the initial data $\mathcal{D}$, with $|\mathcal{D}_{B_i}| = |\mathcal{D}|$, and their

---

Algorithm 1. dCSN-RF($\mathcal{D}$, $\mathbf{X}$, α_f, δ, σ, K)

1: **Input:** instances $\mathcal{D}$ over features $\mathbf{X}$; $\alpha_f \in [0,1]$: ESS factor; δ min num of instances to split, σ min num of features to split, K num of the mixture components
2: **Output:** a mixture of K CNets components $\{\langle \mathcal{G}_i, \gamma_i \rangle\}_{i=1}^K$ with responsibilities $\{\mu_i\}_{i=1}^K$, encoding a pdf over $\mathbf{X}$ learned from $\mathcal{D}$
3: $\alpha \leftarrow \alpha_f |\mathcal{D}|, \hat{\mu} \leftarrow 0$
4: **for** $i = 1, \ldots, K$ **do**
5:      $\mathcal{D}_{B_i} \leftarrow$ bootstrapSample($\mathcal{D}$)
6:      $\langle \mathcal{T}, \boldsymbol{\theta} \rangle \leftarrow$ LearnCLTree($\mathcal{D}_{B_i}, \mathbf{X}, \alpha$)
7:      $\boldsymbol{w} \leftarrow \emptyset$
8:      $\langle \mathcal{G}_i, \gamma_i \rangle \leftarrow$ decompose-RF($\mathcal{D}_{B_i}, \mathbf{X}, \alpha, \mathcal{T}, \boldsymbol{\theta}, \boldsymbol{w}, \delta, \sigma$)
9:      $\mu_i \leftarrow \ell_{\mathcal{D}}(\langle \mathcal{G}_i, \gamma_i \rangle), \hat{\mu} \leftarrow \hat{\mu} + \mu_i$
10: **for** $i = 1, \ldots, K$ **do**
11:      $\mu_i \leftarrow \mu_i / \hat{\mu}$

---

responsibilities are computed as being proportional to the likelihood score they got on $\mathcal{D}_{B_i}$, resulting in the more robust estimate for each instance: $\hat{P}(\xi) = \sum_{i=1}^K \mu_i P(\xi : \langle \mathcal{G}_i, \gamma_i \rangle)$, where $\mu_i = \ell_{\mathcal{D}}(\langle \mathcal{G}_i, \gamma_i \rangle) / \sum_{j=1}^K \ell_{\mathcal{D}}(\langle \mathcal{G}_j, \gamma_j \rangle)$. A sketch of our bagged approach in dCSN-RF is visible in Algorithm 1, where LearnCLTree is the procedure, as already reported in [7], learning leaf trees implementing the Chow-Liu algorithm [4] and smoothing (α) probability estimations.

As the complexity of such approach grows linearly in the number of components, a natural time saving extension would be to consider only a portion of the features while growing the CNet, and then select the best one among them. The extension of the decompose procedure used in dCSN [7], as described in the previous section, to dCSN-RF is outlined in Algorithm 2. The difference now lies in line 5, where only a random subset of $\mathbf{X}$ is evaluated. We choose to sample $\sqrt{|\mathbf{X}|}$ as suggested in [10]. While this can clearly lead to sub optimal structures when a single component is involved, it would derive a better density estimation by forcing each of them to specialize on a particular, partial, view of the joint distribution as the number of components increases. Adopting this strategy within the bagging framework equals to a *random forest* approach [2] translated in a Bayesian generative framework. Random forests have been successfully applied in discriminative tasks like classification and regression, leading to accurate, low variance ensembles [10]. They have also recently being proposed as density estimators in [5]. More in general, our approach is affine to those shown in [1,20], where mixtures of tree structured density estimators are learned by perturbing their components via bagging and random subspace combination. Here we are willingly to trade off accuracy for shorter learning times.

4   Experiments

We evaluated dCSN-RF against the mixture version in our Bayesian framework, using only bagging, dCSN-B, and the mixture extensions of the original single

Algorithm 2. decompose-RF($\mathcal{D}$, $\mathbf{X}$, α, $\mathcal{T}$, $\boldsymbol{\theta}$, $\boldsymbol{w}$, δ, σ)

1: **Input:** instances $\mathcal{D}$ over $\mathbf{X}$; α: ESS; $\mathcal{T}$: a tree structured model and its parameters
   $\boldsymbol{\theta}$; δ min num of instances to split, σ min num of features to split
2: **Output:** a CNet encoding a pdf over $\mathbf{X}$ learned from $\mathcal{D}$
3: **if** $|\mathcal{D}| > \delta$ and $|\mathbf{X}| > \sigma$ **then**
4:    $\ell_{\text{best}} \leftarrow -\infty$, $\mathbf{X}_R \leftarrow$ randomSubset($\mathbf{X}$, $\sqrt{|\mathbf{X}|}$)
5:    **for** $X_i \in \mathbf{X}_R$ **do**
6:       $\mathcal{G}_i \leftarrow \emptyset$, $\boldsymbol{w}_i \leftarrow \emptyset$, $\boldsymbol{\theta}_i \leftarrow \emptyset$, C_i is the OR Node associate to X_i
7:       **for** $x_i^j \in Val(X_i)$ **do**
8:          $\mathcal{D}_j \leftarrow \{\xi \in \mathcal{D} : \xi[X_s] = x_s^j\}$, $w_{ij} \leftarrow |\mathcal{D}_j|/|\mathcal{D}|$
9:          $\langle \mathcal{T}_j, \boldsymbol{\theta}_{ij} \rangle \leftarrow$ LearnCLTree($\mathcal{D}_j$, $\mathbf{X}_{R \setminus s}$, αw_{ij})
10:         $\mathcal{G}_i \leftarrow$ addSubTree(C_i, $\mathcal{T}_j$)
11:         $\boldsymbol{w}_i \leftarrow \boldsymbol{w}_i \cup \{w_{ij}\}$, $\boldsymbol{\theta}_i \leftarrow \boldsymbol{\theta}_i \cup \{\theta_{ij}\}$
12:      $\ell_i \leftarrow \ell_{D_i}(\langle \mathcal{G}_i, \boldsymbol{w}_i \cup \boldsymbol{\theta}_i \rangle)$
13:      **if** $\ell_i > \ell_{\text{best}}$ and $\ell_i > \ell_{D_i}(\langle \mathcal{T}, \boldsymbol{\theta} \rangle)$ **then**
14:         $\ell_{\text{best}} \leftarrow \ell_i$, $X_{\text{best}} \leftarrow X_i$, $\mathcal{G}_{\text{best}} \leftarrow \mathcal{G}_i$, $\boldsymbol{\theta}_{\text{best}} \leftarrow \boldsymbol{\theta}_i$, $\boldsymbol{w}_{\text{best}} \leftarrow \boldsymbol{w}_i$
15:   **if** $\ell_{\text{best}} - \ell_{\mathcal{D}}(\langle \mathcal{T}, \boldsymbol{\theta} \rangle) > (log|\mathcal{D}|)/2$ **then**
16:      substitute $\mathcal{T}$ with $\mathcal{G}_i$ and set $\boldsymbol{w} \leftarrow \boldsymbol{w} \cup \boldsymbol{w}_{\text{best}}$
17:      **for** $x_b^j \in Val(X_{\text{best}})$ **do**
18:         $\mathcal{D}_j \leftarrow \{\xi \in \mathcal{D} : \xi[X_b] = x_b^j\}$
19:         decompose-RF($\mathcal{D}_j$, $\mathbf{X}_{\setminus \text{best}}$, αw_{ij}, $\mathcal{T}_j$, $\boldsymbol{\theta}_j$, $\boldsymbol{w}$, δ, σ)

CNet learning algorithms, as presented in [19]: CNet-B and CNet-RF for the bagged and random forest variants without pruning and CNetP-B and CNetP-RF for those including post pruning, the technique used in [19] to cope with the risk of overfitting. In addition to these, we employed MT [16] as one of the more solid mixture learning competitors [19,21]. We implemented all algorithms in Python[1], while for MT we used the implementation available in the Libra toolkit [15]. We tested them on an array of 18 real world datasets introduced in [12] and [9] as binary variants of frequent itemset mining, recommendation and classification datasets[2].

For all CNet learners, and for each dataset, we bootstrapped 40 dataset samples, then we learned mixture models incrementally, by adding components whose number K ranged from 5 to 40, by steps of 5, looking for the best average log-likelihood on the validation set. For CNet-B, CNet-RF, CNetP-B and CNetP-RF we set $\alpha = 1.0$ as done in [19]. For dCSN-B and dCSN-RF we employed a grid search on the validation sets for the parameters $\alpha_f \in \{5, 10\}$ and $\delta \in \{100, 200, 300, 400, 500, 1000\}$, while fixing $\sigma = 3$. For MT we reproduced the experiment in [21], setting K from 2 to 30 by steps of 2. We found out that while all bagged models achieved the best validation scores with $K = 40$, the best value for this parameter for MT greatly oscillates from 2 to 30, highlighting how EM could lead to different local optima.

---

[1] Source code is available at http://www.di.uniba.it/~ndm/dcsn/.
[2] All experiments have been run on a 4-core Intel Xeon E312xx (Sandy Bridge) @2.0 GHz with 8 Gb of RAM and Ubuntu 14.04.1, kernel 3.13.0-39.

Table 1. Times (in seconds) taken to learn the best models on each dataset for all algorithms.

	CNet-B	CNet-RF	CNetP-B	CNetP-RF	dCSN-B	dCSN-RF	MT
NLTCS	154	111	224	138	144	87	290
MSNBC	2458	1911	2997	1825	2887	1783	8645
Plants	724	345	847	361	2247	441	7414
Audio	1789	571	1903	658	4992	760	6566
Jester	1375	436	1993	691	4798	551	3064
Netflix	2716	861	3604	1184	9681	1177	11402
Accidents	1308	402	1413	405	6844	859	14073
Retail	2822	585	2776	591	1838	924	320
Pumsb-star	2421	495	2480	561	12274	1257	18533
DNA	305	102	388	119	2892	254	228
Kosarek	10565	1927	10119	1888	7248	1616	18782
MSWeb	17667	2331	19425	2028	23168	3509	36076
Book	17209	1488	18028	1552	16111	2385	5918
EachMovie	8056	801	9127	978	18060	941	12100
WebKB	9610	917	11589	843	14109	1195	931
Reuters-52	34170	2381	36106	2392	68296	3256	15082
BBC	8583	415	8473	467	14144	1637	1324
Ad	7499	791	7436	829	42707	1930	6850

Concerning learning times (see Table 1), as expected, random forest variants were up to one magnitude order faster than the bagged versions. For instance for $K = 40$, on EachMovie the times in seconds are: 8056, 801, 9127, 978, 18060, 941 for CNet-B, CNet-RF, CNetP-B CNetP-RF, dCSN-B, dCSN-RF respectively. For all the implemented versions, we run a pairwise Wilcoxon signed rank test to assess the statistical significance of the scores, even if a correction to counteract the problem of multiple comparisons should be used. For instance, adopting the Bonferroni correction and testing the single six hypotheses with a significance level of 0.05 corresponds to use a whole significance level of 0.3. In Table 2, in bold are reported the best values (round off to two decimal places), compared to all others, for each dataset (p-value= 0.05). As expected, the principled Bayesian learning of dCSN-B and dCSN-RF produces more accurate models than entropy based structure learners (those are the only two algorithms performing consistently better than MT, see Table 3). Moreover, it is clearly visible that the random forest approach generally *outperforms* all algorithmic variants using bagging, even in our Bayesian framework, becoming the significantly best performer on 10 datasets.

Table 2. Empirical risk for all algorithms.

	CNet-B	CNet-RF	CNetP-B	CNetP-RF	dCSN-B	dCSN-RF	MT
NLTCS	−6.09	−6.05	−6.02	−6.01	−6.02	−6.01	**-6.01**
MSNBC	−6.06	−6.06	−6.04	−6.05	−6.04	**−6.04**	−6.08
Plants	−12.31	**12.18**	−12.38	−12.25	−12.21	−12.27	−12.93
Audio	−42.14	−41.64	−40.68	−40.06	−40.17	**− 39.96**	− 40.14
Jester	−57.60	−57.43	−53.08	**−52.85**	−52.99	−52.89	−53.06
Netflix	−63.03	−62.15	−57.54	−56.89	−56.63	**−56.53**	−56.71
Accidents	−30.26	−29.75	−30.26	−29.84	**−28.99**	−29.00	−29.69
Retail	−11.00	**−10.84**	−10.88	−10.90	−10.87	**−10.82**	**−10.84**
Pumsb-star	−24.37	−23.98	−24.20	−23.93	−23.32	**−23.32**	−23.70
DNA	−91.13	−87.61	−86.88	−87.26	−84.93	**−84.83**	−85.57
Kosarek	−10.97	−10.68	−10.85	−10.66	−10.85	−10.67	**−10.62**
MSWeb	−9.96	−9.88	−9.91	−9.93	−9.86	**−9.71**	−9.82
Book	−35.91	−35.81	−35.60	−35.92	−35.92	−35.94	**−34.69**
EachMovie	−54.35	−53.35	−54.02	**−53.20**	−53.91	−53.84	−54.51
WebKB	−156.43	−155.47	−156.68	−158.59	**−155.20**	−155.10	−157.00
Reuters-52	−86.36	**−84.83**	−86.89	−85.24	**−85.69**	−84.58	−86.53
BBC	−252.26	−251.12	−257.09	−257.08	−251.14	**−249.56**	−259.96
Ad	−15.98	−16.04	−16.04	−16.04	**−13.73**	−14.35	−16.01

Table 3. Numbers of statistically significant victories for the algorithms on the rows compared to those on columns

	CNet-B	CNet-RF	CNetP-B	CNetP-RF	dCSN-B	dCSN-RF	MT
CNet-B	-	2	3	3	0	1	5
CNet-RF	15	-	9	8	3	3	4
CNetP-B	13	7	-	6	3	1	4
CNetP-RF	12	9	9	-	5	5	5
dCSN-B	16	10	14	11	-	4	11
dCSN-RF	17	14	15	12	9	-	15
MT	12	12	11	11	5	2	-

5   Conclusions

We extended the principled Bayesian approach to learn the structure of CNets proposed in [7] to a random forest ensemble framework with the main aim of reducing learning times considerably while preserving test accuracy. We proved empirically that such an approach leads to significantly more accurate models than a simple bagging scheme with the same number of components on entropy based models as well as on the Bayesian one. This, combined with the great time reduction, substantially proves our approach, making bayesian random Cutset forests an even more attractive tractable PGM.

Acknowledgements. Work supported by the project PUGLIA@SERVICE (PON02 00563 3489339) financed by the Italian Ministry of University and Research (MIUR) and by the European Commission through the project MAESTRA, grant no. ICT-2013-612944.

References

1. Ammar, S., Leray, P., Defourny, B., Wehenkel, L.: Probability density estimation by perturbing and combining tree structured Markov networks. In: Sossai, C., Chemello, G. (eds.) ECSQARU 2009. LNCS, vol. 5590, pp. 156–167. Springer, Heidelberg (2009)
2. Breiman, L.: Random forests. Mach. Learn. **45**(1), 5–32 (2001)
3. Choi, M.J., Tan, V.Y.F., Anandkumar, A., Willsky, A.S.: Learning latent tree graphical models. J. Mach. Learn. Res. **12**, 1771–1812 (2011)
4. Chow, C., Liu, C.: Approximating discrete probability distributions with dependence trees. IEEE Trans. Inf. Theory **14**(3), 462–467 (1968)
5. Criminisi, A., Shotton, J., Konukoglu, E.: Decision forests: a unified framework for classification, regression, density estimation, manifold learning and semi-supervised learning. Found. Trends Comput. Graph. Vis. **7**, 81–227 (2011)
6. Dempster, A.P., Laird, N.M., Rubin, D.B.: Maximum likelihood from incomplete data via the EM algorithm. J. R. Stat. Soc. **39**(1), 1–38 (1977)
7. Di Mauro, N., Vergari, A., Esposito, F.: Learning accurate cutset networks by exploiting decomposability. In: Gavanelli, M., Lamma, E., Riguzzi, F. (eds.) AI*IA 2015: Advances in Artificial Intelligence (2015)
8. Friedman, N., Geiger, D., Goldszmidt, M.: Bayesian network classifiers. Mach. Learn. **29**(2–3), 131–163 (1997)
9. Haaren, J.V., Davis, J.: Markov network structure learning: a randomized feature generation approach. In: Proceedings of the 26th Conference on Artificial Intelligence. AAAI Press (2012)
10. Hastie, T., Tibshirani, R., Friedman, J.: The Elements of Statistical Learning. Springer, New York (2009)
11. Koller, D., Friedman, N.: Probabilistic Graphical Models: Principles and Techniques. MIT Press, Cambridge (2009)
12. Lowd, D., Davis, J.: Learning Markov network structure with decision trees. In: Proceedings of the 10th IEEE International Conference on Data Mining, pp. 334–343. IEEE Computer Society Press (2010)
13. Lowd, D., Domingos, P.: Learning arithmetic circuits. CoRR, abs/1206.3271 (2012)
14. Lowd, D., Rooshenas, A.: Learning Markov networks with arithmetic circuits. In: Proceedings of the 16th International Conference on Artificial Intelligence and Statistics, JMLR Workshop Proceedings, vol. 31, pp. 406–414 (2013)
15. Lowd, D., Rooshenas, A.: The Libra Toolkit for Probabilistic Models. CoRR, abs/1504.00110 (2015)
16. Meilă, M., Jordan, M.I.: Learning with mixtures of trees. J. Mach. Learn. Res. **1**, 1–48 (2000)

17. Pearl, J.: Probabilistic Reasoning in Intelligent Systems: Networks of Plausible Inference. Morgan Kaufmann Publishers Inc., San Francisco (1988)
18. Poon, H., Domingos, P.: Sum-product network: a new deep architecture. In: NIPS 2010 Workshop on Deep Learning and Unsupervised Feature Learning (2011)
19. Rahman, T., Kothalkar, P., Gogate, V.: Cutset networks: a simple, tractable, and scalable approach for improving the accuracy of chow-liu trees. In: Calders, T., Esposito, F., Hüllermeier, E., Meo, R. (eds.) ECML PKDD 2014, Part II. LNCS, vol. 8725, pp. 630–645. Springer, Heidelberg (2014)
20. Ridgeway, G.: Looking for lumps: boosting and bagging for density estimation. Comput. Stat. Data Anal. **38**(4), 379–392 (2002)
21. Rooshenas, A., Lowd, D.: Learning sum-product networks with direct and indirect variable interactions. In: Proceedings of the 31st International Conference on Machine Learning, JMLR Workshop and Conference Proceedings, pp. 710–718 (2014)
22. Roth, D.: On the hardness of approximate reasoning. Artif. Intell. **82**(1–2), 273–302 (1996)

Classifier Fusion Within the Belief Function Framework Using Dependent Combination Rules

Asma Trabelsi[1]($\boxtimes$), Zied Elouedi[1], and Eric Lefèvre[2]

[1] LARODEC, Institut Supérieur de Gestion de Tunis,
Université de Tunis, Tunis, Tunisia
trabelsyasma@gmail.com, zied.elouedi@gmx.fr
[2] EA 3926, Laboratoire de Génie Informatique et d'Automatique de l'Artois
(LGI2A), University Artois, 62400 Béthune, France
eric.lefevre@univ-artois.fr

Abstract. The fusion of imperfect data within the framework of belief functions has been studied by many researchers over the past few years. Up to now, there are some proposed combination rules dealing with dependent information sources. Moreover, the choice of one rule among several alternatives is crucial but the criteria to be based on are still non clear. Thus, in this paper, we evaluate and compare some dependent combination rules for selecting the most efficient one under the framework of classifier fusion.

Keywords: Belief function theory · Combination rules · Dependent information sources · Multiple classifier fusion

1 Introduction

Pattern recognition has been extensively explored in last decades almost always using ensemble classifiers. Thus, several combination approaches have been proposed to combine multiple classifiers such as plurality, Bayesian theory, belief function theory, etc. [6]. This latter has many interpretations such as the Transferable Belief Model (TBM) [6] which offers numerous combination rules. Some of these rules assume the independence of information sources [5,8] while others deal only with dependent information sources [1,2]. The choice of the convenient rule is a crucial task but it has not been yet deeply explored. In this paper, we are interested in the combination of multiple classifiers within the framework of belief functions to evaluate and compare some dependent combination rules in order to pick out the most efficient one among them. Basically, we compare the most known dependent combination rules: the cautious conjunctive rule [2], the normalized cautious rule [2] and the cautious Combination With Adapted Conflict rule [1]. The remaining of this paper is organized as follows: we provide in Sect. 2 a brief overview of the fundamental concepts of the belief function theory. We present three combination rules dealing with dependent sources of information in Sect. 3. Section 4 is devoted to describing our comparative approach. Experiments and results are outlined in Sect. 5. Section 6 draws conclusion.

© Springer International Publishing Switzerland 2015
F. Esposito et al. (Eds.): ISMIS 2015, LNAI 9384, pp. 133–138, 2015.
DOI: 10.1007/978-3-319-25252-0_14

2   Fundamental Concepts of Belief Function Theory

Let Θ be a finite non-empty set of N elementary events related to a given problem, called the frame of discernment. The beliefs held by an agent on the different subsets of the frame of discernment Θ are represented by the so-called basic belief assignment (bba). The bba is defined as follows:

$$m : 2^{\Theta} \rightarrow [0, 1]$$

$$\sum_{A \subseteq \Theta} m(A) = 1 \tag{1}$$

The quantity $m(A)$ states the degree of belief committed exactly to the event A. From the basic belief assignment, we can compute the commonality function (q). It is defined as follows: $q(A) = \sum_{B \supseteq A} m(B)$.

Decision making aims to select the most reasonable hypothesis for a given problem. In fact, it consists of transforming beliefs into probability measure called the pignistic probability denoted by $BetP$ and defined as follows [10]:

$$BetP(A) = \sum_{B \subseteq \Theta} \frac{|A \cap B|}{|B|} \frac{m(B)}{1 - m(\emptyset)} \forall A \in \Theta \tag{2}$$

The dissimilarity between two bbas can be computed. One of the well-known measures is the one proposed by Jousselme [4]:

$$d(m_1, m_2) = \sqrt{\frac{1}{2}(m_1 - m_2)^T D(m_1 - m_2)} \tag{3}$$

where D is the Jaccard similarity measure defined by:

$$D(A, B) = \begin{cases} 1 & \text{if } A = B = \emptyset \\ \dfrac{|A \cap B|}{|A \cup B|} & \forall A, B \in 2^{\Theta} \end{cases} \tag{4}$$

3   Combination of Pieces of Evidence

The TBM framework offers several tools to aggregate a set of bbas induced form dependent information sources:

1. The cautious conjunctive rule, denoted $\oslash$, has been proposed by [2] in order to aggregate pieces of evidence induced from reliable dependent information sources using the conjunctive canonical decomposition proposed by Smets [9]. Let m_1 and m_2 be two non-dogmatic bbas $(m(\Theta) > 0)$ and let $m_1 \oslash m_2$ be the result of their combination. We get:

$$m_1 \oslash m_2(A) = \bigcirc_{A \subset \Theta} A^{w_1(A) \wedge w_2(A)} \tag{5}$$

where $w_1(A) \wedge w_2(A)$ represents the weight function of a bba $m_1 \oslash m_2$ and $\wedge$ denotes the minimum operator. The weights $w(A)$ for every $A \subset \Theta$ can be obtained from the commonalities as follows: $w(A) = \prod_{B \supseteq A} q(B)^{(-1)^{|B|-|A|-1}}$.

2. The normalized version of the cautious conjunctive rule, denoted $\oslash^*$, is obtained by replacing the conjunctive operator $\oslash$ by the Dempster operator $\oplus$ [1] in order to overcome the effect of the value of the conflict generated by the unnormalized version. It is defined by the following equation:

$$m_1 \oslash^* m_2(A) = \underset{\emptyset \neq A \subset \Theta}{\oplus} A^{w_1(A) \wedge w_2(A)} \tag{6}$$

3. The cautious CWAC rule, based on the cautious rule and inspired from the behavior of the CWAC rule, is defined by an adaptive weighting between the unormalized cautious and the normalized ones [1]. The cautious CWAC rule is then defined as follows $\forall A \subseteq \Theta$ and $m_\oslash(\emptyset) \neq 1$:

$$m_\odot(A) = D m_\oslash(A) + (1 - D) m_{\oslash^*}(A) \tag{7}$$

with $D = \underset{i,j}{max}[d(m_i, m_j)]$ is the the maximum Jousselme distance between m_i and m_j.

4 Comparative Study

In our investigation, ensemble classifiers, based on the combination of the outputs of individual classifiers, have been proposed as tools for evaluating and comparing dependent combination rules. Let us consider a pattern recognition issue where $B = \{x_1, ..., x_n\}$ be a data set with n examples, $C = \{C_1, ..., C_M\}$ be a set of M classifiers and $\Theta = \{w_1, ..., w_N\}$ be a set of N class labels. B will be partitioned into train and test sets. The classifiers must be built from the training set and then we apply them to predict the label class $w_j \in \Theta$ of any pattern test x. The outputs from M classifiers should be converted into bbas by taking into account the reliability rate r of each classifier. In fact, for each pattern test we have M bbas obtained as follows:

$$m_i(\{w_j\}) = 1 \text{ and } m_i(A) = 0 \,\forall A \subseteq \Theta \, and \, A \neq \{w_j\} \tag{8}$$

with $r_i = \dfrac{\text{Number of well classified instances}}{\text{Total number of classified instances}}$.

Note that $m_i(\{w_j\})$ denotes the part of belief given exactly to the predicted class w_j by the classifier C_i.

Once the outputs of all classifiers are transformed into bbas, we move to the combination of classifier through dependent fusion rules. The combination results will allow us to evaluate and compare these rules in the purpose of selecting the most appropriate one based on two popular evaluation criteria: the distance and the Percent of Correct Classification (PCC). This is justified by the fact that the cautious conjunctive rule does not keep the initial alarm role of the conflict due to its absorbing effect of the conflictual mass, the normalized cautious rule ignores the value of the conflict obtained by combining pieces of evidence whereas the cautious CWAC rule gives the conflict its initial role as an alarm signal.

– The PCC criterion, representing the percent of the correctly classified instances, was employed to compare the cautious CWAC rule of combination with the normalized cautious rule. Such case requires the use of three variables n_1, n_2 and n_3 which respectively represent the number of well classified, misclassified and rejected instances. Hence, for each combination rule, we propose the following steps:

1. We define a tolerance thresholds $S = \{0.1, 0.2, \ldots, 1\}$. For each threshold $s \in S$, we check the mass of the empty set $m(\emptyset)$ induced by any test instance. If $m(\emptyset)$ is greater than s, our classifier chooses to reject instance instead of misclassifying it. Consequently, we increment n_3. Inversely, we compute the $BetP$ in order to make a decision about the chosen class. Accordingly, we increment n_1 if the current class is similar to the real one else we increment n_2.

2. Once we have calculated our well classified, misclassified and rejected instances, we compute then the PCC for each threshold $s \in S$ as follows:

$$PCC = \frac{n_1}{n_1 + n_2} * 100 \qquad (9)$$

The best rule is the one that has the highest values of $PCC \; \forall \; s \in S$.

– The distance criterion, corresponding to the Jousselme distance between two mass functions [4], was used to compare the cautious CWAC rule of combination with the cautious conjunctive rule. Thus, for each combination rule we proceed as follows:

1. The real class w_j of each pattern test should be converted into a mass function: $m_r(\{w_j\}) = 1$.

2. Then, we calculate for the instance x the Jousselme distance between the mass function corresponding to its real class (m_r) and the mass function produced by combining bbas coming from M classifiers.

3. Finally, we aggregate the Jousselme distances obtained by all test patterns in order to obtain the total distance.

The most appropriate rule is the one that has the minimum total distance.

5   Empirical Evaluation

5.1   Experimental Settings

In order to evaluate our combination rules, we have performed a set of experiments on several real world databases with different number of instances, different number of attributes and different number of classes obtained from the U.C.I repository [7]. We have conducted experiments with four machine learning algorithms implemented in Weka [3]. These learning algorithms including Naive Bayes, k-Nearest Neighbors, Decision tree and Neural Network were run based on a validation approach named leave one out cross validation. This method divides a data set with N instances into N-1 parts for training and the remaining instance for testing. This process should be repeated N times where each instance is used once as a test set. Thus, from each classifier we get N test patterns with their predicted class labels.

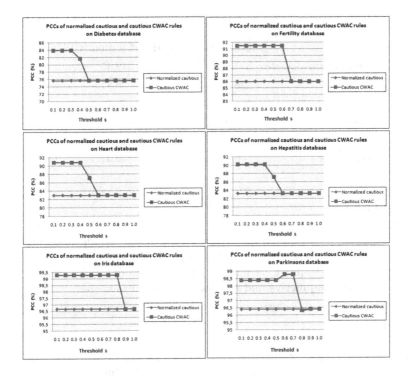

Fig. 1. PCCs for all databases

5.2   Experimental Results

As shown in Table 1, the cautious CWAC rule achieves best results compared with the cautious conjunctive one. In fact, the distance relative to the cautious CWAC rule is lower than that relative to the cautious conjunctive one. Accordingly, we can conclude that the cautious CWAC rule is more efficient than the cautious conjunctive one in term of distance criterion.

Let's lead off by comparing the cautious CWAC rule of combination with the normalized cautious one according to the PCC criterion. Figure 1 presents the PCCs for both the normalized cautious and the cautious CWAC rules relative to all the mentioned databases. From Fig. 1, we can deduce that for the different values of s, the PCC values of the cautious CWAC rule are greater or equal to those relative to the normalized cautious one for all the mentioned databases. So, we can conclude that the cautious CWAC rule is more efficient than the normalized cautious one in term of PCC criterion.

Table 1. Distance results of the cautious conjunctive and cautious CWAC rules.

Datasets	Cautious conjunctive	Cautious CWAC
Pima Indians diabetes	334.87	312.45
Fertility	28.06	26.81
Statlog (Heart)	104.60	96.79
Hepatitis	54.90	51.56
Iris	15.09	14.78
Parkinsons	60.92	50.31

6 Conclusion

In this paper, we have studied some fusion rules when dealing with dependent information sources. Then, we have conducted experimental tests based on multiple classifier systems to judge the efficiency of the cautious CWAC rule compared with the cautious conjunctive and the normalized cautious ones.

References

1. Boubaker, J., Elouedi, Z., Lefèvre, E.: Conflict management with dependent information sources in the belief function framework. In: 14th International Symposium of Computational Intelligence and Informatics (CINTI), vol. 52, pp. 393–398. IEEE (2013)
2. Denœux, T.: Conjunctive and disjunctive combination of belief functions induced by nondistinct bodies of evidence. Artif. Intell. **172**(2), 234–264 (2008)
3. Hall, M., Frank, E., Holmes, G., Pfahringer, B., Reutemann, P., Witten, I.H.: The WEKA data mining software: an update. ACM SIGKDD Explor. Newsl. **11**(1), 10–18 (2009)
4. Jousselme, A., Grenier, D., Bossé, E.: A new distance between two bodies of evidence. Inf. Fusion **2**(2), 91–101 (2001)
5. Lefèvre, E., Elouedi, Z.: How to preserve the conflict as an alarm in the combination of belief functions? Decis. Support Syst. **56**, 326–333 (2013)
6. Mercier, D., Cron, G., Denœux, T., Masson, M.: Fusion of multi-level decision systems using the transferable belief model. In: 7th International Conference on Information Fusion, FUSION 2005, vol. 2, pp. 655–658. IEEE (2005)
7. Murphy, P., Aha, D.: UCI repository databases (1996). http://www.ics.uci.edu/mlearn
8. Pichon, F., Denoeux, T.: The unnormalized dempster's rule of combination: a new justification from the least commitment principle and some extensions. J. Autom. Reason. **45**(1), 61–87 (2010)
9. Smets, P.: The canonical decomposition of a weighted belief. In: 14th International Joint Conference on Artificial Intelligence (IJCAI), vol. 95, pp. 1896–1901 (1995)
10. Smets, P.: The transferable belief model for quantified belief representation. In: Smets, P. (ed.) Quantified Representation of Uncertainty and Imprecision. Handbook of Defeasible Reasoning and Uncertainty Management Systems, vol. 1, pp. 267–301. Springer, Netherlands (1998)

On the Effectiveness of Evidence-Based Terminological Decision Trees

Giuseppe Rizzo$^{(\boxtimes)}$, Claudia d'Amato,
and Nicola Fanizzi

LACAM – Università degli studi di Bari "Aldo Moro", Bari, Italy
{Giuseppe.Rizzo1,Claudia.dAmato,Nicola.Fanizzi}@uniba.it

Abstract. Concept learning methods for Web ontologies inspired by *Inductive Logic Programming* and the derived inductive models for class-membership prediction have been shown to offer viable solutions to concept approximation, query answering and ontology completion problems. They generally produce human-comprehensible logic-based models (e.g. terminological decision trees) that can be checked by domain experts. However, one difficulty with these models is their inability to provide a way to measure the degree of uncertainty of the predictions. A framework for inducing terminological decision trees extended with evidential reasoning has been proposed to cope with these problems, but it was observed that the prediction procedure for these models tends to favor cautious predictions. To overcome this limitation, we further improved the algorithms for inducing/predicting with such models. The empirical evaluation shows promising results also in comparison with major related methods.

1 Introduction

In the context of Web ontologies, which rely on formal representation languages known as Description Logics (DLs) [1], deductive procedures for deciding the membership of an individual are strongly affected by the underlying open world assumption and the possible flaws introduced in the ontology construction (e.g. the lack of disjointness axioms). Consequently, the class-membership cannot be always determined deductively.

To overcome the problem, inductive methods to predict new class-membership assertions have been proposed (e.g. see [2]) so that, as effective alternatives to the standard querying and reasoning services, they may effectively support further applications based on *ontology completion*. An interesting class of such methods can be borrowed from *Inductive Logic Programming*. They offer the advantage of combining effectiveness and comprehensibility of the classification models.

In the literature, various concept learning methods for DLs have been proposed. Most of them rely on a *separate-and-conquer* strategy, such as DL-FOIL [3], an adaptation of the FOIL algorithm, CELOE (*Class Expression Learning for Ontology Engineering*) [4], an accuracy-driven approach for learning concept descriptions, and ELTL (*$\mathcal{EL}$ Tree Learning*) [5] that is based on an

© Springer International Publishing Switzerland 2015
F. Esposito et al. (Eds.): ISMIS 2015, LNAI 9384, pp. 139–149, 2015.
DOI: 10.1007/978-3-319-25252-0_15

ideal downward refinement operator the specific DL. Both CELOE and ELTL are currently implemented in DL-Learner [6], a framework which includes various learning algorithms. On the other hand, adopting a *divide-and-conquer* strategy, *Terminological Decision Trees* (TDTs) [7] have also been proposed, extending the *First Order Logic decision trees* to DL concepts.

However, despite the advantages deriving from logic-based models, these models generally do not provide an epistemic uncertainty measure [4,5,7]. We have started to investigate an extension of TDTs [8] endowed with the operators of the *Dempster-Shafer Theory* (DST) [9]. This model has been devised both to cope with those cases of test resulting in an unknown membership, that is similar to the problem of attributes with missing values in decision trees, and to be able to provide a measure of the uncertainty for the quality of the predicted assertions. Basically, the new classification algorithm follows all branches when an intermediate test is unable to determine a certain value. Once the leaves are reached, the results are merged through DST combination rules [9,10].

The DST has been integrated in various algorithms [11,12] with results that are competitive w.r.t. the classical version. Conversely, the experiments proposed in [8] have shown that the new extension did not perform as well as TDTs. The results suggest that the performance is affected by two factors: 1) the employment of a threshold to control the growth which stops the training phase prematurely; 2) the monotonicity of belief/plausibility/confirmation functions. As a consequence, the resulting models tended not to be able to predict a definite (i.e. positive or negative) membership very easily. Moving from these lessons learned, we have enhanced the DST-TDTs by replacing the previous procedure, that was based on a fixed threshold in the training phase, with a novel one that is more likely to give a definite answer. This has led to a new empirical comparative evaluation with related state-of-the-art methods.

The rest of the paper is organized as follows: Sect. 2 introduces basics concerning the concept learning task in DL knowledge bases and describes the original version of TDTs; in Sect. 3 the algorithm for inducing a TDT based on the DST is proposed while in Sect. 4 the setting and the results of new experiments are described; finally further extensions of this work are proposed in Sect. 5.

2   Background

We will recall the basics of the representation for the Web ontologies and the elements of the Dempster-Shafer theory employed in the proposed classification model.

2.1   Knowledge Bases

Adopting DL languages [1], a domain is modeled through *primitive concepts* (classes) and *roles* (relations), which are used to build complex descriptions regarding *individuals* (resources), employing the specific logic operators of the language of choice.

A *knowledge base* is a couple $\mathcal{K} = (\mathcal{T}, \mathcal{A})$ where the *TBox* $\mathcal{T}$ contains axioms concerning concepts and roles (typically inclusion axioms such as $C \sqsubseteq D$) and the *ABox* $\mathcal{A}$ contains assertions, i.e. axioms regarding the individuals ($C(a)$, resp. $R(a, b)$). The set of individuals occurring in $\mathcal{A}$ is denoted by $\mathsf{Ind}(\mathcal{A})$.

The semantics of concepts/roles/individuals is defined through interpretations. An *interpretation* is a couple $\mathcal{I} = (\Delta^{\mathcal{I}}, \cdot^{\mathcal{I}})$ where $\Delta^{\mathcal{I}}$ is the *domain* of the interpretation and $\cdot^{\mathcal{I}}$ is a *mapping* such that, for each individual a, $a^{\mathcal{I}} \in \Delta^{\mathcal{I}}$, for each concept C, $C^{\mathcal{I}} \subseteq \Delta^{\mathcal{I}}$ and for each role R, $R^{\mathcal{I}} \subseteq \Delta^{\mathcal{I}} \times \Delta^{\mathcal{I}}$. The semantics of complex descriptions descends from the interpretation of the primitive concepts/roles and of the operators employed, depending on the adopted language. $\mathcal{I}$ satisfies an axiom $C \sqsubseteq D$ (C is subsumed by D) when $C^{\mathcal{I}} \subseteq D^{\mathcal{I}}$ and an assertion $C(a)$ (resp. $R(a, b)$) when $a^{\mathcal{I}} \in C^{\mathcal{I}}$ (resp. $(a^{\mathcal{I}}, b^{\mathcal{I}}) \in R^{\mathcal{I}}$). $\mathcal{I}$ is a *model* for $\mathcal{K}$ iff it satisfies each axiom/assertion α in $\mathcal{K}$, denoted with $\mathcal{I} \models \alpha$. When α is satisfied w.r.t. these models, we write $\mathcal{K} \models \alpha$.

We are particularly interested in the *instance-checking* inference service: given an individual a and a concept C, determine if $\mathcal{K} \models C(a)$. Due to the *Open World Assumption* (OWA), answering to a class-membership query is more difficult w.r.t. other settings where the *closed-world reasoning* is the standard. Indeed, it may be not possible to prove either $\mathcal{K} \models C(a)$ or $\mathcal{K} \models \neg C(a)$, as different interpretations that satisfy either cases may be found.

2.2 Dempster-Shafer Theory

In the DST, the *frame of discernment* is a set of exhaustive and mutually exclusive hypotheses $\Omega = \{\omega_1, \omega_2, \ldots, \omega_n\}$ about a domain. A *basic belief assignment* (BBA) is defined as a mapping where $m : 2^{\Omega} \to [0, 1]$ so that $m(\emptyset) = 0$ and $\sum_{A \in 2^{\Omega}} m(A) = 1$. If $m(A) > 0$, A is a *focal element* for m.

Other functions can be derived from BBAs. The *belief* in A, denoted with $\mathrm{Bel}(A)$, represents a measure of the total belief committed to A given the available evidence: $\mathrm{Bel}(A) = \sum_{B \subseteq A} m(B)$. The *plausibility* of A, $\mathrm{Pl}(A)$, represents the amount of belief committed to A when further evidence is available: $\mathrm{Pl}(A) = \sum_{B \cap A \neq \emptyset} m(B)$. In this work we make use of the *confirmation* measure [9] defined $\mathrm{Conf}(A) = \mathrm{Bel}(A) + \mathrm{Pl}(A) - 1$. Note that, differently from m, Bel, Pl and Conf are monotonic. Another useful measure in the context of the DST is *non-specificity*, that conveys the degree of imprecision of a BBA m and is defined as follows:

$$\mathrm{NS}(m) = \sum_{A \in 2^{\Omega}} m(A) \log(|A|) \tag{1}$$

Importantly, *combination rules* are employed in DST for pooling different BBAs. Among the various rules that have been proposed [10], we are interested to associative operators, such as *Dubois-Prade's rule* [13] that derives the combined BBA as follows

$$\forall A \subseteq \Omega \quad m_{1,2}(A) = \sum_{B \cup C = A} m_1(B) m_2(C). \tag{2}$$

3 DST Terminological Trees

3.1 Learning DL Classifiers

The learning task for a knowledge base $\mathcal{K} = (\mathcal{T}, \mathcal{A})$ can be defined as follows:

given

- a target concept C;
- an error threshold ϵ
- a training set Tr of examples for which the correct classification value of $t_C(\cdot) : \mathsf{Ind} \to \{-1, 0, +1\}$ is known, partitioned into
  - $Ps = \{a \in \mathsf{Ind}(\mathcal{A}) \mid \mathcal{K} \models C(a),\ i.e.\ t_C(a) = +1\}$,   *positive ex.'s*
  - $Ns = \{a \in \mathsf{Ind}(\mathcal{A}) \mid \mathcal{K} \models \neg C(a),\ i.e.\ t_C(a) = -1\}$ *negative ex.'s*
  - $Us = Tr \setminus (Ps \cup Ns),\ i.e.\ a \in \mathsf{Ind} : t_C(a) = 0;$ *uncertain-memb. ex.'s*

find a classifier $h_C : \mathsf{Ind} \to \{-1, 0, +1\}$ for C such that

$$\frac{1}{|Tr|} \sum_{a \in Tr} \mathbf{1}[h_C(a) = t_C(a)] > 1 - \epsilon$$

where $\mathbf{1}[\cdot]$ is the *indicator* function returning 1 if the argument is true and 0 otherwise.

A classification model for the problem above, has been defined as follows [7]:

Definition 3.1 (Terminological Decision Tree). *Let $\mathcal{K} = (\mathcal{T}, \mathcal{A})$ a knowledge base, a* Terminological Decision Tree *is a binary tree where each node is labeled with a concept description*

- *leaf node concepts represent the classification to be predicted $(C, \neg C)$*
- *internal node concepts D represent test conditions*
  - *either edge corresponds to the outcome of test for the individual a to be classified: $\mathcal{K} \models D(a)$?. By convention, the left branch corresponds to a successful test outcome.*
  - *if the node labeled with D is linked with a child labeled with E then $E \sqsubseteq D$;*

Given a TDT, deriving the classifier h is straightforward.

TDTs are grown recursively and top-down: candidate concept descriptions are generated by means of a refinement operator and the best one is chosen to be installed as a child node. The best description is the one that maximizes a purity measure respect to the previous level [7], e.g. the information gain.

3.2 DST-TDTs

DST-TDTs extend the TDTs having each node labeled with a couple $\langle D, m \rangle$, with the BBA m related to the membership w.r.t. D. Figure 1 reports a simple example of DST-TDT used for predicting whether a car is to be sent back to the factory (SendBack) or can be repaired. The root concept $\exists hasPart.\top$ is progressively specialized and the refinements are associated to BBAs with a decreasing non-specificity measure.

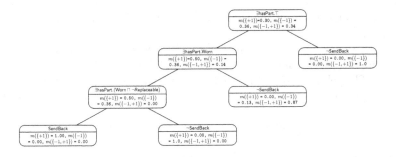

Fig. 1. A simple example of DST-TDT

Algorithm 1. Induction of DST-based TDTs

```
 1  function INDUCEDSTTDTREE(Ps, Ns, Us, C, m): T
 2  input: Ps, Ns, Us: set of training instances; C: concept description; m: BBA
 3  output: T: TDT
 4  const θ ∈ [0, 1] {threshold}
 5
 6  begin
 7      if (|Ps| = 0 and |Ns| = 0) then
 8          if Pr⁺ ≥ Pr⁻ then {distribution over the whole training set}
 9              T_root ← ⟨C, m⟩;
10          else
11              T_root ← ⟨¬C, m⟩;
12          return T;
13
14      if (m({−1}) = 0 and m({+1}) > θ) then
15          begin T_root ← ⟨C, m⟩; return T; end
16      if (m({+1}) = 0 and m({−1}) > θ )then
17          begin T_root ← ⟨¬C, m⟩; return T; end
18
19      S ← GENERATECANDIDATES(Ps, Ns, Us);
20      D ← SELECTBESTCONCEPT(m, S);
21      ⟨⟨P_l, N_l, U_l⟩, ⟨P_r, N_r, U_r⟩⟩ ← SPLIT(D, Ps, Ns, Us);
22      T_root ← ⟨D, m_D⟩;
23      T_left ← INDUCEDSTTDTREE(P_l, N_l, U_l, D, m_D);
24      T_right ← INDUCEDSTTDTREE(P_r, N_r, U_r, ¬D, m_D);
25      return T
26  end
```

Growing DST-Based TDTs. The method for inducing DST-TDTs adopts a *divide-and-conquer* strategy. It requires the target concept C, a partitioned training set $Tr = Ps \cup Ns \cup Us$ and a BBA m associated with C (note that $\Omega = \{+1, -1\}$).

The main learning function (see Algorithm 1) refines the input test concept using the available refinement operator. After candidates are generated, a BBA is computed for each of them. The BBAs for the node concepts are simply estimated by considering the relative frequencies of positive, negative and uncertain instances in the training set.

The set of candidates S is made up of pairs $\langle D, m_D \rangle$ where D is a concept description and m_D is the BBA computed for it. After S has been generated, the algorithm selects the best test concept and the corresponding BBA according to measure computed from the BBA and the best pair $\langle D, m_D \rangle$ is installed as a child node of $\langle C, m \rangle$.

The set S of candidate couples $\langle D, m_D \rangle$ is generated by GENERATECANDI-DATES made up of refinements D of C that can be used as tests and their related BBA m_D. The selection of the best candidate operated by SELECTBESTCANDI-DATE is based on the *non-specificity* of their BBAs (see Eq. (1) in Sect. 2.2):

$$\arg \max_{D \in S} [\mathrm{NS}(m_C) - \mathrm{NS}(m_D)] \tag{3}$$

This strategy is repeated recursively, splitting the examples according to the test concept in each node. Recursion stops when only positive (resp. negative) instances are rooted to a node which becomes a leaf (see the conditions checked in lines 14 and 16). The first conditional (line 7) refers to the case when no positive and negative instances reach the node. In this case the algorithm uses priors, Pr^+ and Pr^-, precomputed for the whole training set.

BBA Creation. The proposed model associates a BBA to each node of a TDT for representing the epistemic uncertainty about the class-membership. The BBA of a child node is created based on the training examples routed to the parent node. When a branch is created together with the related concept description, the membership of the individuals w.r.t. this concept is assessed in order to obtain a BBA whose frame of discernment represents the hypothesis of membership w.r.t. that concept. In addition, when a new node is added as a left or right child, the algorithm knows about the tests performed on the parent node concept for each instance (introducing an implicit kind of conditioning between the corresponding BBAs).

We resort to the DST also to provide the *stop condition* for the tree-growing process. As described above, we add a new child node minimizing the degree of imprecision represented by the non-specificity measure. In this case, the procedure is repeated if a purity condition is verified, i.e. $m(\{-1\}) = 0 \wedge m(\{+1\}) > \theta$ or $m(\{+1\}) = 0 \wedge m(\{-1\}) > \theta$. This condition derives from decision tree induction, where a leaf is formed when only instances that belong to a single class remain. In terms of DST, this idea is well represented by a *certain BBA*, whose only focal element is a singleton (i.e. $\{+1\}$ or $\{-1\}$).

Prediction. The resulting model can be used to predict the class-membership in the usual way. Given an individual $a \in \mathsf{Ind}(A)$, a list with the BBA associated with the reached leaf nodes is produced visiting the DST-TDT (FINDLEAVES function). Paths are followed down recursively the tree according to the results of the test w.r.t. the concept D at each node. If $\mathcal{K} \models D(a)$ the algorithm follows the left branch of the tree. If $\mathcal{K} \models \neg D(a)$ the right branch is followed.

A more interesting case occurs when the result of instance check is unknown, i.e. $\mathcal{K} \not\models D(a)$ or $\mathcal{K} \not\models \neg D(a)$. In this case, both the left and the right branch are followed until the leaves are reached. In this way the algorithm can cope with the OWA. The underlying idea is to collect all the possible classifications

---

Algorithm 2. Evidence-based classification procedure

```
 1  function CLASSIFY(a, T, K): v
 2  input: a: individual,
 3          T: TDT,
 4          K: knowledge base
 5  output: v ⊆ Ω
 6  begin
 7    L ← FINDLEAVES(a, T, K) {list of BBA}
 8    m̄ ← ⊕ m
         m∈L
 9    for v ⊆ Ω do
10      Compute Bel̄_v and Pl̄_v;
11      Conf̄_v ← CONFIRMATION(Bel̄_v, Pl̄_v);
12
13      if (|Conf̄_{+1} − Conf̄_{−1}| ≥ ε) then {forcing procedure}
14        return arg max Conf̄_v;
                  v⊂Ω
15      else
16        return Ω;
17
18  end
```

---

when the result of a test on an internal node is unknown. In these cases the DST seems to be a good framework in order to combine all such results and make a decision on the membership to be assigned.

After the tree exploration, we may have various BBAs in the list (one per reached leaf). Then, these functions are to be pooled according to a combination rule (see Algorithm 2). In this case we are interested to an associative operator for avoiding that the final decision is affected by the order in which the BBAs are collected (in the experiments we used the Dubois-Prade's combination rule [13]). The resulting BBA can be used to compute belief, plausibility or confirmation on the membership hypotheses and the algorithm returns the one that maximize the chosen function. Specifically, in order to deal with the monotonicity of belief and confirmation function, which leads easily to return an unknown membership, the algorithm compares the function values for the positive-membership hypothesis with the negative one and returns the uncertain-membership only when their values are approximately equals. Similarly to our previous works [8,14], we considered the computation of the confirmation in order to balance belief and plausibility in the final decision.

4 Empirical Evaluation

4.1 Setup

We evaluated the effectiveness of the new release of the DST-TDT model also in comparison with related models for inductive classification.

Three Web ontologies were considered (see Table 1): BCO that is a medical ontology, an ontology describing the glycolysis pathway from the EcoCyc database in BioPax format, BIOPAX, and FINANCIAL, an ontology for modeling banking domain.

Table 1. Ontologies employed in the experiments

Ontology	Expressiivity	Concepts	Roles	Individuals
BCO	$\mathcal{ALCIF}(D)$	196	22	112
BioPax	$\mathcal{ALCIF}(D)$	74	70	323
Financial	$\mathcal{ALCIF}(D)$	60	16	1000

For each ontology, 20 query concepts have been randomly generated by combining 2 through 8 concepts, both primitive and universal and existential restrictions on roles, using conjunction and disjunction. The queries have been generated so that at least 20 positive instances and 20 negative individuals can be found in each ontology.

Due to the number of individuals in these datasets, all of them were involved as training/test examples. To best represent the data distribution, a *10-fold stratified cross-validation* procedure was employed in the design of the experiments.

We compared our method both to TDTs [7] and to related approaches in the literature: CELOE and DISJUNCTIVE ELTL algorithms (an extension of ELTL). These methods are implemented as modules of the DL-LEARNER framework, a suite of machine learning algorithms for inducing DL concept descriptions [6]. CELOE performs an accuracy-driven heuristic search [4]. The most important parameter is a measure of the admissible noise (i.e. the rate of false positives). We replicated the experiments for the values 30 %, 40 % and 50 %. DISJUNCTIVE ELTL is an extension of ELTL [5] that allows to learn disjunction of concepts.

The performance of the algorithms has been assessed both in terms of *F-measure*, usually employed for evaluation where closed world can be assumed, and in terms of the following metrics computed through a comparison of the predictions made using the inductive mdel to the gold-standard deductive classifications determined by a reasoner [7]:

- *match* rate: amount of examples for which inductive and deductive classification agree on the membership value, i.e. +1 vs. + 1, −1 vs. − 1, 0 vs. 0;
- *commission* rate: amount of cases for which their classification values are opposite, i.e. +1 vs. − 1, −1 vs. + 1;
- *omission* rate: amount of cases for which inductive classification cannot predict a definite membership while the reasoner does, i.e. 0 vs. − 1| + 1;
- *induction* rate: amount of cases for which the inductive method can predict a definite membership while it is not logically derivable, i.e. −1| + 1 vs. 0.

4.2 Results

Table 2 shows results of the experiments averaged over the replications. We do not report the results for CELOE with a noise rate of 40 % and 50 % because they do not significantly change w.r.t. those made with a noise rate of 30 %. Overall,

Table 2. Average results of the experiments on the considered algorithms

Ontology	Index	TDT	DST-TDT	CELOE	DISJ ELTL
BCO	F_1	67.54 ± 14.36	93.48 ± 09.46	100.0 ± 00.00	64.16 ± 02.15
	M%	68.93 ± 15.87	94.53 ± 07.68	100.0 ± 00.00	47.27 ± 02.35
	C%	06.14 ± 07.20	05.47 ± 07.68	00.00 ± 00.00	52.72 ± 02.35
	O%	16.94 ± 09.74	00.00 ± 00.00	00.00 ± 00.00	00.00 ± 00.00
	I%	00.00 ± 00.00	00.00 ± 00.00	00.00 ± 00.00	00.00 ± 00.00
BIOPAX	F_1	$\mathbf{96.67 \pm 10.54}$	$\mathbf{96.67 \pm 10.54}$	38.02 ± 11.74	93.33 ± 14.05
	M%	$\mathbf{99.67 \pm 01.02}$	$\mathbf{99.67 \pm 01.02}$	70.41 ± 10.97	99.37 ± 01.34
	C%	$\mathbf{00.32 \pm 01.02}$	$\mathbf{00.32 \pm 01.02}$	29.59 ± 10.97	00.32 ± 01.02
	O%	$\mathbf{00.00 \pm 00.00}$	$\mathbf{00.00 \pm 00.00}$	00.00 ± 00.00	00.31 ± 00.99
	I%	$\mathbf{00.00 \pm 00.00}$	$\mathbf{00.00 \pm 00.00}$	00.00 ± 00.00	00.00 ± 00.00
FINANCIAL	F_1	40.00 ± 51.64	$\mathbf{100.0 \pm 00.00}$	$\mathbf{100.0 \pm 00.00}$	$\mathbf{100.0 \pm 00.00}$
	M%	67.06 ± 36.09	$\mathbf{99.70 \pm 00.48}$	$\mathbf{99.70 \pm 00.65}$	$\mathbf{99.70 \pm 00.68}$
	C%	00.00 ± 00.00	$\mathbf{00.00 \pm 00.00}$	$\mathbf{00.00 \pm 00.00}$	$\mathbf{00.00 \pm 00.00}$
	O%	32.94 ± 36.09	$\mathbf{00.00 \pm 00.00}$	$\mathbf{00.30 \pm 00.68}$	$\mathbf{00.30 \pm 00.68}$
	I%	00.00 ± 00.00	$\mathbf{00.30 \pm 00.50}$	$\mathbf{00.00 \pm 00.00}$	$\mathbf{00.00 \pm 00.00}$

the outcomes seem to be promising: our approach appears to be competitive in terms of both F-measure and match rate w.r.t. other approaches.

In the case of BCO, the models resulting from our approach have very good performance, although the F-measure and match rate values are lower than the ones obtained by using CELOE. Despite this, there was a significant improvement of DST-TDTs w.r.t. TDTs and DISJUNCTIVE ELTL. This was mainly due occurrence of node tests that yielded an indefinite membership owing to the OWA, that in the original algorithm hindered the prediction of definite memberships [7]. This behavior explains why the match rate is low while the omission rate is high when TDTs are considered. The visit strategy and the new decision procedure for the DST-TDTs was able to assign a correct membership in most cases. As regards the improvement w.r.t. the DISJUNCTIVE ELTL, this is likely due to the membership assignments w.r.t. very general induced concepts that can be boiled down to a mere random choice.

As regards BIOPAX, the performance of TDTs and DST-TDTs are the same and their models worked considerably better than CELOE, especially in terms of F-measure. The similar performance are motivated by the presence of disjointess axioms available in this ontology. This means that instances more likely had a definite membership w.r.t. the concepts generated by the refinement operator and installed during the training phase. As a consequence, node tests tended to route instances along single paths. Again, the worse performance of CELOE is due to the induction of very general concepts.

Concerning the FINANCE ontology, we observed a general good performance with the DST-TDTs, similarly to CELOE and DISJUNCTIVE ELTL. Besides, the improvement over TDTs was again quite considerable in terms of F-measure and match rate. However, no omission cases were observed with DST-TDTs. Instead we noticed a small percentage of induction, which was absent in the experiments with BCO and BIOPAX.

A final remark concerns the stability of the models induced through the proposed algorithm over the various queries concepts: the outcomes show a lower standard deviation w.r.t. the original version and similar to the other methods considered in the experiments [8,14].

5   Conclusions and Extensions

In this work, we have studied the effectiveness of an improved version of an inductive classification model for Web ontologies, derived from the combination of the evidential measures and supervised learning for DLs, to better cope with the underlying uncertainty due to the inherent incompleteness. We have shown that the new method improves w.r.t. those based on similar models, and the induced models are competitive w.r.t. other concept learning approaches. Specifically, the decision procedure employed in this work allowed to frequently select the correct class-membership value for test individuals. An additional advantage of the new approach is also a higher stability w.r.t. the behavior the original TDTs.

DST-TDTs can be exploited to discover potentially new (non logically derivable) assertions that can be used to complete the extensional part of a Web ontology whose expressiveness allows to represent concepts by means of disjunction and complement operators. However, the validity for these assertions should be decided by a domain expert. Moreover, DST-TDTs can also suggest different versions of the concept definitions to be integrated in the ontologies.

In the next future, we plan to extend the method along various directions. For example, a total uncertainty measure that integrates conflicting evidence [12] can be considered. Additionally, we can study also the effectiveness of further associative combination rules [10] and further refinement operators , as the ones available in DL-Learner [6].

Acknowledgments. This work fulfills the research objectives and has been partially funded by the projects LOGIN project (PII Industry 2015), and VINCENTE project (POR Regione Puglia).

References

1. Baader, F., Calvanese, D., McGuinness, D., Nardi, D., Patel-Schneider, P. (eds.): The Description Logic Handbook, 2nd edn. Cambridge University Press, Cambridge (2007)
2. Rettinger, A., Lösch, U., Tresp, V., d'Amato, C., Fanizzi, N.: Mining the semantic web. Statistical learning for nextgeneration knowledge bases. Data Min. Knowl. Discov. **24**, 613–662 (2012)

3. Fanizzi, N., d'Amato, C., Esposito, F.: DL-FOIL concept learning in description logics. In: Železný, F., Lavrač, N. (eds.) ILP 2008. LNCS (LNAI), vol. 5194, pp. 107–121. Springer, Heidelberg (2008)

4. Lehmann, J., Auer, S., Bühmann, L., Tramp, S.: Class expression learning for ontology engineering. J. Web Sem. **9**, 71–81 (2011)

5. Lehmann, J., Haase, C.: Ideal downward refinement in the EL description logic. In: De Raedt, L. (ed.) ILP 2009. LNCS, vol. 5989, pp. 73–87. Springer, Heidelberg (2010)

6. Lehmann, J.: DL-Learner: learning concepts indescription logics. J. Mach. Learn. Res. **10**, 2639–2642 (2009)

7. Fanizzi, N., d'Amato, C., Esposito, F.: Induction of concepts in web ontologies through terminological decision trees. In: Balcázar, J.L., Bonchi, F., Gionis, A., Sebag, M. (eds.) ECML PKDD 2010, Part I. LNCS, vol. 6321, pp. 442–457. Springer, Heidelberg (2010)

8. Rizzo, G., d'Amato, C., Fanizzi, N., Esposito, F.: Towards evidence-based terminological decision trees. In: Laurent, A., Strauss, O., Bouchon-Meunier, B., Yager, R.R. (eds.) IPMU 2014, Part I. CCIS, vol. 442, pp. 36–45. Springer, Heidelberg (2014)

9. Klir, J.: Uncertainty and Information. Wiley, Hoboken (2006)

10. Sentz, K., Ferson, S.: Combination of evidence in Dempster-Shafer theory. Sandia Report SAND2002-0835 (2002)

11. Denoeux, T.: A k-nearest neighbor classification rule based on Dempster-Shafer theory. IEEE Trans. Syst. Man Cybern. **25**, 804–813 (1995)

12. Sutton-Charani, N., Destercke, S., Denoeux, T.: Classification trees based on belief functions. In: Denoeux, T., Masson, M.-H. (eds.) Belief Functions: Theory and Applications. LNCS, vol. 164, pp. 77–84. Springer, Heidelberg (2012)

13. Dubois, D., Prade, H.: On the combination of evidence in various mathematical frameworks. In: Flamm, J., Luisi, T. (eds.) Reliability Data Collection and Analysis. Eurocourses, vol. 3, pp. 213–241. Springer, Heidelberg (1992)

14. Rizzo, G., d'Amato, C., Fanizzi, N., Esposito, F.: Assertion prediction with ontologies through evidence combination. In: Bobillo, F., et al. (eds.) URSW 2008-2010/UniDL 2010. LNCS, vol. 7123, pp. 282–299. Springer, Heidelberg (2013)

Clustering Classifiers Learnt from Local Datasets Based on Cosine Similarity

Kaikai Zhao[1]([✉]) and Einoshin Suzuki[2]

[1] Graduate School of Systems Life Sciences, Kyushu University, Fukuoka, Japan
cyoukaikai@gmail.com
[2] Department of Informatics, ISEE, Kyushu University, Fukuoka, Japan

Abstract. In this paper we present a new method to measure the degree of dissimilarity of a pair of linear classifiers. This method is based on the cosine similarity between the normal vectors of the hyperplanes of the linear classifiers. A significant advantage of this method is that it has a good interpretation and requires very little information to exchange among datasets. Evaluations on a synthetic dataset, a dataset from the UCI Machine Learning Repository, and facial expression datasets show that our method outperforms previous methods in terms of the normalized mutual information.

Keywords: Clustering classifiers · Distributed database classification

1 Introduction

Many real-world problems involve multiple related classification tasks, and there is a growing interest in measuring the similarity of the classifiers or the relatedness of classification [1–3]. One typical example is classifying physically distributed databases. Collecting distributed data for centralized processing is often unrealistic due to storage, transmission costs and privacy issues, e.g., in medicine, in finance. A more efficient way to deal with this problem is to mine the local datasets and combine the resulting models [4]. However, viewing the local datasets as a part of a single global model is inappropriate because they might differ conceptually [4]. For example, patients with the same disease in hospitals of different regions might be influenced by different living styles and levels.

A feasible way to solve this problem is to first utilize a specific dissimilarity measure to group similar classifiers, and then use only the group of similar classifiers to predict a new example [4]. The resulting dissimilarity among classifiers could be also used to find a common structure or difference among datasets.

Another example is clustered multi-task learning (MTL) [5]. A single task leaning where each database is classified independently will not exploit the relatedness of tasks. MTL learns multiple related tasks simultaneously using a shared

---

E. Suzuki—A part of this research was supported by Grant-in-Aid for Scientific Research 25280085 and 15K12100 from the Japanese Ministry of Education, Culture, Sports, Science and Technology.

F. Esposito et al. (Eds.): ISMIS 2015, LNAI 9384, pp. 150–159, 2015.
DOI: 10.1007/978-3-319-25252-0_16

representation: what is learnt from one task could help learn other tasks better. Existing research assumes all the tasks are related, however, it is usually the case in a practical application that the models of tasks from a group are more similar to each other than those from different groups. Clustered multi-task learning is developed to improve the generalized model by grouping the related tasks. Measuring the degree of the relatedness between a pair of tasks is a significant problem.

In this paper, we propose a dissimilarity measure based on the cosine similarity to estimate the degree of the dissimilarity between a pair of classifiers. It is simple and efficient, and requires very little information exchange among classifiers. The rest of the paper is organized as follows. Section 2 introduces the related works. Section 3 describes our method in detail. Section 4 describes the testing datasets and the evaluation results. Section 5 concludes.

2   Previous Work

Several distance measures have been proposed based on the associations among different datasets. For instance, Parthasarathy et al. [6] proposed to measure the degree of the similarity between a pair of databases based on how their attributes are correlated. They restricted themselves to frequent occurring associations. This method is very sensitive to the parameters such as the minimum support. McClean et al. [7] proposed a Euclidean distance and a Kullback-Leibler information divergence between two homogeneous datacubes, each of which is a contingency table that displays multi-attribute frequency distributions. However, aggregate count data on categorical attributes are required in advance, and this method is difficult to be extended to data with numerical attributes. Chen et al. [8] also used aggregated data that summarize each distributed database to measure the distance between different datasets. In their method, a subset of local data needs to be transmitted to a central site, which requires a lot of transmission costs. In [9], Flores et al. proposed a procedure to create a meta-datasets with several complexity measures of classification problems (proposed in [10]), train several semi-naive Bayesian Networks classifiers for each dataset, and automatically select the best one for each test data. This process could lead to a good classification performance among datasets but can not give the dissimilarity among the datasets.

Tsoumakas et al. [4] measured the similarity between classifiers derived from two separate distributed datasets based on the prediction difference on another independent dataset. Suppose N is the number of overall distributed datasets, and M is the number of tuples of each dataset. To obtain the similarity, their method averages the prediction difference of two classifiers on all other $N - 2$ independent datasets. The complexity of the algorithm is $O(MN^3)$. However, N will be reasonably large in some kind of situations, e.g., communication systems which treat each user's data as a local dataset or multi-task learning systems which involve sequential tasks. In such a case, this method is inappropriate.

For multi-task learning, the mutual relatedness of tasks are estimated in [11], and the performance of transferring knowledge from one task to another task is used as the similarity measure. However, this method uses the k-nearest-neighbor as the classifier for each data set, and thus the transfer among datasets requires a lot of data exchange. Evgeniou et al. [12] proposed the task clustering regularization, however, it requires the group structure as a prior. Xue et al. [13] introduced the Dirichlet process prior to identifying the subgroup of tasks. This method assumes that the parameters of the tasks within the same cluster share a common probabilistic prior. It will fail to take advantage of the negatively related tasks, since they should be clustered into different groups.

3   Our Methods

A real-world classification problem usually has some learnable structures [10]. For instance, tuples from the same category may have some common characteristics and lie in the same density regions, while tuples from other different categories tend to lie in other different regions. In a general case, the entire feature space is divided into different regions by hyperplanes, and a region which is given different predictions based on two different hyperplanes represents a *disagreement region*. The volume of the regions seemed to be an ideal dissimilarity measure.

Figure 1 shows an example that illustrates the idea of a dissimilarity measure based on the disagreement area. The datasets D_1 and D_2 have more similar data distributions than those of D_1 and D_3 or D_2 and D_3. In the rightmost plot, the volume of the disagreement areas between $hyper1$ and $hyper2$ (area 1) is much smaller than those between $hyper1$ and $hyper3$ (area 2).

However, relying on the volume of the disagreement area is likely to cause overfitting because the training examples is typically scarce for the large number of features[1]. An observation in Fig. 1 shows that disagreement volume is positively correlated with the angle between the hyperplanes, i.e., the larger the

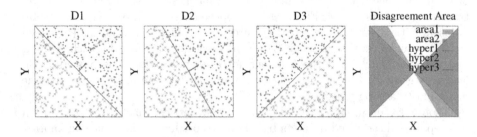

Fig. 1. The illustration of a similarity measure based on the disagreement area between a pair of linear classifiers. From left to right, the datasets D_1, D_2, and D_3, estimated hyperplanes, normal vectors and disagreement areas of the estimated hyperplanes, where red points are positive examples and green points are negative examples. Blue lines are the estimated hyperplanes and magenta lines represent their normal vectors. Area 1 is the disagreement area between $hyper1$ and $hyper2$. Area 2 is the disagreement areas between $hyper1$ and $hyper3$ (Color figure online).

---

[1] The same risk exists for relying on the density in the example space.

volume of the disagreement regions is, the larger the angle between the relevant hyperplanes that shape the disagreement regions. For instance, the angle between *hyper*1 and *hyper*2 (area 1) is much smaller than that between *hyper*1 and *hyper*3 (area 2).

Cosine similarity [14] is frequently used in information retrieval and text mining. We adopt it as the similarity measure between different classifiers. Since training a classifier and estimating its hyperplanes can be performed locally, we only need to exchange the hyperplane information of each dataset, which can be done very efficiently and saves a lot of communication costs.

The cosine similarity of two vectors A and B is given by

$$cos(\theta) = \frac{A \cdot B}{\|A\|\|B\|} \tag{1}$$

where θ is the angle between A and B. For two binary classifiers C_x and C_y, suppose $\mathbf{n}$ and $\mathbf{v}$ are the normal vectors of their estimated hyperplanes, respectively, then the dissimilarity between C_x and C_y can be estimated by

$$dissimilarity(C_x, C_y) = 1 - \frac{\frac{\mathbf{n}\cdot\mathbf{v}}{\|\mathbf{n}\|\|\mathbf{v}\|} + 1}{2} \tag{2}$$

where $\frac{\mathbf{n}\cdot\mathbf{v}}{\|\mathbf{n}\|\|\mathbf{v}\|}$ is the cosine similarity derived from Eq. (1), and $\frac{\frac{\mathbf{n}\cdot\mathbf{v}}{\|\mathbf{n}\|\|\mathbf{v}\|}+1}{2}$ is used to map the original cosine similarity to [0,1]. So $dissimilarity(C_x, C_y)$ is in the range of [0,1] as well. If we want to cluster the negatively related classifier into one group, we just need to modify the above equation to

$$dissimilarity(C_x, C_y) = 1 - \left|\frac{\mathbf{n}\cdot\mathbf{v}}{\|\mathbf{n}\|\|\mathbf{v}\|}\right| \tag{3}$$

here we use the absolute value of $\frac{\mathbf{n}\cdot\mathbf{v}}{\|\mathbf{n}\|\|\mathbf{v}\|}$ to eliminate the influence of the directions of two hyperplanes. In this paper, we use Eq. (2) as our disagreement measure between two classifiers.

Let Ω be the labeled n datasets, $\Omega = \{D_1, D_2, \ldots, D_n\}$, where D_i represents the ith datasets. Each D_i has m features. Let X_i^k and y_i^k be the feature vector and the class label of the kth tuple of D_i, respectively. Let $X_i^{k,j}$ be the jth attribute value of X_i^k.

For linear classification problems, a hyperplane is represented by $\boldsymbol{\Theta}^\tau \alpha = 0$, where $\alpha = [\alpha_0, \alpha_1, ..., \alpha_m]^\tau$ is an augmented feature vector (with $\alpha_0 = 1$) for the parameter vector $\boldsymbol{\Theta} = [\theta_0, \theta_1, ..., \theta_m]^\tau$. We denote the normal vector $\boldsymbol{v}$ by $\boldsymbol{v} = (\theta_1, ..., \theta_m)$. The estimation of the parameter vector $\boldsymbol{\Theta}_i$ of the hyperplane of D_i can be easily obtained by using logistic regression [15]. Algorithm 1 lists the pseudo code of the dissimilarity measure of two binary classifiers derived from D_1 and D_2. Lines 1 - 8 train the linear classifiers for each of D_1 and D_2. Firstly, $\boldsymbol{\Theta}_i$ is initialized to the zero vector, then an iterative procedure is used to keep updating $\boldsymbol{\Theta}_i$ until convergence. Line 5 describes how to update $\theta_{i,j}$ (the jth element of $\boldsymbol{\Theta}_i$) in each iteration, where $h_{\boldsymbol{\Theta}}(X) = g(\boldsymbol{\Theta}^T X) = \frac{1}{1 + \exp^{-\boldsymbol{\Theta}^T X}}$

Algorithm 1. Dissimilarity Measure between Two Binary Classifiers

1: **for** each dataset D_i of D_1 and D_2 **do**
2:    $\Theta_i \leftarrow 0$;
3:    **repeat**
4:       **for** $j = 0$ to m **do**
5:          $\theta_{i,j} = \theta_{i,j} + \alpha \dfrac{1}{m} \sum\limits_{k=1}^{N} [(h_{\Theta_i}(X_i^k) - y_i^k)X_i^{k,j}]$;
6:       **end for**
7:    **until** convergence
8: **end for**
9: $n \leftarrow \Theta_1\{1 : m\}$;
10: $v \leftarrow \Theta_2\{1 : m\}$;
11: $dissimilarity(D_1, D_2) \leftarrow 1 - (\frac{n \cdot v}{\|n\|\|v\|} + 1)/2$;

is the sigmoid function. $\dfrac{1}{m} \sum\limits_{k=1}^{N} [(h_{\Theta_i}(X_i^k) - y_i^k)X_i^{k,j}]$ is the gradient of the cost function at the current Θ_i and α is the learning rate. Lines 9 - 10 assign the respective parameter vectors Θ_1 and Θ_2 to the normal vectors. Line 11 calculates the dissimilarity based on Eq. (2).

For a multiple classification problem in which more than 2 classes are involved, we propose two strategies to deal with it: one-vs-one and one-vs-all. For one-vs-one strategy, we need to estimate the hyperplanes between any pair of classes, calculate their mutual dissimilarities based on Eq. (2), and then average the results. So the dissimilarity measure between two datasets is given by,

$$dissimilarity(D_x, D_y) = \frac{\sum\limits_{i=1}^{L-1} \sum\limits_{j=i+1}^{L} dissimilarity(C_x(i,j), C_y(i,j))}{\binom{L}{2}} \qquad (4)$$

where $C_x(i,j)$ is the binary classifier derived from the tuples that belong to class i or j in dataset D_x. L is the total number of classes. $\binom{L}{2} = \frac{L(L-1)}{2}$ is the total number of pairs of binary classifiers and is used here to normalize the dissimilarity. For N distributed datasets, we need to compare the $\binom{L}{2}$ binary classifiers with respect to all the combinations of pairs of datasets (total $\binom{N}{2}$), so the complexity of the algorithm is $O(L^2N^2)$.

For one-vs-all strategy, we obtain the dissimilarity measure by

$$dissimilarity(D_x, D_y) = \frac{\sum\limits_{i=1}^{L} dissimilarity(C_x(i,\bar{i}), C_y(i,\bar{i}))}{L} \qquad (5)$$

where $C_x(i,\bar{i})$ is the binary classifier that assigns tuples that belong to class i the label 0 and all other tuples the label 1. Here only L hyperplanes are involved for each dataset, so the complexity of the algorithm is $O(LN^2)$.

Compared with Tsoumakas' method [4] (with a complexity of $O(MN^3)$, where M is the number of tuples of each dataset), the complexities of one-vs-all and one-vs-one strategies increase as L increase, but are less influenced by the number N of datasets, and has nothing to do with M, which is usually much greater than L in practice.

4 Experiments

In [4], Tsoumakas et al. have shown that grouping similar classifiers and using only the group of similar classifiers to predict a new example that belongs to one of the dataset in the group could result in a better classification result. Since our focus in this paper is the dissimilarity measure between two classifiers, we did not measure the accuracy of classifiers.

4.1 Datasets

We use a synthetic dataset, a dataset from the UCI Machine Learning Repository [16], and two facial expression datasets [17] to test our method. We generated the synthetic dataset based on a technique described in [18] as follows. Firstly, we generated 18 sub-datasets, each of which contains 12 features $a_1, a_2, \ldots, a_{12}$. Each feature follows a multi-variate Gaussian distribution with mean 0 and varying covariances either 0, ± 0.1, ± 0.2, ± 0.3, ± 0.4, ± 0.5. Then we manually labeled them based on three different class boundaries, $a_1 + a_{12} > 0$, $a_1 + a_2 + a_{11} + a_{12} > 0$, and $a_1 a_2 + a_{12} > 0$. Each hyperplane represents the class boundary of 6 sub-datasets and each sub-dataset includes 10,000 examples.

The second dataset is the vowel dataset from the UCI Machine Learning Repository [16]. We divided it into 6 sub-datasets based on the concept of contextual features [19] (gender in this example). Here we firstly divided the dataset into two parts, based on the gender information (male or female), then we deleted the gender feature, and divided each part into three sub-datasets randomly. Finally, we obtained six sub-datasets that represent two different concepts, 3 for male concept, 3 for female concept.

The third and fourth datasets are facial expression datasets collected using Kinect[2]. We collected these datasets to recognize human facial expressions for building a human monitoring system. 17 features are involved, 6 animation units (AUs), and 11 shape units (SUs). Here the AUs represent facial movements from the neutral face, and SUs represent the particular shapes of the user's head, the neutral position of their mouth, brows, eyes, and so on[3].

Undeliberate facial expression data are more desirable. However, some expressions, e.g., fear, are difficult to collect in real life [20]. So as a first step, we collected the deliberate facial expression from December 2013 to January 2014 [17] (the third dataset). The labeled datasets include 100 individuals, each of them has 25 expressions, and each expression includes at least 25 tuples. The feature values of a same expression are very different for different individuals.

---

[2] http://www.microsoft.com/en-us/kinectforwindows/.
[3] http://msdn.microsoft.com/en-us/library/jj130970.aspx.

Because the magnitude of the facial movements of a same expression and the shapes of the heads of different individuals may vary significantly.

Two experiments were designed using the third dataset. In the first experiments (the *facial3* dataset), we randomly chose 3 individuals, and randomly divided each of their data into 4 sub-datasets. In the second experiments (the *facial100* dataset) we divided 100 individuals' facial expression data into 30 groups with respect to the person IDs and each group includes a varying number of individuals (from 1 to 5). For each of the 30 groups, we further randomly divided all its tuples into 3 sub-datasets.

Then we collected two individuals' (named P1 and P2) as many undeliberate expressions (P1-U, P2-U) as possible from May 2014 to August 2014. We also collected their deliberate 25 expressions (P1-D, P2-D) at once (without interruption). Here we extracted the common 4 expressions {*Happy, Pleased, Tired, Serious*} in each of the four sub-datasets as our test dataset (the fourth dataset).

4.2   Clustering Classifiers

Once we obtain the pairwise distances among classifiers, we can perform the agglomerative hierarchical clustering, which works by merging the most similar two objects each time until all objects are merged. We chose it because the pairwise distances of the classifiers have already been computed, and it doesn't require specifying the number of clusters in advance. Complete linkage is used here as the linkage criteria.

Figure 2 shows the clustering results of the first two datasets using our methods. We can see that our methods successfully clustered similar classifiers. In the left plot, 18 sub-datasets are clustered into three groups, and each group has a different boundary. In the right two plots, male and female sub-datasets are clustered into separate groups by one-vs-all and one-vs-all strategies. Here the classifiers that are clustered into the same group are marked in a same color.

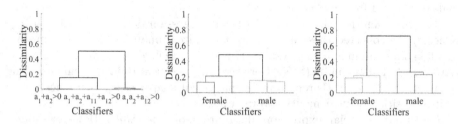

Fig. 2. Clustering results of synthetic dataset (left) and vowel dataset using our method with one-vs-all (middle) and one-vs-one (right) strategies

Figures 3 show the clustering results of the two experiments using the third dataset by applying Tsoumakas' method [4] and our methods separately. We see that both methods succeeded to cluster separate groups of classifiers. However, our resulting clusters have relatively smaller within group distances and greater inter-group distances than Tsoumakas' method.

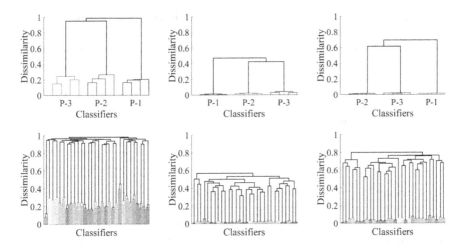

Fig. 3. Clustering results of 3 individuals' facial expression dataset (first row) and 100 individuals' facial expression dataset (second row) using Tsoumakas' method [4] (left) and our method with one-vs-all (middle) and one-vs-one (right) strategies.

For the last facial expression dataset, it is natural to believe that facial expressions of the same individual are more similar than those of the different individuals. Thus P1-D and P1-U should be clustered into one group and P2-D and P2-U be clustered into another group. The left three plots in Fig. 4 show the clustering results of the fourth dataset. We see that our methods successfully clustered the deliberate and undeliberate facial expressions of the same individual into the same group, while Tsoumakas' method failed. Since Tsoumakas' methods [4] rely on prediction difference on the tuples of the local datasets, it will be influenced by how well the datasets fit the true distributions behind them. Our methods use the boundary difference, and thus is more robust to data. The NMI of the resulting clusters are shown in the rightmost of Fig. 4.

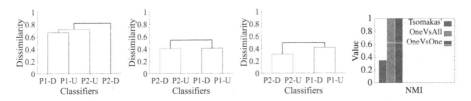

Fig. 4. Clustering results of the fourth datasets using Tsoumakas' method [4] (leftmost), our method with one-vs-all (middle left) and one-vs-one (middle right) strategies. The evaluation results of the fourth dataset (rightmost).

Table 1 shows the learning time spent to compute dissimilarities between pairs of the sub-datasets using Matlab on a Windows system. The benchmarking platform was an Intel(R) Core(TM) i7-3930K CPU @3.2 GHZ system with

Table 1. Time computation (second) of estimating dissimilarities

Dataset	one-vs-all	one-vs-one	Tsoumakas	No. Class	No. Dataset	No. AvgTuple
synthetic	0.011	0.011	2.55	2	18	10,000
vowel	0.009	0.047	0.010	11	4	165
facial3	0.09	1.03	0.24	25	12	645
facial100	5.10	65.3	346.7	25	90	2923
facial4	0.0041	0.0052	0.0079	4	4	915

32 GB memory. The recorded time did not include the time spent on training the local classifiers (here we chose logistic regression classifiers) using Tsoumakas' methods, and estimating the hyperplanes of local classifiers using our methods. Because those processes were performed in local datasets, they did not need to communicate with other datasets. We see that our methods are much faster than Tsoumakas' method when the number of datasets increases significantly, e.g., dataset *facial100*. One-vs-all strategy and one-vs-one strategy have similar performance in terms of clustering but the former costs less time.

We believe the proposed methods could be used to find similar datasets, since the dissimilarity measure is based on the normal vectors of the hyperplanes of the classifiers derived from the different datasets. A limitation of our methods is that it is not suitable for multi-modal datasets where multiple dense areas of classes exist in the attribute space. The examples may not be linearly separable because examples of one class lie in several different dense regions and nonlinear feature transfer may also fail. In such a case, we suggest to find a latent space to represent the original feature space by, for example, sparse coding technique, to learn a set of basis vectors to represent the original dataset, so that the dissimilarity measure could be performed in the new basis space.

5   Conclusions

In this paper, we proposed a novel method to measure the similarity between two classifiers derived from distributed databases. The resulting clusters of classifiers could be used to find similar datasets or difference among physically distributed datasets, and measure the relatedness among tasks for clustered multi-task learning. A significant advantage of this method is that it has a good interpretation and requires very few information to exchange among datasets. In our future work we will investigate similarity measures between more general classifiers.

References

1. Ben-David, S., Schuller, R.: Exploiting task relatedness for multiple task learning. In: Schölkopf, B., Warmuth, M.K. (eds.) COLT/Kernel 2003. LNCS (LNAI), vol. 2777, pp. 567–580. Springer, Heidelberg (2003)

2. Pedersen, T., Pakhomov, S.V.S., Patwardhan, S., Chute, C.G.: Measures of Semantic Similarity and Relatedness in the Biomedical Domain. J. Biomed. Inf. **40**(3), 288–299 (2007)
3. Li, Y., Tian, X., Song, M., Tao, D.: Multi-task proximal support vector machine. Pattern Recogn. **48**(10), 3249–3257 (2015)
4. Tsoumakas, G., Angelis, L., Vlahavas, I.P.: Clustering classifiers for knowledge discovery from physically distributed databases. Data Knowl. Eng. **49**(3), 223–242 (2004)
5. Jacob, L., Bach, F.R., Vert, J.P.: Clustered multi-task learning: a convex formulation. In: NIPS 2008, pp. 745–752 (2009)
6. Parthasarathy, S., Ogihara, M.: Clustering distributed homogeneous datasets. In: Zighed, D.A., Komorowski, J., Żytkow, J.M. (eds.) PKDD 2000. LNCS (LNAI), vol. 1910, pp. 566–574. Springer, Heidelberg (2000)
7. McClean, S.I., Scotney, B.W., Morrow, P.J., Greer, K.: Knowledge discovery by probabilistic clustering of distributed databases. Data Knowl. Eng. **54**(2), 189–210 (2005)
8. Chen, R., Sivakumar, K., Kargupta, H.: Collective mining of Bayesian networks from distributed heterogeneous data. Knowl. Inf. Syst. **6**(2), 164–187 (2004)
9. Flores, M.J., Gmez, J.A., Martnez, A.M.: Meta-prediction of semi-naive Bayesian network classifiers based on dataset complexity characterization. In: Proceedings of Sixth European Workshop on Probabilistic Graphical Models, pp. 107–114 (2012)
10. Ho, T.K., Basu, M.: Complexity measures of supervised classification problems. IEEE Trans. Pattern Anal. Mach. Intell. **24**(3), 289–300 (2002)
11. Thrun, S., O'Sullivan, J.: Clustering learning tasks and the selective cross-task transfer of knowledge. In: Thrun, S., Pratt, L. (eds.) Learning To Learn, pp. 235–257. Kluwer, New York (1998)
12. Evgeniou, T., Micchelli, C.A., Pontil, M.: Learning multiple tasks with kernel methods. J. Mach. Learn. Res. **6**, 615–637 (2005)
13. Xue, Y., Liao, X., Carin, L., Krishnapuram, B.: Multi-task learning for classification with Dirichlet process priors. J. Mach. Learn. Res. **8**, 35–63 (2007)
14. Singhal, A.: Modern information retrieval: a brief overview. IEEE Data Eng. Bull. **24**(4), 35–43 (2001)
15. Hosmer, D.W., Lemeshow, S.: Introduction to the logistic regression model. In: Applied Logistic Regression, 2 edn., pp. 1–30. Wiley (2005)
16. Lichman, M.: UCI Machine Learning Repository. University of California, Irvine, School of Information and Computer Sciences (2013). http://archive.ics.uci.edu/ml
17. Erna, A., Yu, L., Zhao, K., Chen, W., Suzuki, E.: Facial expression data constructed with Kinect and their clustering stability. In: Ślęzak, D., Schaefer, G., Vuong, S.T., Kim, Y.-S. (eds.) AMT 2014. LNCS, vol. 8610, pp. 421–431. Springer, Heidelberg (2014)
18. Scott, P.D., Wilkins, E.: Evaluating data mining procedures: techniques for generating artificial data sets. Inf. Softw. Technol. **41**(9), 579–587 (1999)
19. Widmer, G., Kubat, M.: Learning in the presence of concept drift and hidden contexts. Mach. Learn. **23**(1), 69–101 (1996)
20. Sebe, N., Lew, M., Sun, Y., Cohen, I., Geners, T., Huang, T.: Authentic facial expression analysis. Image Vis. Comput. **25**(12), 1856–1863 (2007)

HC-edit: A Hierarchical Clustering Approach to Data Editing

Paul K. Amalaman[(✉)] and Christoph F. Eick

Department of Computer Science, University of Houston,
Houston, TX 77204-3010, USA
{pkamalam, ceick}@uh.edu

Abstract. Many nearest neighbor based classification approaches require that atypical and mislabeled examples be removed from the training dataset in order to achieve high accuracy. Current editing approaches often remove excessive amount of examples from the training set which does not always lead to optimum accuracy rate. We introduce a new editing method, called HC-edit,—for the k-nearest neighbor classifier —that recognizes areas in the dataset of high purities and removes minority class examples from those regions. The proposed method takes hierarchical clusters with purity information as its input. To edit the data, clusters with purities above a user-defined purity threshold are selected and minority class examples are removed from the selected clusters. Experiments carried out on real datasets using trees generated by traditional agglomerative hierarchical approaches, and trees generated by a supervised taxonomy method which incorporates class label information in the clustering process show that the new approach leads to improved accuracy and does well in comparison to other editing methods.

Keywords: Supervised clustering · Classification · Hierarchical clustering · Predictive model · Data editing

1 Introduction

One of the most popular classification techniques is the k-nearest neighbor method (k-NN) which assigns the majority class label in the k nearest neighbor set to a point that needs to be classified [1]. The basic k-NN has the advantage of being easy to implement but also requiring a large memory to store the model — which is the training set. Additionally, the classifier is sensitive to atypical examples whose presence in the training set may lead to poor accuracy and unnecessary storage of examples [7]. Consequently, most k-NN based approaches deal with these two issues using a technique known as "condensing" and improve accuracy by applying a technique known as "editing" [2]. Condensing aims at reducing a classifier training time while achieving no degradation in classification accuracy. Editing, on the other hand, seeks to remove noise examples from the original dataset with the goal of improving classification accuracy. Most editing approaches, in addition to cleaning up the data, have the advantage of cleaning up overlapping areas between clusters.

This paper focuses on a new editing method, called HC-edit which takes as input an agglomerative clustering —a tree of clusters— to "clean up" the training set. The tree

© Springer International Publishing Switzerland 2015
F. Esposito et al. (Eds.): ISMIS 2015, LNAI 9384, pp. 160–170, 2015.
DOI: 10.1007/978-3-319-25252-0_17

stores in its nodes the purities of the node clusters. The tree can be generated by traditional agglomerative hierarchical clustering algorithm such as the popular UPGMA (Un-weighted Pair Group Method with Arithmetic Mean) which merges the closest pairs first, and the distant examples/clusters last or by a supervised taxonomy algorithm such as STAXAC (Supervised Taxonomy Agglomerative Clustering) which incorporates purity information in the tree construction process. Due to its novelty, the STAXAC algorithm —which attempts to maximize purity while merging the examples/clusters will be explained in Sect. 3; it represents a "side" contribution of the paper. When traditional agglomerative method is used, purity information for the node datasets must be computed and stored after the tree is built. To edit the training set, user specifies a minimum purity threshold β and the algorithm retrieves the clusters whose purities are greater or equal to β such that the union of all the retrieved clusters equals the input dataset and each example appears only in one cluster. Then the minority class examples are removed from the clusters. A k-NN rule is then used to classify unlabeled examples using the edited dataset. The advantage of using a hierarchical clustering is that hierarchical clustering computes all needed clusters in advance, whereas an ordinary clustering algorithm has to be rerun for different purity thresholds. Figure 1 illustrates the steps of HC-edit editing process. Figure 1a-1 illustrates a hypothetical dataset and selected clusters which purities are above a user-defined threshold β. Figure 1a-2 shows the resulting clusters after the minority examples have been removed. Different purity thresholds yield different decision boundaries. Figure 1b-1 illustrates another cluster selection for a different user-defined purity threshold from the same training set and Fig. 1b-2 the corresponding result after editing.

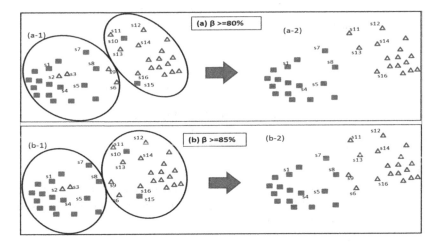

Fig. 1. Different purity thresholds yields different decision boundaries

The contributions of this paper are the following:

1. The data editing method of previous approaches visits each example and analyzes local information, namely the k-nearest neighbors of the example to decide if the

example should be removed or kept. HC-edit, on the other hand, is a more 'regional' approach in that it looks for regions of high purity in the dataset and removes the minority class examples from the regions.

2. The proposed method has the capacity to remove fewer points than widely known editing approaches which can be beneficial to the classifier accuracy rate. Many editing approaches tend to smooth the boundaries between clusters even if the clusters do not overlap. Excessive examples removal, especially, in the boundary regions may lead to a decrease in classifier accuracy since the chances of mis-classification increase due to a widened gap between the clusters.

3. A novel supervised hierarchical clustering approach is introduced.

To illustrate the second claim, let us consider the dataset in Fig. 1a and b with border points s7, s8, s9, and s6 where s8 and s9 are nearest neighbor of each other. Figure 1a-1 and b-1 show non overlapping clusters obtained from HC-edit of same dataset. If k = 1, current approaches will remove s8 and s9. On the other hand HC-edit would discard either s9 or s8, or none but not both. In Fig. 1a-2 s9 is removed but s8 is kept. In Fig. 1b-2 s8 and s9 are kept in the edited dataset.

2   Editing Algorithms

The motivation for editing the training set relies on the fact that the k-NN classifier achieves higher accuracy when the training set is rid of atypical and mislabeled examples; examples belonging to minority class label in comparison to other examples in the local region [7]. Editing also gets rid of the overlapping region examples.

Given an example x, a training dataset D, the k-neighborhood of x, noted $N_k(x)$, consisting of its k nearest neighbors can formally be defined as

$$N_k(x) \subseteq D; \ |N_k(x)| \ = \ k \ (k > 0)$$

$$\forall s \in N_k(x), \ q \in D \backslash N_k(x) \rightarrow d(s, x) \leq \ d(q, x) \, (q \in D \wedge q \notin N_k(x))$$

Given $N_k(x)$, the k-NN rule for x can be defined as follows:
Assign to x the class label of the majority class label in $N_k(x)$.
    This rule gives equal weight to each example's vote.

Given $N_k(x)$, we define a weighted k-NN rule as follows:
Assign a weight to each nearest neighbor's vote which is inversely proportional to its distance to x. Sum the vote by class label. The class label with the highest score is assigned to x. Each class score is computed as

$$V(x, c) = \sum_{j=1}^{k} \frac{p(x_j, c)}{(1+d(x, x_j))}$$

where $x_j \in N_k(x)$ and $p(x_j, c) = 1$ if x_j is labeled with c and $P(x_j, c) = 0$ otherwise.

Several editing approaches have been proposed; the most important ones are briefly discussed in the remainder of this section. The pseudo code for each approach is provided in Fig. 2.

<table>
<tr><td>

Wilson
Preprocessing:
Given a set of labeled examples $x_i \in D$,
$S \leftarrow D$
 For each x in D
   Discard x from S if it is misclassified using the k-NN
   rule with examples in D- {x}.

</td><td>

Multi-Edit
Preprocessing:
Repeat until no example is discarded
   Divide D into m random subsets $D_1,..,D_m$ with m>2
   Classify Di using $D_{(i+1)}$ as training set (i=1,..,m)
   Discard all examples in D_i incorrectly classified
   Replace D with union of $D_1,..D_m$

</td></tr>
<tr><td>

Supervised Clustering Editing
Preprocessing:
Use supervised clustering to cluster the data
Compute representative of each cluster
Delete all non-representative examples

</td><td>

WilsonProb
Preprocessing:
Given a set of labeled examples $x_i \in D$,
$S \leftarrow D$
For each x in D
   Discard x from S if it is misclassified using the
   weighted k-NN rule with examples in D- {x}.

</td></tr>
</table>

Fig. 2. Editing algorithms

Wilson Editing. Wilson editing [8] relies on the idea that if an example is erroneously classified using the k-NN rule it has to be eliminated from the training set.

Multi-Edit. Devijver and Kittler [3] proposed the Multi-edit technique. The algorithm repeatedly applies Wilson editing to m random subsets of the original dataset until no more examples are removed.

Supervised Clustering Editing. In supervised clustering editing [4], a supervised clustering algorithm is used to cluster a dataset D. Then D is replaced with a subset consisting of cluster representatives. Supervised clustering [5] deviates from traditional clustering in that it is applied on classified examples with the objective of identifying clusters with high probability density with respect to a single class. However, it does not organize the clusters in hierarchical fashion as HC-edit; it output ordinary clusters maximizing a fitness function. When the parameters to the fitness function change the clustering algorithm needs to regenerate the clusters. With HC-edit, the tree generation is independent of the parameter β —which is only used to select the clusters.

WilsonProb. This method [9] edits the training set based on a probability of an example to belong to a certain class in its neighborhood. The estimated probability is the weighted k-NN rule.

After the preprocessing, a new example is classified using k-NN rule.

3 The HC-edit Approach

The proposed method uses a supervised taxonomy clustering algorithm, called STAX-AC, to construct a binary tree which stores the purities of the clusters in the nodes. A user-defined purity threshold β is then used to retrieve clusters which purities are above the threshold. Due to its novelty STAXAC is explained in the following section.

3.1 The STAXAC Algorithm

Traditional hierarchical agglomerative clustering recursively merges the two closest clusters into a larger cluster until the last two clusters are merged. Thus proximity is the only criterion for merging. Supervised taxonomy algorithms such as STAXAC, on the other hand, incorporate class label information in the tree building process. STAXAC maximizes purity (minimizes impurity growth) by merging clusters in which a majority of the examples have the same class label. It uses distance information as a constraint so that only neighboring clusters are merged (Algorithm 1).

Algorithm 1: *STAXAC*

Input: examples with class labels and their distance matrix D.
Output: Hierarchical clustering
1. Start with a clustering X of one-object clusters.
2. $\forall C, C' \in X$; merge-candidate(C',C) $\Leftrightarrow$ (1-NN$_x$(C') = C or 1-NN$_x$(C)=C')
3. WHILE there are merge-candidates (C,C*) left
  BEGIN
    a.   Merge the pair of merge-candidates (C,C^*) obtaining a
        new cluster C'=C$\cup$C* and a new clustering X' for which Purity(X') has the largest
        value
    b.   Update merge-candidates:
        $\forall C'' \in X$; merge-candidate $(C', C'') \Leftrightarrow$ (merge-candidate(C, C'') or
        merge-candidate (C^*, C''))
    c.   Extend dendrogram by drawing edges from C and C* to C'
  END
4. Return constructed dendrogram

STAXAC identifies/updates merge candidates—clusters that potentially could be merged—and then creates a new cluster by choosing the merge candidate that maximizes the purity objective function. It continues this process until no more clusters can be merged. It starts off with single-object clusters (line 1) and identifies pairs of objects such that one is nearest-neighbor of the other (merge candidates) from the initial set (line 2)[1]. Then STAXAC merges the best merge-candidate (C, C*), creating a new

---

[1] It should be noted that the initial set of merge candidates is a subset of the actual merge candidates; in general, clusters could be neighboring with more than 2 other clusters. However, determining all clusters which are neighboring is only feasible in 2D-space by computing the Voronoi tessellation for the points in the dataset; unfortunately, the Voronoi tessellation cannot be computed for higher dimensional spaces.

cluster C' = C∪C* and computes the merge-candidates of the newly created cluster C' as the union of the merge candidates of clusters C and C* that were merged (line 3a and b). In summary, STAXAC conducts a wider search when constructing the tree than do traditional agglomerative hierarchical clustering algorithms, as it merges neighboring clusters, but it does not necessarily merge the pair of clusters that are closest to each other.

Illustration. Figure 3 shows a two-class dataset with green and white examples. The binary tree in Fig. 3a has been generated by STAXAC-like algorithm and the tree in Fig. 3b by a traditional agglomerative hierarchical clustering on the same dataset.

The arrows represent the "1-NN" relationship among the examples. Example x_3 and x_4 are the closest pair examples. However, because merging them will result into a cluster of 50 % purity, STAXAC chooses to merge x_2, and x_3 (into cluster C_1) to maximize purity because 1-NN(x_2) = x_3 (Fig. 3a). The traditional agglomerative method which does not incorporate class label information in its merging procedure merges x_3, and x_4 (into cluster C_1) (Fig. 3b). The HC-edit pseudo code is as follows:

Preprocessing:
Inputs:
  T, tree generated by hierarchical clustering algorithm such as STAXAC with purity information for each node cluster
  β, user-defined cluster purity
Output: Ds, dataset that is a subset of the original dataset
EDIT_TREE (T, root node)

Function EDIT_TREE (T, *node*)
BEGIN
 IF T=NULL exit
 IF *node* (purity) >= β
  Remove minority examples
  Add the surviving examples to Ds
  T←T\ sub-tree rooted at *node*; remove sub-tree rooted at node *node* from tree T
  EDIT_TREE(T, left_node)
  EDIT_TREE(T, right_node)
 Return Ds
END

Classification Use k-NN weighted vote rule to classify an example
 Because the algorithm starts the search from the root, it returns the largest cluster with purity equal or greater than the threshold in the selected branch of the tree. It can be observed that if Wilson editing is applied on the dataset presented in Fig. 3 x_3 and x_4 will be removed creating a wider gap between the clusters which may lead to potential misclassification. Secondly, although there are unlimited choices for β, most values return identical cluster sets. The set, β_set, composed of all the node purities, contains potential values for β. Figure 4 illustrates the trees obtained for the dataset depicted in Fig. 3 by both clustering approaches (β_set = {60,100} for STAXAC tree and β_set = {60, 66.67, 50, 100} for HAC.

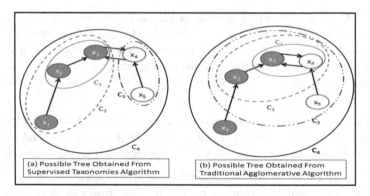

Fig. 3. STAXAC generated tree vs traditional hierarchical agglomerative tree (HAC)

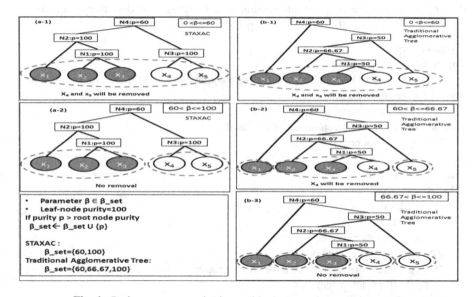

Fig. 4. Purity parameter selection and its impact on the editing result

If user queries the tree with $0 < \beta <= 60$ both trees return the entire dataset (Fig. 4a-1 and b-1); in that case minority examples x_4 and x_5 will be removed. On the hand, if $\beta > 60$, the tree generated by STAXAC returns two 100 % purity clusters (Fig. 4a.2). If $60 < \beta <= 66.67$, the search in the traditional tree returns three clusters (Fig. 4b.2); only x_4 will be removed by editing. If $\beta > 66.67$ five singleton clusters are returned; in this case the edited dataset is identical to the original dataset (Fig. 4b-3). Overall, the editing result is influenced by the choice of β. The value 100 is always element of β_set (leaf-nodes have 100 % purities). When $\beta = 100$ the edited dataset is identical to the original dataset (no removal since all clusters are pure). As β decreases, more minority examples are removed. (An optimum solution for the choice of β is not investigated in this paper).

4   Experimental Evaluation

HC-edit by default uses as input, trees generated by STAXAC to edit the dataset. We implemented a variant that uses tree generated by the widely known traditional agglomerative hierarchical method UPGMA. To distinguish both we denote them by HC-edit-STAXAC and HC-edit-UPGMA throughout the remaining of the paper. Additionally, we implemented the Wilson editing, and the WilsonProb algorithms. The performances of the algorithms with respect to accuracy were evaluated on seven real world datasets.

4.1   Datasets Description

We used the datasets summarized in Table 1 throughout the experiments.

Table 1. Datasets

Dataset	Description	Size	Number of Cl. Labels
Joint Genome Institue Ecosystem Datasets			
E.coli	Niche breadth	82	3
BEE	Bacteria ecosystem class: engineered environment	120	4
BEV	Bacteria ecosystem class: environmental	311	4
AEV	Archaea ecosystem type: environmental	70	4
UCI Repository			
Bos	Boston Housing	506	3
Bld	Bupa Liver Disease	345	2
Vot	Congress	435	2

Joint Genome Institute Datasets: Distance used for distance matrices was the patristic distance.

E. coli: This dataset was obtained by measuring the growth of 82 strains of *E. coli* in 10 distinct environments. Strains were then characterized as specialists, intermediate, or generalists depending on arbitrary divisions of the standard deviation of their growth in the environments.

Ecosystem datasets: Datasets characterize principle ecosystem type of bacteria (engineered environment, BEE; environmental, BEV) and archaea (environmental, AEV).

Ecosystem type and sequence information were downloaded from the Joint Genome Institute website [10].

UCI Datasets [6]: All datasets were preprocessed into dissimilarity matrices before the experiments. Dissimilarity matrices were generated using z-scores (except for the Vot dataset which has attributes with binary values).

4.2   Accuracy Results

The 10-fold cross-validation method (90 % of the original instances have been used as the training set and 10 % for test purposes) has been employed to estimate the overall classification accuracy. Table 2 reports on the experimental results obtained by the different algorithms over the 7 datasets. These results have been averaged over the ten partitions.

Table 2. Accuracy results

	E. Coli				
	HC-edit-STAXAC	HC-edit-UPGMA	WilsonProb	Wilson	K-NN
K=1	56.1 (50) [43.20]	47.56 (100)[0.00],(90)[0.00]	57.32 *[49.73]*	57.32 *[49.73]*	51.22
K=3	56.10 (100)[0],(90)[1.06],(80)[8.13]	52.44 (100)[0.00],(90)[0.00]	57.32 [44.53]	57.32 [45.73]	56.10
K=5	58.54 (100)[0.00],(90)[10.67]	53.66 (100)[0.00],(90)[0.00]	51.22 [42.40]	51.22 [41.86]	54.88
K=7	**63.41** (100)[0.00],(90)[10.67]	62.20 (80)[4.00]	52.44 [43.20]	54.88[46.80]	56.1
	BEV				
	HC-edit-STAXAC	HC-edit-UPGMA	WilsonProb	Wilson	K-NN
K=1	81.35 (90)[6.00]	82.64 (80)[14.17]	**83.6** *[19.96]*	**83.6** *[19.96]*	79.1
K=3	81.03 (90)[6.00]	81.99 (80)[14.17]	82.64 [16.92]	81.79 [16.75]	81.67
K=5	81 (100)[0.00]	83.28 (80)[14.17]	81.67[16.50]	79.74 [18.60]	80.39
K=7	81.35 (90)[6.00]	83.28 (80)[14.17]	81.35[16.25]	80.06 [18.60]	80.39
	Vot				
	HC-edit-STAXAC	HC-edit-UPGMA	WilsonProb	Wilson	K-NN
K=1	91.89 (100)[0.00]	90.09 (90)[5.80],(80)*[10.9]*	90.95 *[9.00]*	90.95 *[9.00]*	89.66
K=3	91.38 (100)[0.00]	91.38 (90)[5.80]	90.95 [8.00]	90.95 [8.00]	92.24
K=5	92.67 (100)[0.00],(90)[5.38]	92.24 (100)[0.00],(90)[5.80]	90.95 [8.00]	90.95 [8.00]	91.38
K=7	**93.10** (100)[0.00]	92.67 (100)[0.00]	90.95 [8.00]	90.95 [8.00]	92.67
	Bld				
	HC-edit-STAXAC	HC-edit-UPGMA	WilsonProb	Wilson	K-NN
K=1	61.11 (100)[0.00]	62.09 (80)[7.15]	61.44 *[39.26]*	61.44 *[39.26]*	60.13
K=3	59.80 (100)[0.00]	59.80 (90),(80)[7.15]	59.48 [35.64]	59.48 [35.60]	59.48
K=5	63.07 (90)[3.21]	62.75 (100)[0.00],(90)[0.6]	59.8 [32.00]	59.8 [31.73]	63.4
K=7	64.05 (90)[3.21]	**64.38** (100)[0.00],(90)[0.6]	61.44 [29.48]	61.44 [29.37]	64.05
	AEV				
	HC-edit-STAXAC	HC-edit-UPGMA	WilsonProb	Wilson	K-NN
K=1	**80** (80)[10.50]	78.57 (80)[5.6]	77.14 *[25.61]*	77.14 *[25.61]*	77.14
K=3	78.57 (90)[2.50],(80)[10.50]	78.57 (100)[0.00],(90)[0.7],(80)[5.6]	72.86 [18.25]	74.29 [16.84]	78.57
K=5	**80** (90)[2.5]	77.14 (100)[0.00],(90)[0.7],(80)[5.6]	72.86 [18.60]	71.43 [17.89]	78.57
K=7	78.57 (100)[0.00],(90)[2.5],(80)[10.5]	78.57 (100)[0.00],(90)[0.7],(80)[5.6]	71.43 [18.25]	70 [18.94]	74.29
	BEE				
	HC-edit-STAXAC	HC-edit-UPGMA	WilsonProb	Wilson	*K-NN*
K=1	80.83 (90)[7.31]	84.17 (80)[15.83]	78.33 [17.68]	78.69 [17.68]	78.33
K=3	81.67 (90)[7.31]	84.17 (80)[15.83]	82.5 [19.07]	85 *[20.09]*	82.5
K=5	82.50 (100)[0.00],(90)[7.31]	84.17 (80)[15.83]	82.5 [17.96]	76.15 [17.96]	80.83
K=7	84.17 (90)[7.31]	84.17 (80)[15.83]	84.17 [16.67]	**85.83** [16.57]	83.33
	Bos				
	HC-edit-STAXAC	HC-edit-UPGMA	WilsonProb	Wilson	K-NN
K=1	69.57 (90)[37.28]	72.33 (80)[7.14]	67.00 [23.24]	65.5 [23.24]	67
K=3	70.55 (90)[37.28]	71.94 (80)[7.14]	70.36 [23.68]	70.55 [23.48]	67.39
K=5	70.75 (100)[0.00]	70.36 (90)[1.03]	67.46 [23.61]	66.4 [24.16]	70.55
K=7	**73.12** (100)[0.00]	**73.12** (100)[0.00]	69.37 *[25.83]*	69.57 [24.18]	**73.12**

Bold figures indicate the best method in terms of classification accuracy for each dataset. The largest compression values are in italic. All k values tried out during classification and training phase are reported (HC-edit uses k for classification only). The result are presented as "x (y)[z]" where x is the accuracy rate, y is the purity parameter β (for HC-edit) and z the compression rate. For HC-edit multiple values of β were run for a given k and best results are reported. Whenever more than one (y)[z] are reported they are separated by commas. With respect to accuracy rate, we observe that the plain k-NN approach has overall best performance on one dataset (Bos). Wilson has best performance on two datasets (BEV, and BEE). WilsonProb wins best performance on one dataset (BEV). HC-edit-STAXAC wins 4 times (Ecoli, AEF,Vot and Bos) and HC-edit-UPGMA wins 2 times (Bld, and Bos). Overall, the result suggests a superiority of HC-edit over the traditional k-NN, WilsonProb and Wilson approaches. With respect to compression rate, as expected, HC-edit removes fewer examples than other methods.

5 Conclusion

Editing improves the accuracy of the k-NN classifier; however, current editing methods tend to remove too many examples from the training set which does not always lead to optimum accuracy rate. We proposed a new editing algorithm, called HC-edit that identifies regions in the dataset of high purity and removes minority examples from the identified regions. HC-edit takes as input hierarchical clusters augmented with purity information in the nodes; which facilitates clusters retrieval—based on user-defined purity values. Traditional k-nearest neighbor methods used the k parameter for both editing and classification. By allowing two parameters for modeling —purity for editing, and k for the classification—, HC-edit provides greater landscape for a model selection. Experiments over seven data sets have been carried out in order to evaluate the performance of the new editing approach. HC-edit's performance has been compared with that of other traditional techniques. The experiments reveal that the HC-edit has improved accuracy while removing less of examples.

References

1. Dasarathy, B.V.: Nearest Neighbor (NN) Norms: NN Pattern Classification Techniques. Mc Graw-Hill Computer Science Series. IEEE Computer Society Press, Los Alamitos (1991)
2. Devijver, P.A., Kittler, J.: Pattern Recognition: A Statistical Approach. Prentice Hall, Englewood Cliffs (1982)
3. Devijver, P., Kittler, J.: On the edited nearest neighbor rule. In: IEEE 1980 Pattern Recognition, vol. 1, pp. 72–80 (1980)
4. Eick, C.F., Zeidat, N., Vilalta, R.: Using representative-based clustering for nearest neighbor dataset editing. In: ICDM, pp. 375–378 (2004)
5. Eick, C.F., Zeidat, N., Zhenghong, Z.: Supervised clustering – algorithms and benefits. In: Proceedings of the 16th IEEE International Conference on Tools with Artificial Intelligence (ICTAI 2004), Boca Raton, Florida, pp. 774–776, November 2004
6. UCI Repository of Machine Learning. http://archive.ics.uci.edu/ml/
7. Sánchez, J.S., Barandela, R., Marqués, A.I., Alejo, R., Badenas, J.: Analysis of new techniques to obtain quality training sets. Pattern Recogn. Lett. **24**, 1015–1022 (2003)
8. Wilson, D.L.: Asymptotic properties of nearest neighbor rules using edited data. IEEE Trans. Syst. Man Cybern. **2**, 408–420 (1972)

9. Vázquez, F., Sánchez, J., Pla, F.: A stochastic approach to wilson's editing algorithm. In: Marques, J.S., Pérez de la Blanca, N., Pina, P. (eds.) IbPRIA 2005. LNCS, vol. 3523, pp. 35–42. Springer, Heidelberg (2005)
10. http://jgi.doe.gov/. Accessed May 2015

Ontology-Based Topic Labeling
and Quality Prediction

Heidar Davoudi$^{(\boxtimes)}$ and Aijun An

Department of Electrical Engineering and Computer Science,
York University, Toronto, Canada
{davoudi,aan}@cse.yorku.ca

Abstract. Probabilistic topic models based on Latent Dirichlet Allocation (LDA) are increasingly used to discover hidden structure behind big text corpora. Although topic models are extremely useful tools for exploring and summarizing large text collections, most of time the inferred topics are not easy to understand and interpret by human. In addition, some inferred topics may be described by words that are not much relevant to each other and are thus considered low quality topics. In this paper, we propose a novel method that not only assigns a label to each topic but also identifies low quality topics by providing a reliability score for the label of each topic. Our rationale is that a topic labeling method cannot provide a good label for a low quality topic, and thus predicting label reliability is as important as topic labeling itself. We propose a novel measure (Ontology-Based Coherence) that can assess coherence of topics with respect to an ontology structure effectively. Empirical results on a real dataset and our user study show that the proposed predictive model using the defined measures can predict the label reliability better than two alternative methods.

Keywords: Topic modeling · Topic labeling · Labeling reliability

1 Introduction

Despite the fact that topic models (e.g., LDA [1]) are quite useful tools to explore and summarize the corpus, most of time the resulting latent (topic) space is not easily to interpret/evaluate by human. One approach to address this problem is to assign a label to each topic automatically [2–4]. However, most of automatic topic labeling methods do not provide information about the reliability for assigned labels. In other words, while resulting topics consist of some fused (i.e., more than one concept [5]) and random topics (i.e., no clear connection between the words [6]), automatic topic labeling schemes only deal with topic assignment and have nothing to do with these kinds of topics. Intuitively, *an automatic topic labeling method cannot assign a good label to a low quality topic.* As such, predicting reliability for the assigned labels is as important as the label assignment. Following this idea, we propose a novel automatic topic labeling

© Springer International Publishing Switzerland 2015
F. Esposito et al. (Eds.): ISMIS 2015, LNAI 9384, pp. 171–179, 2015.
DOI: 10.1007/978-3-319-25252-0_18

technique that not only assigns a label to each topic but also predicts the reliability of assigned labels. We define a novel coherence measure (Ontology-Based Coherence) based on the external ontology with a Directed Acyclic Graph (DAG) structure. The proposed predictive model utilizes this measure as the external labeling coherence criterion and average Normalized Pairwise Mutual Information (NPMI) as the internal topic coherence measure to predict the reliability of assigned labels. The major contributions of this study can be summarized as follows:

- Predicting topic labels and their reliabilities based on a DAG-structured ontology.
- Defining an appropriate measure to calculate the coherence of assigned label with respect to the external source (ontology).
- Designing a model based on both internal topic coherence (average NPMI) and the external labeling coherence (based on specificity) to calculate the reliability of predicted labeled.
- Empirical studies in a real dataset and respective user study show the predictive power of our proposed model.

2   Related Work

Topic labeling and evaluation tasks are conceptually related even though little studies have been done on topic labeling. For example, Mei et al. proposed a probabilistic framework for automatic topic labeling [2]. Their model extracted the meaningful phrases as the labels and then ranked them based on an scoring function. They defined the scoring function based on notion of good label (i.e., understandable, semantically relevant and distinguishable), but their model still cannot produce the labels which are the hypernyms of topic words [7]. Lau et al. [3] generated label candidates form Wikipedia (by querying the top topic words). They used a set of existing features (e.g., PMI, student t-test) and supervised and unsupervised methods to rank the candidates. In the unsupervised approach, each individual score (feature) between the label and topic words served as the ranking measure while in the supervised method, combination of features and users ratings were utilized to build a regression model. Meggati et al. [4] labeled topics based on alignment with hierarchal ontology (Google Directory). They used 6 similarity measures to find the best nodes in the tree. The final algorithm tried to find the best node in the tree by some heuristic rules even though the experiment was preliminary and did not show different aspects of the proposed algorithm. In another study, Newman et al. [7] conducted a large scale user study and investigated the correlation between PMI and human score for each topics. They showed that although top-10 topic words were reasonably good to be used as a part of label, they were not generally sufficient for the entire label. Musat et al. [8] suggested some coherence measures (based on coverage and specificity) with respect to an external ontology. The primary purpose of the paper was to evaluate the topics (and not labeling) and their measurements were based on a tree structure ontology. While all these studies dealt with either evaluation or labeling tasks, our proposed method assigns labels and predicts their reliability based on the novel measures defined on a DAG structured ontology.

3 Problem Formulation

In this section, we formally define *topic, ontology,* and *topic labeling*. Given document collection D and vocabulary V, a topic t extracted by LDA is multinomial distribution over words. Usually top k probable words are considered as representative keywords for topic t.

Definition 1 (Topic): Topic t is defined by set $T_t = \{(w_{it}, p_{it}) : w_{it} \in V, p_{it} = P(w_{it}|t)\}$ where $P(w_{it}|t)$ is a multinomial distribution over words.

The basic idea behind topic models is that documents are mixtures of topics and each topic is a distribution over words.

Definition 2 (Topic Ontology): A topic ontology $O = (C, R)$ is a *Directed Acyclic Graph (DAG)*, where each vertex $c_i \in C$ represents a concept and each edge $r_{ij} \in R$ (from c_i to c_j) represents that concept c_j is a subtopic for concept c_i. Unlike a tree structure, an ontology may have more than one root (i.e., a node without parents). The root nodes can be considered as the most generic concepts.

Given top k keywords for topic t, a topic labeling procedure aims to assign the most semantically meaningful label to topic t.

Definition 3 (Topic Labeling): Suppose $Topics_U = \{T_t : t = 1, 2, ..., n\}$ represents the set of topics produced by the topic modeling method (i.e., LDA), topic labeling produces a set $Topics_L = \{(l_t, a_t) : t = 1, 2, ..., n\}$ where l_t is the label assigned to topic T_t and a_t is a reliability score for l_t.

4 The Proposed Technique

4.1 Labeling Score

We utilized Normalized *PMI (NPMI)*, which has been used successfully in a wide range of applications [2,3,7,9], to define the coherence score between the top words (based on the probability of the words in the topic) in each topic and the concepts in the ontology. Suppose that p_{it} is the probability of w_{it} (the $i'th$ word in topic t) and c is a concept in the ontology $O = (C, R)$, the coherence score between topic t and concept c can be defined as follows:

$$Score(t, c) = \frac{1}{k\,|label(c)|} \sum_{i=1}^{k} \sum_{q \in label(c)} p_{it}\, NPMI(w_{it}, q) \qquad (1)$$

where k is the number top words in each topic used for calculating the score. The label for topic t can be calculated as follows:

$$l_t = \arg\max_{c \in C} Score(t, c) \qquad (2)$$

Intuitively, we use the probability of each word in a topic as the weight in computation of semantics similarity between topics and ontology concepts. While LDA explores the co-occurrence between the words in a particular topic, the proposed score tries to measure the co-occurrence between the words in a topic and the words in a particular ontology.

4.2   Ontology-Based Specificity (OBS)

Intuitively, the specificity of a concept can be defined based on the ontology structure. Suppose $G = (C, R)$ is an ontology and $c \in C$ is a concept reachable from some roots by following a set of edges in R. We define Ontology-Based Specificity (OBS) of concept c as:

$$OBS(c) = \frac{1}{|R|} \sum_{i \in R} depth_i(c) \qquad (3)$$

where $depth_i(c)$ is the length of path from root i to concept c added to 1 (OBS of each root node equals to 1).

4.3   Ontology-Based Coherence (OBC)

Assume $G = (C, R)$ is an ontology and $P(t) \subseteq C$ is the set of top m labels (i.e., concepts) for topic t (that is, for each $c \in P(t)$, $Score(t, c)$ is among the m highest scores for t). We define the cohesion of topic t based on the specificity of the most specific common ancestor of its top m labels in the ontology. The Ontology-Based Coherence (OBC) of topic t is defined as:

$$OBC(t) = \max_{cp \in CP(t)} OBS(cp) \qquad (4)$$

where $CP(t)$ is a set of ancestors that are common among the members of $P(t)$:

$$CP(t) = \bigcap_{p \in P(t)} Ancestor(p) \qquad (5)$$

Note that in case of $CP(t) = \emptyset$, the OBS measure is defined as 0.

4.4   Average NPMI

We use the average NPMI among the top k words in a topic as the internal measure of topic coherence. The average NPMI can be written as follows:

$$AvgNPMI(t) = \frac{1}{k\,(k-1)} \sum_{i=1}^{k} \sum_{\substack{j=1 \\ j \neq i}}^{k} Npmi(w_{it}, w_{jt}) \qquad (6)$$

where w_{it}/w_{jt} are the $i'th/j'th$ words of topic t respectively.

Input: $Topics_U, O = (C, R)$	$\triangleright$ $Topics_U$ is a set of unlabeled topics
	$\triangleright$ O is an ontology
Output: $Topics_L$	$\triangleright$ $Topics_L$ is a set of labeled topics

```
 1: function TOPICLABELING(Topics_U, O)
 2:     Topics_L ← ∅
 3:     for each topic t ∈ Topics_U do
 4:         Calculate Score(t, c) using Equation (1)
 5:         l_t ← arg max Score(t, c)                    ▷ Label assignment
                 c∈C
 6:         P(t) ← top m concepts c ∈ C based on Score(t, c)
 7:         f_1(t) ← OBC(t)                              ▷ (Equation 4)
 8:         f_2(t) ← Avg NPMI(t)                         ▷ (Equation 6)
 9:         Calculate a_t using Equation (7)            ▷ Label accuracy prediction
10:         Topics_L ← Topics_L ∪ {(l_t, a_t)}
11:     end for
12:     Return Topics_L
13: end function
```

Fig. 1. Proposed topic assignment algorithm

4.5 Building Predictive Model

We have defined two coherence measures as quality measures for a topic. In this section, we show how these measures can be combined to estimate the reliability of a topic and its label. Let $f_1(t) = OBC(t)$ and $f_2(t) = AvgNPMI(t)$. We can define the reliability score of the assigned label (l_t) to topic t using the linear combination of these features (*i.e.*, $\lambda f_1(t) + (1 - \lambda)f_2(t)$) where $\lambda \in [0, 1]$ can be set by the user to indicate the relative importance of each measure.

Alternatively, we can use a supervised approach that learns the importance of each measure using a set of training data if it is available (which is our method). The training data contain a set of topics, their labels, their $OBCs$, their $AveNPMIs$ and the average user evaluation scores (given to each topic by a group of users, ranging from 0 to 1). We use the following logistic regression model to predict the reliability of the assigned label to topic t:

$$Pr(l_t \text{ is ``reliable''} | w_1, w_2, w_3) = \frac{1}{1 + \exp^{-(w_1 f_1(t) + w_2 f_2(t) + w_3)}} \tag{7}$$

The model parameters can be learned from the training data using the maximum likelihood estimation method [10].

Figure 1 illustrates the overall algorithm of our proposed method. Step 5 determines the best label for topic t according to score calculated in (step 4). In step 6, we find the top m concepts $(P(t))$ in the ontology for topic t, then the OBC and average PMI for topic (t) are calculated respectively (step 7–8). Finally, the accuracy of predicted label is calculated using the trained predictive model described by (7). In particular, our proposed labeling approach not only assigns a label to each topic but also estimates the reliability of labels.

5   Empirical Evaluation

5.1   Dataset and Experimental Setup

For evaluation, we use Dow Jones Intelligent Indexing System[1] as the ontology. The structure of Dow Jones ontology is similar to what is described in Definition 2. The corpus used in our experiments is 359145 on-line news published by The Globe and Mail[2] newspaper between the year 2010 to 2014. In the user-study, 5 users are hired and asked to classify each topic into tree levels (i.e., 1, 0.5 and 0 correspond to *good, neutral* and *bad*). For the predictive model training, average rates greater than/less than or equal to 0.5 are considered as good/bad labels respectively. Topics are extracted by Mallet software [11] using default parameter setting (k=19) and for OBS measure we considered top 3 words ($m=3$). The performance of predictive model evaluated using 10-fold cross validation method.

Table 1. Top-10 words of first five topics extracted from the globe and mail dataset

Topic 1	drug, health, drugs, people, disease, treatment, cases, hiv, virus, flu
Topic 2	debt, bond, bonds, credit, investors, interest, government, rates, yield, yields
Topic 3	event, events, royal, day, king, queen, prince, wedding, visit, couple
Topic 4	people, don, ve, lot, things, make, thing, kind, ll, doesn
Topic 5	hotel, travel, park, mountain, beach, resort, trip, town, visit, tourism

Table 2. Top labels of topics illustrated in Table 1

Human Rate	Avg NPMI	OBC	1'st Label	2'nd label	3rd label
0.90	**0.27**	**0.13**	Health	AIDS/HIV	Hepatitis
1.00	**0.24**	**0.38**	Debt	Corporate Debt	Government Debt
0.80	0.70	0.00	Calendar of Events	Festivals	Dance/Ballet
0.40	0.09	0.00	People Profile	Society/Community/Work	School
1.00	**0.21**	**0.25**	Travel	All Inclusive Resorts	Cruises

---

[1] http://new.dowjones.com/dj-intelligent-indexing/.
[2] http://www.theglobeandmail.com.

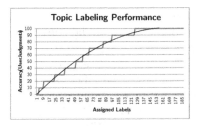

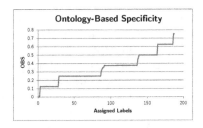

Fig. 2. Users judgments on assigned labels

Fig. 3. OBS of assigned labels

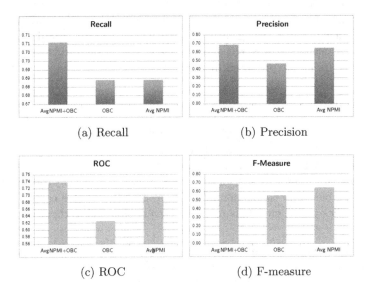

(a) Recall

(b) Precision

(c) ROC

(d) F-measure

Fig. 4. Performance of proposed predictive model

5.2 Experimental Results

Table 2 shows the top-3 labels assigned to the topics illustrated in Table 1 respectively. While topics 1, 2 and 5 have reasonably good labels (i.e., Health, Debt, Travel), topics 3 and 4 are not coherent enough and as a result assigned labels are not semantically meaningful. It can be seen that average NPMI and Ontology-Based Coherent (OBC) measures are quite appropriate measures to assess the coherence of topics and assigned labels (OBC is zero for topic 3 and 4) and are correlated to average human judgments (column 1). Figure 2 demonstrates the average users judgments (between 0 to 100 % corresponding to *bad* to *good* label) about the labels. We observe that only 27 % of assigned labels have less than 50 % accuracy. Moreover, 30 % of labeled is evaluated by all user as good labels. Figure 3 shows the assigned label specificities distribution according to the OBS measure. As it can be seen, proposed method assigns labels with variety of specificities and in fact, the label assignment method does not have a bias towards the generic labels (generic labels have a good coverage and are hyperonym of many topics, so it is very likely to be chosen by coverage-based topic labeling methods). Figure 4 illustrates the performance of proposed label reliability prediction method based on different criteria. As it can be seen, designed predictive models using internal topic coherence (average NPMI) and external label assignment coherence (OBC) is consistently better than the case when one measure is utilized. While in terms of recall average PMI is slightly better than OBC (Fig. 4a), average NPMI precision is 18 % better than OBC precision (Fig. 4b). Interestingly, in both cases, proposed method (Average NPMI + OCE) is the best (with recall/precision equal to 0.71/0.69 respectively). Moreover, analysis using Roc curve demonstrates that proposed method (with ROC equal to 0.74) provides the best balance between true positive and false positive rates. Similarly, the best F-measure performance (at 0.69) signifies that best trade-off between precision and recall occurs in the proposed method.

6   Conclusions

In this paper, we proposed a novel approach for both topic labeling and labeling reliability prediction with respect to an external ontology. In particular, we defined some measures based on the notion of coherence which can serve as the both labeling and reliability assessment criteria. We showed that the logistic prediction model based on the combination of proposed measures can evaluate the quality of topics and their labels more effectively than using them individually.

Acknowledgement. This research is supported by the Center for Innovation in Information Visualization and Data Drive Design (CIVDDD), a CRD Grant from Natural Sciences and Engineering Research Council of Canada (NSERC) and The Globe and Mail. We thank The Globe and Mail for providing the dataset and ontology used in this research

References

1. Blei, D.M., Ng, A.Y., Jordan, M.I.: Latent dirichlet allocation. J. Mach. Learn. Res. **3**, 993–1022 (2003)
2. Mei, Q., Shen, X., Zhai, C.: Automatic labeling of multinomial topic models. In: Proceedings of the 13th ACM SIGKDD International Conference on Knowledge Discovery and Data Mining, pp. 490–499. ACM (2007)
3. Lau, J.H., Grieser, K., Newman, D., Baldwin, T.: Automatic labelling of topic models. In: Proceedings of the 49th Annual Meeting of the Association for Computational Linguistics: Human Language Technologies, vol. 1, pp. 1536–1545. Association for Computational Linguistics (2011)
4. Magatti, D., Calegari, S., Ciucci, D., Stella, F.: Automatic labeling of topics. In: 2009 Ninth International Conference on Intelligent Systems Design and Applications. ISDA'09, pp. 1227–1232. IEEE (2009)
5. Chuang, J., Gupta, S., Manning, C., Heer, J.: Topic model diagnostics: assessing domain relevance via topical alignment. In: Proceedings of the 30th International Conference on Machine Learning (ICML-13), pp. 612–620 (2013)

6. Mimno, D., Wallach, H.M., Talley, E., Leenders, M., McCallum, A.: Optimizing semantic coherence in topic models. In: Proceedings of the Conference on Empirical Methods in Natural Language Processing, pp. 262–272. Association for Computational Linguistics (2011)

7. Newman, D., Noh, Y., Talley, E., Karimi, S., Baldwin, T.: Evaluating topic models for digital libraries. In: Proceedings of the 10th Annual Joint Conference on Digital Libraries, pp. 215–224. ACM (2010)

8. Musat, C., Velcin, J., Trausan-Matu, S., Rizoiu, M.A.: Improving topic evaluation using conceptual knowledge. In: 22nd International Joint Conference on Artificial Intelligence (IJCAI), vol. 3, pp. 1866–1871 (2011)

9. Newman, D., Lau, J.H., Grieser, K., Baldwin, T.: Automatic evaluation of topic coherence. In: Human Language Technologies: The 2010 Annual Conference of the North American Chapter of the Association for Computational Linguistics, pp. 100–108. Association for Computational Linguistics (2010)

10. Murphy, K.P.: Machine learning: a probabilistic perspective. MIT Press, Cambridge (2012)

11. McCallum, A.K.: Mallet: a machine learning for language toolkit (2002). http://mallet.cs.umass.edu

Tweets as a Vote: Exploring Political Sentiments on Twitter for Opinion Mining

Muhammed K. Olorunnimbe[(✉)] and Herna L. Viktor

School of Information Technology and Engineering, University of Ottawa,
Ottawa, ON K1N 6N5, Canada
{molor068,hviktor}@uottawa.ca

Abstract. Twitter feeds provide data scientists with a large repository for entity based sentiment analysis. Specifically, the tweets of individual users may be used in order to track the ebb and flow of their sentiments and opinions. However, this domain poses a challenge for traditional classifiers, since the vast majority of tweets are unlabeled. Further, tweets arrive at high speeds and in very large volumes. They are also suspect to change over time (so-called concept drift). In this paper, we present the PyStream algorithm that addresses these issues. Our method starts with a small annotated training set and bootstraps the learning process. We employ online analytic processing (OLAP) to aggregate the opinions of the individuals we track, expressed in terms of the votes they would cast in a national election. Our results indicate that we are able to capture the sentiments of individuals as they evolve over time.

Keywords: Sentiment analysis · Entity analytics · Adaptive learner · Semi-supervised learning · Concept drift

1 Introduction

Twitter has become an important social media platform for political activism and for the expression of opinions, 140 characters at a time. With the #Occupy-WallStreet movement and the #ArabSpring proliferated through twitter, there is no doubt that this platform has empowered many ordinary citizens of various countries around the world to express their opinions. As such, it provided a source for role players and data scientists to study and to predict user sentiments. This paper concerns the use of sentiment analysis to predict the votes of individuals in a country-wide election, through exploring their aggregated sentiments on particular candidates, as collected over time. Our political opinion mining case study concerns the presidential election in Nigeria which was concluded on March 28th, 2015. Although there was a total of fourteen candidates that vied for this office, the two main candidates were Goodluck Ebele Jonathan (GEJ) and General Muhammadu Buhari (GMB); we focus our attention on these two.

Analyzing such twitter data is a challenging task. It follows that the contents of twitter traffic are not expected to remain constant. This is a phenomena

© Springer International Publishing Switzerland 2015
F. Esposito et al. (Eds.): ISMIS 2015, LNAI 9384, pp. 180–185, 2015.
DOI: 10.1007/978-3-319-25252-0_19

referred to as concept drift and is commonplace in such data streams. In order to facilitate for such changes, one needs to update the training set to keep the feature set relevant for current analysis. To this end, we developed a semi-supervised approach in order to bootstrap the learning process and to facilitate incremental learning. Our PyStream algorithm starts with a small training set, using a subset of the entire dataset. This training dataset is annotated by a domain expert, and used to build initial classification models. The models are used to make predictions on each example of the unlabeled dataset. Next, prediction labels with a probability that is higher than a predefined threshold value are used to augment the training data. The newly updated training dataset is subsequently used to update the classification model, in order to facilitate incremental learning. This paper is organized as follows. Section 2 discusses the data capturing process. Section 3 details our PyStream method and Sect. 4 presents our results. Section 5 concludes the paper.

2 Data Capturing

During the period under survey, our PyStream system connected to Twitter through the streaming API, and the tweets to be analyzed were cached to an OLAP database. A total of 11,944,368 tweets were captured between January 14th and April 2nd, 2015. During this period, we collected all tweets with the words 'GMB', 'Buhari', 'GEJ', 'Jonathan'. The majority of these tweets, however, were found to be irrelevant for our purpose. In this research, emphasis is placed on unique tweets from a person relevant to the study. We started by removing all retweets, all tweets generated automatically for marketing or campaign purposes, and all users that are not registered as English language users.

By the time we were done with all of our preprocessing tasks, we were left with 1,205,444 tweets, representing 256,236 unique users. We noticed that there are significantly more users, and in effect, more tweets, between 30th and 31st of March, when the result was announced.

From this set of tweets, the number of our search terms can be seen in Table 1. We used 1020 tweets as training data from the dataset. For an adequate representation of the timeline, these tweets were taken from the total tweets, ordered by time of creation, for each week in the period under survey. Each of these test tweets were manually annotated, for independent sentiments for each of the candidates. As the expert for this task, we consulted with a social media content consultant, based in Nigeria (Table 2).

3 PyStream Methodology

Semi-supervised learning is a form of classification task that is employed in situations with limited number of labeled examples, relative to the unlabeled set [4]. This is particularly useful in a situation where the cost of labeling adequate amounts of training data is high, due to factors such as time, computation or monetary value, as is the case in our task. In our work, we used the Naive Bayes

Table 1. Count of search terms

Term	tweets #
Jonathan	553,529
GEJ	123,507
Buhari	593,035
GMB	160,822

Table 2. Sentiment count

Annotated	Count
Jonathan +	38
Jonathan -	105
Jonathan	874
Buhari +	162
Buhari -	58
Buhari	874

Table 3. Total changes in training dataset over time

Sentiment	Initial set	Total after
Jonathan +	90	44,840
Jonathan -	253	44,841
Jonathan	674	0
Buhari +	254	67,398
Buhari -	114	67,397
Buhari	650	0

classifier, available in the Natural Language Toolkit, NLTK as base learner, since it has been shown to work well in text classification tasks [1]. The Naive Bayes classifier builds a probabilistic model, such that the hypothesis is represented as $p(l_i|t)$, i.e. the probability of obtaining label l_i from the classified text, x. The value of $p(l_i|t)$ is compared to a pre-defined threshold value, and the training data is updated with the new labeled example if the condition is met.

The first step of the feature extraction stage is tokenization. For each tweets, we considered the url (http, https, www.), hashtag (#string), twitter username (@string), abbreviation, ellipsis, currency, amongst others. This list is precisely expressed as regular expressions that we defined for our tokenization. For the tokens of each tweet, uni-gram, bi-gram and tri-gram are created as feature set in the training dataset. After this, the training set is fed to the classifier to generate a model. Once the model is created, the unlabeled tweets from the database are tokenized, and a prediction, which is the label with the highest probability distribution, as generated by the model, is assigned as the sentiment of the tweet. Algorithm 1 shows the pseudocode of our algorithm. Our algorithm proceeds as follows:

Step 1: Recall that 1,020 examples (S) were taken from the dataset ($T \rightarrow \{t_1, t_2, ..., t_n\}$). These examples, S, is given to the domain expert, to annotate as 'Positive', 'Negative' or 'Neutral' ($l_i \in l$). The annotation is done independently (S_1, S_2) for each of the candidates. A different training set is used for each search term, independent of the others. In this study, the search terms {'Buhari', 'GMB'} and {'Jonathan', 'GEJ'} are treated as one for each candidate.

Step 2: Initial values are set for the probability threshold (τ_p), length threshold (τ_l) and reset threshold (τ_r). All threshold values were obtained through preliminary experimentation. Details of these threshold values are shown in below.

Step 3: Based on the above annotation of S, the Naive Bayes classifier is applied, and models $\{h_{m_1}, h_{m_2}\} \in h_m$ are generated for each of the candidates. Having separate models, based on their individual annotations, aids to address the issue of entity based sentiment analysis.

Step 4: For each candidate, the respective model is used to predict the label of each of the tweets in the dataset $\{t_1, t_2, ..., t_n\} \in T$, using a probability.

Algorithm 1

Input: T is the input stream of tweets $\{t_1, t_2, ..., t_n\}$;
   C is text classification algorithm (Naive Bayes);
   S is the initial training set drawn from T, each with label $l_i \in l$, as annotated
   by domain expert (1,020 tweets), separately (S_1, S_2) for each candidate;
   S_τ is a time sequenced, training set update, initialized to $\varnothing$ (unlabeled),
   separately (S_{τ_1}, S_{τ_2}) for each candidate;;
   $x \ni x_i$ is a set of counters $\forall\, l_i \in l$, all initialized to 0;
   τ is a set of threshold parameters $\{\tau_p = probability, \tau_l = length, \tau_r = reset\}$;

```
 1: for each candidate do
 2:     initialize h_m = model induced by C from S
 3:     for each tweet t ∈ T do
 4:         l_i(t) = predicted label for t using model h_m, with probability p(l_i|t)
 5:         if length(S_τ) ≤ τ_l then
 6:             S_τ ← S_τ + (t, l_i)
 7:         else
 8:             remove first index in S_τ
 9:             S_τ ← S_τ + (t, l_i)
10:         x_i ← x_i + 1
11:         if sum(x) mod τ_r = 0 then
12:             Reset h_m with S + S_τ
```

Output: label: $l_i(t)$

Step 5: We set the value of $\tau_p = 99\,\%$, and any tweet with label prediction probability, $p(l_i|t)$, higher than τ_p is stored in the update set, S_τ.

Step 6: We also used the length threshold, τ_l, as the size of S_τ for each candidate. For our experimentation, we set the value of τ_l to 4000. If the value of τ_l is met, the first item in S_τ is removed, and a new one is appended to the end of the list. Since the tweets in T are time ordered, this enables us to keep the training set updates in S_τ relevant to the current discussions.

Step 7: The reset threshold, τ_r, is used to initiate a model reset, when the update window of 20 is met. This means that whenever 20 additional updates have been added to both sets of S_τ, a new model is generated from S and S_τ. That is, when updates to $S_{\tau_1} + S_{\tau_2} = 20$, models h_{m_1}, h_{m_2} are reset with $S_1 + S_{\tau_1}$ and $S_2 + S_{\tau_2}$ respectively, for each candidate.

Step 8: In this experimentation, we are primarily interested in 'Positive' and 'Negative' labels, so those are the ones we keep in the update set, S_τ, when the τ_p condition is met. To ensure an unbiased representation of these labels, we ensure that they are equally updated in S_τ. These enable the labels to over-weight the bias of the 'Neutral' label from the training dataset S. The expertly annotated training examples in S are always kept as part of the model reset. Table 3 shows the distribution of the initial training set, S, as well as the count of the total updates added to S_τ, for the duration of the process. Of this count, a full update set, S_τ, only keeps 2000 for each label, at any particular time.

4   Results and Analysis

Our primary objective in this work was to detect the most likely candidate each individual user will vote for based on their tweets. The label assigned to each of the tweets are aggregated for each candidate. We used an OLAP engine for exploring the results, using Tableau for the OLAP presentation. The OLAP engine enables us to treat each entity independently, and it further allows us to aggregate the user sentiments in relation to each candidate. Based on our objective of detecting each user's overall sentiment on each candidate, we identify each unique user by the maximum polarity of his/her overall tweets.

The distributions of the sentiments per candidate is depicted in Fig. 1. Our first observation is that, although we started with a larger set of neutral labels in the training dataset, a fewer number of tweets were labeled neutral for both models. This can be attributed to the fact the we eventually have more positive and negative labels as a result of the bootstrap method we implemented in our algorithm, which models the evolution of opinions about a specific candidate. Recall that the tweets that are identified as neutral are those with unknown sentiment. This means that their polarity, or 'vote' for the candidates cannot be determined, with regards to the task of aggregating the users' overall sentiments.

The reader should notice that the above values, however, do not indicate the actual user votes, as defined in this experiment. Recall that our aim is to aggregate all the labels for each user, such that the overall sentiment expressed by the classification of a user's tweets, is assumed to be the his/her vote. This means that if a user has a total of 100 Positive and Negative labels during the experimentation timeline, the label with more that 50 % counts is assumed to be the user's opinion, and it is used as the user's vote. Considering the fact that a large number of additional users tweeted after the results were announced, we also limit the aggregation timeline till the election day. Thus we only consider the overall user sentiments, from the start until the date of the election.

From the overall 256,236 users that sent all the tweets across the timeline, only 190,954 were sending messages before the election was concluded. The aggregated result for these users are shown in Fig. 2. As observed, a total of 92.42 % has Positive sentiments towards Buhari, while only 14.01 % has Positive sentiments towards Jonathan. This is quite different from the outcome of the elections, with 54.96 % and 44.96 % for Buhari and Jonathan respectively, from a total of 28,278,083 voters. We believe that this result is due to a number of factors. Firstly, we initially postulated that the entire population is not adequately represented on twitter and that the vocal minority may have a very different profile than the silent majority. This is, indeed, the case and our results thus confirms the observations of [3]. That is, while there is a correlate between activities on twitter and election results, the correlation is not enough to estimate the distribution of the vote. Elections are guided by complicated political and

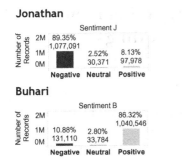

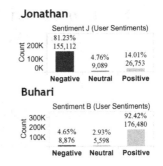

Fig. 1. Polarity of tweets

Fig. 2. Overall user sentiments (till date of election)

socio-economical factors, and our results seem to reinforce this fact. What our results do indicate, however, is that the support of the vocal minority was generally directed towards Buhari, who ended up winning the election. The reader should also notice that the demographic profiles of twitter users indicate that they mostly belong to the younger generation [2]. It would thus be of interest to conduct a longitudinal study, in order to see how these users' opinions evolve.

5   Conclusion and Future Work

In this study, we introduced an algorithm for entity based sentiment analysis. Our incremental learning method employs the aggregated value of user sentiments, based on multiple independent models, in order to predict the political opinions of individuals, as their opinions evolve over time. We annotated a corpus using expert knowledge, and also formulated a method for improving on correctly annotated dataset, by way of bootstrapping the classification results. In this work, we did not weight the tweets based on the date they were created. Our future work will explore ways to give weights to individual tweets, based on the timeline and other socio-political events. We will also formulate a framework that is based on this approach, and can be used on any future topic of interest.

References

1. Medhat, W., Hassan, A., Korashy, H.: Sentiment analysis algorithms and applications: a survey. Shams Eng. J. **4**(5), 1093–1113 (2014)
2. Tumasjan, A., Sprenger, T.O., Sandner, P.G., Welpe, I.M.: Predicting elections with twitter: what 140 characters reveal about political sentiment. In: Proceedings 4th International AAAI Conference on Weblogs & Social Media, pp. 178–185 (2010)
3. Skoric, M., Poor, N., Achananuparp, P., Lim, E-P., Jiang, J.: Tweets and votes: a study of the 2011 Singapore general election. In: Proceedings of the 45th Hawaii International Conference on System Sciences (2012)
4. Chen, Y., Zhang, X., Li, Z., Ng, J.: Search engine reinforced semi-supervised classification and graph-based summarization of microblogs. Neurocomputing **25**(152), 274–286 (2015)

Sentiment Dictionary Refinement
Using Word Embeddings

Aleksander Wawer[(✉)]

Institute of Computer Science, Polish Academy of Science, ul. Jana Kazimierza 5,
01-238 Warszawa, Poland
axw@ipipan.waw.pl

Abstract. Previous works on Polish sentiment dictionaries revealed
the superiority of machine learning on vectors created from word contexts (concordances or word co-occurrence distributions), especially compared to the SO-PMI method (semantic orientation of pointwise mutual
information). This paper demonstrates that this state-of-the-art method
could be improved upon when extending the vectors by word embeddings, obtained from skip-gram language models. Specifically, it proposes a new method of computing word sentiment polarity using feature
sets composed of vectors created from word embeddings and word co-
occurrence distributions. The new technique is evaluated in a number of
experimental settings.

Keyword: Sentiment analysis

1 Introduction

Despite the recent progress in other fields, such as machine learning , dictionaries
of sentiment still occupy an important place in automated sentiment recognition.
Their initial usage can be probably traced to measuring psycholinguistic dimensions of text, as implemented in tools such as LIWC (Linguistic Inquiry and
Word Count) or The General Inquirer [1].

The usability of sentiment lexicons has been recently confirmed in multiple contexts and settings, for instance in tasks such as feature-level Sentiment analysis and
fine-grained sentiment extraction [2]. Also message-level sentiment classification
benefits from sentiment lexicons. In the recent SemEval twitter-level sentiment
recognition task, top performing systems apply machine learning to feature space
composed in part from features computed using sentiment lexicons [3].

The scope of this paper concerns methods of automated generation of sentiment lexicons, and more specifically: how to increase their quality in terms
of precision of automated predictions, provided the most recent state-of-the-art
methods of word embeddings and combining them with existing approaches that
focus on machine learning and corpus-based data.

This work was funded by the National Science Centre of Poland grant nr UMO-
2012/05/N/ST6/03587.

F. Esposito et al. (Eds.): ISMIS 2015, LNAI 9384, pp. 186–193, 2015.
DOI: 10.1007/978-3-319-25252-0_20

The paper is organized as follows. Section 2 discusses the gold standard Polish sentiment lexicon with manual labellings, which is used to evaluate proposed methods. Section 3 is dedicated to word embeddings. Section 3 provides a short overview of this concept, while Sect. 3.1 discusses the corpora used in our work. Section 3.2 presents the design of experiments on using word embedding vectors for predicting sentiment. Their results are presented and discussed in Sect. 3.3, where performance of various models is compared. Section 4 concludes the paper.

2 Polish Sentiment Lexicon with Manual Scorings

This section describes the baseline for the efforts described in our work, namely the sentiment lexicon and methods of automated labeling of word sentiments.

The lexicon was described in [4]. It consists of 1204 word forms, obtained by lexico-syntactic patterns submitted to a search engine. The goal was to acquire words that have high likelihood of carrying sentiment. The words were then labeled for sentiment by two independent linguists. Each of them used a five point scale (strongly negative, negative, neutral, positive, strongly positive).

The experiments described in [4] were focused on comparing two automatic methods of word sentiment estimation. The first was an unsupervised one, based on the well-known SO-PMI formula (proposed by [5] and then extended by [6]). It requires a corpus and two small sets of paradigm words, representatives of each polar class (positive and negative). The second approach was a supervised one, based on the idea of training machine learning classifiers on vectors generated from word contexts. The comparison took into account different context (window) sizes, samples of contexts of various lengths and investigated the impact of morphosyntactic data.

The results reported in [4] indicate superiority of the supervised method. For this reason, the alternative approach based on SO-PMI formula is not considered in our work. Therefore, steps are undertaken to improve the supervised machine learning approach by introducing new data source: vectors of word embeddings.

3 Word Embeddings

Word embeddings is a name for a set of techniques in language modeling where words from the vocabulary are mapped to vectors of real numbers in a low dimensional space, relative to the vocabulary size. Methods of generating this mapping include neural networks, dimensionality reduction of the word co-occurrence matrix and explicit representation of the contexts in which words appear.

Their recent popularity is related to a growing list of natural language processing tasks where they are successfully applied, often achieving state-of-the-art performance.

One of the most popular techniques to obtain word embeddings that outperformed other approaches on analogy tasks is based on skip-gram models with negative sampling as in [7] and [8]. In Sentiment analysis, word embeddings are used to model compositional effects between words. One notable example of application of these techniques to Sentiment analysis (by modeling phrase-level sentiment compositions) has been described in [9].

Skip-grams are a generalization of n-grams in which in addition to allowing adjacent sequences of words, we allow tokens to be skipped. The method of skip-gram modeling is in this case a neural network language model with the training objective to learn word vector representations that predict the nearby words given the current word. One interesting property of this network is that jointly learns not only the statistical language model but also word vector representations (called also word embeddings).

As demonstrated by [10], skip-gram models with negative sampling are implicitly factorizing a word-context matrix, whose cells are the pointwise mutual information (PMI) of the respective word and context pairs. In fact, PMI has been used explicitly in the task of word sentiment estimation as demonstrated in multiple papers.

3.1 Corpora

The most popular implementation of skip-gram models with negative sampling, also used in this paper, is word2vec. It automatically learns word embedding vectors using large amounts of unannotated plain text. The output is a set of vectors (one vector per word) that can be viewed as matrices of size #words * #length. The length of vectors is a parameter of word2vec. In word analogy tasks, length of 100 was demonstrated to perform well.

In our experiments on word embeddings, we used two corpora. The first one is a subset of Polish wikipedia of 2.5 million tokens. Its advantage is lexical richness and mostly clean language, of little errors. The second corpus is perhaps more relevant to our goals due to its more emotional and less formal character. It is a 3-year period subset (2011, 2012 and 2013) of OpenSubtitles data for Polish movies, of around 90 MB of plain text. It has been created as described in [11] – from a free on-line resource of user uploads (www.opensubtitles.org). It contains various genres and time periods and combines features from spoken language corpora and narrative texts, including dialogs, idiomatic expressions, dialectal expressions and slang.

3.2 Predicting Word Sentiment from Word Embedding Vectors

This section describes the result of predicting word sentiment from vectors obtained using skip-gram models with negative sampling. The experiments were carried using two corpora described above: a 2.5 million tokens subset of Wikipedia and a subset of Polish subtitles.

The question that needs to be addressed when creating word embedding vectors is how to treat various morphological forms of the same lexeme, as is the case in morphosyntactically rich languages such as Polish. The problem arises when many surface (orthographical) forms of a word map to one lemma. It becomes a problem when creating word embedding matrices, as neither of the two corpora, nor the word2vec implementation, are designed to preserve mappings between word forms and lemmas. Word embedding vectors are created for every orthographical word form encountered in corpus. However, sentiment information in the existing gold standard Polish sentiment dictionary is assigned to lemmas, not to orthographical word forms. Therefore, the issue that needs to be resolved is how to create word embedding vectors on the level of lemmas that also include the information from all related forms of a word.

In the experiment described in this section, we employ two distinct approaches to this problem:

- Use word embedding only of the lemma form (L).
- Use matrices of all orthographical word forms for a lemma (word forms pulled from the Polimorf morphological dictionary for a given lemma, including lemma itself) and then compute a mean matrix (M).

We carry all experiments in a 10-fold cross-validation scenario and apply Maximum Entropy classifier. We use the liblinear library [12] due to its maturity and overall high performance in text-related classification tasks.

The most notable difference between [4] and our study, that also influences the results, was the size of the data set. We limited the amount of lexemes in our golden sentiment lexicon to 943 by removing those that either occur less than 20 times in the subset of the National Corpus of Polish or do not occur in Wikipedia and subtitles corpora (therefore, no word embedding vectors could be generated).

Table 1. Class frequencies in the golden sentiment lexicon.

Classes	Positive	Neutral	Negative
Frequency	371	284	288

For this distribution, baseline estimation method of the most frequent class, indicates the accuracy at 0.39. The other baseline method of accuracy calculation, stratified random prediction (each classification decision is random sampled with probability corresponding to class frequencies from Table 1) can be estimated at 0.33.

3.3   Results

The results reported in Table 2 were computed on mean matrices of all orthographical word forms for a lemma (M). Table shows number of lemmas, vector sizes, micro precision[1] and macro F1[2] (measured in 10-fold cross-validation).

---

[1] *Micro* means to calculate metrics globally by counting the total true positives, false negatives and false positives. Thus, it takes into account label imbalance. The formula is: $F1 = 2 * (precision * recall)/(precision + recall)$.

[2] *Macro* means to calculate metrics for each label, and find their unweighted mean. This does not take label imbalance into account.

Table 2. Sentiment estimation from matrices of word embeddings (M).

corpus	lemmas	vector size	micro prec	macro F1
subs	831	50	0.528	0.517
subs	831	100	0.555	0.545
subs	831	200	0.577	0.568
subs	831	300	0.565	0.553
wiki	942	50	0.545	0.543
wiki	942	100	0.533	0.533
wiki	942	200	0.544	0.540
wiki	942	300	0.562	0.559
subs+wiki	942	50	0.544	0.541
subs+wiki	942	100	0.581	0.577
subs+wiki	942	200	0.592	0.587
subs+wiki	942	300	0.598	0.595

Reported results outperform by far the random stratified baseline. This demonstrates that word embeddings contain information usable to discriminate word sentiment.

The best results were obtained using a mixed corpora composed of Wikipedia (wiki) and subtitles (subs). The most likely reason behind the best performance of mixed corpora as it was already discussed, it can be explained by lexical richness and varied language types. The optimum vector length was 300. The length could be extended further, for instance to vectors of 500 and more, what may be a promising direction to raise the sentiment recognition quality even further. However, due to their size, Wiki+subs models for vectors longer than 1000 could be difficult to use on typical hardware (as even the vectors of length 300 exceed 9 GB memory).

The Table 3 compares two types of matrices, mean (M) computed from individual matrices of all word forms associated with a lemma, with a matrix of only lemma (L). The table shows matrix type (M or L), and as before: number of lemmas, vector sizes, micro precision and macro F1 (both measured in 10-fold cross-validation). For each corpus type only best performing models selected from Table 2 are presented and contrasted with lemma-only variants.

As expected, the results demonstrate that lemma-only matrices (L) are significantly inferior to those obtained as mean values (M) from all orthographical word form matrices associated to a lemma.

3.4   Sentiment from Word Embeddings and Word Contexts

Finally, we extend the best performing method of estimations from word embeddings by vectors of word contexts as in supervised scenario in [4], where vectors

Table 3. Sentiment estimation from mean (M) and lemma-only (L) matrices of word embeddings.

corpus	type	lemmas	vector size	micro prec	macro F1
subs	M	831	200	0.577	0.568
subs	L	831	200	0.524	0.494
wiki	M	942	300	0.562	0.559
wiki	L	942	300	0.521	0.504
subs+wiki	M	942	300	0.598	0.595
subs+wiki	L	942	300	0.552	0.544

Table 4. Sentiment estimation from mixed feature space of word embeddings and word contexts.

corpus	lemmas	micro prec	macro F1
subs+wiki + contexts	942	0.647	0.638
subs+wiki	942	0.598	0.595

are created from word co-occurrences within a given context of a word that is about to be predicted.

The examination of the performance of word contexts alone falls out of the scope of this paper and has also been described in [4] (however, using a different subset of the golden standard sentiment dictionary).

We apply the optimal values of parameters reported in [4]. Namely, context (window) size of 3 tokens left and right from the target word (whose sentiment was to be predicted), samples of contexts consisting of 300 random concordances. As in [4], we used the balanced 300 million-tokens subset of the National Corpus of Polish and the Poliqarp query engine.

Table 4 compares the best performing word-embedding model (mean (M) computed from individual matrices of all word forms associated with a lemma, mixed corpus type, vectors of 300 elements) with this feature space extended by word context vectors. Table 4 presents number of lemmas, vector sizes, micro precision and macro F1 (both measured in 10-fold cross-validation).

The increase of prediction quality is notable. The results demonstrate clear superiority of the method that utilized all available information: word embeddings and word contexts.

4   Conclusions

The novelty of this paper is to apply word embedding matrices, obtained from the skip-gram models with negative sampling, to the task of word sentiment prediction. Indeed, the experiments demonstrate the advantage of this idea. Even word embedding matrices alone, used as features for classification, reach promising performance in predicting word sentiment. Yet the best results were obtained using a feature space created from two sources, word embedding vectors from mixed corpus of Wikipedia (probably due to lexical richness) and subtitles (less formal, spoken language) merged into one feature space with bag-of-words word contexts, as in [4] (typical word context representations). This is by far the best performing feature space for predicting word sentiment in Polish.

In the future, we plan to continue the work by testing other vector methods of representing word meaning, especially GLOVE [13] and CBOW (continuous bag-of-words) representation and comparing these to word2vec embeddings, tested in this paper. We also intend to investigate the possibility of computing word embeddings on lemmatized corpora. If successful, this could avoid the issue of matching vectors of different orthographic word forms to their corresponding lemma vectors, examined in this paper.

References

1. Stone, P.J., Dunphy, D.C., Ogilvie, D.M., Smith, M.S.: The General Inquirer: A Computer Approach to Content Analysis. MIT Press, Cambridge (1966)
2. Liu, B.: Sentiment Analysis and Opinion Mining. Morgan and Claypool Publishers (2012)
3. Zhu, X., Kiritchenko, S., Mohammad, S.: Nrc-canada-2014: Recent improvements in the sentiment analysis of tweets. In: Proceedings of the 8th International Workshop on Semantic Evaluation (SemEval 2014), Dublin, Ireland, Association for Computational Linguistics and Dublin City University, pp. 443–447, August 2014
4. Wawer, A., Rogozinska, D.: How much supervision? Corpus-based lexeme sentiment estimation. In: 2012 IEEE 12th International Conference on Data Mining Workshops (SENTIRE 2012), Los Alamitos, CA, USA, pp. 724–730. IEEE Computer Society (2012)
5. Turney, P., Littman, M.: Measuring praise and criticism: Inference of semantic orientation from association. ACM Trans. Inf. Syst. **21**, 315–346 (2003)
6. Grefenstette, G., Qu, Y., Evans, D.A., Shanahan, J.G.: In: Validating the Coverage of Lexical Resources for Affect Analysis and Automatically Classifying New Words along Semantic Axes. Springer, Netherlands (2006)
7. Mikolov, T., Sutskever, I., Chen, K., Corrado, G., Dean, J.: Distributed representations of words and phrases and their compositionality. In: Proceedings of NIPS (2013)
8. Mikolov, T., Chen, K., Corrado, G., Dean, J.: Efficient estimation of word representations in vector space. In: Proceedings of Workshop at ICLR (2013)
9. Socher, R., Perelygin, A., Wu, J., Chuang, J., Manning, C., Ng, A., Potts, C.: Recursive deep models for semantic compositionality over a sentiment treebank. In: Conference on Empirical Methods in Natural Language Processing (2013)
10. Levy, O., Goldberg, Y.: Neural word embedding as implicit matrix factorization. In: The Twenty-Eighth Annual Conference on Neural Information Processing Systems (NIPS 2014) (2014)
11. Tiedemann, J.: Parallel data, tools and interfaces in opus. In: Chair, N.C.C., Choukri, K., Declerck, T., Dogan, M.U., Maegaard, B., Mariani, J., Odijk, J., Piperidis, S., (eds.): Proceedings of the Eight International Conference on Language Resources and Evaluation (LREC 2012), Istanbul, Turkey, European Language Resources Association (ELRA), May 2012

12. Fan, R.E., Chang, K.W., Hsieh, C.J., Wang, X.R., Lin, C.J.: LIBLINEAR: A library for large linear classification. J. Mach. Learn. Res. **9**, 1871–1874 (2008)
13. Pennington, J., Socher, R., Manning, C.: Glove: global vectors for word representation. In: Proceedings of the 2014 Conference on Empirical Methods in Natural Language Processing (EMNLP), Association for Computational Linguistics, pp. 1532–1543 (2014)

Knowledge Representation,
Semantic Web

The Cube of Opposition and the Complete Appraisal of Situations by Means of Sugeno Integrals

Didier Dubois[1], Henri Prade[1], and Agnès Rico[2]([⊠])

[1] IRIT, Université Paul Sabatier, 118 route de Narbonne,
31062 Toulouse cedex 9, France
[2] ERIC, Université Claude Bernard Lyon 1, 43 bld du 11 novembre,
69100 Villeurbanne, France
agnes.rico@univ-lyon1.fr

Abstract. The cube of opposition is a logical structure that under-lies many information representation settings. When applied to multiple criteria decision, it displays various possible aggregation attitudes. Sit-uations are usually assessed by combinations of properties they satisfy, but also by combinations of properties they do not satisfy. The cube of opposition applies to qualitative evaluation when criteria are weighted as well as in the general case where any subset of criteria may be weighted for expressing synergies between them, as for Sugeno integrals. Sugeno integrals are well-known as a powerful qualitative aggregation tool which takes into account positive synergies between properties. When there are negative synergies between properties we can use the so-called desintegral associated to the Sugeno integral. The paper investigates the use of the cube of opposition and of the if-then rules extracted from these integrals and desintegrals in order to better describe acceptable situations.

1 Introduction

The description of situations (or objects, or items) is usually based on the degrees to which they satisfy properties (or criteria). Sugeno integrals [12,13] are quali-tative integrals first used as aggregation operators in multiple criteria decision. They deliver a global evaluation between the minimum and the maximum of the partial evaluations. The definition of the Sugeno integral is based on a monotonic set function, called capacity or fuzzy measure, which represents the importance of the subsets of criteria.

More recently Sugeno integrals have been used as a representation tool for describing more or less acceptable objects [11] under a bipolar view. In such a context the properties are supposed to be positive, i.e., the global evaluation increases with the partial ratings. But some objects can be accepted because they do not satisfy some properties. So we also need to consider negative properties, i.e., the global evaluation increases when the partial ratings decreases. Hence a pair of evaluations made of a Sugeno integral and a reversed Sugeno integral are

© Springer International Publishing Switzerland 2015
F. Esposito et al. (Eds.): ISMIS 2015, LNAI 9384, pp. 197–207, 2015.
DOI: 10.1007/978-3-319-25252-0_21

used to describe acceptable objects in terms of properties they must have and of properties they must avoid. This reversed integral is a variant of Sugeno integral, called a desintegral. Their definition is based on a decreasing set function called anti-capacity.

Moreover, it was proved that the Sugeno integrals and the associated desintegrals can be encoded as a possibilistic logic base [5,7]. These results have been used for extracting decision rules from qualitative data evaluated on the basis of Sugeno integrals [1]. This paper extends these results to the extraction of decision rules from qualitative data using qualitative desintegrals. These decision rules should help completing the results presented in [11].

Besides, we can distinguish the optimistic part and the pessimistic part of any capacity [8]. It has been recently indicated that Sugeno integrals associated to these capacities and their associated desintegrals form a cube of opposition [9], the integrals being present on the front facet and the desintegrals on the back facet of the cube (each of these two facets fit with the traditional views of squares of opposition). As this cube summarizes all the evaluation options, we may consider the different Sugeno integrals and desintegrals present on the cube in the selection process of acceptable situations.

The paper is organized as follows. Section 2 introduces the cube of opposition and discusses its relevance for multiple criteria aggregation. Section 3 restates the main results on Sugeno integrals, desintegrals, and their logical rule counterparts, before presenting the cube of opposition for Sugeno integrals and desintegrals in Sect. 4. Section 5 takes advantage of the cube for discussing the different aggregation attitudes and their relations.

2   Square and Cube of Opposition in Multiple Criteria Evaluation

The traditional square of opposition [10] is built with universally and existentially quantified statements in the following way. Consider a statement ($\mathbf{A}$) of the form "all P's are Q's", which is negated by the statement ($\mathbf{O}$) "at least one P is not a Q", together with the statement ($\mathbf{E}$) "no P is a Q", which clearly expresses a form of opposition to the first statement ($\mathbf{A}$). These three statements, together with the negation of the last statement, namely ($\mathbf{I}$) "at least one P is a Q" can be displayed on a square whose vertices are traditionally denoted by the letters $\mathbf{A}$, $\mathbf{I}$ (affirmative half) and $\mathbf{E}$, $\mathbf{O}$ (negative half), as pictured in Fig. 1 (where $\overline{Q}$ stands for "not Q").

As can be checked, noticeable relations hold in the square:

- (i) $\mathbf{A}$ and $\mathbf{O}$ (resp. $\mathbf{E}$ and $\mathbf{I}$) are the negation of each other;
- (ii) $\mathbf{A}$ entails $\mathbf{I}$, and $\mathbf{E}$ entails $\mathbf{O}$ (it is assumed that there is at least one P for avoiding existential import problems);
- (iii) together $\mathbf{A}$ and $\mathbf{E}$ cannot be true, but may be false;
- (iv) together $\mathbf{I}$ and $\mathbf{O}$ cannot be false, but may be true.

Changing P into $\neg P$, and Q in $\neg Q$ leads to another similar square of opposition **aeoi**, where we also assume that the set of "not-P's" is non-empty.

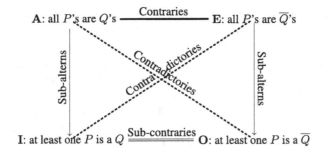

Fig. 1. Square of opposition

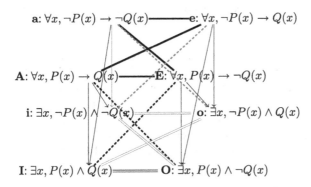

Fig. 2. Cube of opposition of quantified statements

Then the 8 statements, **A, I, E, O, a, i, e, o** may be organized in what may be called a *cube of opposition* [4] as in Fig. 2 (where → denotes material implication).

The front facet and the back facet of the cube are traditional squares of opposition. In the cube, if we also assume that the sets of "Q's" and "not-Q's" are non-empty, then the thick non-directed segments relate contraries, the double thin non-directed segments sub-contraries, the diagonal dotted non-directed lines contradictories, and the vertical uni-directed segments point to subalterns, and express entailments. Stated in set vocabulary, **A, I, E, O, a, i, e, o**, respectively means $P \subseteq Q$, $P \cap Q \neq \emptyset$, $P \subseteq \overline{Q}$, $P \cap \overline{Q} \neq \emptyset$, $\overline{P} \subseteq \overline{Q}$, $\overline{P} \cap \overline{Q} \neq \emptyset$, $\overline{P} \subseteq Q$, $\overline{P} \cap Q \neq \emptyset$. In order to satisfy the four conditions of a square of opposition for the front and the back facets, we need $P \neq \emptyset$ and $\overline{P} \neq \emptyset$. In order to have the inclusions indicated by the diagonal arrows in the side facets, we need $Q \neq \emptyset$ and $\overline{Q} \neq \emptyset$ as further normalization conditions.

Suppose P denotes a set of important properties, Q a set of satisfied properties (for a considered object). Vertices **A, I, a, i** correspond respectively to 4 different cases: (i) all important properties are satisfied, (ii) at least one important property is satisfied, (iii) all satisfied properties are important, (iv) at least one non satisfied property is not important.

Note also the cube is compatible with an understanding having a bipolar flavor [3]. Suppose that among possible properties for the considered objects, some are desirable (or requested) and form a subset R and some others are excluded (or undesirable) and form a subset E. Clearly, one should have $E \subseteq \overline{R}$. For a considered object the set of properties is partitioned into the subset of satisfied properties S and the subset $\overline{S}$ of properties not satisfied. Then vertex **A** corresponds to $R \subseteq S$ and **a** to $\overline{R} \subseteq \overline{S}$. Then **a** also corresponds to $E \subseteq \overline{S}$.

More generally, satisfaction of properties may be graded, and importance (both with respect to desirability and undesiraribility) is also a matter of degree. It is the case in multiple criteria aggregation where objects are evaluated by means of a set $\mathcal{C}$ of criteria i (where $1 \leq i \leq n$). Let us denote by x_i the evaluation of a given object for criterion i, and $x = (x_1, \cdots, x_i, \cdots, x_n)$. We assume here that $\forall i, x_i \in [0,1]$. $x_i = 1$ means that the object fully satisfies criterion i, while $x_i = 0$ expresses a total absence of satisfaction. Let $\pi_i \in [0,1]$ represent the level of importance of criterion i. The larger π_i the more important the criterion.

Simple qualitative aggregation operators are the weighted min and the weighted max [2]. The first one measures the extent to which all important criteria are satisfied; it corresponds to the expression $\bigwedge_{i=1}^{n} (1 - \pi_i) \vee x_i$, while the second one, $\bigvee_{i=1}^{n} \pi_i \wedge x_i$, is optimistic and only requires that at least one important criterion be highly satisfied. Under the hypothesis of the double normalization ($\exists i, \pi_i = 1$ and $\exists j, \pi_j = 0$) and the hypothesis $\exists r, x_r = 1$ and $\exists s, x_s = 0$, weighted min and weighted max correspond to vertices **A** and **I** of the cube on Fig. 3.

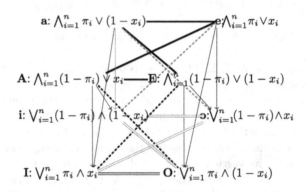

a: $\bigwedge_{i=1}^{n} \pi_i \vee (1 - x_i)$ ——— e: $\bigwedge_{i=1}^{n} \pi_i \vee x_i$

A: $\bigwedge_{i=1}^{n} (1 - \pi_i) \vee x_i$ ——— E: $\bigwedge_{i=1}^{n} (1 - \pi_i) \vee (1 - x_i)$

i: $\bigvee_{i=1}^{n} (1 - \pi_i) \wedge (1 - x_i)$ ——— o: $\bigvee_{i=1}^{n} (1 - \pi_i) \wedge x_i$

I: $\bigvee_{i=1}^{n} \pi_i \wedge x_i$ ——— O: $\bigvee_{i=1}^{n} \pi_i \wedge (1 - x_i)$

Fig. 3. Cube of weighted qualitative aggregations

There is a correspondence between the aggregation functions on the right facet and those on the left facet, replacing x with $1 - x$.

Suppose that a fully satisfactory object x is an object with a global rating equal to 1. Vertices $\mathbf{A}, \mathbf{I}, \mathbf{a}$ and $\mathbf{i}$ correspond respectively to 4 different cases: x is such that

(i) **A**: all properties having some importance are fully satisfied (if $\pi_i > 0$ then $x_i = 1$ for all i).

(ii) **I**: there exists at least one important property i fully satisfied ($\pi_i = 1$ and $x_i = 1$),

(iii) **a**: all somewhat satisfied properties are fully important (if $x_i > 0$ then $\pi_i = 1$ for all i)

(iv) **i**: there exists at least one unimportant property i that is not satisfied ($\pi_i = 0$ and $x_i = 0$). These cases are similar to those presented in the cube on Fig. 2.

Example 1. We consider $\mathcal{C} = \{1, 2, 3\}$ and $\pi_1 = 0$, $\pi_2 = 0.5$ and $\pi_3 = 1$; see Fig. 4.

- on vertex A (resp. I) a fully satisfied object is such that $x_2 = x_3 = 1$ (resp. $x_3 = 1$),
- on vertex a (resp. i) a fully satisfied object is such that $x_1 = x_2 = 0$ (resp. $x_1 = 0$).

The operations of the front facet of the cube of Fig. 3 merge positive evaluations that focus on the high satisfaction of important criteria, while the local ratings x_i on the back could be interpreted as negative ones (measuring the intensity of faults). Then aggregations yield global ratings evaluating the lack of presence of important fault. In this case, weights are tolerance levels forbidding a fault to be too strongly present. Then the vertices a and i in the back facet are interpreted differently: a is true if all somewhat intolerable faults are fully absent; i is true if there exists at least one intolerable fault that is absent. This framework this involves two complementary points of view, recently discussed in a multiple criteria aggregation perspective [6].

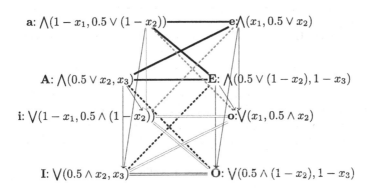

Fig. 4. Example of a cube of weighted qualitative aggregations

3   Sugeno Integrals, Desintegrals and Rules

In the definition of Sugeno integrals the relative weights of the set of properties are represented by a capacity (or fuzzy measure) which is a set function $\mu : 2^{\mathcal{C}} \to L$ that satisfies $\mu(\emptyset) = 0$, $\mu(\mathcal{C}) = 1$ and $A \subseteq B$ implies $\mu(A) \leq \mu(B)$.

In order to translate a Sugeno integral into rules we also need the notions of conjugate capacity, qualitative Moebius transform and focal sets: The conjugate capacity of μ is defined by $\mu^c(A) = 1 - \mu(\overline{A})$ where $\overline{A}$ is the complementary of A.

The inner qualitative Moebius transform of a capacity μ is a mapping $\mu_\# : 2^\mathcal{C} \to L$ defined by

$$\mu_\#(E) = \mu(E) \text{ if } \mu(E) > \max_{B \subset E} \mu(B) \text{ and } 0 \text{ otherwise.}$$

It contains the minimal information characterizing μ ($\mu(A) = \max_{E \subseteq A} \mu_\#(E)$). A set E for which $\mu_\#(E) > 0$ is called a focal set. The set of the focal sets of μ is denoted by $\mathcal{F}(\mu)$. The Sugeno integral of an object x with respect to a capacity μ is originally defined by [12,13]:

$$S_\mu(x) = \max_{\alpha \in L} \min(\alpha, \mu(\{i|x_i \geq \alpha\})). \qquad (1)$$

There are two equivalent expressions used in this article:

$$S_\mu(x) = \max_{A \in \mathcal{F}(\mu)} [\min(\mu(A), \min_{i \in A} x_i)] = \min_{A \in \mathcal{F}(\mu^c)} [\max(\mu(\overline{A}), \max_{i \in A} x_i)]. \qquad (2)$$

The first expression in (2) is the generalisation of a normal disjunctive form from Boolean functions to lattice-valued ones. The second is the generalisation of a normal conjunctive form. Clearly using the first form, $S_\mu(x) = 1$ if and only if there is a focal set E of μ for which $\mu(E) = 1$ and $x_i = 1$ for all i in E. Likewise, using the second form $S_\mu(x) = 0$ if and only if there is a focal set F of μ^c for which $\mu^c(F) = 1$ and $x_i = 0$ for all i in F. The Sugeno integral can then be expressed in terms of if-then rules that facilitate the interpretation of the integral, when it has been derived from data (see [1] for more details).

Selection Rules. Each focal T of μ corresponds to the selection rule:

$$\text{If } x_i \geq \mu(T) \text{ for all } i \in T \text{ then } S_\mu(x) \geq \mu(T).$$

The objects selected by such rules are those satisfying to a sufficient extent all properties present in the focal set appearing in the rule.

Elimination Rules. Each focal set of the conjugate μ^c with level $\mu^c(F)$ corresponds to the following elimination rule:

$$\text{If } x_i \leq 1 - \mu^c(F) \text{ for all } i \in F \text{ then } S_\mu(x) \leq 1 - \mu^c(F).$$

The objects rejected by these rules are those that do not satisfy enough the properties in the focal set of μ^c of some such rule.

When Sugeno integrals are used as aggregation functions for selecting acceptable objects, the properties are considered positive: the global evaluation increases with the partial ratings. But generally, we have also negative properties: the global evaluation increases when the partial ratings decrease. In such a context we can use a desintegral associated to the Sugeno integral. We now presents this desintegral.

In the case of negative properties, weights are assigned to sets of properties by means of an anti-capacity (or anti-fuzzy measure) which is a set function

$\nu : 2^{\mathcal{C}} \to L$ such that $\nu(\emptyset) = 1$, $\nu(\mathcal{C}) = 0$, and if $A \subseteq B$ then $\nu(B) \leq \nu(A)$. Clearly, ν is an anti-capacity if and only if $1 - \nu$ is a capacity. The conjugate ν^c of an anti-capacity ν is an anti-capacity defined by $\nu^c(A) = 1 - \nu(\overline{A})$, where $\overline{A}$ is the complementary of A. The desintegral is defined from the corresponding Sugeno integral, by reversing the direction of the local value scales (x becomes $1 - x$), and by considering a capacity induced by the anti-capacity ν, as follows:

$$S_\nu^\downarrow(x) = S_{1-\nu^c}(1 - x). \tag{3}$$

Based on this identity, we straightforwardly obtain the following rules associated to the desintegral $S_\nu^\downarrow$ from those derived from the integral $S_{1-\nu^c}$:

Proposition 1. Selection Rules. *Each focal T of $1 - \nu^c$ corresponds to the selection rule:*

$$\text{If } x_i \leq \nu^c(T) \text{ for all } i \text{ in } T, \text{ then } S_\nu^\downarrow(x) \geq 1 - \nu^c(T).$$

The objects selected by these rules are those that do not possess, to a high extent (less than $\nu^c(T)$), faults present in the focal set of the capacity $1 - \nu^c$.
Elimination Rules. *Each focal set of $1 - \nu$ corresponds to the elimination rule:*

$$\text{If } x_i \geq 1 - \nu(F) \text{ for all } i \in F \text{ then } S_\nu^\downarrow(x) \leq \nu(F).$$

The objects rejected by these rules are those possessing to a sufficiently large extent the faults in the focal sets of the capacity $1 - \nu$.

Proof. The first result is obvious. For the second, notice that $1 - (1 - \nu^c)^c(A) = \nu(A)$. $\quad\blacksquare$

4   The Cube of Opposition and Sugeno Integrals

When we consider a capacity μ, its pessimistic part is $\mu_*(A) = \min(\mu(A), \mu^c(A))$ and its optimistic part is $\mu^*(A) = \max(\mu(A), \mu^c(A))$ [8]. We have $\mu_* \leq \mu^*$, $\mu_*{}^c = \mu^*$ and $\mu^{*c} = \mu_*$. We need these notions in order to respect the fact that, in the square of opposition, the vertices $\mathbf{A}, \mathbf{E}$ express stronger properties than vertices $\mathbf{I}, \mathbf{O}$.

Proposition 2. *A capacity μ induces the following square of opposition for the associated Sugeno integrals*

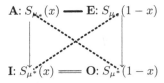

Proof. **A** entails **I** and **E** entails **O** since $\mu_* \leq \mu^*$. **A** and **O** (resp. **E** and **I**) are the negation of each other since for all capacities μ we have the relation:

$$S_\mu(x) = 1 - S_{\mu^c}(1 - x). \tag{4}$$

Let us prove that expressions at vertices **A** and **E** cannot be both equal to 1. Consider x such that $S_{\mu_*}(x) = 1$, hence using Eq. (4), $1 = 1 - S_{\mu^*}(1 - x)$. So $S_{\mu_*}(1 - x) \leq S_{\mu^*}(1 - x) = 0$ which entails $S_{\mu_*}(1 - x) = 0$.

Similarly we can prove that **I** and **O** cannot be false together.

Remark 1. In the above square of opposition, **A** and **E** can be false together and **I** and **O** can be true together. For instance, if μ is a non-fully informed necessity measure N (for instance $\mathcal{F}(N) = \{E\}$, with weight 1, where E is not a singleton), it comes down to the known fact that we can find a subset A such that $N(A) = N(\overline{A}) = 0$, and for possibility measure $\Pi(A) = 1 - N(\overline{A})$ it holds that $\Pi(A) = \Pi(\overline{A}) = 1$.

Note that $S_{\mu_*}(1 - x) = S_{1-\mu^*}(x)$ and $S_{\mu^*}(1 - x) = S_{1-\mu_*}(x)$ where $1 - \mu^*$ $1 - \mu_*$ are anti-capacities. Hence a capacity μ defines a square of opposition where the decision rules on the vertices **A** and **I** (resp. **E** and **O**) are based on a Sugeno integral (resp. desintegral).

In order to present the cube associated to Sugeno integrals we need to introduce the negation of a capacity μ, namely the capacity $\overline{\mu}$ defined as follows: $\overline{\mu}_{\#}(E) = \mu_{\#}(\overline{E})$ and $\overline{\mu}(A) = \max_{\overline{E} \subseteq A} \mu_{\#}(E)$. A square of opposition **aieo** can be defined with the capacity $\overline{\mu}$. Hence we can construct a cube **AIEO** and **aieo** as follows: [9]:

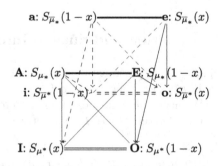

Fig. 5. Cube of opposition associated to μ

Proposition 3. *If $\exists i \neq j \in \mathcal{C}$ such that $x_i = 0, x_j = 1$ and $\{i\}, \{j\}, \mathcal{C}\backslash\{i\}, \mathcal{C}\backslash\{j\}$ are focal sets of μ, then the cube in Fig. 5 is a cube of opposition.*

Proof. **AIEO** and **aieo** are squares of opposition (due to $\mu_* \leq \mu^*$ and $\overline{\mu}_* \leq \overline{\mu}^*$). Let us prove that $S_{\mu_*}(x) \leq S_{\overline{\mu}^*}(1 - x)$ (the arrow **A** → **i** on the left side facet). $S_{\mu_*}(x) = \min_A[\max(\min(\mu(\overline{A}), 1 - \mu(A)), \max_{i \in A} x_i]$ and $S_{\overline{\mu}^*}(1 - x) =$

$\max_A[\min(\max(\overline{\mu}(A), 1 - \overline{\mu}(\overline{A})), \min_{i \in A}(1 - x_i)]$. So we just need to find sets E and F such that one term inside $\min_A$ is less that one term inside $\max_A$. Let us consider $A = \{i\}$ where i is such that $x_i = 0$. We have

$\max(\min(\mu(\overline{A}), 1 - \mu(A)), \max_{i \in A} x_i) = \min(\mu(\mathcal{C}\backslash\{i\}), 1 - \mu(\{i\}))$ and
$\min(\max(\overline{\mu}(A), 1 - \overline{\mu}(\overline{A})), \min_{i \in A}(1 - x_i)) = \max(\overline{\mu}(\{i\}), 1 - \overline{\mu}(\mathcal{C}\backslash\{i\}))$.

Note that $\mu(\mathcal{C}\backslash\{i\}) = \mu_\#(\mathcal{C}\backslash\{i\})$, since $\mathcal{C}\backslash\{i\}$ is focal for μ and $\overline{\mu}(\{i\}) = \overline{\mu}_\#(\{i\}) = \mu_\#(\mathcal{C}\backslash\{i\})$ by definition. So $\min(\mu(\mathcal{C}\backslash\{i\}), 1 - \mu(\{i\})) \leq \mu(\mathcal{C}\backslash\{i\}) = \mu_\#(\mathcal{C}\backslash\{i\}) \leq \max(\overline{\mu}(\{i\}), 1 - \overline{\mu}(\mathcal{C}\backslash\{i\}))$. The inequality $S_{\overline{\mu}_*}(1 - x) \leq S_{\mu^*}(x)$ linking $\mathbf{a}$ with $\mathbf{I}$, is obtained, under the condition $\exists j \in \mathcal{C}$ such that $x_j = 1$, and $\mathcal{C}\backslash\{j\}$ must be a focal set of $\overline{\mu}$, i.e., $\{j\}$ is focal for μ using the Eq. (4). For arrow $\mathbf{E} \rightarrow \mathbf{o}$ and $\mathbf{e} \rightarrow \mathbf{O}$, we need by symmetry to exchange i and j in the above requirements. This concludes the proof.

The cube of opposition is reduced to the facet **AEIO** if and only if $\mu = \overline{\mu}$. So the cube is degenerated if and only if for each focal set $T \neq \mathcal{C}$ of μ, the complement $\overline{T}$ is also focal and $\mu_\#(T) = \mu_\#(\overline{T})$, in other words, capacities expressing ignorance in the sense that $\mu(A) = \mu(\overline{A})$ for all $A \neq \emptyset, \mathcal{C}$. Likewise the cube is reduced to the top facet **AEea**, if μ is self-conjugate, that is, $\mu = \mu^c$, i.e., $\mu(A) + \mu(\overline{A}) = 1$.

5 Discussing Aggregation Attitudes with the Cube

In the following, we characterize situations where objects get a global evaluation equal to 1 using aggregations on the side facet. According to the selection rules, we can restrict to focal sets of the capacity with weight 1.

Proposition 4. *The global evaluations at vertices **AIai** of a cube associated to a capacity μ are maximal respectively in the following situations pertaining to the focal sets of μ:*

A: The set of totally satisfied properties contain a focal set with weight 1 and overlaps all other focal sets.

I: The set of satisfied properties contains a focal set with weight 1 or overlaps all other focal sets.

a: The set of totally violated properties contains no focal set and its complement is contained in a focal set with weight 1.

i: The set of totally violated properties contains no focal set or its complement is contained in a focal set with weight 1.

Proof. **A:** $S_{\mu_*}(x) = 1$ iff there exists a set A such that $\mu_*(A) = 1$ and for all i in $A, x_i = 1$. $\mu_*(A) = 1$ is equivalent to $\mu(A) = 1$ and $\mu^c(A) = 1$, that is $\mu(\overline{A}) = 0$. So A contains a focal set of μ with weight 1 and overlaps all focal sets.

I: $S_{\mu^*}(x) = 1$ iff $\exists A$ such that $\mu^*(A) = 1$ and for all i in $A, x_i = 1$. Since $\mu^*(A) = \max(\mu(A), \mu^c(A)) = 1$, either $\mu(A) = 1$ or $\mu^c(A) = 1$, i.e., $\mu(\overline{A}) = 0$. So A contains a focal set of μ with weight 1 or it overlaps all focal sets.

a: $S_{\overline{\mu}_*}(1 - x) = 1$ if and only if there exists a set A such that $\overline{\mu}_*(A) = 1$ and for all i in $A, x_i = 0$. The condition $\overline{\mu}_*(A) = 1$ reads $\overline{\mu}(A) = 1$ and $\overline{\mu}^c(A) = 1$. The first condition says that $\overline{\mu}_\#(B) = 1$ for some subset B of A that is $\mu_\#(\overline{B}) = 1$, where $\overline{B}$ contains $\overline{A}$. The second condition says that $\overline{\mu}(\overline{A}) = 0 = \mu(A)$. It means that there is no focal set of μ contained in A. So $S_{\overline{\mu}_*}(1 - x) = 1$ if and only if there exists A such that for all i in $A, x_i = 0$, and $\overline{A}$ is contained in a focal set of μ with weight 1, and A contains no focal set of μ.

i: $S_{\overline{\mu}^*}(1 - x) = 1$ iff $\exists A$ such that $\overline{\mu}^*(A) = 1$ and for all i in $A, x_i = 0$. Since $\overline{\mu}^*(A) = \max(\overline{\mu}(A), \overline{\mu}^c(A)) = 1$, either $\overline{\mu}(A) = 1$ or $\overline{\mu}^c(A) = 1$, i.e., $\mu(\overline{A}) = 0$. So, from the previous case we find that $S_{\overline{\mu}^*}(1 - x) = 1$ if and only if for all i in $A, x_i = 0$, and $\overline{A}$ is contained in a focal set of μ with weight 1, or A contains no focal set of μ.

Example 2. Assume $\mathcal{C} = \{p, q, r\}$. We want to select objects with properties p and q or p and r. Hence we have $\mathcal{F}_\mu = \{\{p, q\}, \{p, r\}\}$ with $\mu_\#(\{p, q\}) = \mu_\#(\{p, r\}) = 1$. We can calculate the useful capacities:

capacity	{p}	{q}	{r}	{p,q}	{p,r}	{q,r}	{p,q,r}
$\mu_\#$	0	0	0	1	1	0	0
μ	0	0	0	1	1	0	1
μ^c	1	0	0	1	1	1	1
$\mu^c_\#$	1	0	0	0	0	1	0
$\overline{\mu}_\#$	0	1	1	0	0	0	0
$\overline{\mu}$	0	1	1	1	1	1	1
$\overline{\mu}^c_\#$	0	0	0	0	0	1	0
$\overline{\mu}^c$	0	0	0	0	0	1	1

$\mu^c \geq \mu$ so $\mu_* = \mu$ and $\mu^* = \mu^c$
$\overline{\mu} \geq \overline{\mu}^c$ so $\overline{\mu}^* = \overline{\mu}$ and $\overline{\mu}_* = \overline{\mu}^c$
Note that $\overline{\mu}$ is a possibility measure.

The aggregation functions on the vertices are:
A : $S_\mu(x) = \max(\min(x_p, x_q), \min(x_p, x_r))$
I : $S_{\mu^c}(x) = \max(x_p, \min(x_q, x_r))$
a : $S_{\overline{\mu}^c}(1 - x) = \min(1 - x_q, 1 - x_r)$
i : $S_{\overline{\mu}}(1 - x) = \max(1 - x_q, 1 - x_r)$.

Note that for vertex **A**, the two focal sets overlap so that the first condition of Proposition 4 is met when $S_\mu(x) = 1$. For vertex **I**, one can see that $S_{\mu^c}(x) = 1$ when $x_p = 1$ and $\{p\}$ does overlap all focal sets of μ; the same occurs when $x_q = x_r = 1$. For vertex **a**, $S_{\overline{\mu}^c}(1 - x) = 1$ when $x_q = x_r = 0$, and note that the complement of $\{q, r\}$ is contained in a focal set of μ, while $\{q, r\}$ contains no focal set of μ. For vertex **i**, $S_{\overline{\mu}}(1 - x) = 1$ when, $x_q = 0$ or $x_r = 0$, and clearly, neither $\{q\}$ not $\{r\}$ contain any focal set of μ, but the complement of each of them is a focal set of μ.

6   Concluding Remarks

This paper has shown how the structure of the cube of opposition extends from ordinary sets to weighted min- and max-based aggregations and more generally to Sugeno integrals, which constitute a very important family of qualitative aggregation operators, which moreover have a logical reading. The cube exhausts all the possible aggregation attitudes. Moreover, as mentioned in Sect. 2, it is compatible with a bipolar view where we distinguish between desirable properties and rejected properties. It thus provides a rich theoretical basis for multiple criteria aggregation.

References

1. Dubois, D., Durrieu, C., Prade, H., Rico, A., Ferro, Y.: Extracting decision rules from qualitative data using Sugeno integral: a case study. In: Destercke, S., Denoeux, T. (eds.) ECSQARU 2015. LNCS, vol. 9161, pp. 14–24. Springer, Heidelberg (2015)
2. Dubois, D., Prade, H.: Weighted minimum and maximum operations. An addendum to 'A review of fuzzy set aggregation connectives'. Inf. Sci. **39**, 205–210 (1986)
3. Dubois, D., Prade, H.: An introduction to bipolar representations of information and preference. Int. J. Intel. Syst. **23**(8), 866–877 (2008)
4. Dubois, D., Prade, H.: From Blanché's hexagonal organization of concepts to formal concept analysis and possibility theory. Logica Univers. **6**, 149–169 (2012)
5. Dubois, D., Prade, H., Rico, A.: Qualitative integrals and desintegrals – towards a logical view. In: Torra, V., Narukawa, Y., López, B., Villaret, M. (eds.) MDAI 2012. LNCS, vol. 7647, pp. 127–138. Springer, Heidelberg (2012)
6. Dubois, D., Prade, H., Rico, A.: Qualitative integrals and desintegrals: how to handle positive and negative scales in evaluation. In: Greco, S., Bouchon-Meunier, B., Coletti, G., Fedrizzi, M., Matarazzo, B., Yager, R.R. (eds.) IPMU 2012, Part III. CCIS, vol. 299, pp. 306–316. Springer, Heidelberg (2012)
7. Dubois, D., Prade, H., Rico, A.: The logical encoding of Sugeno integrals. Fuzzy Sets Syst. **241**, 61–75 (2014)
8. Dubois, D., Prade, H., Rico, A.: On the informational comparison of qualitative fuzzy measures. In: Laurent, A., Strauss, O., Bouchon-Meunier, B., Yager, R.R. (eds.) IPMU 2014, Part I. CCIS, vol. 442, pp. 216–225. Springer, Heidelberg (2014)
9. Dubois, D., Prade, H., Rico, A.: The cube of opposition. A structure underlying many knowledge representation formalisms. In: Proceedings of the 24th International Joint Conf. on Artificial Intelligence (IJCAI 2015), Buenos Aires, 25–31 July, pp. 2933–2939 (2015)
10. Parsons, T.: The traditional square of opposition. In: Zalta, E.N. (ed.) The Stanford Encyclopedia of Philosophy (2008)
11. Prade, H., Rico, A.: Describing acceptable objects by means of Sugeno integrals. In: Proceedings of the 2nd IEEE Internatonal Conference on Soft Computing and Pattern Recognition (SoCPaR 2010), Cergy Pontoise, Paris, 7–10 December, pp. 6–11 (2010)
12. Sugeno, M.: Theory of Fuzzy Integrals and its Applications, Ph.D. Thesis, Tokyo Institute of Technology, Tokyo (1974)
13. Sugeno, M.: Fuzzy measures and fuzzy integrals. A survey. In: Gupta, M.M., Saridis, G.N., Gaines, B.R. (eds.) Fuzzy Automata and Decision Processes, pp. 89–102. North-Holland (1977)

Model Checking Based Query and Retrieval in OpenStreetMap

Tommaso Di Noia, Marina Mongiello$^{(\boxtimes)}$, and Eugenio Di Sciascio

Dipartimento di Ingegneria Elettrica e Dell'informazione, Politecnico di Bari,
Via E. Orabona, 4-70125 Bari, Italy
{tommaso.noia,marina.mongiello,eugenio.sciascio}@poliba.it

Abstract. OpenStreetMap (OSM) is a crowd source geographical database that gives users a wide range of tools for searching and locating points of interest and to support the user in navigation on a map. This paper proposes to define a query language for OSM specifying user requests about the route to select between a source and a destination. To this purpose we use Uppaal (http://www.uppaal.org/) model checker: the user poses her query specifying desired points of interest via temporal logic. the method model checks the negation of the desired property, whose counterexample will retrieve the desired path.

1 Introduction

Open Street Map [11] is an initiative to create and provide free geographic data, such as street maps for the user who is looking for a path tailored to her needs. Anyway it is not always easy to detect an optimized route. For example the user locates her current position and searches on the map her destination, but along the path she needs to find some desired points of interest. Even though all the points of interest are easily localized on the map, a structured and automated search of them in the region of interest bounded by starting and arrival points is not available.

In this paper we introduce an approach to query OSM by specifying user's requests with regard to the route to take between a source and a destination together with some potential point of interests. To this purpose the area of interest on the map is modeled in a graph containing the potential POIs for the user. The region of interest is extracted from the map and encoded in XML graph of the Uppall model checker's syntax [2]. More precisely we model check the negation of the desired property, whose counterexample will retrieve the path satisfying the user's needs. The main contributions are: (a) a translation of OSM map to UPPAAL; (b) a novel application of model checking technology; (c) temporal logic as a query languages on a map; (d) a monitoring automaton for temporal logics formulae verification. In the remaining of this paper we start by introducing some background notions on temporal logics and Uppaal Model checker in Sect. 2. We then propose our approach and describe the architecture of the developed system in Sect. 3. Conclusion and Future Work close the paper.

© Springer International Publishing Switzerland 2015
F. Esposito et al. (Eds.): ISMIS 2015, LNAI 9384, pp. 208–213, 2015.
DOI: 10.1007/978-3-319-25252-0_22

2   Background

Temporal Logics and Uppaal Model Checker. A Temporal Logic provides
a formal system for reasoning about the truth values of assertions that change
over time. In this section we briefly describe main categories of temporal logics
The interested reader may refer to the surveys for temporal logics in [1,4–6,10].

Linear Temporal Logics (LTL). In a linear temporal logic the non-temporal
portion of the logic is defined according to classical propositional logic. The
formulae of LTL are built up from atomic propositions, truth-functional connectives ($\vee, \wedge, \neg$, etc.) and temporal operators. The basic temporal operators
of this system are Fp ("sometime p" also read as "eventually p"), Gp ("always
p" also read as "henceforth p"), Xp ("next time p"), and p U q ("p until q").
Branching Temporal Logics. A Branching Temporal Logic allows the use of basic
temporal operators such as a path quantifier either A "for all futures" or E "for
some future", followed by a single one of the usual linear temporal operators G
"always", F "sometime", X "next time", or U "until". Computation Tree Logic
(CTL) is a Branching TL. *Real time temporal logics.* Real-time temporal logics
are different in terms of *expressiveness, order, time metric, temporal modalities, time model* and *time structure.* They also have different capabilities for the
specification and verification of real-time systems. TCTL is a real-time extension of CTL. It extends CTL with hidden clock bounded operators. Uppaal[1]
[2,9,12] is a toolbox for verification of real-time systems jointly developed by
Uppsala University and Aalborg University. It has been applied successfully in
several application problems ranging from communication protocols to multimedia applications. The tool is designed to verify systems that can be modelled
as networks of timed automata extended with integer variables, structured data
types, user defined functions, and channel synchronisation. Uppaal uses a simplified version of TCTL that consists of path formulae and state formulae. State
formulae describe individual states and are expressions that can be evaluated for
a state without looking at the behaviour of the model. Path formulae quantify
over paths or traces of the model and can be classified into reachability, safety
and liveness.

3   Approach

In this section we start by defining the mathematical model we use for managing
the map in our method.

Definition 1. *Region map. The region map is formally defined as a directed
graph $G = < N, E >$ that models the information about a geographical region. The
nodes are point on the map that belong to the set $N = \{n_0, n_1, ...n_n\}$; each n_i is a
tuple $n_i = (id, latitude, longitude)$; the edges belong to the set $E = \{e_0, e_1, ...e_m\}$
where each e_j is a path between two nodes.*

---

[1] http://www.uppaal.org/.

A subregion of the map is a subgraph $S =< SN, SE >$ with $SN \subseteq N$ and $SE \subseteq E$ respectively.

The experimental framework we use adopts the Uppaal model checker and hence translates the region map in the XML-based internal encoding of the model checker.

Definition 2. *Uppaal region map. A UPPAAL region map is a XML structure* $M = \{l\}$ *where location* l *is the tuple* $l = (location, id, x, y, name)$.

With respect to other approaches, planning or other existing query languages [3,7,8], the approach we propose is strengthened by the use of TL and by the possibility to check the property expressed in TL. The choice of temporal logics was motivated by the possibility to give a sequence of points of interest or of events that the users desires to find. to check the property satisfaction we use model checking that gives the user as response the retrieved path if it exists. To retrieve the path the formula is negated and the path is obtained as counterexample of the property to verify.

The choice of Uppaal model checker was motivated by implementation issue concerned with its interface and data management. Even though Uppaal is mainly a Real-time system verifier it has been easily integrated in our approach to model a direct graph and to model a context in which systems are not necessarily time based, by using a monitoring automata. By the way, among other existing model checkers – Spin, NuSMv, Maude and so on – it proved several advantages as performance, usability, efficiency, and was really suitable for the purposes of our approach. Internal representation of data is instead supported by XML modeling thus enabling an easy reduction of OSM map to Uppaal model. At last with respect to other model checkers more obsolete and less flexible, Uppaal architectural model is much more open and interoperable to be integrated in even complex component based systems. Thanks to all these advantages we opted for using Uppaal instead of other model checkers and solved the well-known limitation in term of TL verification by using an ad hoc implemented module (the monitoring automata).

Monitoring Automaton. One of the main drawback of using Uppaal is the temporal logic verifier it embeds. In fact, it is able to check only a subset of the temporal logic without nested operator. The query: *The user is leaving on the state* START, *she wants to reach the state* END *and wants to pass a* POI *before reaching her destination.* is not directly executable with Uppaal because it is not possible to nest more than one temporal operator. To solve this drawback we use a Monitoring Automaton, i.e. an automaton whose state transition depends on the occurrence of properties in a secondary automaton. To make sure that required properties are verified on the primary automaton it is sufficient to check that the secondary automaton, the monitor, has reached a certain state. So it is sufficient to design the monitor automaton to move to a new state each time a specific condition changes and then formulate query on that automaton. The term monitoring automaton is derived from the presence of a robot that monitors that POI will be crossed.

3.1 Prototype System

The architectural framework of the navigation system we developed is made up of several integrated components:

- A Graphical interface where the user poses her query through a simple selection of icons or points of interest.
- Overpass turbo APIs for accessing and manipulating OSM data.
- A XML parser for extracting and filtering data from OSM and to select intermediate points of the route.
- The XML file extracted from the parser that models the Uppaal's model checker graph.
- A CTL/LTL package for modeling user's query in temporal logics specifications.
- A Monitoring Automaton (see below) used to verify some complex Temporal logics specifications with nested operators or quantifiers not directly verifiable by Uppaal verifier.

The Overpass API[2] (or OSM3S) is a read-only API that serves up custom selected parts of the OSM map data. Unlike the main API, which is optimized for editing, Overpass API is optimized for data consumers that need a few elements selected by search criteria like e.g. location, type of objects, tag properties, proximity, or combinations of them.

In a typical scenario, the user selects a starting point on the map and a destination, specifying the characteristics of the path in terms of POIs. For example, if the user requires a place to sleep she will be looking for POIs classified as hotels, B&B, etc. If she requires a place to eat the categories of course include restaurant, pizzeria, snack bar. The graphical interface allows the user to enter the specifications using a text box or by selecting the icon of the required point of interest. The iconic or textual query formulated by the user is translated into a Temporal Logic formula and provided as input to the Uppaal verifier.

To retrieve the path satisfying the user's needs we model check the negation of the desired property, whose counterexample will give us the desired path. Let us now briefly consider some query examples in LTL that the user can pose to the framework. Suppose the user is unaware of the syntax of the logical language. The application scenario is the following: *the user is in Bari in the University campus and starting from this point she wants to go downtown, specifically to have lunch. She also needs to find an ATM to withdraw some money. She is going by car so she wants to avoid pedestrian street and before arriving to the restaurant she wants to find a car parking. As a final remark, she also wants to avoid peripheral areas in order to have the opportunity to visit the city even while driving.* The logical form of the query can be specified as follows using specification patterns as:

---

[2] http://overpass-api.de/api/.

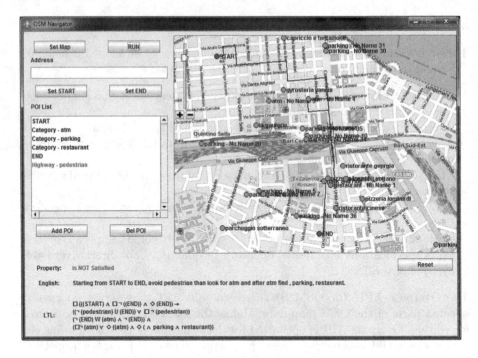

Fig. 1. An example of retrieved path

$Between((START) \wedge (END))$
$Absence((Peripheralzones) \wedge (Pedestrian))$
$Existence((ATM) \wedge (Parking) \wedge (Restaurant))$
$After(Parking) \ Existence(Restaurant)$
Starting from **Start** to **End**, avoid **pedestrian** then look for **ATM** and after **parking** find **restaurant** that in LTL is formalized as:
$(G(START) \wedge G(\neg(END) \wedge F(END)) \rightarrow (\neg((Pedestrain) \wedge (\neg(Peripheral) \wedge F(ATM) \wedge (Parking) \wedge (Restaurant)) \wedge (G\neg(ATM) \vee (F(Restaurant) \wedge (F(Parking) \wedge F(Restaurant)))))$
Figure 1 shows the retrieved path after the query evaluation.

4   Conclusion and Future Work

In this paper we proposed a model-based approach to query Open Street Map. We used temporal logics to specify user's requests about the map and Uppaal model checker, properly embedded in a Java navigation system, to check and verify the queries posed by the user. We modeled the area of interest on the map as a graph containing the potential POIs for the user and then we pose the query as a temporal logic formula thus allowing the system to solve the overall problem via model checking techniques. We are currently working to develop a Domain Specific Language to hide the LTL operators for normal users. At last, we are extending our proposal to build an app working in a mobile environment, to allow the user to get access to the map using his smartphone.

References

1. Alur, R., Henzinger, T.A.: Logics and models of real time: a survey. In: de Bakker, J.W., Huizing, C., de Roever, W.P., Rozenberg, G. (eds.) Real-Time: Theory in Practice. LNCS, vol. 600, pp. 74–106. Springer, Heidelberg (1992)
2. Bengtsson, J., Larsen, K., Larsson, F., Pettersson, P., Yi, W.: UPPAAL - a tool suite for automatic verification of real-time systems. In: Alur, R., Sontag, E.D., Henzinger, T.A. (eds.) HS 1995. LNCS, vol. 1066, pp. 232–243. Springer, Heidelberg (1996)
3. Colombo, A., Fontanelli, D., Legay, A., Palopoli, L., Sedwards, S.: Motion planning in crowds using statistical model checking to enhance the social force model. In: 2013 IEEE 52nd Annual Conference on Decision and Control (CDC), pp. 3602–3608. IEEE (2013)
4. Emerson, E.A.: Handbook of Theoretical Computer Science. North-Holland publisher, Amsterdam (1995)
5. Katoen, J.P.: Stochastic model checking. In: Cassandras, C.G., Lygeros, J. (eds.) Stochastic Hybrid Systems, Control Engineering Series. CRC Press inc, Boca Raton (2007)
6. Konur, S.: Real-time and probabilistic temporal logics: An overview. CoRR, abs/1005.3200 (2010)
7. Lahijanian, M., Almagor, S., Fried, D., Kavraki, L.E., Vardi, M.Y.: This time the robot settles for a cost: Quantitative approach to temporal logic planning with partial satisfaction (2015)
8. Lahijanian, M., Kavraki, L.E., Vardi, M.Y.: A sampling-based strategy planner for nondeterministic hybrid systems. In: 2014 IEEE International Conference on Robotics and Automation (ICRA), pp. 3005–3012. IEEE (2014)
9. Larsen, K.G., Pettersson, P., Yi, W.: Model-checking for real-time systems. In: Karpinski, M. (ed.) FCT 1977. LNCS, vol. 56, pp. 62–88. Springer, Heidelberg (1977)
10. Clarke, E.M., Grumberg, O., Peled, D.A.: Model Checking. MIT press, Cambridge (1999)
11. OpenStreetMap (OSM). https://www.openstreetmap.org/#map=5/51.509/0.044 (2004)
12. Yi, W., Pettersson, P., Daniels, M.: Automatic Verification of real-time communicating systems by constraint-solving. In: Hogrefe, D., Leue, S. (eds.) Proceedings of the 7th Inernational Conference on Formal Description Techniques, pp. 223–238. North-Holland (1994)

Granular Rules and Rule Frames for Compact Knowledge Representation

Antoni Ligęza[✉]

AGH University of Science and Technology,
al. Mickiewicza 30, 30-059 Kraków, Poland
ligeza@agh.edu.pl

Abstract. Efficient management of big Rule-Based Systems constitutes an important challenge for Knowledge Engineering. This paper presents an approach based on Granular Sets and Granular Relations. Granules of data replace numerous low-level items and allow for concise definition of constraints over a single attribute. Granular Relations are used for specification of preconditions of rules. A single Granular Rule can replace numerous rules with atomic preconditions. By analogy to Relational Databases, a complete Granular Rule Frame consists of Rule Scheme and Rule Specification. Such approach allows for efficient and concise specification of powerful rules at the conceptual level and makes analysis of rule set easier. The detailed specifications of Granular Rules are much more concise than in the case of atomic attribute values, but still allow for incorporating all necessary details.

1 Introduction

Rule-Based Systems [3] are one of the most visible result of application of Artificial Intelligence tools in various domains of technology, business and administration. Knowledge representation in the form of *if-then rules* is not only transparent and intuitive, but it constitutes a powerful, generic, and universal approach to transitional knowledge specification. However, definition of hundreds of rules is a very tedious task, and rule-base management becomes a non-trivial task.

In a monograph [6] a novel approach to knowledge representation in Rule-Based Systems was put forward. The main focus was on (i) concise representation of groups of similar rules and (ii) transitions among such groups rather than single, flat rule-base. Moreover, clear logical foundations and systematic design and analysis methodology were proposed as well. Some further details concerning attributive logic for knowledge representation and tools developed within the proposal were reported in [7,11] and [12]. A thorough study of the methodological issues of developing rule-based systems is presented in [9].

This paper presents the continuation of those research motivated by the need of efficient representation of *potentially large bodies of detailed knowledge* with

A. Ligęza—AGH University of Science and Technology; Research Contract No. 18.18.120.859.

F. Esposito et al. (Eds.): ISMIS 2015, LNAI 9384, pp. 214–223, 2015.
DOI: 10.1007/978-3-319-25252-0_23

compact, abstract tools. An attempt at applying a *granular* approach and two-level representation scheme is reported. In fact, one idea of capturing the plethora of data and knowledge chunks is to organize them and manipulate through more abstract items, e.g. granules. Such approaches, often emerging from Fuzzy Set Theory [18], or Rough Sets [14], are becoming more and more popular in various domains of computer science [1,15].

In this paper we use a purely algebraic approach based on simple concepts of granular sets and granular relations, as introduced in [4] and followed in some consecutive works [5,7]. Instead of perceiving and manipulating single elements, we group them into naturally emerging subsets of the universe of discourse. A selection of such subsets form a *granular set*, where each of the subsets defines a *granule* of data. Such a set is conceived as a tool for dealing with numerous detailed data at some more abstract level. Further consideration of join granular sets leads in a natural way to the concept of *granular relation*. And such a relation can be used as specification of preconditions of *granular rules* for covering numerous input cases.

For intuition, consider a specification of time instant when some activity must be performed. Every month with 31 days, every 1-st, 11-th, 21-st and 31-st one has to repeat some action **a**. Instead of 28 flat rules, we can provide a granular specification of the form:

```
-----------------------------------------------------
| Month is:          | Day is:        || Action is: |
=====================================================
| {1,3,5,7,8,10,12}  | {1,11,21,31}  ||      a      |
-----------------------------------------------------
```

which covers (through the Cartesian Product) all the individual cases. The subsets of the universe of months {1,3,5,7,8,10,12} and the universe of days {1,11,21,31} are *granules*. The Cartesian Product of them forms a *tube* or a *box* covering all the individual combinations *month-day*. More similar tubes would form a *granular relation*. And finally, the specification as above is a *granular rule frame*; it is combined of *granular rule scheme* (the heading) and *granular rule specification* (the second row). In fact, the specification can be composed of as many rows, as necessary.

Such a granular rule scheme can cover a number of rule specifications. An important, practical question is how to manage irregular (*defective*) cases, where some elements of the Cartesian Product of granules are to be excluded or they have different specification w.r.t the action/conclusion? Well, there is no single elegant way to solve such cases. What is proposed in this paper is to (i) keep the concise granular rule specification, (ii) specify as many rules in the capacious way with explicit granules (or constraints defining them), and (iii) all the necessary individual cases provide as a set of records of a relational database table identified with a foreign key corresponding to the key of the particular rule scheme.

The proposed approach is influenced by the theory and work on relational databases [2] and tries to keep things as simple as possible, with the ultimate goal of efficiency in mind.

1.1 Related Work

The notion of a *granule* has been omnipresent in Data and Knowledge Representation for some period of time now. It naturally appears in the calculus in the form of an interval [10]. A number of recent works put forward various concepts concerning more efficient knowledge representation with levels of abstraction, and granularity is perhaps one of the most prominent directions [15]. Most of the promoted ideas have its roots in Fuzzy Set Theory [18], Rough Sets [14,16] or following concepts such as *coverings, tolerance spaces* [16], or neighborhood [17]. Some two books try to present a survey on concepts concerning Granular Computing [1] and [15].

The main differences between majority of such approaches and the presented paper are: (i) the main focus here is on *compact knowledge representation*; fuzzy and rough-based approaches usually deals with imprecise character of knowledge, (ii) the assumptions here are as weak as possible — the approach is oriented towards practical applications (e.g. incomplete, overlapping partitions), and (iii) use of standard tools, both with respect to theory (algebra) and applications (relational databases) is insisted on. Some further comments on relationship to rough sets were presented in [8].

2 Granular Sets: Motivation and Basic Concepts

In this section we re-introduce the notion of *Granular Set* (GS for short). The presented idea of granular sets and granular relations was introduced in [4], and further developed in [5]. A *granular attributive logic* was a further step toward efficient knowledge representation in rule-based systems [7]. In this paper we extend these ideas over representation of rules with the *Granular Rules* and *Granular Rules Frames*.

For intuition, a *Granular Set* is a set composed of several subsets of a universe under consideration. Such subsets, usually defined in a natural way, allow for more efficient dealing with to numerous elements of the universe by referring to *small number* of distinguished groups of items rather than to each of *numerous* detailed elements of the universe. A formal presentation follows.

Consider a set V and several subsets of it, say $V_1, V_2, \ldots, V_k$.

Definition 1. *A semi-partition of V is any collection of its subsets $V_1, V_2, \ldots, V_k$. A semi-partition is normalized (in normal form) iff $V_i \cap V_j = \emptyset$ for all $i \neq j$.*

A semi-partition is called an *incomplete partition*, or a σ-partition, and it will be denoted as $\sigma(V)$. If not stated explicitly, the considerations will refer to normalized σ-partitions, however, overlapping granules V_i can also be considered.

A question can be raised why to consider semi-partitions instead of classical partitions (which are complete and disjoint)? The answer is that the presented research are practice-driven; partitions are a *fine mathematical concept*, while semi-partitions are what can be met in practice by a knowledge engineer, where some groups of items can be named, and there are still individual items not belonging to any group (the so-called *residual elements*). There can be also *defective groups*, i.e. well-defined groups with the necessity to exclude several individuals.

Definition 2. *Let U denote a set of objects under consideration, i.e. the so-called universe and let σ be a mapping of the form:*

$$\sigma : U \to 2^{(2^U)},$$

where $\sigma(U) = \{U_1, U_2, \ldots, U_n\}$; U_i denote some subsets of U. The pair $G(U) = (U, \sigma(U))$ will be called granular space over U induced by σ.

The sets U_i are called *granules*; $\sigma(U)$ defines the *level of granularity*; it can be *fine* (small granules) or *coarse* (big granules).

As an example consider a calendar year as the universe U; it is composed of 365 days. Its granular representation is by splitting it into 12 months, say $M_1, M_2, \ldots M_{12}$, and it can be considered as a fine granulation. A more coarse granulation can be by splitting it into quarters of the year, say Q_1, Q_2, Q_3, Q_4.

The set of granules can also better or worse cover the universe. We introduce the notion of the sum of elements covered as the *support*.

Definition 3. *The support of an σ-partition $\sigma(U)$ is defined as $[\![\sigma(U)]\!] = U_1 \cup U_2 \cup \ldots \cup U_k$.*

A σ-partition is *complete* iff $U = [\![\sigma(U)]\!]$. Both the above granulation — into months and into quarters — are complete. Note that the granules defined over U are not necessarily assumed to cover all the elements of U, and so the partition may remain *incomplete*. Obviously, a clear case is when $\sigma(U)$ defines a partition (it is by definition complete).

Definition 4. *Let $S \subseteq U$. A granular set $G(S)$ in granular space $(U, \sigma(U))$ is defined as a pair:*

$$G(S) = (S, \sigma(S)) \tag{1}$$

where

$$\sigma(S) = \{S_i : S_i = S \cap U_i, S_i \neq \emptyset, i = 1, 2, \ldots, n\}$$

For the sake of being concise, if $S = [\![\sigma(S)]\!]$ (the granular representation of S is complete), we shall use a shorthand notation $G(S) = \sigma(S)$ (no residual, uncovered elements exist).

For intuition, let us return to the example of year and months being the predefined granules. Let S be a set of days defining summer holidays as from June 15-the to September 15-th (hopefully). The induced granular set is $\sigma(S) = \{June\_2\_half, July, August, September\_1\_half\}$, and so it is composed of two original granules (July, August) and two being intersections of S and the original granules (second half of June and first half of September).

For intuition, a granular set $G(S)$ defined in a granular space $G(U)$ is represented by the set S itself and:

– all the *entire* granules U_i, such that $U_i \subseteq S$ (most wanted),
– all the granules U_i having non-empty intersection with S, restricted to elements of S only.

The idea behind the definition is that (i) we keep the original domain S and we find as many predefined granules within S to cover it as possible; if, by chance, there are still granules overlapping with S we restrict them to the elements of S and adjoin these new granules to improve covering of S. Finally, all the remaining elements, the so-called *residuals* (or *orphans*) (if any) are left uncovered (but specified with the definition of S). Obviously, the most elegant case could consist of S being exactly composed of some elements of σ being a regular partition.

3   Granular Relations

Having in mind the idea of granular set, a *granular relation* can be defined in a straightforward way.

Consider some family of sets $Q^1, Q^2, \ldots, Q^m$; for intuition these sets can constitute domains of a set of m attributes of interest used to describe system state. Let there be defined some granular sets on them with predefined semi partitions, i.e. $\sigma_1(Q^1), \sigma_2(Q^2), \ldots, \sigma_m(Q^m)$ are given. For simplicity of further considerations, let us assume that $Q^i = [\![\sigma_i(Q^i)]\!]$, i.e. all the semi partitions are complete[1].

Definition 5. *A granular relation is any set R_G such that $R_G \subseteq U_G$ where*

$$U_G = \sigma_1(Q^1) \times \sigma_2(Q^2) \times \ldots \times \sigma_m(Q^m). \tag{2}$$

The set U_G will be referred to as Granular Cartesian Product. If at least one of the granular subsets was unnormalized, the relation is also said to be unnormalized one.

The elements (rows) of a granular relation will be called *boxes* or *tubes*. Note that in fact a granular relation defines a kind of meta-relation, i.e. one based on subsets instead of single elements of the universe. If R is a relation defined as $R \subseteq Q^1 \times Q^2 \times \ldots \times Q^m$, then any tuple of R is like a *thread* in comparison to elements of R_G which are like a tube. In fact, any tube of R_G may contain numerous elements of R. Hence, the representation may be very concise.

4   Granular Rules and Rule Frames

Granular relations allow for very compact specification of 'covers' for Cartesian Product elements. In fact, every element of granular relation represents the whole Cartesian Product of the composite granules (which are de facto sets). Let R_G

---

[1] If this is not the case, it is enough to create a separate, single element granule for any remaining residual element.

be a granular relation in the Cartesian Product of U_G, as defined by Eq. 2, i.e.
$R_G \subseteq \sigma_1(Q^1) \times \sigma_2(Q^2), \times \ldots, \sigma_m(Q^m)$.

For simplifying the notation let $(\overline{Q}_j) \in R_G$ denote a certain tube — a single element of granular relation (a single *tube*); $\overline{Q}_j$ is in fact composed of granules belonging to each $\sigma(Q^i$, where $i = 1, 2, \ldots, m$. Now, a *granular rule* is a rule with preconditions specified by some granular relation. Such preconditions are composed of one or more such tubes. The way of definition of conclusion (typically a setting of a parameter) is not important here.

Definition 6. *Let* $\overline{Q}_1, \overline{Q}_2, \ldots, \overline{Q}_k$ *be some tubes of Granular Relation* R_G *defined by Definition 5. A Granular Rule Specification GR is a pair of preconditions specified with set of these tubes, and conclusion defined by variable assignment of the form:*

$$GR = (\{\overline{Q}_1, \overline{Q}_2, \ldots, \overline{Q}_k\}, h) \tag{3}$$

To understand the meaning of a granular rule the definition must be completed by the *scheme* of a rule.

Definition 7. *A granular Rule Scheme RS is a sequence of attributes*

$$RS = (A_1, A_2, \ldots, A_m, H)$$

where $A_1 - A_m$ *are precondition attributes (corresponding to sets* $Q^1, Q^2, \ldots, Q^m$, *which, as a matter of fact, can be considered to specify the domains of these attributes), and H is the decision attribute.*

Finally, for specification of a single rule, or mostly a set of granular rules of the same scheme we need the notion of *rule frame*.

Definition 8. *A Granular Rule Frame GFR is a pair*

$$GFR = (RS, \{GR_1, GR_2, \ldots, GR_k\})$$

where all the rules GR_i, $i = 1, 2, \ldots k$ *have the same scheme RS.*

For illustration, consider the Granular Rule Frame as below.

Student in:	Time in:	Rooms is:	Topic is:	Professor is:
S1	[T1-T2]	b1	c1	p1
S2	[T3-T4]	b2	c2	p2
S3	[T5-T6]	b2	c3	p3

In the specification above the first row defines the Rule Schema for a university schedule (common to all the rules), and the following rows are specifications of three Granular Rules. Note that in fact the three granular rules cover perhaps hundreds of assignments of professors for single students and time instants covered by granules S1, S2, S3 and the time intervals of interest.

5 An Example

In this section an example of knowledge representation with granular rules and rule frames is presented. The example is based on the specification of 18 rules defining *set-point* for air condition system. The rules were initially described in [13] and comprehensively analyzed in [6].

The selection of the particular temperature set-point value depends on the season (SE), whether it is a working day or not (TD=wd or TD=wk), and the time of the day. The original rules can be grouped into four separate schemes corresponding to four frame rules:

- there are two rules for determining if the operation is during working day (wd) or weekend (wk),
- there are four rules for determining if the operation is during working hours (dbh) or not (ndbh),
- there are four rules for defining the season: Summer (sum), Autumn (aut), Winter (win), or Spring (spr),
- finally, there are eight rules for defining the temperature set-point depending on the season and the working hours parameter. Below we present the representation of the rules using rule frames and granular rules. The specification of the first two rules is as follows:

```
-----------------------------------------------------------
| Day in:                                      || TD is: |
===========================================================
| {monday,tuesday,wednesday,thursday,friday} || wd      |
-----------------------------------------------------------
| {saturday,sunday}                           || wk      |
-----------------------------------------------------------
```

In this rule frame we have two granular rules; the meaning of them is straightforward. The rules are very simple since in fact the preconditions of them are specified with single granules. The 'Day' is an input value and TD is a variable denoting 'today'.

The next four rules are specified with the following rule frame:

```
-----------------------------------
| TD is:  | TM in:  || OP is:  |
===================================
| wd      | [9,17]  || dbh     |
-----------------------------------
| wd      | [0,9)   ||         |
| wd      | (17,24) || ndbh    |
| wk      | _       ||         |
-----------------------------------
```

The rule frame above covers four original rules of [13], but thanks to granular specification we have only two granular rules; the first (single row) says that if today is working day (TD=wd) and the time is between 9 and 17 (TM in [9,17]) then the operation is during business hours (OP=dbh). The second granular rule is specified in three lines; there are three separate tubes specifying alternative preconditions, and there is a single conclusion (OP=ndbh). The three tubes form a proper granular relation.

The following four rules are specified as follows:

MO in:		SE is:
{january,february,december}		sum
{march,april,may}		aut
{june,july,august}		win
{september,october,november}		spr

The meaning of the four simple one-line rules is obvious: depending on the month (MO; input value) the season (SE) is established.

Finally, the last eight rules is specified with the following rule frame:

SE is:	OP is:		THS is:
spr	dbh		20
spr	ndbh		15
sum	dbh		24
sum	ndbh		17
aut	dbh		20
aut	ndbh		16
win	dbh		18
win	ndbh		14

Depending on current season (SE) and operation (OP; dbh – during business hours, ndbh – not during business hours) the current temperatures (THS) is set to a specific set-point. Due to detailed specification, in fact no granularity is observed in this last table. However, note that for rules 1 and 5, THS=20; this gives rise to possible more concise, granular representation of these two rules with a single granular. Moreover, assuming some tolerance allowing for slight adjustment of the set-points, one can easily form a set of compact granular rules *almost* equivalent to the initial specification, e.g.:

```
----------------------------------------
| SE is:          | OP is: || THS is: |
========================================
| {spr,aut,win}  | dbh     || 20      |
----------------------------------------
| sum            | dbh     || 24      |
----------------------------------------
| _              | ndbh    || 16      |
----------------------------------------
```

A workflow can be used to visualize the flow of the inference process; a simple BPMN diagram of the process is presented in Fig. 1. After start, the process is split into two independent paths of activities. The upper path is aimed at determining the current season (SE) and the lower path is used to determine the weekday (WD) and then — the operation hours (OP). Finally, the last rectangle covers the rules for determining the set-point (THS).

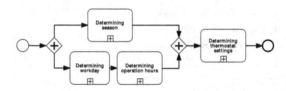

Fig. 1. An example BPMN flow diagram — top-level specification of rule frames

6   Conclusions and Further Work

The main aim of this paper was to present the possibility of applying granular knowledge representation to specification of rules. There are three straightforward goals of using granular specifications instead of atomic ones: (i) assuring compact knowledge specification, (ii) making knowledge transparent and readable, and (iii) improving efficiency of inference and knowledge management.

The ultimate success of granular specification of knowledge in general, and rules in particular, depends much on the *language* for representing granules and tools for efficient management of them. It seems efficient knowledge specification and management can be done with help of traditional relational databases. Than appropriate level of granularity should allow to hide the detailed atomic specification and allow for easier analysis by human operator.

References

1. Bargieła, A., Pedrycz, W.: Granular Computing: An Introduction. Kluwer Academic Publishers, Dordrecht (2003)
2. Conolly, T., Begg, C., Strachan, A.: Database Systems. A Practical Approach to Design, Implementation and Management. Addison-Wesley, Melbourne (1998)
3. Liebowitz, J. (ed.): The Handbook of Applied Expert Systems. CRC Press, Boca Raton (1998)
4. Ligęza, A.: Granular sets and granular relations. An algebraic approach to knowledge representation and reasoning. In: AIMETH 2002 Methods of Artificial intelligence, pp. 47–54, Gliwice (2002)
5. Ligęza, A.: Granular sets and granular relations: towards a higher abstraction level in knowledge representation. In: Kłopotek, M.A., Wierzchoń, S.T., Michalewicz, M. (eds.) Intelligent Information Systems 2002. Advances in Soft Computing, vol. 17, pp. 331–340. Physica-Verlag HD, Springer-Verlag Berlin Heidelberg (2002)
6. Ligęza, A.: Logical Foundations for Rule-Based Systems. In the series: Studies in Computational Intelligence, vol. 11. Springer-Verlag, Berlin, Heidelberg (2006)
7. Ligęza, A., Nalepa, G.J.: Knowledge representation with granular attributive logic for XTT-based expert systems. In: Proceedings of the 20th Florida Artificial Intelligence Research Society Conference FLAIRS, pp. 530–535 (2007)
8. Ligęza, A., Szpyrka, M.: A note on granular sets and their relation to rough sets. In: RSEISP 2007 International Conference Rough Sets and Emerging Intelligent Systems Paradigms, pp. 251–260 (2007)
9. Ligęza, A., Nalepa, G.J.: A study of methodological issues in design and development of rule-based systems: proposal of a new approach. Wiley Interdisc. Rev. Data Mining Knowl. Discov. 1, 117–137 (2011)
10. Moore, R.: Interval Analysis. Prentice-Hall, Upper Saddle River (1966)
11. Nalepa, G.J., Ligęza, A.: The HeKatE methodology: hybrid engineering of intelligent systems. Int. J. Appl. Math. Comput. Sci. 20(1), 35–53 (2010)
12. Nalepa, G.J., Bobek, S., Ligęza, A., Kaczor, K.: HalVA - rule analysis framework for XTT2 rules. In: Bassiliades, N., Governatori, G., Paschke, A. (eds.) RuleML 2011 - Europe. LNCS, vol. 6826, pp. 337–344. Springer, Heidelberg (2011)
13. Negnevitsky, M.: Artificial Intelligence. A Guide to Intelligent Systems. Addison-Wesley, Harlow (2002)
14. Pawlak, Z.: Rough Sets. Theoretical Aspects of Reasoning about Data. Kluwer Academic Publishers, Dordrecht (1991)
15. Pedrycz, W., Skowron, P.W.A., Kreinovich, W.: Handbook of Granular Computing. Wiley, New York (2008)
16. Ślęzak, D., Wasilewski, P.: Granular sets – foundations and case study of tolerance spaces. In: An, A., Stefanowski, J., Ramanna, S., Butz, C.J., Pedrycz, W., Wang, G. (eds.) RSFDGrC 2007. LNCS (LNAI), vol. 4482, pp. 435–442. Springer, Heidelberg (2007)
17. Yao, Y.Y.: Granular computing using neighborhood systems. In: Roy, R., Furuhashi, T., Chawdhry, P.K. (eds.) Engineering Design and Manufacturing, pp. 539–553. Springer-Verlag, London (1999)
18. Zadeh, L.A.: Towards a theory of fuzzy information granulation and its centrality in human reasoning and fuzzy logic. Fuzzy Sets Syst. 19, 111–127 (1997)

FIONA: A Framework for Indirect Ontology Alignment

Marouen Kachroudi[(✉)], Aymen Chelbi, Hazem Souid,
and Sadok Ben Yahia

Université de Tunis El Manar, Faculté des Sciences de Tunis,
LIPAH Programmation Algorithmique et Heuristique, 2092 Tunis, Tunisia
{marouen.kachroudi,sadok.benyahia}@fst.rnu.tn

Abstract. Ontology alignment process is seen as a key mechanism for reducing heterogeneity and linking the diverse data and ontologies arising in the Semantic Web. In such large infrastructure, it is inconceivable to assume that all ontologies dealing with a particular knowledge domain are aligned in pairs. Furthermore, the high performance of the alignment techniques is closely related to two major factors, *i.e.*, time consumption and resource machine limitations. Indeed, good quality alignments are valuable and it would be appropriate to harness. This paper introduces a new indirect ontology alignment method. The proposed method implements a strategy of indirect ontology alignment based on a smart direct alignments composition and reuse. Results obtained after extensive carried experiments are very encouraging and highlight many useful insights about the new proposed method.

Keywords: Semantic web · Interoperability · Indirect ontology alignment · Alignment algebra

1 Context and Motivations

Semantic Web active actors publish and share their data sources in their own respective languages or formalisms. Moreover, the explicitation of the associated concepts related to a particular domain of interest takes advantage of ontologies, considered as the Semantic Web cornerstone [1]. In addition, the open and dynamic resources of the Semantic Web endow it with a heterogeneous aspect, which reflects at once the formats or the conceptual varieties of its description. Indeed, the informative volume reachable via the Semantic Web stresses needs of techniques guaranteeing the share, reuse and interaction of all resources. Based on the fact that ontologies are likely to be authored by different actors using different terminologies, structures and natural languages, ontology alignment has emerged as a way to achieve semantic interoperability. The operation of finding correspondences is called *ontology alignment* and its result is a set of correspondences called an alignment. Obviously, these alignments are subject of extensive evaluation, improvement and finally integrated across multiple

© Springer International Publishing Switzerland 2015
F. Esposito et al. (Eds.): ISMIS 2015, LNAI 9384, pp. 224–229, 2015.
DOI: 10.1007/978-3-319-25252-0_24

environments before being used by several applications. This paper meets challenges strictly bound at the interoperability level. Indeed, it proposes a new idea about a minimal framework for indirect ontology alignment called FIONA (A Framework for Indirect ONtology Alignment). The main idea is to capitalize on existing alignments expressed in the RDF format and which already exist within such a semantic environment. The remainder of this paper is as follows. Section 2 reviews the existing methods in the field of indirect ontology alignment and defines some terminologies and notations for the rest of this paper. Section 3 thoroughly describes the FIONA method, its foundation and its various steps as the main contribution of this work. Section 4 reports the experimental encouraging results obtained with the considered test base. Finally, Sect. 5 draws the conclusion and sketches future issues of this paper.

2 Key Notions and Background

Ontology alignment is considered as an evaluation of the degrees of resemblance or the differences detected on the considered ontological entities [2]. Besides, the process of alignment can be defined as follows: being given two ontologies $\mathcal{O}_1$ and $\mathcal{O}_2$, an alignment between $\mathcal{O}_1$ and $\mathcal{O}_2$ is a set of correspondences, (*i.e.*, a quadruplet): $< e_1, e_2, r, Conf(n) >$, with $e_1 \in \mathcal{O}_1$ and $e_2 \in \mathcal{O}_2$, r is a relation between two given entities e_1 and e_2, while $Conf(n)$ represents the confidence level in this relation. In a wider context, produced alignments in the Semantic Web can be reused. Indeed, the main advantage of the reuse process reduces the cost of conventional alignment operations. In what follows, e_1, e_2 and e_3 are three entities belonging respectively to $\mathcal{O}_1$, $\mathcal{O}_2$ and $\mathcal{O}_3$. Indeed, alignment composition determines the semantic relations which labels the link between e_1 to e_3, based on already existing links between e_1, e_2 and e_2, e_3. Secondly, the composition process determines the confidence value on the new composed relation between and e_1 to e_3. In fact, relation composition for two given entities is carried out according to several rules [3], where *"are equivalent"* ($=$), *"includes"* ($>$), *"is-included-in"* ($<$), *"overlaps"* ($\between$), and *"disjoint"* ($\perp$). The composed relation should be supported in turn by a confidence value (n or n') that reflects the correspondence degree between the two considered entities. The latter value can be derived according to several ways [3]:

- Multiplication : $Conf(n, n') = n \times n'$
- Normalization : $Conf(n, n') = \frac{n \times n'}{2}$
- Maximization (or Minimization) : $Conf(n, n') = max(n, n')$ or $min(n, n')$

In the literature, several approaches that tried to exploit and implement the notions previously introduced. These approaches proposed mechanisms that enable them to connect data fragments initially isolated. The interconnection operation involves treatments based on reasoning and/or semantic transitivity into the considered data amount space. A theoretical idea was also presented for building indirect alignments between multilingual ontologies [4]. The basic principle of this method is the reuse of already existing and stored alignment files.

An intermediary alignment should be carried out between source and target ontology to compose a new alignment using such objects. Beforehand, equivalence between multilingual entities belonging in both considered distinct ontologies should be discovered and established by a human expert. Then, a process of alignment composition is applied using alignment algebra introduced in [3]. To sum up, there are several methods that have addressed the indirect alignment issue. Each of these methods has its own vision and its own account for indirect alignment. In addition, the only work that has methodically addressed this issue is that of Jung et al. [4] although the work remains limited, primitive and without thorough tests on some concrete cases. In the next section, we introduce FIONA, as the new indirect ontology alignment method.

3   The FIONA Method

In the following, a detailed description of the FIONA method that revolves around four major components.

3.1   Step 1 : Ontology Network Examiner (ONEX)

The driving idea is to examine the provided space formed by n ontologies aligned in pairs, what we call *ontology network*. The purpose of this module is to determine the number of deducible indirect alignments from the provided direct ones.

3.2   Step 2 : Alignment Parsing and Pretreatment (APP)

After defining the available direct alignments between ontologies involved in the ONEX component, the APP module initiates the next phase, for parsing and preprocessing. This module operates on the supplied alignment `rdf` files.

3.3   Step 3 : Indirect Alignment Composition (IACOMP)

The IACOMP module is the kernel of the FIONA method. This module encompasses three complementary sub modules. This can be achieved through the alignment algebra sketched in Sect. 2, then we formally define a composed alignment, denoted by $\mathcal{A}_{ik}$, as follows:

Definition 1. *Given two alignments $\mathcal{A}_{ij}$ and $\mathcal{A}_{jk}$, if it exists a bridging entity interlinking two existing direct correspondences, then the composed alignment $\mathcal{A}_{ik}$ is defined by a set of composed correspondences:*

$$\mathcal{A}_{ik} = \mathcal{A}_{ij} \oplus \mathcal{A}_{jk} = \{\langle e_1, e_3, \mathcal{F}_{rel}(r, r'), \mathcal{F}_{conf}(n, n')\rangle | e_1 \in \mathcal{O}_1, e_3 \in \mathcal{O}_3\}$$

where:

- $e_1 \in \mathcal{O}_1$, $e_2 \in \mathcal{O}_2$ *and* $e_3 \in \mathcal{O}_3$.
- $\langle e_1, e_2, r, Conf(n)\rangle \in \mathcal{A}_{ij}$, $\langle e_2, e_3, r', Conf(n')\rangle \in \mathcal{A}_{ij}$.

– the bridge entity e_2.
– $\mathcal{F}_{rel}$, $\mathcal{F}_{conf}$, are functions designed for composing two relations and two confidence values, respectively.

In the following, a detailed description of the sub modules constituting the IACOMP module.

3.3.1   Bridge Entity Detection

In this step, from developed parsed entity lists obtained by the APP module, we determine the most important element of the composition, which is the *bridge entity*. This *bridge entity* ensures and guarantees navigability as well as semantic transitivity between any three alignments.

3.3.2   Relation Composition

New relation composition that label the link between two given entities e_1 and e_3, as previously described is achieved through the composition algebra [3].

3.3.3   Confidence Value Computation

Confidence value composition is performed according to one of four manners (*i.e.* namely, multiplication, normalization, maximization and minimization) introduced in Sect. 2.

3.4   Step 4 : Indirect Alignment Generation (IAG)

The alignment file generation is the last step in the FIONA method process. At this level, an alignment file is built from the aligned entities, as well as the relation and its confidence value. This step is performed by means of the alignment API[1].

4   Experimental Study

In what follows we will present the experimental study, based on the metrics of Precision[2] (P) and Recall[3] (R).

4.1   Test Cases

The carried out experimental evaluation uses alignments file-battery provided by the OAEI campaign (Ontology Alignment Evaluation Initiative Campaign)[4]. These files are alignment methods outputs that participated in the 2014 campaign. Tests were conducted on alignments produced by three pioneering methods of OAEI campaign, namely AML [5], LOGMPLT [6] and XMAP++ [7]. The chosen base is *Conference*. Indeed, the choice of this test base is well studied, considering its evaluation specificity.

---

[1] http://alignapi.gforge.inria.fr/.
[2] $Precision = \frac{|N_{correct}|}{|N_{found}|}$.
[3] $Recall = \frac{|N_{correct}|}{|N_{expected}|}$.
[4] http://oaei.ontologymatching.org/.

Table 1. Precision and Recall values for the considered methods and alignments

I.C.A[a]	AML				LogMpLt				Xmap++			
	ByEq		Sem		ByEq		Sem		ByEq		Sem	
	P	R	P	R	P	R	P	R	P	R	P	R
1	0.20	0.50	0.27	0.60	0.15	0.36	0.18	0.40	0.14	0.42	0.18	0.50
2	0.17	0.40	0.23	0.50	0.22	0.45	0.30	0.57	0.27	0.60	0.30	0.66
3	0.66	0.65	0.75	0.75	0.55	0.88	0.75	1.00	0.44	0.60	0.50	0.66
4	0.15	0.55	0.18	0.60	0.14	0.68	0.18	0.75	0.14	0.70	0.18	0.75
5	0.33	0.65	0.41	0.71	0.35	0.76	0.41	0.83	0.28	0.76	0.33	0.80
6	0.25	0.88	0.33	1.00	0.25	0.92	0.33	1.00	0.25	0.90	0.33	1.00
7	0.27	0.75	0.33	0.83	0.18	0.57	0.26	0.66	0.26	0.78	0.33	0.83
8	0.20	0.85	0.28	1.00	0.16	0.93	0.21	1.00	0.18	0.90	0.21	1.00
9	0.28	0.88	0.36	1.00	0.14	0.59	0.18	0.66	0.20	0.88	0.27	1.00
10	0.15	0.55	0.20	0.71	0.08	0.30	0.08	0.33	0.07	0.54	0.12	0.60
11	0.15	0.88	0.13	0.60	0.12	0.95	0.13	1.00	0.10	0.57	0.13	1.00
12	0.20	0.56	0.28	0.66	0.18	0.92	0.28	1.00	0.28	1.00	0.42	1.00
13	0.14	0.66	0.14	1.00	0.10	0.90	0.14	1.00	0.21	1.00	0.24	1.00
14	0.20	0.60	0.26	0.66	0.18	0.90	0.26	0.80	0.20	0.78	0.26	0.80
15	0.17	0.65	0.21	0.80	0.22	0.77	0.15	0.60	0.10	0.70	0.15	0.75

[a]I.C.A stands for Indirect Composed Alignment as shown in Table 1

4.2   Results and Discussion

The bridging entity detection is fitted through two scenarios through which we are interested in, to assess their impact on indirect alignment quality. The first scenario uses a string based similarity technique abbreviated (ByEq), which aims to detect the bridge entity by a simple equality test between two strings, each belonging to a certain alignment, and so, representing a given entity. The second scenario uses an external resource called Semilar[5][8] abbreviated (Sem). In this second case, we seek to palliate the drawbacks of String-Based Similarity techniques, that can induce processing errors as well as a loss in the meaning or deviate the meaning of some bridge entities, thus altering the composed alignments. Table 1 highlights that the use of Semilar API provides better Precision and Recall values, this means that by using Semilar, bridge entity detection supplied a good impact and boosted the composition task. Indeed, we note that the obtained values, during the second scenario, outperform those of the first scenario by around 14 % for the metric Recall and 10 % Precision (as maximum values).

---

[5] http://www.semanticsimilarity.org/.

5   Conclusions and Outlooks

This paper we introduced the FIONA method for indirect ontology alignment. The obtained results are very promising and accentuate other aspects related to ontology alignment and reuse. In this respect, it is important to highlight the external resource contribution. The proposed method showed a good performance, but still requires some improvements. In the near future, we also intend also to enhance the performance of the FIONA so that it can handle a wider range of alignments and under other constraints. Furthermore, the integration of new external resources can provide a wider choice of semantic equivalents that could be of benefit to the task of bridge entity detection. Besides, a graphical user interface (GUI) is needed to assist ordinary users.

References

1. Berners-Lee, T.: Designing the web for an open society. In: Proceedings of the 20th International Conference on World Wide Web (WWW2011), pp. 3–4, Hyderabad (2011)
2. Ehrig, M.: Ontology Alignment: Bridging the Semantic Gap. Springer-Verlag, New-York (2007)
3. Euzenat, J.: Algebras of ontology alignment relations. In: Sheth, A.P., Staab, S., Dean, M., Paolucci, M., Maynard, D., Finin, T., Thirunarayan, K. (eds.) ISWC 2008. LNCS, vol. 5318, pp. 387–402. Springer, Heidelberg (2008)
4. Jung, J.J., Håkansson, A., Hartung, R.: Indirect alignment between multilingual ontologies: a case study of korean and swedish ontologies. In: Håkansson, A., Nguyen, N.T., Hartung, R.L., Howlett, R.J., Jain, L.C. (eds.) KES-AMSTA 2009. LNCS, vol. 5559, pp. 233–241. Springer, Heidelberg (2009)
5. Faria, D., Martins, C., Nanavaty, A., Taheri, A., Pesquita, C., Santos, E., Cruz, I., Couto, F.: Agreement maker light results for Oaei 2014. In: Proceedings of the 11th International Workshop on Ontology Matching (OM-2014), Colocated with the 13th International Semantic Web Conference (ISWC-2014), **1317** of CEUR-WS, pp. 105–112, Trentino (2014)
6. Jiménez-Ruiz, E., Grau, B.C., Xia, W., Solimando, A., Chen, X., Cross, V., Yuan Gong, a.S.Z., Chennai-Thiagarajan, A.: Logmap family results for Oaei 2014. In: Proceedings of the 11th International Workshop on Ontology Matching (OM-2014,) Colocated with the 13th International Semantic Web Conference (ISWC-2014), **1317** of CEUR-WS, pp. 126–134, Trentino (2014)
7. Djeddi, W.E., Khadir, M.T.: Xmap++: results for Oaei 2014. In: Proceedings of the 11th International Workshop on Ontology Matching (OM-2014), Colocated with the 13th International Semantic Web Conference (ISWC-2014), **1317** of CEUR-WS, pp. 163–169, Trentino (2014)
8. Rus, V., Lintean, M., Banjade, R., Niraula, N., Stefanescu, D.: Semilar: the semantic similarity toolkit. In: Proceedings of the 51st Annual Meeting of the Association for Computational Linguistics, pp. 163–168, Sofia (2013)

Safe Suggestions Based on Type Convertibility to Guide Workflow Composition

Mouhamadou Ba[1](✉), Sébastien Ferré[2], and Mireille Ducassé[1]

[1] IRISA/INSA Rennes, 20 Avenue des Buttes de Coesmes,
35708 Rennes Cedex, France
mouhamadou.ba@irisa.fr
[2] IRISA/Université de Rennes 1, 263 Avenue Général Leclerc,
35042 Rennes Cedex, France

Abstract. This paper proposes an interactive approach that guides users in the step-by-step composition of services by providing safe suggestions based on type convertibility. Users specify the points of the workflow (called the focus) they want to complete, and our approach suggests services and connections whose data types are compatible with the focus. We prove the safeness (every step produces a well-formed workflow) and the completeness (every well-formed workflow can be built) of our approach.

1 Introduction

In a number of domains, particularly in bioinformatics, there is a need for complex data analysis. To that end, elementary data analysis operations, called services, are composed as workflows. Due to the distributed and heterogeneous resources, it is difficult to compose services. Some systems rely on experts who manually develop workflows as scripts. The approach has limitations in usability and scalability [1]. For example, even for experts, the development of scripts quickly becomes heavy and time-consuming. Moreover, developing scripts is, in general, beyond the capacities of domain end-users. Automated approaches are proposed to generate workflows [2]. For example, Kashlev and al. [3] generate executable workflows by matching and transforming input and output data. AI planning techniques are also used to automatically generate compositions of services [4]. Although fully automatic approaches seem appealing, in the context of bioinformatics they are not adapted because users are scientists who generally wish to control the construction of workflows. Furthermore, those approaches require *a priori* complete specifications that are almost impossible to produce in a bioinformatics context. In order to integrate users in the process of composing services, interactive approaches propose support along two axes: verification and suggestions. For example, Kim et al. [5] provide an error scan algorithm that detects errors and provides suggestions to fix them. Their suggestion mechanism relies on global properties (e.g., no cycle, no redundancy) of the workflow but does not focus suggestions to particular points of the workflow the user wants to manage. DiBernado et al. [6] are able to decompose data structures in order

© Springer International Publishing Switzerland 2015
F. Esposito et al. (Eds.): ISMIS 2015, LNAI 9384, pp. 230–236, 2015.
DOI: 10.1007/978-3-319-25252-0_25

to suggest services that can consume parts of those data. However, they do not enable suggestions on multiple points, and do not recompose input data from several parts of the output. The approach of Kim et al. has the same drawback.

In this paper, we propose a guided approach of service composition based on convertibility between input and output types of services. We formally define a workflow model, a suggestion mechanism and possible user actions. The contributions of the paper are manifold. Firstly, users can specify several points where they want to connect a new service. Those points form what we call the focus. The suggestions give only the services that can be safely connected to the focus, and all of them. Hence the number of suggestions that users have to consider for an insertion is significantly reduced. Secondly, all user actions are safe with respect to type convertibility: they lead to a well-formed workflow. There is therefore no need for *a posteriori* type verification. Thirdly, the approach is complete: all workflows that are well-formed with respect to type convertibility can be built. The construction steps can be applied backward or forward (from inputs to outputs, or from outputs to inputs) at any time. Furthermore, the convertibility algorithm, defined in Ba et al. [7], allows to take into account in a fine way the composite structure of data types. A proof-of-concept implementation exists.

Section 2 summarizes our previous work on convertibility between inputs and outputs of services. Section 3 defines in a formal way the principles of our workflow composition approach. Section 4 states properties of safeness and completeness, and Sect. 5 shows a use case.

2   Type Convertibility

Type convertibility is used to assess compatibility between inputs and outputs of services. It relies on a rule system based on type abstractions More details about convertibility can be found in Ba et al. [7].

Services and Data Type Abstractions. In this paper, we concentrate on the data types of the main parameters of services. The inputs and outputs of services are represented by abstract types defined from an open set of primitive types and a fixed set of type constructors. The type constructors are *Tag, Optional, Tuple, List* and *Union*. *Tag* denotes XML elements, *Optional* denotes not required elements, *Tuple* denotes sequences of elements that have different types, *List* denotes sequences of elements that have the same type, and *Union* denotes union of types. Supported types represent data through annotations, hierarchy and composition, the elementary types being primitive types such as *integer* and *string*.

Rule System. The rule system decides whether types representing service inputs and outputs are convertible. Convertibility is defined as a judgement function $f : S \rightarrow T$, where S is a source type and T a target type. A judgement $f : S \rightarrow T$ holds true if and only if it is possible to build a proof tree with that judgement at the root, and where each node instantiates a rule. This is done by a deduction system that works by structural induction on couples of types (S, T), covering all combinations of type constructors for which convertibility is possible. It uses base judgements on tags and on primitive types, defined according to the application domain. Our system, by applying convertibility on a library of services, detects all possible conversions between input and output types of services. Furthermore it generates executable converters.

3   Our Workflow Composition Approach

Our approach guides a user to construct a workflow step by step, using background knowledge about the services and type convertibility presented in the previous section. To define a workflow, the addition of a service consists of the following 3 steps. First, users indicate the points of the workflow that they want to grow. Second, our system makes suggestions related to those points, using background knowledge and preserving workflow properties. Third, users can select a suggestion to extend the workflow. They can also remove an element of the workflow at any time. In the following, we introduce the components used in our approach, and formalize them.

3.1   Knowledge Base

The knowledge base is composed of types $(\mathcal{T})$, ports (Π), domain services (Σ) and type convertibilities (Λ).

Definition 1. *An abstract port $\pi = (name, type) \in \Pi$ has a name and a type $\in \mathcal{T}$. A service $\sigma = (name, \Pi^{in}, \Pi^{out}) \in \Sigma$ has a name, a set of abstract input ports Π^{in}, and a set of abstract output ports Π^{out}. A convertibility $\lambda = (type_1, type_2) \in \Lambda$ has a couple of types such that $type_1$ is convertible to $type_2$.*

In the following, we use dot notation to designate tuple components in definitions, for example $\sigma.\Pi^{out}$ is the set of output ports of service σ.

3.2   Workflow

A workflow is defined by instantiating services and ports, and by linking them.

Definition 2 (Workflow). *A workflow is a 3-tuple $W = (P, T, L)$, where*

P is a set of concrete ports, namely instances of abstract ports. Each concrete port $p = (id, \pi)$ is uniquely identified by id and refers to the port π it instantiates.

T is a set of tasks, namely instances of services. Each task $t = (id, \sigma, P^{in}, P^{out})$ is an instance of the σ service, it is uniquely identified by id. It defines input ports and output ports $(P^{in}, P^{out} \subset P)$ that are, respectively, instances of the input and output ports of σ $(\sigma.\Pi^{in}, \sigma.\Pi^{out})$. T contains two special tasks $(t_{setter}$ and $t_{getter})$ handling the global inputs and outputs of the workflow.

L is a set of links between ports. Each link $l = (p_{out}, p_{in})$ is defined from an output concrete port to an input concrete port.

We note $W.IP$ and $W.OP$, respectively, the set of input ports and the set of output ports of the tasks of a workflow. The empty workflow is the workflow that has no task and no links, apart the special tasks t_{getter} and t_{setter} and their ports. A workflow W can be abstracted as a dataflow graph.

Definition 3 (Dataflow graph). *Given a workflow $W = (P, T, L)$, its dataflow graph $DF(W)$ is defined by the graph $G = (V, E)$, where $V = W.P$ and $E = W.L \cup \{(p_{in}, p_{out}) \mid \exists t \in W.T : p_{in} \in t.P^{in} \wedge p_{out} \in t.P^{out}\}$*

3.3 Workflow Properties

Properties define constraints on the workflow that have to be satisfied during composition. To define these properties, we use port dependencies from the dataflow graph. In a well-formed workflow (see Definition 4), there is no cycle, and an input only consumes data from one output. However, a data from an output may feed in several inputs. For each link, data consumed by the input must be convertible from the output data. We also consider fully-defined workflows where all inputs are connected.

Definition 4. *Considering the dataflow graph, a port p_2 depends on a port p_1 if and only if there is a path from p_1 to p_2 in the dataflow graph. A workflow is well-formed if and only if (1) no port of the workflow depends upon itself, i.e., there is no cycle in $DF(W)$, (2) every input port of the workflow is, at most, the target of one link and (3) every link (p_{out}, p_{in}) verifies convertibility, i.e., $(p_{out}.\pi.type, p_{in}.\pi.type) \in \Lambda$.*

3.4 Focus and Suggestions

Suggestions are components proposed to users during composition. Given a workflow $W = (P, T, L)$, we define a focus $F \subseteq P$ as a set of ports of the tasks composing the workflow. The focus defines points of the workflow from which tasks and links can be added. We define $F^{in} = F \cap W.IP$ the input ports of the focus and $F^{out} = F \cap W.OP$ the output ports of the focus. Asking for suggestions with reference to the focus is equivalent to requesting services that can be connected to all focus ports. Suggestions concern on the one hand services whose input and output types are convertible to/from the port types of the focus, and on the other hand links verifying convertibility between ports of the workflow. Suggestions must guarantee that the workflow after insertion is well-formed.

Function $sugg^\sigma$ (see Definition 5) returns suggested services and the associated links. In the definition of $sugg^\sigma$, detected convertibilities between the ports of the focus and ports of a service are described by mappings (φ^{in}) from the set of output ports of the focus to the set of input ports of the service, and (φ^{out}) from the set of input ports of the focus to the set of output ports of the service. The two mappings define links between focus ports and ports of suggested services. For a suggestion to be valid, it is necessary that (1) no input port of the focus depends on a port of the workflow, and (2) no output port of the focus depends on an input port of the focus. The first condition prevents multiple links on input ports, and the second prevents cycles. The function for suggestion of links between existing tasks is not shown but it is similar to the function for suggestion of tasks, it is based on the same properties.

Definition 5 (Service suggestions). *Provided that $\forall p_o \in F^{out}, p_i \in F^{in}$, p_o does not depend on p_i, and p_i does not depend on any port of the workflow, the suggested services are defined as follows: $sugg^\sigma(W, F) = \{(\sigma, \varphi^{in}, \varphi^{out}) \mid \sigma \in \Sigma \wedge \varphi^{in} \in F^{out} \rightarrow_{injective} \sigma.\Pi^{in} \wedge \forall p \in F^{out} : (p.\pi.type, \varphi^{in}(p).type) \in \Lambda \wedge \varphi^{out} \in F^{in} \rightarrow \sigma.\Pi^{out} \wedge \forall p \in F^{in} : (\varphi^{out}(p).type, p.\pi.type) \in \Lambda\}$*

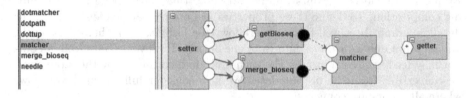

Fig. 1. Insertion of the suggested service 'matcher' in a workflow under construction.

3.5 Transformations

Transformations define authorized actions. The main transformations during workflow construction are insertion and removal of services and links. The insertion of a service corresponds to the addition of a new task to the workflow, and the addition of links between ports of the task and ports of the focus. Insertion of links is similar to insertion of tasks and removal is a simpler case. The services and links to insert in the workflow are chosen from the suggestions. Every insertion of suggested service or link, and every removal produces a well-formed workflow, when starting with a well-formed workflow (see Sect. 4). We assume that functions $inst^\pi$ and $inst^\sigma$ define the instantiation of ports and services. Function $insert^\sigma$ uses the focus and returns the new workflow after insertion. From a practical point of view, links actually contain converters when needed. Note that users may have to make a choice, when several converters are generated.

Definition 6 (Task insertion). *Let $(\sigma, \varphi^{in}, \varphi^{out}) \in sugg^\sigma(W, F)$, insertion is defined by: $insert^\sigma(W, F, (\sigma, \varphi^{in}, \varphi^{out})) = W'$, where $t = inst^\sigma(\sigma)$,*
$W'.P = W.P \cup t.P^{in} \cup t.P^{out}, \quad W'.T = W.T \cup \{t\},$
$W'.L = W.L \cup \{(p, p') \mid p \in F^{out} \wedge p' = inst^\pi(\varphi^{in}(p))\}$
$\cup \{(p', p) \mid p \in F^{in} \wedge p' = inst^\pi(\varphi^{out}(p))\}$

4   Safeness and Completeness

We state the theorems about the safeness and completeness of suggestions and transformations. Due to lack of space, we only sketch proofs here.

Theorem 1 (Safeness & completeness). *Starting from $W_\emptyset$ (empty workflow), every sequence of transformations leads to a well-formed workflow. For every well-formed workflow W, there is a finite sequence of transformations starting from $W_\emptyset$.*

Proof (Sketch). The empty workflow is well-formed because it does not contain links (e.g., no dependencies). Each transformation on a well-formed workflow produces a well-formed workflow. Thus, by induction, any sequence of transformations leads to a well-formed workflow. We prove the completeness by induction on well-formed workflows with decreasing size, i.e. number of tasks and links. ∎

5   Use Case

Figure 1 shows an example of service insertion in a well-defined workflow during composition. Black circles represent the focus. It is composed of two output ports of tasks *getBioseq* and *merge_bioseq* that have the same type *bioseq*. Considering the focus, the user is interested in services having at least two input ports such that each input is convertible from *bioseq*. According to our system, suggested services at the left of Fig. 1 verify that condition. Service *matcher* is one of them and can be inserted. The resulting workflow is well-formed. In practice, ports composing the focus are added one by one. For example, in Fig. 1, if we suppose that the user initially adds the output port of *getBioseq*, the initial suggestions contain more suggestions, including services with only one input. When the user adds the output port of *merge_bioseq*, some services are filtered out from the suggestions. The more ports are added to the focus, the more services are filtered out from the suggestions. The use case is created from our proof-of-concept implementation. Our implementation uses about 120 services from the version 5 of Emboss [8]. The GUI enables users to select suggestions and to apply transformations. It is still basic but it is reactive and already complete for the composition of well-formed workflows. A perspective is to take into account others aspects of services and provide ranking using Logical Information Systems [9].

6   Conclusion

This paper provides a guided approach to compose services. The approach is based on convertibility between input and output data types of services. Our system enables users to incrementally specify the focus (a set of ports) of a workflow to be completed; it suggests services and links whose data are compatible with the focus. Our approach produces a well-formed workflow at every step, and enables to produce every well-formed workflow.

References

1. Gil, Y.: Workflow composition: semantic representations for flexible automation. In: Taylor, I.J., Deelman, E., Gannon, D.B., Shields, M. (eds.) Workflows for e-Science, pp. 244–257. Springer, Heidelberg (2007)
2. Romano, P.: Automation of in-silico data analysis processes through workflow management systems. Brief Bioinform **9**(1), 57–68 (2008)
3. Kashlev, A., Lu, S., Chebotko, A.: Coercion approach to the shimming problem in scientific workflows. In: IEEE International Conference on Services Computing, pp. 416–423 (2013)
4. Rao, J., Su, X.: A survey of automated web service composition methods. In: Cardoso, J., Sheth, A.P. (eds.) SWSWPC 2004. LNCS, vol. 3387, pp. 43–54. Springer, Heidelberg (2005)
5. Kim, J., Spraragen, M., Gil, Y.: An intelligent assistant for interactive workflow composition. In: International Conference on Intelligent User Interfaces, pp. 125–131 (2004)
6. DiBernardo, M., Pottinger, R., Wilkinson, M.: Semi-automatic web service composition for the life sciences using the biomoby semantic web framework. J. Biomed. Inform. **41**(5), 837–847 (2008)
7. Ba, M., Ferré, S., Ducassé, M.: Generating data converters to help compose services in bioinformatics workflows. In: Decker, H., Lhotská, L., Link, S., Spies, M., Wagner, R.R. (eds.) DEXA 2014, Part I. LNCS, vol. 8644, pp. 284–298. Springer, Heidelberg (2014)
8. Rice, P., Longden, I., Bleasby, A.: Emboss: the european molecular biology open software suite. Trends Genet. **16**(6), 276–277 (2000)
9. Ferré, S.: Camelis: a logical information system to organise and browse a collection of documents. Int. J. Gen. Sys. **38**(4), 379–403 (2009)

MUSETS: Diversity-Aware Web Query Suggestions for Shortening User Sessions

Marcin Sydow[1,2]([✉]), Cristina Ioana Muntean[3], Franco Maria Nardini[3],
Stan Matwin[1,4], and Fabrizio Silvestri[5]

[1] Polish Academy of Sciences, Warsaw, Poland
msyd@ipipan.waw.pl
[2] Polish-Japanese Institute of Information Technology, Warsaw, Poland
[3] ISTI-CNR, Pisa, Italy
nardini@isti.cnr.it
[4] Big Data Institute, Dalhousie University, Halifax, Canada
[5] Yahoo Labs, London, UK

Abstract. We propose MUSETS (multi-session total shortening) – a novel formulation of the query suggestion task, specified as an optimization problem. Given an ambiguous user query, the goal is to propose the user a set of query suggestions that optimizes a diversity-aware objective function. The function models the expected number of query reformulations that a user would save until reaching a satisfactory query formulation. The function is diversity-aware, as it naturally enforces high coverage of different alternative continuations of the user session. For modeling the topics covered by the queries, we also use an extended query representation based on entities extracted from Wikipedia. We apply a machine learning approach to learn the model on a set of user sessions to be subsequently used for queries that are under-represented in historical query logs and present an evaluation of the approach.

Keywords: Web query suggestions · Diversity · Session shortening · Query logs · Learning to rank

1 Introduction

We consider the problem of generating search query suggestions triggered by an ambiguous or underspecified user query in order to shorten the subsequent iterative query refinement process while the actual information need is unknown.

Imagine a web search session s started by a user with a query q. Typically, if the query q is ambiguous or underspecified, then, once started, such query session can be continued in many possible ways, each representing a different information need behind the starting query. For example, the query "windows" has many possible interpretations. Some users would reformulate it after seeing the first result page to $q_1 =$ "windows 7" or, alternatively, $q_2 =$ "big windows". These reformulations could be subsequently even further reformulated until the user gets to a satisfactorily refined level and finds satisfactory results, for example

© Springer International Publishing Switzerland 2015
F. Esposito et al. (Eds.): ISMIS 2015, LNAI 9384, pp. 237–247, 2015.
DOI: 10.1007/978-3-319-25252-0_26

q_1 can be further reformulated to $q_{11} =$ "windows 7 download" or $q_{12} =$ "windows 7 manual" and q_2 to $q_{21} =$ "big picture windows", etc. Notice that the problem we consider is different (more general) than a simple automatic query completion since the reformulations can have different wording.

Our MUSETS (MUlti-SEssion Total Shortening) approach has the following goal. Given the initial query q, the goal is to present the user a *set of query suggestions* S_q satisfying the following two conditions:

- it *shortens* maximally the subsequent possible session continuation to lead the user faster to the satisfactory level of refinement of the query
- it is *diversity-aware*, i.e. potentially covers many possible aspects or interpretations of q;

We treat each *potential session* starting with q and continued with a particular sequence of reformulations, e.g. $q, q_1, q_{12}, \ldots$, or $q, q_2, q_{21}, \ldots$, etc. as a basic mean of representing a *separate aspect or interpretation* of the initial query q.

Contributions of the Paper: The paper proposes a novel general approach to diversity-aware query suggestion and some particular initial implementations of this idea. In particular it proposes: a novel form of specification of the query suggestion problem as an optimization problem; a new diversity-aware user intent model based on sessions from query logs; several variants of utility measure for query suggestions; a machine learning approach to the problem, and experimental results that would help in conceptualizing the sequel of this work.

2   Related Work

Various techniques proposed during last years for query recommendation are very different, yet they have in common the exploitation of usage information recorded in query logs. One subset of these recommendation algorithms uses clustering to determine groups of similar queries that lead users to related documents [1]. Deng *et al.* propose the use of frequency-inverse query frequency (UF-IQF) [6], based on entropy models. An efficient approach for generating suggestions is "Search Shortcuts" by Broccolo *et al.* [4]. Recently, Bonchi *et al.* [3] present a recommendation method based on the well-known concept of the center-piece subgraph that allows for time/space efficient generation of suggestions, also for rare, i.e., long-tail queries. Furthermore, Vahabi *et al.* [12] propose a novel approach that is not based on the keywords of the original query. Authors intentionally seek out (build) orthogonal queries, which are related queries that have (almost) no common terms with the user's query. This allows an orthogonal query to satisfy the user's informational need when small perturbations of the original keyword set are insufficient.

In Boldi *et al.* [2], the basic unit is a query and query refinement is modeled as a stochastic process over the queries. In our work we take a very different approach, where each separate *session* is treated as a mean to model a separate information need or interpretation of a given initial query. Shortening the query sessions was studied in [8], where the aim was to predict the *final query* of a started session. That approach did not explicitly address the diversity-awareness, while ours does. Moreover, as explained in Sect. 3, the optimal suggestions in MUSETS *do not have to be final queries*. In [11] diversified query suggestions were generated based on a very different approach: a pair-wise dissimilarity model between queries.

Work similar to ours is Ozertem *et al.* [9]. Authors propose a machine learning approach to learn the probability that a user may find a follow-up query both useful and relevant, given his initial query. The approach is based on a machine learning model which enables generalization of queries that have never occurred in the logs. The approach we are proposing is different, as we are using a machine learning framework to approximate the expected number of steps carried out by a user until reaching a satisfactory query formulation.

3   Problem Description

Each user session can be represented as a sequence of queries in this session. Let $sessions(q)$ denote the set of all sessions (i.e. sequences of queries) recorded in the logs that start with query q, and $sessions(q, q')$ - the subset of $sessions(q)$ that contain query q'. Notice that a given session (i.e. a particular sequence of queries) can repeatedly occur in the logs. Thus, each session $s \in sessions(q)$ has an associated multiplicity $mult_q(s)$. Let $len(s)$ denote the number of reformulations in s (the number of queries in the sequence minus 1). Let $pre(s, q')$ denote the number of queries that precede the earliest[1] occurrence of q' in s, that can be interpreted as how much the session would be *shortened* if q' was submitted by the user immediately after the initial query of the session s.

Hence, it is natural to define the degree of shortening of the session s with query q' in this way; in this paper we propose the following two variants:

- **"absolute shortening"**:

$$sessionShort(s, q') = pre(s, q')$$

- **"normalised shortening"**:

$$sessionShort(s, q') = pre(s, q')/len(s)$$

Example. Let assume that the initial ambiguous user query is a and that the logs contain the following set of sessions starting with query a: $s = (a, b, c, d)$, $s' = (a, e, f)$ (repeated twice in the logs), $s'' = (a, e, g, h)$, $s''' = (a, e, h)$. According to this: $sessions(a) = \{s, s', s'', s'''\}$, $sessions(a, f) = \{s'\}$, $mult_a(s) = 1$, $mult_a(s') = 2$, $len(s) = len(s'') = 3$, $len(s') = 2$, $pre(s, d) = 3$, $pre(s'', e) = 1$.

We generalize the shortening utility from a single query suggestion q' to a *set* of queries S_q as follows:

$$sessionShortSet(s, S_q) = \max_{q' \in S_q} sessionShort(s, q') \tag{1}$$

---

[1] Optionally, it is also possible to consider the latest occurence thereof.

Maximum operator is justified by the assumption that the user presented with a set of suggestions will exactly know which suggestion would maximally shorten their sequence of reformulations[2].

Now, we define the following objective set function to be maximised:

$$f(S_q) = \sum_{s \in sessions(q)} P(s|q) \cdot sessionShortSet(s, S_q) \tag{2}$$

where $P(s|q)$ is the likelihood that s will be the subsequent continuation of q.

We propose several variants for modeling $P(s|q)$[3]:

- **"simplistic likelihood"**:
$$P(s|q) = 1$$

- **"cardinality-based likelihood"**:
$$P(s|q) = mult_q(s)/(\sum_{s' \in sessions(q)} mult_q(s'))$$

- **"length-weighted likelihood"**:
$$P(s|q) = (len(s) * mult_q(s))/\sum_{s' \in sessions(q)} (len(s') * mult_q(s'))$$

The Eq. 2 can be interpreted as the expectation (over the possible session continuations) of the shortening utility of the set of query suggestions S_q presented to the user.

Example, cont. To illustrate the concepts, let's use our working example. The variants of $P(s'|a)$ are equal to: 2/5 ("cardinality-based likelihood"), $4/12 = 1/3$ ("weighted likelihood"), respectively. The value of $sessionShort(s, c)$ is equal to: 2 for the "absolute shortening" variant (and 2/3 for "normalized shortening").

Finally, we define the MUSETS problem as an optimization problem as follows:

INPUT: query q, number k of suggestions to be generated[4] and a set of recorded sessions $sessions(q)$ that start with q

OUTPUT: a k−element set S_q of query suggestions that *maximises* the objective function presented in Eq. 2.

Example cont. For the combination of the absolute shortening variant and cardinality-based likelihood variant, the optimal values of the objective function

---

[2] While this is an "optimistic" assumption, other variants can be also considered, in particular min operator ("pessimistic" variant) or some other aggregation operators.

[3] However other options are possible, what is envisaged in the continuation work.

[4] The choice of the value of k is an independent interesting problem that depends on the level of ambiguity of the query and some other external conditions such as the space available for presenting the suggestions in the front-end application. We do not study this problem here and treat k as an external parameter.

on our example are as follows: $f(\{e\}) = 5$, $f(\{d, e\}) = 7$ and $f(\{d, f, g\}) = 10$, for $k = 1, 2, 3$, respectively.

Let us now highlight some important and interesting properties of the objective function $f(\cdot)$:

- *(inherent diversity-awareness)*: the definition naturally supports *diversity* understood as the awareness of many possible alternative interpretations of q expressed by possible continuations of the query. Technically, the definition of f *promotes* the selection of suggestions that shorten as many different possible session continuations as possible.
- *(nonfinal queries)*: the optimal set S_q may contain query suggestions that are not necessarily "final" in their sessions. In our example, for $k = 1$ and the simplistic-absolute combination variant, the optimum set is $S_q = \{e\}$ i.e. the best suggestion is not a final query in any session. This makes our approach different from that in [8].
- *(non-monotonicity)*: i.e., the solution for a higher value of k does not necessarily subsume the solution for lower value of k. In addition, sometimes an earlier query in a session can be preferred than a later one. Non-monotonicity is a typical feature of set-wise approach common in diversity-aware applications, implying that a simple greedy approach might be sub-optimal here.

4 Solving the Problem

If the initial query q and sessions started by q are sufficiently represented in query logs, the problem could be treated as a standard optimization problem and approached directly by optimizing the objective function[5].

However, in practice, the sessions starting with q might be insufficiently represented in historical logs. For this reason, in this paper we propose an approximate efficient approach to solve the optimisation problem defined in Sect. 3. Namely, instead of directly optimising the set-wise objective function 2 we greedily select suggestions and use machine learning approach to approximate shortening utility value of each candidate suggestion. More precisely, we proceed in two phases. In the first (training) phase we learn the session model with some pre-computed, session-independent features of queries that are well represented in the historical logs. Then, in the second phase, for an incoming query q, we apply the model to predict the shortening utility $shortening(q, q')$ of each potential suggestion q' out of a set of some pre-filtered candidates by defining it simply as follows:

$$shortening(q, q') = f(\{q'\})$$

Finally, we construct S_q out of top-k candidate suggestions. The proposed approach turned out to be computationally feasible but we are aware that such

---

[5] Such a direct approach can be of high computational complexity though.

a simplification is sub-optimal and that our current approach leaves room for improvement that can be tackled in future work.

Machine Learning Approach. Given a user query q, we propose to learn a ranking function $h : X \rightarrow Y$, where the input space X is a vector representation of a candidate suggestion q' for q. We employ the query features shown in Table 1 to map the value of $shortening(q, q')$ in the input space X. We compute a series of features that best describe the pair regarding type, popularity and topics covered. Firstly we consider the tokens from which the queries are composed and calculate union, intersection, difference and symmetric difference. We also take into account the queries they co-occur with, whether one query is part of the other and whether they tend to be initial or inner session queries. Moreover, we also calculate entropy- and probability-related features.

A challenging task is to represent the queries from a topic point of view. In order to do this, we use entity linking techniques presented in [5]. For each query, we compute the list of spots (parts of the query that are present as anchors in Wikipedia) and the list of entities for each of the spots. By doing so, we are able to find all the potential meanings, without disambiguation. For the initial queries we also compute the extended representation of entities from annotated final queries with which they co-occur in clicked sessions. With these, we compute features like union, intersection, difference and symmetric difference, as presented in Tables 1 and 2. Note that due to space limitations some features calculated for spots (similarly to the ones for entities) are not shown.

The output space Y for the learning problem consists of a set of ground-truth labels determining the relevance of each query prediction. As our aim here is to learn a model that maximizes a set-wise function, we select as training examples for each session queries actually co-occurring in the user sessions of training set. For such candidate q', we compute their value of $shortening(q, q')$. Moreover, we add to the training set some negative examples by selecting random queries from sessions not starting from q and assigning them a target label equal to zero. Target labels for the ambiguous query q are positive and negative examples assigned according to:

$$y_{q'} = \begin{cases} shortening(q, q'), & \text{if } q' \text{ is in a session starting with } q; \\ 0, & \text{otherwise.} \end{cases}$$

We employ Multiple Additive Regression Trees (MART) [7], state-of-the-art in web search ranking, as the learning to rank algorithm. It consists of an ensemble of regression trees for determining the predicted value of the label assigned to each candidate query. The loss function used for optimizing the learning is RMSE between the label assigned to each example in the training set, i.e., $y_{q'}$, and its predicted value, $h(q')$.

In the training phase, examples for a set of training sessions are used to learn the MART. The model is then employed in the test phase to score test candidates. The input of the test phase is a list of candidate queries for any given test session. The output of the test phase is a re-ranked list of candidates sorted by decreasing probability of being the suggestion query of the test session.

Table 1. List of query-related features used to model a *shortening*(q, q').

Features	
qi-tokens	The number of tokens in the initial query
qc-tokens	The number of tokens in the candidate query
token-intersection	The intersection of tokens for the two queries
token-union	The union of tokens for the two queries
token-difference1	The difference of tokens between the initial query and the candidate query
token-difference2	The difference of tokens between the candidate query and the initial query
token-symmetric-difference	The symmetric difference of tokens
coocurring-queries-union	The union of co-occurring queries with the initial and the candidate query
cooccuring-queries-intersection	The intersection of co-occurring queries with the initial and the candidate query
difference-qi-qc	The portion of text where the two queries differ, more precisely, the remainder of the candidate query, starting from where it's different from the initial query
qi-substring-of-qc	Reflects whether the initial query is a substring of the candidate query
type-of-query-qc	Reflects whether the candidate query is preponderantly an initial query or an inner query
type-of-query-qi	Reflects whether the initial query is preponderantly an initial query or an inner query
edit-distance-for-queries	Computes the Levenshtein Distance between the initial and the candidate query
entropy-qi	The entropy of the initial query
entropy-qc	The entropy of the candidate query
probability-qi	The probability of the initial query
probability-qc	The probability of the candidate query
qi-as-qf-probability	The probability of the initial query of being a final query

5   Experiments

We evaluate MUSETS on sessions devised from the publicly available MSN RFP 2006 query logs [10]. We preprocess the log by converting all the queries to lowercase, and by removing stop-words and punctuation/control characters. We then apply a session splitting technique based on the Query Flow Graph [2]. From

Table 2. List of entity-related features used to model a *shortening*(q, q').

Features	
entities-qi	The number of entities found for the initial query
entities-qi-extended	The number of entities for the initial queries computed from annotated co-occurring queries
entities-qc	The number of entities found for the candidate query
entities-union	The union of entities of initial and candidate query
entities-intersection	The union of entities of initial and candidate query
entities-difference1	The difference of entities between the initial query and the candidate query
entities-difference2	The difference of entities between the candidate query and the initial query
entities-symmetric-difference	The symmetric difference of entities between the candidate query and the initial query
entities-union-extended	The union of entities between the extended entity representation of the initial query and the entities of the candidate query
entities-intersection-extended	The intersection of entities between the extended entity representation of the initial query and the entities of the candidate query
entities-difference1-extended	The difference of entities between the extended entity representation of the initial query and the entities of the candidate query
entities-difference2-extended	The difference of entities between the entities of the candidate query and the extended entity representation of the initial query
entities-symmetric-difference-extended	The symmetric difference of entities between the extended entity representation of the initial query and the entities of the candidate query
probability-most-frequent-entity	The probability of the most frequent entity of the initial query in respect to the other entities from the extended entity representation of the initial query
probability-second-most-frequent-entity	The probability of the second most frequent entity of the initial query in respect to the other entities from the extended entity representation of the initial query
probability-third-most-frequent-entity	The probability of the third most frequent entity of the initial query in respect to the other entities from the extended entity representation of the initial query
probability-avg-3-most-frequent-entity	The average probability of the top three most frequent entities of the initial query in respect to the other entities from the extended entity representation of the initial query
entities-with-freq-1	The number of entities with frequency equal to one in the extended entity representation of the initial query

Table 3. Example of Suggestions derived for the query "nemo"

Query	Candidate Suggestions	$shortening(q, q')$
nemo	finding nemo	0.65
	sea otter	0.07
	great white shark	0.06
	dolphins pictures	0.04
	sea creatures pictures	0.04
	whale sharks	0.03
	nemo pictures	0.03
	finding nemo video clip	0.02
	nemo and friends lamp	0.02
	nemo video	0.02

the set of sessions obtained, we filter out sessions with 3 or less queries assuming their starting query is ambiguous. The resulting set of sessions is then divided in training (30, 000 sessions) and test set (2, 000 sessions), used to measure the performance of our model.

To do so we need to extract, build and evaluate the performance of a machine-learned model. Given a list of sessions starting with the same ambiguous query, we model the pair (q, q') by means of the features presented before Table 1. Moreover, for pairs in the training and test sets we compute the value of the $shortening(q, q')$ to be the target label of the pair as in the previous section.

We employ the training set above to learn the MART by optimizing RMSE. To so, we use RankLib[6]. As a preliminary evaluation, we are reporting below an example of suggestions produced with a MART model learned by using the "simplistic" strategy for modeling $P(s|q)$ and the "absolute shortening" strategy for modeling $sessionShort(s, q')$. First of all, the top-1 recommendation (the one with the highest value of $shortening(q, q')$) reveals the correct meaning of the query "nemo", i.e., the well-know cartoon from Disney. Moreover, the nature of the recommendations in the list is twofold. The method returns relevant suggestions w.r.t the topic and it is also able to diversify them between aspects related to the movie (e.g., "nemo pictures", "finding nemo video clips", etc.) and aspect related to the sea and nature that are however closely related with the movie (e.g., "sea otter", "great white shark", "whale sharks", etc.).

We also test the effectiveness of the MART model in terms of NDCG (Normalised Discounted Cumulative Gain) at different k by scoring the test file built as described above. Table 4 provides some results. NDCG@10 is above 0.8. It means that the MART we learn is able to reproduce the target labels (i.e., $shortening(q, q')$) for the test set and thus, to build in a effective way the final list of suggestions in response to a given query. Moreover, NDCG@2 and NDCG@5 are also very high confirming that we can decrease the number of suggestions returned (k) without losing quality of the returned list (Table 4).

---

[6] http://people.cs.umass.edu/~vdang/ranklib.html.

Table 4. Performance on the test set for values of $k \in \{2, 5, 10\}$.

$P(s\|q)$	Metric	Score
Simplistic	NDCG@2	0.7836
	NDCG@5	0.8011
	NDCG@10	0.8214

6   Conclusions

In this paper we proposed a novel approach to generate a set of web search query suggestions to a user who submitted a potentially ambiguous query. The basic idea is to model the unkown user information need as the whole sequence of queries that consitute the potential session continuation and to achieve the diversity-awareness by modeling the objective function as an expectation over many potential session continuations. We proposed several variants for the components of the objective function. An efficient, approximate machine-learning approach was proposed to practically solve the problem. Preliminary results on selected variants of the problem show that MUSETS is able to produce recommendations that are both relevant and diverse with respect to the query of the user. It would be interesting to consider other variants and study its important theoretical properties (e.g. submodularity) and improve the practical method of computing the solution as well as continue experiments on larger samples, including some user evaluation experiments.

Acknowledgments. The work is partially supported by Polish National Science Centre 2012/07/B/ST6/01239 "DISQUSS" grant.

References

1. Baeza-Yates, R., Hurtado, C.A., Mendoza, M.: Query recommendation using query logs in search engines. In: Lindner, W., Fischer, F., Türker, C., Tzitzikas, Y., Vakali, A.I. (eds.) EDBT 2004. LNCS, vol. 3268, pp. 588–596. Springer, Heidelberg (2004)
2. Boldi, P., Bonchi, F., Castillo, C., Donato, D., Gionis, A., Vigna, S.: The query-flow graph: model and applications. In: Proceedings of CIKM'08. ACM (2008)
3. Bonchi, F., Perego, R., Silvestri, F., Vahabi, H., Venturini, R.: Efficient query recommendations in the long tail via center-piece subgraphs. In: Proceedings of SIGIR'12. ACM (2012)
4. Broccolo, D., Marcon, L., Nardini, F.M., Perego, R., Silvestri, F.: Generating suggestions for queries in the long tail with an inverted index. Inf. Process. Manage. **48**(2), 326–339 (2012)
5. Ceccarelli, D., Lucchese, C., Orlando, S., Perego, R., Trani, S.: Dexter: an open source framework for entity linking. In: Proceedings of ESAIR'13 (2013)
6. Deng, H., King, I., Lyu, M.R.: Entropy-biased models for query representation on the click graph. In: Proceedings of SIGIR'09. ACM (2009)

7. Friedman, J.H.: Greedy function approximation: a gradient boosting machine. Ann. Stat. **29**, 1189–1232 (2001)
8. Muntean, C.I., Nardini, F.M., Silvestri, F., Sydow, M.: Learning to shorten query sessions. In: Proceedings WWW '13 Companion (2013)
9. Ozertem, U., Chapelle, O., Donmez, P., Velipasaoglu, E.: Learning to suggest: a machine learning framework for ranking query suggestions. In: Proceedings of SIGIR '12. ACM (2012)
10. Silvestri, F.: Mining query logs: turning search usage data into knowledge. Found. Trends Inf. Retrieval **1**(1–2), 1–174 (2010)
11. Sydow, M., Ciesielski, K., Wajda, J.: Introducing diversity to log-based query suggestions to deal with underspecified user queries. In: Bouvry, P., Kłopotek, M.A., Leprévost, F., Marciniak, M., Mykowiecka, A., Rybiński, H. (eds.) SIIS 2011. LNCS, vol. 7053, pp. 251–264. Springer, Heidelberg (2012)
12. Vahabi, H., Ackerman, M., Loker, D., Baeza-Yates, R., Lopez-Ortiz, A.: Orthogonal query recommendation. In: Proceedings of RecSys'13. ACM (2013)

Encoding a Preferential Extension of the Description Logic $\mathcal{SROIQ}$ into $\mathcal{SROIQ}$

Laura Giordano[1]([✉]) and Valentina Gliozzi[2]

[1] DISIT - University Piemonte Orientale, Alessandria, Italy
laura.giordano@uniupo.it
[2] Dipartimento di Informatica - University di Torino, Torino, Italy
gliozzi@di.unito.it

Abstract. In this paper we define an extension of the description logic $\mathcal{SROIQ}$ based on a preferential semantics to introduce a notion of typicality in the language which allows defeasible inclusions to be represented in a knowledge base. We define a polynomial encoding of the resulting language into $\mathcal{SROIQ}$, thus showing that reasoning in the preferential extension of $\mathcal{SROIQ}$ has the same complexity as reasoning in $\mathcal{SROIQ}$.

1 Introduction

Recently, many different approaches have been proposed to extend the basic formalism of Description Logics (DLs) with nonmonotonic reasoning features [1,3–6,9–11,13,18,19,23,25]; the purpose of these extensions is that of allowing reasoning about *prototypical properties* of individuals or classes of individuals, as well as combining DLs with nonmonotonic rule-based languages, such as Datalog under the answer set semantics. The most well known semantics for nonmonotonic reasoning have been used to the purpose, from default logic [1], to circumscription [3], to Lifschitz's nonmonotonic logic MKNF [9,19,23], to Answer set Semantics [10], to preferential reasoning [4,5,11,13], to rational closure [5,6,8,15].

In particular, preferential extensions of description logics allow defeasible inclusions to be added to a knowledge base, to model typical, defeasible and non-strict properties of individuals. Their semantics extends DL semantics with a preference relation among domain individuals, along the lines of the preferential semantics introduced by Kraus, Lehmann and Magidor [20,22] (KLM for short) for the preferential logic P and for the rational logic R. In particular, in [11,12] a preferential extension of the description logic $\mathcal{ALC}$ has been introduced, while in [4] a rational extension of $\mathcal{ALC}$ is developed. Preferential description logics have been used as the bases of stronger non-monotonic formalisms, such as the rational closure defined by Lehmann and Magidor [22]. In particular, in [15] a rational closure construction for $\mathcal{ALC}$ has been presented which is based on a preferential extension of $\mathcal{ALC}$.

In this paper, we develop a preferential extension of the expressive DL $\mathcal{SROIQ}$ [16], called $\mathcal{SROIQ}^P\mathbf{T}$, that we regard as a first step towards the

© Springer International Publishing Switzerland 2015
F. Esposito et al. (Eds.): ISMIS 2015, LNAI 9384, pp. 248–258, 2015.
DOI: 10.1007/978-3-319-25252-0_27

definition of a stronger non-monotonic extension of $\mathcal{SROIQ}$, the description logic which is at the basis of the fragment of OWL called OWL 2 DL [21]. Generalizing [12], we introduce in the language of $\mathcal{SROIQ}$ a typicality operator $\mathbf{T}$, that selects the most typical instances of a concept C and we attribute to $\mathbf{T}$ the properties of KLM preferential consequence relation P [20]. We show that the preferential logic $\mathcal{SROIQ}^P\mathbf{T}$ can be polynomially encoded into $\mathcal{SROIQ}$, which means that reasoning in $\mathcal{SROIQ}^P\mathbf{T}$ has the same complexity as reasoning in $\mathcal{SROIQ}$ and that the optimized proof methods developed for $\mathcal{SROIQ}$ can be used for reasoning in $\mathcal{SROIQ}^P\mathbf{T}$ as well.

2   The Description Logic $\mathcal{SROIQ}$

We follow [16,21] for the description of the syntax and the semantics of $\mathcal{SROIQ}$.

Definition 1. *Consider an alphabet of concept names* $\mathcal{C}$, *role names* $\mathcal{R}$ *and individual constants* $\mathcal{O}$. *Let* U *be the universal role,* $A \in \mathcal{C}$, $S \in \mathcal{R}$, $a \in O$, *and* $n \in \mathbb{N}$. *We define:*

$$C := A \mid \top \mid \bot \mid \neg C \mid C \sqcap C \mid C \sqcup C \mid \forall S.C \mid \exists S.C \mid (\geq nS.C \mid (\leq nS.C) \mid \exists S.Self \mid \{a\}$$
$$S := U \mid R \mid R^-$$

A KB is a triple (TBox, RBox, ABox). TBox contains a finite set of *general concept inclusion axioms* (GCI) $C_1 \sqsubseteq C_2$; RBox contains a finite set of *complex role inclusion axioms* of the form $S_1 \circ \ldots \circ S_n \sqsubseteq S$, as well as a set of *roles assertions* of the form: *Ref(R), Irr(R)* and *Dis(R,S)* (for $R, S \neq U$), to require a role R to be reflexive, irreflexive and disjoint from role S, respectively[1]. ABox contains *individual assertions* of the form $C(a)$ and $R(a,b)$, where $a,b \in \mathcal{O}$.

As usual, we assume that only *simple roles* (see [16,21]) can occur in number restrictions, in the concept $\exists S.Self$, in individual assertions as well as in irreflexive and disjointness role assertions. Also, we assume the *regularity* requirement on role inclusions on the RBox, for which, again, we refer to [16].

Definition 2 (Interpretations in $\mathcal{SROIQ}$). *An interpretation in* $\mathcal{SROIQ}$ *is any structure* $I = (\Delta^I, \cdot^I)$ *where:* Δ^I *is a domain;* $\cdot^I$ *is an interpretation function that maps each atomic concept* A *to set* $A^I \subseteq \Delta^I$, *each atomic role* R *to a binary relation* $R^I \subseteq \Delta^I \times \Delta^I$, *and each individual name* a *to an element* $a^I \in \Delta^I$. *I is extended to complex concepts, to inverse roles and to the universal role as follows:*

- $\top^I = \Delta$ $\bot^I = \emptyset$
- $(C \sqcap D)^I = C^I \cap D^I$ $(C \sqcup D)^I = C^I \cup D^I$
- $(\neg C)^I = \{x \in \Delta \mid x \notin C^I\}$ $(\{a\})^I = \{a^I\}$
- $(\exists R.C)^I = \{x \in \Delta \mid \exists y.(x,y) \in R^I \text{ and } y \in C^I\}$
- $(\forall R.C)^I = \{x \in \Delta \mid \forall y.(x,y) \in R^I \text{ implies } y \in C^I\}$

---

[1] The symmetric and transitive role assertions *Sym(R)* and *Tra(R)* can be replaced by suitable role inclusion axioms, respectively, $R^- \sqsubseteq R$ and $R \circ R \sqsubseteq R$ [16].

- $(\geq nR.C)^I = \{x \in \Delta \mid \sharp\{y \mid (x,y) \in R^I \text{ and } y \in C^I\} \geq n\}$
- $(\leq nR.C)^I = \{x \in \Delta \mid \sharp\{y \mid (x,y) \in R^I \text{ and } y \in C^I\} \leq n\}$
- $(\exists R.Self)^I = \{x \in \Delta \mid (x,x) \in R^I\}$
- $(R^-)^I = \{(x,y) \mid (y,x) \in R^I\}$ $U^I = \Delta \times \Delta$

Definition 3 (Model of a KB). *We say that an interpretation* $I = (\Delta^I, \cdot^I)$ *satisfies a generalized concept inclusion axiom* $C \sqsubseteq D$ *if* $C^I \subseteq D^I$; *it satisfies a role inclusion axiom* $S_1 \circ \ldots \circ S_n \sqsubseteq S$ *if* $S_1^I \circ \ldots \circ S_n^I \subseteq S^I$; *it satisfies an assertion* $C(a)$ *if* $a^I \in C^I$; *it satisfies an individual assertion* $R(a,b)$ *if* $(a^I, b^I) \in R^I$. *I satisfies a role assertion* $Refl(R)$ *if the relation* R^I *is reflexive, and similarly for all the other role assertions.*

Given a KB=(TBox,RBox,ABox), we say that $I = (\Delta^I, \cdot^I)$ *is a model of KB if I satisfies all GCIs in TBox, all role inclusion axioms and role assertions in RBox, and all individual assertions in ABox.*

3   A Preferential Extension of $\mathcal{SROIQ}$

In this section we define an extension of the description logic $\mathcal{SROIQ}$ by allowing concepts of the form $\mathbf{T}(C)$, where C is a $\mathcal{SROIQ}$ concept, whose instances are intended to be the *typical* instances of a concept C. We call $\mathbf{T}$ the *typicality operator*. The concept $\mathbf{T}(C)$ can be used on the left hand side of GCIs to express typicality inclusions (or defeasible inclusions) of the form $\mathbf{T}(C) \sqsubseteq D$, meaning that the *typical* instances of a concept C are also instances of the concept D. We can therefore distinguish between properties that hold for all instances of C, expressed by *strict inclusions* $(C \sqsubseteq D)$, and those that only hold for the typical instances of C, expressed by *typicality* or *defeasible inclusions* $(\mathbf{T}(C) \sqsubseteq D)$.

As a difference with [12,15], where a typicality operator was introduced for $\mathcal{ALC}$, here we do not require that the typicality operator only occurs on the left hand side of concept inclusions. For instance, we can say that the friends of mary are all typical italians by introducing the following axiom expressing the requirement that all elements which are in a friendship relation with mary are tipical italians: $\exists friendOf.\{mary\} \sqsubseteq \mathbf{T}(Italian)$ (where $friendOf \equiv hasFriends^-$).

We introduce a notion of extended concept C_E in order to take into account the typicality operator $\mathbf{T}$ as follows:

$$C_E := C \mid \mathbf{T}(C) \mid \neg C_E \mid C_E \sqcap C_E \mid C_E \sqcup C_E \mid \forall S.C_E \mid \exists S.C_E \mid (\geq nS.C_E) \mid (\leq nS.C_E)$$

where C is a $\mathcal{SROIQ}$ concept as defined in Sect. 2. Observe that the typicality operator $\mathbf{T}$ cannot be nested. We assume that concept inclusion axioms in TBox have the form $C_E \sqsubseteq C_E$ and that extended concepts can occur in individual assertions in ABox. Also, we assume that the restrictions to simple roles, which hold for $\mathcal{SROIQ}$ concepts, hold for extended concepts as well. Hence, a $\mathcal{SROIQ}^P\mathbf{T}$ knowledge base is defined as a $\mathcal{SROIQ}$ knowledge base: a triple KB=(TBox,RBox,ABox), where extended concepts can occur in TBox and ABox.

Consider the following example of knowledge base, stating that typical Italians have black hair, typical students have at least 20 friends, except if they are shy (in which case they have at most 5 friends), that all Mary's friends are typical italians, and that Mary has at most 5 Italian friends. We also have the assertions stating that Mary is a student, that Bob is a Student, Italian and with Blond hair, and that they are friends.

Example 1. TBox:

$\mathbf{T}(Italian) \sqsubseteq hasBlackHair$

$\mathbf{T}(Student) \sqsubseteq\, \geq 20\ hasFriend.\top$

$\mathbf{T}(Student \sqcap Shy) \sqsubseteq\, \leq 5\ hasFriend.\top$

$hasBlackHair \sqcap hasBlondHair \sqsubseteq \bot$

$\exists friendOf.\{mary\} \sqsubseteq \mathbf{T}(Student)$

$\{mary\} \sqsubseteq\, \leq 5\ hasFriend.Italian$

RBox: $\{friendOf \equiv hasFriend^-, hasFriend^- \equiv hasFriend\}$

ABox: $\{Student(mary), (Student \sqcap Italian \sqcap hasBlondHair)(bob), hasFriend$ $(mary,\ bob)\}$, where, as usual, $R \equiv R'$ stands for $R \sqsubseteq R'$ and $R' \sqsubseteq R$.

In $\mathcal{SROIQ}^P\mathbf{T}$ we can infer all what can be inferred in $\mathcal{SROIQ}$: for instance we can derive that Bob is not a typical italian (from the first inclusion, as he is blond), but he is a typical student and has more than 20 friends (from the fifth inclusion, since he is Mary's friend). As a further step with respect to $\mathcal{SROIQ}$, by virtue of the $\mathbf{T}$ operator here we can simultaneously have the second and third inclusions (exceptional with respect to each other) without having to conclude that $Student \sqcap Shy$ is an empty concept with no instances (as it would be the case if the two inclusions were strict, without $\mathbf{T}$). Notice also that in this preferential extension of $\mathcal{SROIQ}$ we can make some other extra weak inferences: if we further added an inclusion concerning students, such as $\mathbf{T}(Student) \sqsubseteq Young$, it would follow that $\mathbf{T}(Student \sqcap Young) \sqsubseteq\, \geq 20\ hasFriend.\top$. This is a start, and indeed in this paper we prepare the ground to a nonmonotonic extension of $\mathcal{SROIQ}$ that supports stronger inferences, in the spirit of what done for $\mathcal{ALC}$ and $\mathcal{SHIQ}$ in [13,14], although here there are some extra difficulties, see Sect. 5 below. In the long run, we would like to define a nonmonotonic extension of $\mathcal{SROIQ}$ that allows us to conclude that for instance *mary* is a typical student with more than 20 friends (although at most 5 will be Italians), and that all typical italian students have black hair.

To define a preferential extension of $\mathcal{SROIQ}$ we attribute to $\mathbf{T}$ the properties of preferential consequence relation P [20,22]. As it has been done for the preferential extension of the logic $\mathcal{ALC}$ [12], we define the semantics of $\mathcal{SROIQ}^P\mathbf{T}$ in terms of preferential models, extending ordinary models of $\mathcal{SROIQ}$ with a *preference relation* $<$ on the domain, whose intuitive meaning is to compare the "typicality" of domain elements, that is to say, $x < y$ means that x is more typical than y. The typical instances of a concept C (the instances of $\mathbf{T}(C)$) are the instances x of C that are minimal with respect to the preference relation $<$ (no other instance of C is preferred to x).

Definition 4 (Models in $\mathcal{SROIQ}^P\mathbf{T}$). *A $\mathcal{SROIQ}^P\mathbf{T}$ interpretation $\mathcal{M}$ is any structure $\langle \Delta, <, \cdot^I \rangle$ where:*

- *Δ and $\cdot^I$ are a domain and an interpretation function, as in Definition 2;*
- *$<$ is an irreflexive, transitive and well-founded relation over Δ;*
- *the interpretation of concept $\mathbf{T}(C)$ is defined as follows:*

$$(\mathbf{T}(C))^I = Min_<(C^I)$$

where $Min_<(S) = \{u : u \in S$ and $\nexists z \in S$ s.t. $z < u\}$.

Furthermore, we say that an irreflexive and transitive relation $<$ is well-founded if, for all $S \subseteq \Delta$, for all $x \in S$, either $x \in Min_<(S)$ or $\exists y \in Min_<(S)$ such that $y < x$.

Definition 5 (Satisfiability and entailment in $\mathcal{SROIQ}^P\mathbf{T}$). *Given a $\mathcal{SROIQ}^P\mathbf{T}$ interpretation $\mathcal{M} = \langle \Delta, <, \cdot^I \rangle$, the notion of satisfiability of concept inclusion axioms, role inclusion axioms, role assertions and individual assertions is defined as in Definition 3, but for extended concepts instead of concepts.*

Given a $\mathcal{SROIQ}^P\mathbf{T}$ knowledge base KB=(TBox,RBox,ABox), we say that the interpretation $\mathcal{M} = \langle \Delta, <, \cdot^I \rangle$ is a model of KB if $\mathcal{M}$ satisfies all GCIs in TBox, all role inclusion axioms and role assertions in RBox, and all individual assertions in ABox.

Given a KB, let a query F be either a concept inclusion $C \sqsubseteq D$, where C and D are extended concepts, or an individual assertion. We say that F is entailed by KB, written $KB \models_{\mathcal{SHIQ}^R\mathbf{T}} F$, if for all models $\mathcal{M} = \langle \Delta, <, \cdot^I \rangle$ of KB, $\mathcal{M}$ satisfies F.

The logic $\mathcal{SROIQ}^P\mathbf{T}$, as well as the underlying $\mathcal{SROIQ}$, does not enjoy the finite model property [17]. The fact that irreflexive and transitive relation $<$ on Δ is well-founded guarantees that there are no infinite descending chains of elements of Δ.

As observed in [12], the meaning of $\mathbf{T}$ can be split into two parts: for any element $x \in \Delta$, $x \in (\mathbf{T}(C))^I$ when (i) $x \in C^I$, and (ii) there is no $y \in C^I$ such that $y < x$. In order to isolate the second part of the meaning of $\mathbf{T}$, one can introduce a new modality $\square$ and interpret the preference relation $<$ as an accessibility relation of this modality. Well-foundedness of $<$ ensures that typical elements of C^I exist whenever $C^I \neq \emptyset$, by avoiding infinitely descending chains of elements. The interpretation of $\square$ in $\mathcal{M}$ is as follows: $(\square C)^I = \{x \in \Delta \mid$ for every $y \in \Delta$, if $y < x$ then $y \in C^I\}$. It is easy to observe that x is a typical instance of C if and only if it is an instance of C and $\square \neg C$, that is to say:

Proposition 1 ([12]). *Given a model $\mathcal{M}$, given a concept C and an element $x \in \Delta$, we have that $x \in (\mathbf{T}(C))^I$ iff $x \in (C \sqcap \square \neg C)^I$*

In the following section we exploit this definition of the typicality operator $\mathbf{T}$ in terms of a Gödel-Löb modality $\square$ to define an encoding of $\mathcal{SROIQ}^P\mathbf{T}$ into $\mathcal{SROIQ}$.

4   Encoding of $\mathcal{SROIQ}^P\mathbf{T}$ in $\mathcal{SROIQ}$

In this section we show that reasoning in $\mathcal{SROIQ}^P\mathbf{T}$ can be reduced polynomially to reasoning in $\mathcal{SROIQ}$. This result generalizes the result given for $\mathcal{SHIQ}$ in [14] showing that a similar result holds for $\mathcal{SROIQ}$ and for a more general language which does not impose restrictions on the occurrences of $\mathbf{T}$.

The encoding exploits the definition of the typicality operator $\mathbf{T}$ in terms of a Gödel-Löb modality $\square$ as follows: $\mathbf{T}(C)$ is defined as $C \sqcap \square \neg C$ where the accessibility relation of the modality $\square$ is the preference relation $<$ in preferential interpretations.

Let KB=(TBox, RBox, ABox) be a $\mathcal{SROIQ}^P\mathbf{T}$ knowledge base. We define the encoding KB'=(TBox', RBox', ABox') of KB in $\mathcal{SROIQ}$ as follows. First, RBox'= RBox. TBox' is defined as follows:

For each $\mathbf{T}(A)$ occurring in the TBox, we introduce a new atomic concept $\square_{\neg A}$ and, for each inclusion $C \sqsubseteq D \in$ TBox, we let in TBox' the inclusion $C' \sqsubseteq D'$, where C' and D' are obtained from C and D, respectively, by replacing each occurrence of a concept $\mathbf{T}(A)$ with the concept $A \sqcap \square_{\neg A}$.

To capture the properties of the $\square$ modality, a new role name $P_<$ is introduced to represent the relation $<$ in preferential models, and the following concept inclusion axioms are also introduced in TBox' (for all concepts A such that $\mathbf{T}(A)$ occurs in KB):

$$\square_{\neg A} \sqsubseteq \forall P_<.(\neg A \sqcap \square_{\neg A}) \tag{1}$$

$$\neg \square_{\neg A} \sqsubseteq \exists P_<.(A \sqcap \square_{\neg A}) \tag{2}$$

The first inclusion accounts for the transitivity of $<$. The second inclusion accounts for the smoothness (see [22]): the fact that if an element is not a typical A element then there must be a typical A element preferred to it.

We define ABox' as follows. We replace each occurrence of the concept $\mathbf{T}(A)$ in any individual assertions $C(d) \in$ ABox, with the concept $A \sqcap \square_{\neg A}$, and include the resulting assertion in ABox'. All the assertions of the form $R(a, b) \in$ ABox are included unaltered in ABox'. Concerning queries, a query $C_1 \sqsubseteq C_2$ can be encoded as the inclusion $C_1' \sqsubseteq C_2'$, as done for TBox concept inclusions. A query $A(c)$ or $R(a, b)$ can be encoded as done for ABox assertions. It is clear that the size of KB' and of F' are both polynomial in the size of the KB, when the size of F is polynomial as well.

Proposition 2. *Let* **KB=(TBox,RBox,ABox)** *be a knowledge base and F a query. It holds that* $KB \models_{\mathcal{SROIQ}^P\mathbf{T}} F$ *iff* $KB' \models_{\mathcal{SROIQ}} F'$, *where KB' and F' are, respectively, the polynomial encodings in $\mathcal{SROIQ}$ of KB and F.*

Proof. We prove the two directions.

(If) By contraposition, let us assume that $KB \not\models_{\mathcal{SROIQ}^P\mathbf{T}} F$. We want to prove that $KB' \not\models_{\mathcal{SROIQ}} F'$. From the hypothesis, there is a preferential model $\mathcal{M} = \langle \Delta, <, \cdot^I \rangle$ of KB such that F is not satisfied in $\mathcal{M}$. We build a $\mathcal{SROIQ}$ model $\mathcal{M}' = (\Delta', \cdot^{I'})$ satisfying KB' as follows: $\Delta' = \Delta$; for all concept names C in $\mathcal{C}$, $C^{I'} = C^I$; for all role names R in $\mathcal{R}$, $R^I = R^{I'}$; for all individual names a in $\mathcal{O}$, $a^{I'} = a^I$; $(x, y) \in P_<^{I'}$ if and only if $y < x$ in the model $\mathcal{M}$; $(\square_{\neg A})^{I'} = \{x : \text{for all } y < x \text{ in } \mathcal{M}, y \in (\neg A)^I\}$.

We want to show that $\mathcal{M}'$ is a model of KB. First, observe that, as $<$ is transitive and antisymmetric, by construction the relation $P_<^{I'}$ is transitive and antisymmetric as well. By construction, it is clear as well that, for all concepts $C \in \mathcal{C}$ which do not include occurrences of the operator $\mathbf{T}$, $y \in C^I$ if and only if $y \in (C')^{I'}$ and that, for all the roles $R \in \mathcal{R}$, $(x, y) \in R^I$ if and only if $(x, y) \in R^{I'}$. As a consequence, all the role inclusion axioms in RBox and all role assertions in RBox which are satisfied in $\mathcal{M}$ are also satisfied in $\mathcal{M}'$, and, hence, RBox' is satisfied by $\mathcal{M}'$.

For all concepts A such that $\mathbf{T}(A)$ occurs in KB it holds that:

$$(\mathbf{T}(A))^I = (A \sqcap \square_{\neg A})^{I'}. \tag{3}$$

In fact, a concept A occurring in $\mathbf{T}(A)$ cannot contain the $\mathbf{T}$ operator. Let $x \in (\mathbf{T}(A))^I$. Then $x \in min_<(A^I)$, hence: $x \in A^I$ and for all $y < x$, $y \notin A^I$. Thus, $x \in A^{I'}$ and for all $y < x$ in $\mathcal{M}$, $y \notin A^{I'}$. Hence, $x \in (A \sqcap \square_{\neg A})^{I'}$. The viceversa holds as well.

Given equality (3), it follows easily by induction that, for all the (extended) concepts C in KB (possibly containing occurrences of the $\mathbf{T}$ operator),

$$y \in C^I \text{ if and only if } y \in (C')^{I'} \tag{4}$$

From (4) it follows that, for each concept inclusion $C \sqsubseteq D$ in TBox, $\mathcal{M}'$ satisfies the corresponding concept inclusion $C' \sqsubseteq D'$ in TBox'.

It is easy to see that the new concept inclusion axioms (1) and (2) added in TBox' are also satisfied in $\mathcal{M}'$. For axiom (1), observe that, by construction:

$$(\square_{\neg A})^{I'} = \{x : \text{ for all } y < x, \, y \in (\neg A)^I\} \text{ by definition of } (\square_{\neg A})^{I'}$$
$$= \{x : \text{ for all } (x, y) \in P_<^{I'}, \, y \in (\neg A)^{I'}\} \text{ by definition of } P_<^{I'}.$$

Hence, if $x \in (\square_{\neg A})^{I'}$, then $x \in (\forall P_<.(\neg A))^{I'}$. Let y be such that $(x, y) \in P_<^{I'}$. We want to show that $y \in (\square_{\neg A})^{I'}$. If not, by construction, there is a z such that $z < y$ and $z \notin (\neg A)^I$. As $<$ is transitive, $z < x$ and $z \notin (\neg A)^I$ which contradicts the hypotesis that $x \in (\square_{\neg A})^{I'}$. Therefore, $x \in (\forall P_<.(\square_{\neg A}))^{I'}$, and hence $x \in (\forall P_<.(\neg A \sqcap \square_{\neg A}))^{I'}$. Thus, inclusion axiom (1) is satisfied by $\mathcal{M}'$. In a similar way, we can show that axiom (2) is satisfied in $\mathcal{M}'$, and we can conclude that all the inclusion axioms in TBox' are satisfied in $\mathcal{M}'$.

For all the individual assertions $C(b) \in$ ABox', if $b^I \in C^I$ then, by property (4) and the fact that $b^{I'} = b^I$, we get $b^{I'} \in (C')^{I'}$.

Therefore, $\mathcal{M}'$ is a model of KB'. To conclude the (If) direction it suffices to show that $\mathcal{M}' \not\models_{SROIQ} F'$. From the hypothesis that $\mathcal{M}$ does not satisfy F, it follows that $\mathcal{M}'$ does not satisfy F': if the query F is $A(c)$ and $x \notin (A)^I$, then also $x \notin (F')^{I'}$ by (4). If the query F is $C \sqsubseteq D$ and, for some $x \in C^I$, $x \notin D^I$ then by, (4), also $x \in (C')^{I'}$ and $x \notin (D')^{I'}$, so that $\mathcal{M}'$ does not satisfy $C' \sqsubseteq D'$. Hence, $KB' \not\models_{SROIQ} F'$.

(Only if) By contraposition, let us assume that $KB' \not\models_{SROIQ} F'$. We want to prove that $KB \not\models_{SROIQ^P\mathbf{T}} F$. From the hypothesis, we know there is a model $\mathcal{M}' = (\Delta', I')$ of KB' in $SROIQ$ such that the query F' is not satisfied. We build a model $\mathcal{M} = (\Delta, <, I)$ of KB in $SROIQ^P\mathbf{T}$ such that F is not satisfied in $\mathcal{M}$. We let: $\Delta = \Delta'$; for all individual names $a \in \mathcal{O}$, $a^I = a^{I'}$; for all concept names $C \in \mathcal{C}$, $C^I = C^{I'}$; for all roles names $R \in \mathcal{R}$, $R^I = R^{I'}$.

It would be natural to define the relation $<$ in $\mathcal{M}$ as the transitive closure $(P_<^{I'})^+$ of $P_<^{I'}$. However, this would not guarantee that the resulting relation $<$ is well-founded and irreflexive. To define the relation $<$ in $\mathcal{M}$ so that it is transitive, well-founded and irreflexive we introduce some more notation. For all

$y \in \Delta$, we let $\Box_y = \{\Box_{\neg A} \mid \Box_{\neg A}$ occurs in KB and $y \in (\Box_{\neg A})^I\}$. Observe that for the elements x_i in a descending chain $x_0 > \ldots > x_i > x_{i+1}, \ldots$ the set $\Box_{x_i}$ is monotonically increasing: $\Box_{x_i} \subseteq \Box_{x_{i+1}}$.

In the model $\mathcal{M} = (\Delta, <, I)$, we define the preference relation $<$ as follows: $y < x$ if and only if $(x, y) \in (P_<{}^{I'})^+$ and $\Box_x \subset \Box_y$. In essence, for a pair of elements (x, y) such that $(y, x) \in (P_<{}^{I'})^+$ but x and y are instances of exactly the same boxed concepts ($\Box_x = \Box_y$) we do not include the pair (x, y) in $<$. In this way, for the elements x_i in a descending chain is strictly increasing, i.e. $\Box_{x_i} \subset \Box_{x_{i+1}}$. As the number of formulas $\Box_{\neg A}$ in KB is finite, there cannot be infinite descending chains. It is easy to see that $<$ is transitive, irreflexive and well-founded. By construction, we can show that $\mathbf{T}(A)^I = (A \sqcap \Box_{\neg A})^{I'}$, and that $C^I = C^{I'}$, for all concepts C. Hence, $\mathcal{M}$ is a model of KB, and the query F is not satisfied in $\mathcal{M}$, i.e., $KB \not\models_{\mathcal{SROIQ}^P\mathbf{T}} F$. $\qquad\Box$

The relevance of the above result is that standard and optimized proof methods for $\mathcal{SROIQ}$ can be exploited for reasoning in $\mathcal{SROIQ}^P\mathbf{T}$. As the above encoding only exploits constructs of $\mathcal{ALC}$, it also works for encoding the preferential extension of $\mathcal{ALC}$, called $\mathcal{ALC} + \mathbf{T}$ [11,12], into $\mathcal{ALC}$.

5 A Minimal Model Semantics for $\mathcal{SROIQ}^P\mathbf{T}$?

Preferential logics, although rather weak non-monotonic logics, have been used as the basis of the rational closure construction both for classical logic, for which rational closure was originally introduced by Lehmann [22], and for description logics as proposed, for instance, for $\mathcal{ALC}$ and $\mathcal{SHIQ}$ in [14,15]. The semantics of rational closure corresponds to a strengthening of the preferential semantics through a minimal model construction [15,22]. In this section we discuss some problems emerging when we want to define a minimal model semantics for more expressive preferential DLs.

Although the typicality operator $\mathbf{T}$ is nonmonotonic ($\mathbf{T}(C) \sqsubseteq D$ does not imply $\mathbf{T}(C \sqcap E) \sqsubseteq D$), the logic $\mathcal{SROIQ}^P\mathbf{T}$ is monotonic. In particular, it does not allow to deal with irrelevance: In Example 1, for instance, one cannot infer that typical blond students have more than 20 friends, even if all typical students have more than 20 friends and being blond is irrelevant with respect to the number of friends (has nothing to do with the number of friends).

To get this kind of inferences one can restrict the consideration to those models which maximize the typicality of all domain element. In such models, a student (like mary) can be assumed to be a typical student, when consistent. These models are said minimal as the typicality of domain elements is maximized by minimizing the distance of each element from the most preferred ones. The approach is skeptical as all the possible minimal models are considered.

For the logics $\mathcal{ALC}$ and $\mathcal{SHIQ}$ it has been shown that a minimal canonical model semantics can be defined [13,14] (where a canonical model must contain an instance for each satisfiable concept). A natural question arises, whether a minimal canonical model semantics can be defined for stronger DLs like $\mathcal{SROIQ}$

as well. The answer is negative in the general case. Due to the interaction of nominals with number restriction, a consistent $\mathcal{SROIQ}$ (as well as a $\mathcal{SHOIQ}$) knowledge base may have no canonical models (whence no minimal canonical ones). Let us consider the following example:

Example 2. Consider a KB, where ABox= $\{Student(mary), Italian(mary)\}$ and TBox contains:

$\mathbf{T}(Student) \sqsubseteq\ \geq 20\ HasFriends.\top$

$\{mary\} \sqsubseteq\ \leq 1HasFriend^-.\top$

$\neg\{mary\} \sqsubseteq \exists HasFriend.\{mary\}$

A model of KB can at most contain two elements. Although $Student \sqcap Italian$, $Student \sqcap \neg Italian$, $\neg Student \sqcap Italian$, $\neg Student \sqcap \neg Italian$ are all concepts that are consistent with KB, there is no model of the KB which contains an instance for each one of these concepts. Hence, there is no canonical model for KB.

This example shows that the notion of minimal canonical model is too strong for defining a minimal model semantics for $\mathcal{SROIQ}$. A suitable refinement of the minimal model semantics is needed as well as a refinement of the rational closure construction.

6   Conclusions and Related Works

In this work we have presented a preferential extension $\mathcal{SROIQ}^P\mathbf{T}$ of the description logic $\mathcal{SROIQ}$ and we have shown that reasoning in $\mathcal{SROIQ}^P\mathbf{T}$ can be polynomially reduced to reasoning in $\mathcal{SROIQ}$. This is the first step of the development of a stronger non-monotonic extension of $\mathcal{SROIQ}$. As future work, we aim to extend to $\mathcal{SROIQ}$ the notion of rational closure [22] and to develop its semantic characterization in therms of a minimal models semantics along the line of the notion of rational closure proposed for $\mathcal{ALC}$ [15]. There are a number of works which are closely related to our proposal.

In [12,13] nonmonotonic extensions of DLs based on the $\mathbf{T}$ operator have been proposed. These extensions focus on the basic DL $\mathcal{ALC}$ and the semantics of $\mathbf{T}$ is based on preferential logic P[20]. In [4] a preferential extension of $\mathcal{ALC}$ based on rational logic R is proposed.

The first notion of rational closure for DLs was defined by Casini and Straccia [6], Their rational closure construction for $\mathcal{ALC}$ directly uses entailment in $\mathcal{ALC}$ over a materialization of the KB. A variant of this notion of rational closure has been studied in [5], and a semantic characterization for it has been proposed. To overcome the limitations of rational closure, in [8] an approach is introduced based on the combination of rational closure and *Defeasible Inheritance Networks*, while in [7] a lexicographic closure is proposed.

The typicality concepts $\mathbf{T}(C)$ used in this paper as well as in [12], might appear to be related with the notion of lower approximation $\underline{C}$ of a concept in Rough DLs [24]. However, a major difference between the two approaches is that the semantics of typicality, as that in [11], is based on a preference relation (a total preorder) on domain elements, while the semantics of Rough DLs uses an equivalence (indiscernibility) relation. This difference also reflects in the encoding. Given the preference-based semantics, our notion of typicality satisfies a set of postulates (studied in [12] for $\mathcal{ALC}$) that are essentially a reformulation of the postulates of preferential consequence relation [22]. A deeper analysis and comparison with Rough DLs will be subject of future work.

Among the recent nonmonotonic extensions of DLs are the formalisms for combining DLs with logic programming rules, such as for instance, [10] and [23]. In [2] a non monotonic extension of DLs is proposed based on a notion of overriding, supporting normality concepts and enjoying nice computational properties. In [19] a general DL language is introduced, which extends $\mathcal{SROIQ}$ with nominal schemas and epistemic operators as defined in [23], and encompasses some of the most prominent nonmonotonic rule languages, including Datalog under the answer set semantics.

Acknowledgement. We thank the anonymous referees for their helpful comments. This research is partially supported by INDAM - GNCS Project 2015 *Description Logics and Nonmonotonic reasoning* and by Compagnia di San Paolo Project *GINSENG*.

References

1. Baader, F., Hollunder, B.: Priorities on defaults with prerequisites, and their application in treating specificity in terminological default logic. J. Autom. Reasoning (JAR) **15**(1), 41–68 (1995)
2. Bonatti, P.A., Faella, M., Petrova, I., Sauro, L.: A new semantics for overriding in description logics. Artif. Intell. **222**, 1–48 (2015)
3. Bonatti, P.A., Lutz, C., Wolter, F.: The complexity of circumscription in DLs. J. Artif. Intell. Res. (JAIR) **35**, 717–773 (2009)
4. Britz, K., Heidema, J., Meyer, T.: Semantic preferential subsumption. In: Brewka, G., Lang, J. (eds.) Principles of Knowledge Representation and Reasoning: Proceedings of the 11th International Conference (KR 2008), pp. 476–484. AAAI Press, Sidney, September 2008
5. Casini, G., Meyer, T., Varzinczak, I.J., Moodley, K.: Nonmonotonic Reasoning in Description Logics: Rational Closure for the ABox. In: DL 2013 CEUR Workshop Proceedings, vol. 1014, pp. 600–615 (2013)
6. Casini, G., Straccia, U.: Rational closure for defeasible description logics. In: Janhunen, T., Niemelä, I. (eds.) JELIA 2010. LNCS, vol. 6341, pp. 77–90. Springer, Heidelberg (2010)
7. Casini, G., Straccia, U.: Lexicographic Closure for defeasible description logics. In: Proceedings of the Australasian Ontology Workshop vol. 969, pp. 28–39 (2012)
8. Casini, G., Straccia, U.: Defeasible inheritance-based description logics. J. Artif. Intell. Res. (JAIR) **48**, 415–473 (2013)
9. Donini, F.M., Nardi, D., Rosati, R.: Description logics of minimal knowledge and negation as failure. ACM Trans. Comput. Logic (ToCL) **3**(2), 177–225 (2002)
10. Eiter, T., Lukasiewicz, T., Schindlauer, R., Tompits, H.: Combining answer set programming with description logics for the semanticweb. In: Proceedings of the

Ninth International Conference on Principles of Knowledge Representation and Reasoning (KR 2004). AAAI Press, Whistler, Canada, June 2004

11. Giordano, L., Gliozzi, V., Olivetti, N., Pozzato, G.L.: Preferential description logics. In: Dershowitz, N., Voronkov, A. (eds.) LPAR 2007. LNCS (LNAI), vol. 4790, pp. 257–272. Springer, Heidelberg (2007)

12. Giordano, L., Gliozzi, V., Olivetti, N., Pozzato, G.L.: ALC+T: a preferential extension of description logics. Fundam. informaticae **96**, 1–32 (2009)

13. Giordano, L., Gliozzi, V., Olivetti, N., Pozzato, G.L.: A nonmonotonic description logic for reasoning about typicality. Artif. Intell. **195**, 165–202 (2013)

14. Giordano, L., Gliozzi, V., Olivetti, N., Pozzato, G.L.: Rational closure in SHIQ. In: DL2014 CEUR Workshop Proceedings, vol. 1193, pp. 1–13 (2014)

15. Giordano, L., Gliozzi, V., Olivetti, N., Pozzato, G.L.: Semantic characterization of rational closure: From propositional logic to description logics. Artif. Intell. **226**, 1–33 (2015)

16. Horrocks, I., Kutz, O., Sattler, U.: The even more irresistible SROIQ. In: Proceedings, Tenth International Conference on Principles of Knowledge Representation and Reasoning, Lake District of the United Kingdom, pp. 57–67, 2–5 June 2006

17. Horrocks, I., Sattler, U., Tobies, S.: Practical reasoning for very expressive description logics. Logic J. IGPL **8**(3), 239–263 (2000)

18. Ke, P., Sattler, U.: Next steps for description logics of minimal knowledge and negation as failure. In: Baader, F., Lutz, C., Motik, B. (eds.) Proceedings of Description Logics. CEUR Workshop Proceedings, vol. 353, CEUR-WS.org, Dresden, May 2008

19. Knorr, M., Hitzler, P., Maier, F.: Reconciling owl and non-monotonic rules for the semantic web. In: ECAI 2012, pp. 474–479 (2012)

20. Kraus, S., Lehmann, D., Magidor, M.: Nonmonotonic reasoning, preferential models and cumulative logics. Artif. Intell. **44**(1–2), 167–207 (1990)

21. Krötzsch, M., Simančík, F., Horrocks, I.: A description logic primer. CoRR abs/1201.4089 (2012). http://arxiv.org/abs/1201.4089

22. Lehmann, D., Magidor, M.: What does a conditional knowledge base entail? Artif. Intell. **55**(1), 1–60 (1992)

23. Motik, B., Rosati, R.: Reconciling description logics and rules. J. ACM **57**(5), 93–154 (2010)

24. Schlobach, S., Klein, M.C.A., Peelen, L.: Description logics with approximate definitions - precise modeling of vague concepts. In: IJCAI 2007, Hyderabad, pp. 557–562 (2007)

25. Straccia, U.: Default inheritance reasoning in hybrid kl-one-style logics. In: Bajcsy, R. (ed.) Proceedings of IJCAI 1993, pp. 676–681. Morgan Kaufmann, Chambéry, August 1993

iQbees: Towards Interactive Semantic Entity Search Based on Maximal Aspects

Grzegorz Sobczak[1](✉), Mateusz Chochół[2], Ralf Schenkel[3],
and Marcin Sydow[1,2]

[1] Institute of Computer Science, Polish Academy of Sciences, Warsaw, Poland
gsobczak@piwstk.edu.pl
[2] Polish-Japanese Institute of Information Technology, Warsaw, Poland
[3] Universität Passau, Passau, Germany

Abstract. Similar entity search by example is an important task in the area of retrieving information from semantic knowledge bases. In this paper we define a new interactive variant of this problem that is called iQbees for "Interactive Query-by-Example Entity Search" and is an extension of a previous QBEES approach. We also present a working on-line prototype demo which implements the proposed approach.

Keywords: Similar entity search · Query by example · Semantic knowledge base · Interactive search · QBEES · Demo · Diversity

1 Introduction

Imagine a situation when one wants to find a list of entities which share a specific property, for example countries which economies are based on oil export. From a user's point of view, a convenient way would be to provide a semantic description of such a property. Another possibility is to supply a system with a few examples which share a desirable property.

The problem of finding entities close to the given examples is known as similar entities search. Recent work [7] presents QBEES ("Query By Entity Examples") – one of the possible approaches to solve this problem. It is based on an idea of maximal aspects that is briefly described in Sect. 3.

In this paper we propose how this solution can be extended to **iQbees** – for "Interactive QBEES". In original QBEES, a user provided k examples of entities and the system, based on an underlying semantic knowledge graph returned a list of similar entities. The solution presented in this paper is more interactive - a user provides one example at the beginning and iteratively converges to the desired result by selecting next hint entities that refine the original query.

We also present a working on-line prototype demo of the system which follows the described approach. We adapted software that supports QBEES framework, implemented a graphical interface and integrated into one system. The prototype is available under the following URL: http://webmining.pjwstk.edu.pl/qbees-dev/.

F. Esposito et al. (Eds.): ISMIS 2015, LNAI 9384, pp. 259–264, 2015.
DOI: 10.1007/978-3-319-25252-0_28

Motivations. The main motivation behind this improvement is its usefulness on any system which shows entities to its users, like Wikipedia or internet shops. For example, when the user sees information about Qatar on Wikipedia it might be convenient to him to navigate directly to other oil exporters.

Another reason is to increase the functionality of the QBEES framework and similar. The main drawback of such systems is the difficulty in providing many good input examples by the user. We believe that making it possible for the user to iteratively extend the input example list by selecting some suggested entities returned by the system is much more convenient for the user and potentially increases the performance of the system.

Contributions. The main contributions of this paper are: (1) introduction of the new paradigm of interactive similar entity search based on the previously proposed maximal aspect approach (2) providing a working on-line demo prototype implementation of the proposed approach to illustrate it and providing a proof-of-concept and experimentation tool for future research.

Outline. In Sect. 2 related work is discussed, in Sect. 3 a solution presented in [7] is introduced. Section 4 presents the interactive search solution and describes a typical scenario how to use the system. The prototype of the system is presented in Sect. 5 and future work is described in Sect. 6.

2   Related Work

Entity search has been considered extensively in the past, with a focus on finding related entities and list completion, and with extensive evaluation campaigns at TREC [2] and INEX [4]. We consider the specific scenario where entities from a knowledge graph are searched. Existing systems usually build on entity similarity measures, exploiting the graph structure (e.g., SimRank [6]), the context of entities in the graph (e.g., Albertoni and De Marino [1]), or additional context outside the graph (e.g., Bron et al. [3], which combines a term-based language model with a simple structural model).

The problem of example-based entity search has seen some recent work. Yu et al. [11] solve a slightly different problem where entities similar to a single query entity are computed, exploiting a small number of example results. Focusing on heterogeneous similarity aspects, they propose to use features based on so-called meta paths between entities and several path-based similarity measures, and apply learning-to-rank methods for which they require labeled test data.

Mottin et al. [8,9] introduce the concept of exemplar queries. Similar to our setting, an example result is used instead of a query. However, the setting in their XQ system is strictly different since it consideres examples in the form of a connected subgraph of entities, not single entities, and determines result subgraphs based on their similarity to the query graph. The problem is therefore in some sense easier, as more information can be exploited for identifying query results.

The GQBE system by Jayaram et al. [5] is similar to XQ, but does not use connected subgraphs, but just entities that form a query result as input; the meaningful connections between those entities are explored by the system. Again, the main difference to our system is that we consider only single entities as results, not combinations, and hence have fewer information for identifying relevant results.

Diversity-aware entity summarization was considered in [10], but we are not aware of any work on entity search that takes diversity into account.

3 Aspect-Based Model

In this section, basic ideas of aspect-based model are presented. More detailed description is included in [7].

Knowledge Graph. A Knowledge Graph KG is a directed multi-graph that consists of three basic components, a *Fact Graph FG*, an *Ontology Tree O*, and a set of type assignment arcs TA connecting the two. Arcs in KG are labelled. We will use the notation relation(arg1,arg2) for any directed arc with label relation in KG that points from node arg1 to arg2.

The *Fact Graph FG=(E,F)* is a directed multigraph where nodes in E represent *entities* (e.g. Warsaw, Poland) and edges in F represent *facts* about the entities. For example, an arc isCapitalOf(Warsaw,Poland) represent the fact "Warsaw is the capital of Poland".

The *Ontology Tree O=(C,S)* is a graph where each node (class) $c \in C$ represents some *type* of entities (e.g. person). The class nodes are connected by directed arcs labelled as subClassOf. For instance, subClassOf(composer,musician) indicates that every composer is also a musician.

The *Type Assignment TA* is a set of arcs labelled hasType which connect entities from the Fact Graph and classes from the Ontology Tree. For example the arc hasType(Chopin,composer) means that "Chopin is a composer".

Basic Aspects. In this section the concept of a *basic aspect* is introduced with an example. A basic aspect represents an "atomic property" that describes a specific entity q. We distinguish three types of basic aspects: type aspects, relational aspects and factual aspects.

Type aspects are obtained by replacing the particular entity q in an arc that represents a type with a variable. For example, a type arc hasType(Chopin, composer) naturally induce predicates of the form hasType(.,composer) that represents the "basic property" of this entity of "being a composer".

Fact aspects are created from arcs that represent a fact similarly as type aspects. The arc bornIn(Chopin,Poland) induces the fact aspect bornIn(.,Poland).

Relational aspects are also obtained from arcs that represents facts. Additionaly to fact aspects we also replace the remaining argument of a factual aspect by a free variable ? such that actedIn(.,?) indicates that an entity acted in at least one movie.

Maximal Aspects. A set of basic aspects is called a *compound aspect*. For example, a property "being a composer born in Poland", which consists of two "atomic properties" - "being a composer" and "being born in Poland", are represented by a compound aspect {bornIn(.,Poland), hasType(.,composer)}.

It is easy to see that each entity can be treated as a set of basic aspects. Let A_e be a set of all basic aspects of $e \in E$. Let q be an query example and A_q be its set of basic aspects. For all $e \in E, e \neq q$ consider set of all basic aspects common with q, that is $A'_e = A_e \cap A_q$. These compound aspects naturally form a lattice (with inclusion as an operation). *Maximal aspects* are those compound aspects which are maximal in the lattice.

This definition naturally extends to a set Q of query entities ($A_Q = \bigcap_{q \in Q} A_q$).

List of Entities. Based on the initial single user-provided query entity the initial list of suggested k similar entities is computed by the underlying QBEES approach. The detailed description of the ranking method is outside of the scope of this paper and is presented in [7].

In short, based on a set of query examples, maximal aspects are calculated and one of them is selected as the most promising one. Among the entities which satisfy the chosen maximal aspect one entity is picked. The procedure is repeated k times taking into consideration the list of already selected entities and their aspects.

4   Interactive QBEES

Here we shortly describe the interactive QBEES paradigm. The procedure is as follows:

1. A user provides an initial example entity as the input.
2. The system returns a list of similar entities based on the QBEES approach.
3. If the results do not satisfy user information need, the user can treat the returned entities as *refinement suggestions* and select one of them as a hint for the system to refine his query. This entity is appended to the list of previously selected query entities. The user can go back to the point 2 until she finds the result successful or wishes to restart the search.

In this approach a user interactively, step by step, converges to the expected result. She exploits partial results given by the system. An important property of the list of returned entities is its diversity-awareness. Since the user interests are not known in advance, diversified results make it possible for the user to select one of possible alternative branches to further explore.

Scenario: Zbigniew Religa. An example of an ambiguous query is a famous Polish surgeon and, in addition, a politician – Zbigniew Religa. The branch of surgeons in the results returned by the current setting of our system is represented by "Victor Chang" and "Christiaan Barnard". The political part of the returned result list in our system for that query is represented by a post-communist Polish politician Leszek Miller - a former Polish prime minister. Further exploration of this branch leads to other Polish politicians.

5   iQbees: A Prototype On-line Demo Implementation of the Approach

We adapted QBEES with YAGO2 as the underlying semantic knowledge base for the purpose of the prototype demo implementation. The prototype is available under the following URL: http://webmining.pjwstk.edu.pl/qbees_dev/.

The goals of this implemention are multi-fold: as a proof-of-concept, as an illustration and, most importantly as an important tool for experimenting with the system and tuning its parameters in order to improve it in future.

The provided graphical interface has the following functionality:

- providing by the user one or more initial entity examples
- providing by the user the desired number k of the entities to be returned
- browse through the returned entities
- see the maximal aspect set of any returned entity
- select any of the returned/suggested entities as a further refinement hint to the system.

Exemplary use case of the GUI involves:

1. Typing a chosen entity, for example "Arnold_Schwarzenegger"[1]
2. Pressing the "Calculate" button. It takes a few seconds to output the result as the current implementation is not highly optimised for efficiency.
   (a) viewing the list of basic aspects constituting the maximal aspect set for the corresponding entity: by clicking the "Show debug" button
   (b) refining the query by appending an entity from the returned results/suggestions to the previously provided input query entities: by pressing the "GO" button to the left of the selected entity.

6   Conclusions and Future Work

We proposed a novel interactive extension of the problem of semantic search of similar entities approach that was based on the concept of maximal aspects.

This is an ongoing work and below we describe some planned future directions of the development of the approach and its prototype implementation. Concerning the model, we plan to focus more on the ranking algorithm. Furthermore, we plan to work on providing more control on the diversity-awareness of the

---

[1] In the current version of the system the white-space separator is disabled and is represented by the '_' symbol. This is planned to be changed in future.

returned list of the results. Concerning the functionality, we envisage introduction of the "back-button" functionality or, more generally, to provide the "delete entity" option and provide some control on the order of the presented entities. Among other improvements we plan to implement typing suggestions, make the browsing through entities more handy, plug other semantic database, in particular DBPedia, design and execute a series od semantic experiments, speed up our system.

Acknowledgements. We would like to thank Steffen Metzger, who provided the substantial part of the code of the back-end of the current state of the demo implementation. This work is partially supported by the Polish National Science Centre grant 2012/07/B/ST6/01239 and partially supported by the European Union under the European Social Fund Project PO KL "Information technologies: Research and their interdisciplinary applications", Agreement UDA-POKL.04.01.01-00-051/10-00.

References

1. Albertoni, R., Martino, M.D.: Asymmetric and context-dependent semantic similarity among ontology instances. J. Data Semant. **10**, 1–30 (2008)
2. Balog, K., Serdyukov, P., de Vries, A.P.: Overview of the TREC 2011 entity track. In: TREC (2011)
3. Bron, M., Balog, K., de Rijke, M.: Example based entity search in the web of data. In: Serdyukov, P., Braslavski, P., Kuznetsov, S.O., Kamps, J., Rüger, S., Agichtein, E., Segalovich, I., Yilmaz, E. (eds.) ECIR 2013. LNCS, vol. 7814, pp. 392–403. Springer, Heidelberg (2013)
4. Demartini, G., Iofciu, T., de Vries, A.P.: Overview of the INEX 2009 entity ranking track. In: Geva, S., Kamps, J., Trotman, A. (eds.) INEX 2009. LNCS, vol. 6203, pp. 254–264. Springer, Heidelberg (2010)
5. Jayaram, N., Gupta, M., Khan, A., Li, C., Yan, X., Elmasri, R.: GQBE: querying knowledge graphs by example entity tuples. In: IEEE 30th International Conference on Data Engineering, Chicago, ICDE 2014, IL, USA, 31 March - 4 April, pp. 1250–1253 (2014)
6. Jeh, G., Widom, J.: SimRank: a measure of structural-context similarity. In: KDD, pp. 538–543 (2002)
7. Metzger, S., Schenkel, R., Sydow, M.: Qbees: query by entity examples. In: Proceedings of the 22nd ACM International Conference on Conference on Information #38; Knowledge Management, CIKM 2013, pp. 1829–1832. ACM, New York, NY, USA (2013)
8. Mottin, D., Lissandrini, M., Velegrakis, Y., Palpanas, T.: Exemplar queries: give me an example of what you need. PVLDB **7**(5), 365–376 (2014)
9. Mottin, D., Lissandrini, M., Velegrakis, Y., Palpanas, T.: Searching with XQ: the exemplar query search engine. In: International Conference on Management of Data, SIGMOD 2014, Snowbird, UT, USA, 22–27 June 2014, pp. 901–904 (2014)
10. Sydow, M., Pikula, M., Schenkel, R.: The notion of diversity in graphical entity summarisation on semantic knowledge graphs. J. Intel. Inf. Syst. **41**, 109–149 (2013)
11. Yu, X., Sun, Y., Norick, B., Mao, T., Han, J.: User guided entity similarity search using meta-path selection in heterogeneous information networks. In: CIKM, pp. 2025–2029 (2012)

Emotion Recognition, Music Information Retrieval

Emotion Detection Using Feature Extraction Tools

Jacek Grekow[✉]

Faculty of Computer Science, Bialystok University of Technology,
Wiejska 45A, 15-351 Bialystok, Poland
j.grekow@pb.edu.pl

Abstract. This paper presents an analysis of the effect of features obtained from 3 different audio analysis tools on classifier accuracy during emotion detection. The research process included constructing training data, feature extraction, feature selection, and building classifiers. The obtained results indicated leaders among feature extraction tools used during classifier building for each emotion. An additional result of the conducted research was obtaining information on which features are useful in the detection of particular emotions.

Keywords: Music emotion recognition · Audio feature extraction · Audio analysis tools · Music information retrieval

1 Introduction

It cannot be denied that listening to music has an emotional character. Detecting the emotions contained in music is one of the main causes of listening to it [1]. In the era of the Internet, searching music databases for emotions has become increasingly important. Automatic emotion detection enables indexing files in terms of emotions [2].

This paper presents the use of 3 different audio analysis tools (Marsyas, jAudio and Essentia) during emotion detection. The positives of this experiment were: we gained experience using these tools; we learned their strengths and weaknesses; we got insight into their construction and terms of use; and we checked their usefulness in emotion detection. Another result of this experiment was we extracted information on which features are useful during the detection of each emotion.

Music emotion detection studies are mainly based on two popular approaches: categorical or dimensional. The categorical approach [3–5] describes emotions with a discrete number of classes - affective adjectives. In the dimensional approach [6,7], emotions are described as numerical values of valence and arousal.

There are several other studies on the issue of emotion detection with the use of different audio tools for musical feature extraction. Studies [4,8] used a collection of tools that use the Matlab environment called MIR toolbox [9]. Feature extraction library jAudio [10] was used in studies [5,7]. Feature sets

© Springer International Publishing Switzerland 2015
F. Esposito et al. (Eds.): ISMIS 2015, LNAI 9384, pp. 267–272, 2015.
DOI: 10.1007/978-3-319-25252-0_29

extracted from PsySound [11] were used in study [6], while study [12] used the Marsyas framework [13]. The Essentia [14] library for audio analysis was used in study [15].

There are also papers devoted to the evaluation of audio features for emotion detection within one program. Song et al. [4] explored the relationship between musical features extracted by MIR toolbox and emotions. They compared the emotion prediction results for four sets of features: dynamic, rhythm, harmony, and spectral features. A comprehensive review of the methods that have been proposed for music emotion recognition was prepared by Yang et al. [16].

2   Music Data Sets

In this research, we use four emotion classes: e1 (energetic-positive), e2 (energetic-negative), e3 (calm-negative), e4 (calm-positive). They cover the four quadrants of the two-dimensional Thayer model of emotion [17]. They correspond to four basic emotion classes: happy, angry, sad, and relaxed.

To conduct the study of emotion detection, we prepared two sets of data. One set was used for building one common classifier for detecting the 4 emotions, and the other data set for building four binary classifiers of emotion in music. Both data sets consisted of six-second fragments of different genres of music: classical, jazz, blues, country, disco, hip-hop, metal, pop, reggae, and rock. The tracks were all 22050 Hz mono 16-bit audio files in .wav format. Music samples were labeled by the author of this paper, a music expert with a university musical education. Six-second music samples were listened to and then labeled with one of the emotions (e1, e2, e3, e4). In the case when the music expert was not certain which emotion to assign, such a sample was rejected. In this way, each file was associated with only one emotion.

The first training data set for emotion detection consisted of 324 files, with 81 files for each emotion (e1, e2, e3, e4). We obtained the second training data from the first set. It consisted of 4 sets of binary data. For example, data set for binary classifier e1 consisted of 81 files labeled as e1 and 81 files labeled as not e1 (27 files each from e2, e3, e4). In this way, we obtained 4 binary data sets (consisting of examples of e and not e) for 4 binary classifiers e1, e2, e3, e4.

3   Feature Extraction Using Audio Analysis Tools

With the Marsyas [13], the following features can be extracted: Zero Crossings, Spectral Centroid, Spectral Flux, Spectral Rolloff, Mel-Frequency Cepstral Coefficients (MFCC), and chroma features. For each of these basic features, Marsyas calculates four statistic features (the mean of the mean, the mean of the standard deviation, the standard deviation of the mean, the standard deviation of the standard deviation).

The following features are implemented in jAudio [10]: Zero Crossing, Root Mean Square, Fraction of Low Amplitude Frames, Spectral Centroid, Spectral

Flux, Spectral Rolloff, Spectral Variability, Compactness, Mel-Frequency Cepstral Coefficients (MFCC), Beat Histogram, Strongest Beat, Beat Sum, Strength of Strongest Beat, Linear Prediction Coefficients (LPC), Method of Moments (Statistical Method of Moments of the Magnitude Spectrum), Area Method of Moments. jAudio also calculates metafeatures, which are the feature templates that automatically produce new features from existing ones. jAudio provides three basic metafeature classes (mean, standard deviation, and derivative), which are also combined to produce two more metafeatures (derivative of the mean and derivative of the standard deviation).

We used version 2.0.1 of Essentia [14], which contains a number of executable extractors computing music descriptors for an audio track: spectral, time-domain, rhythmic, tonal descriptors. Essentia also calculates many statistic features: the mean, geometric mean, power mean, median of an array, and all its moments up to the 5th-order, its energy and the root mean square (RMS). To characterize the spectrum, flatness, crest and decrease of an array are calculated. Variance, skewness, kurtosis of probability distribution, and a single Gaussian estimate are calculated for the given list of arrays.

The previously prepared, labeled by emotion, music data sets served as input data for 3 tools used for feature extraction. The obtained lengths of feature vectors, dependent on the package used, were as follows: Marsyas - 124 features, jAudio - 632 features, and Essentia - 471 features.

4   Results

4.1   The Construction of Classifiers

We built classifiers for emotion detection using the WEKA package [18]. During the construction of the classifier, we tested the following algorithms: J48, RandomForest, BayesNet, IBk (K-nn), SMO (SVM). The classification results were calculated using a cross validation evaluation CV-10.

The first important result was that during the construction of the classifier for 3 data sets obtained from Marsyas, jAudio and Essentia, the highest accuracy among all tested algorithms was obtained for SMO algorithm. SMO was trained using polynominal kernel. The second best algorithm was RandomForest.

Table 1. Accuracy obtained for SMO algorithm

	Marsyas	jAudio	Essentia
Accuracy before attribute selection	55.24 %	58.95 %	59.26 %
Accuracy after attribute selection	58.95 %	**67.90 %**	**64.50 %**

The results obtained for SMO algorithm are presented in Table 1. The result (classifier accuracy) improved after applying attribute selection (attribute evaluator: WrapperSubsetEval, search method BestFirst). The best results were obtained after applying attribute selection for data from jAudio (67.90 %) and Essentia (64.50 %).

Table 2. The most important features obtained from jAudio and Essentia

Tool	Selected features
jAudio	Strongest Beat, MFCC, LPC (Linear Prediction Coefficients), Statistical Method of Moments of the Magnitude Spectrum
Essentia	Energy of the Erbbands, MFCC, Onset Rate, Beats Loudness Band Ratio, Key Strength, Chords Histogram

The most important features obtained from jAudio and Essentia are presented in Table 2. In both cases, such features as MFCC and those pertaining to rhythm confirmed their usefulness in emotion detection; they are present in the obtained features from jAudio as well as Essentia. What distinguishes Essentia's set of features is the use of tonal features (Key Strength, Chords Histogram); and what distinguishes the features selected from jAudio is the use of statistical moments of the magnitude spectrum.

4.2   The Construction of Binary Classifiers

Once again the best results were obtained for SMO algorithm. The results are presented in Fig. 1. The best results were obtained for emotion e2 (91 %) regardless of the type of audio analysis tools. It is difficult to unequivocally select the best audio analysis tools used for features extraction. For detection of emotion e1, the best results were obtained using Essentia (80.86 %). For detection of emotion e2, all tools had the same results (approx. 91 %). For detection of emotion e3, the best results were obtained using the tools jAudio and Essentia (87 %), and for emotion e4 - Marsyas (82.71 %).

Essentia achieved the best results since in three cases the obtained classifier accuracy was the highest (for e1, e2, e3); the remaining tools achieved the

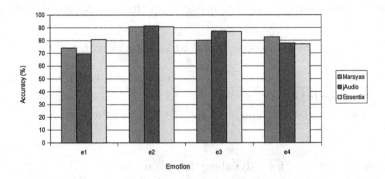

Fig. 1. Classifier accuracy for emotions e1, e2, e3, and e4 obtained for SMO

best results in two cases: Marsyas (e2, e4) and jAudio (e2, e3). The obtained binary classifier accuracy results were higher (15–24 percentage points) than the accuracy of one classifier recognizing four emotions. Table 3 presents the most important features obtained after feature selection for Essentia features for each emotion. In each features set, we had representatives of low-level, rhythm features, even though we had different sets for each emotion. Only in the case of classifier e4, tonal features were not used. The energies of the bands are important for e1, e2, and e4 classifiers, but they differ in to which bands they pertain: e1 - Barkbands, e2 - Erbbands, and Melbands, e4 - Barkbands, and Erbbands. High Frequency Content, which is characterized by the amount of high-frequency content in the signal is important for e3 and e4 classifiers. Beats Loudness Band Ratio (the beat's energy ratio on each band) is very important for emotion detection because it is used in all sets. Another important feature was the tonal feature: Chords Histogram, which was used by e2 and e3.

Table 3. Selected features used for building binary classifiers using Essentia

Classifier	Selected features
e1	Energy of the Barkbands, Onset Rate, Beats Loudness Band Ratio, Key Strength
e2	Average Loudness, Dissonance, Energy of the Erbbands and the Melbands, MFCC, Beats Loudness Band Ratio, Chords Changes, Chords Histogram
e3	High Frequency Content, Silence Rate, Spectral Energy Band Middle Low, Beats Loudness Band Ratio, Key Strength, Chords Histogram
e4	Energy of the Barkbands, Energy of the Erbbands, High Frequency Content, Pitch Salience, Beats Loudness Band Ratio

5   Conclusions

This paper presents an analysis of the effect of features obtained from different audio analysis tools on classifier accuracy during emotion detection. The research process included constructing training data, feature extraction, feature selection, and building classifiers. The collected data allowed comparing different tools during emotion detection. The obtained results indicated leaders among feature extraction tools used during classifier building for each emotion. Only the use of several different tools achieves high classifier accuracy (80–90 %) for all basic emotions (e1, e2, e3, e4). An additional result of the conducted research was obtaining information on which features are useful in the detection of particular emotions. The obtained results present a new and interesting view of the usefulness of different feature sets for emotion detection.

Acknowledgments. This paper is supported by the S/WI/3/2013.

References

1. Krumhansl, C.L.: Music: a link between cognition and emotion. Am. Psychol. Soc. **11**(2), 45–50 (2002)
2. Grekow, J., Raś, Z.W.: Emotion Based MIDI Files Retrieval System. In: Raś, Z.W., Wieczorkowska, A.A. (eds.) Advances in Music Information Retrieval. SCI, vol. 274, pp. 261–284. Springer, Heidelberg (2010)
3. Grekow, J., Raś, Z.W.: Detecting emotions in classical music from MIDI files. In: Rauch, J., Raś, Z.W., Berka, P., Elomaa, T. (eds.) ISMIS 2009. LNCS, vol. 5722, pp. 261–270. Springer, Heidelberg (2009)
4. Song, Y., Dixon, S., Pearce, M.: Evaluation of musical features for emotion classification. In: Proceedings of the 13th International Society for Music Information Retrieval Conference (2012)
5. Xu, J., Li, X., Hao, Y., Yang, G.: Source separation improves music emotion recognition. In: ACM International Conference on Multimedia Retrieval (2014)
6. Yang, Y.-H., Lin, Y.-C., Su, Y.-F., Chen, H.H.: A regression approach to music emotion recognition. IEEE Trans. Audio, Speech, Language Process. **16**(2), 448–457 (2008)
7. Lin, Y., Chen, X., Yang, D.: Exploration of music emotion recognition based on midi. In: Proceedings of the 14th International Society for Music Information Retrieval Conference (2013)
8. Saari, P., Eerola, T., Fazekas, G., Barthet, M., Lartillot, O., Sandler, M.: The role of audio and tags in music mood prediction: a study using semantic layer projection. In: Proceedings of the 14th International Society for Music Information Retrieval Conference (2013)
9. Lartillot, O., Toiviainen, P.: MIR in Matlab (II): A toolbox for musical feature extraction from audio. In: International Conference on Music Information Retrieval, pp. 237–244 (2007)
10. McKay C., Fujinaga I., Depalle P.: jAudio: a feature extraction library. In: Proceedings of the 6th International Conference on Music Information Retrieval (ISMIR 05), pp. 600–603 (2005)
11. Cabrera, D.: PSYSOUND: a computer program for psychoacoustical analysis. In: Proceedings of the Australian Acoustical Society Conference, pp. 47–54 (1999)
12. Grekow, J.: Mood tracking of radio station broadcasts. In: Andreasen, T., Christiansen, H., Cubero, J.-C., Raś, Z.W. (eds.) ISMIS 2014. LNCS, vol. 8502, pp. 184–193. Springer, Heidelberg (2014)
13. Tzanetakis, G., Cook, P.: Marsyas: a framework for audio analysis. Organized Sound **10**, 293–302 (2000)
14. Bogdanov, D., Wack N., Gomez E., Gulati S., Herrera P., Mayor O., Roma G., Salamon J., Zapata J., Serra X.: ESSENTIA: an audio analysis library for music information retrieval. In: Proceedings of the 14th International Conference on Music Information Retrieval, pp. 493–498 (2013)
15. Laurier, C.: Automatic Classification of Musical Mood by Content-Based Analysis. Ph.D. thesis, UPF, Barcelona, Spain (2011)
16. Yang, Y.-H., Chen, H.H.: Machine recognition of music emotion: a review. ACM Trans. Intell. Sys. Technol. **3**(3), 61 (2012). Article No. 40
17. Thayer, R.E.: The Biopsychology Arousal. Oxford University Press, Cambridge (1989)
18. Witten, I.H., Frank, E.: Data Mining: Practical Machine Learning Tools and Techniques. Morgan Kaufmann, San Francisco (2005)

Improving Speech-Based Human Robot Interaction with Emotion Recognition

Berardina De Carolis[✉], Stefano Ferilli, and Giuseppe Palestra

Dipartimento di Informatica, Università di Bari, 70126 Bari, Italy
berardina.decarolis@uniba.it

Abstract. Several studies report successful results on how social assistive robots can be employed as interface in the assisted living domain. In this domain, a natural way to interact with robots is to use a speech. However, humans often use particular intonation in the voice that can change the meaning of the sentence. For this reason, a social assistive robot should have the capability to recognize the intended meaning of the utterance by reasoning on the combination of linguistic and acoustic analysis of the spoken sentence to really understand the user's feedback. We developed a probabilistic model that is able to infer the intended meaning of the spoken sentence from the analysis of its linguistic content and from the output of a classifier able to recognise the valence and arousal of the speech prosody starting from dataset. The results showed that reasoning on the combination of the linguistic content with acoustic features of the spoken sentence was better than using only the linguistic component.

Keywords: Speech based interaction · Emotion recognition · Social robotics

1 Introduction and Motivation

In the Ambient Assisted Living (AAL) domain, a Smart Environment should support people in their daily activities in order to increase their quality of life [1], by assisting and facilitating users when interacting with environment services in a natural and easier way. The required assistance may be provided to the user in different modalities. The choice of an assistive robot agent as interface is supported by several considerations. First of all, the robot has a physical presence and it may participate in the user's daily life, assistive robots can move around and execute actions, follow and observe the user in the environment, which is fundamental when designing supportive technologies for elderly people [2]. In addition to typical service-oriented features, social assistive robots can be equipped with social and conversational capabilities, thus improving the naturalness and effectiveness of the interaction between users and smart environment services.

Several studies report successful results on how social assistive robots can be employed as interface in the assisted living domain. For instance, ROBOCARE [3], Nursebot [4], Care-o-bot [5], CompaniAble [6] and Ksera [7] are projects aiming to create assistive intelligent environments for the elderly in which robots offer support to the user at home. However, in order to make the robot able to perform accurately the required tasks, it is strictly necessary for them to understand correctly the command

© Springer International Publishing Switzerland 2015
F. Esposito et al. (Eds.): ISMIS 2015, LNAI 9384, pp. 273–279, 2015.
DOI: 10.1007/978-3-319-25252-0_30

given by the human user. Since speech is a natural way for humans to interact with robots [8], in this phase of the project we focused on the problem of interpreting the user's utterances in order to understand correctly the intended meaning.

In order to express their intentions humans use words and transfer emotions and emphasis by modulating their voice tone. In fact, speech conveys two main types of information: it carries linguistic information according to the rules of the used language and paralinguistic information that are related to acoustic features, such as variations in pitch, intensity and energy [9]. Usually the first component conveys information about the content of the communication and the second one about the user's attitude or affective state.

In this paper we present a module to be used by a social assistive robot, embodied in Aldebaran Robotics H25 NAO in our project, for interpreting the communicative intention in spoken utterances when interacting with a robot acting as a mediator in smart environments. In this module, we have coupled a linguistic parser, formalized in the grammar of an Automatic Speech Recogniser (ASR), that allows interpreting the spoken sentence in terms of speech act [10], with an acoustic analyser able to extract the prosodic features of the user spoken input in order to classify the conveyed emotion. Then, using a probabilistic model based on a Bayesian Network (BN) [11], this module infers the user's intention by combining these two knowledge sources. Results showed that the combination of the two information sources allowed the recognition of the intended meaning with a higher accuracy that linguistic content only.

2 The Communicative Intention Recognition Module

In order to endow a social assistive robot with the previously described capability, we use an annotated dataset of spoken sentences collected in a previous phase of the project and it is composed by 592 spoken utterances. Each utterance (the correspondent transcript and the related 16-bit single channel, 8 kHz signal (in a .wav format)) was stored in a database. More details about the Corpus and the annotation process can be found in [12].

We have designed the architecture shown in Fig. 1 to endow the Aldebaran NAO robot with this capability. The system is composed by two fundamental units: the NAO humanoid robot and the workstation connected with NAO robot. Audio files in wav format, recorded from 4 microphones located in the head of the NAO, are collected by the Application Programming Interface (API) provided with NAO Software Development Kit (SDK). Captured audio files are sent to the Communicative Intention Recognition Module in order to allows a translation of voice commands into written sentences and to recognize the valence and arousal. An Automatic Speech Recognition Module (ASR) performs the first task whereas the second task is accomplished by Voice Classifier for Emotions (VOCE) Module. Then, the results of the recognition intention module are sent to Behavior Decision Module that choose the appropriate behavior and send it to the robot to be executed. Communication between the robot and the workstation has been performed using the NAOqi API, currently available in 8 programming languages.

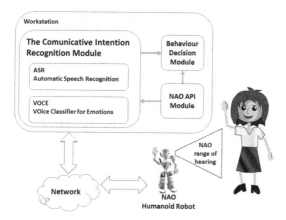

Fig. 1. Overview of the proposed system

The system is composed by two fundamental units: the NAO humanoid robot and the workstation connected with NAO robot. Audio files in wav format, recorded from 4 microphones located in the head of the NAO, are collected by the Application Programming Interface (API) provided with NAO Software Development Kit (SDK). Captured audio files are sent to the Communicative Intention Recognition Module in order to allows a translation of voice commands into written sentences and to recognize the valence and arousal. An Automatic Speech Recognition Module (ASR) performs the first task whereas the second task is accomplished by Voice Classifier for Emotions (VOCE) Module. Then, the results of the recognition intention module are sent to Behavior Decision Module that choose the appropriate behavior and send it to the robot to be executed. Communication between the robot and the workstation has been performed using the NAOqi API, currently available in 8 programming languages.

2.1 VOCE: VOice Classifier for Emotions

The analysis of emotional content of the spoken utterance is made by using VOCE: a classifier to recognize the valence and arousal. VOCE can be used to analyze recorded files or as a web service that can be invoked by any application that needs to recognize affect from voice. As explained previously, each audio file in the corpus was labeled with the corresponding level of arousal and valence and was analyzed using PRAAT [13] for extracting acoustic features related to: (i) the variation of the fundamental frequency (pitch minimum, mean, maximum and standard deviation, slope); (ii) the variation of energy and harmonicity (min, max and standard deviation); (iii) the central spectral moment, standard deviation, gravity centre, skewness and kurtosis; (iv) the speech rate.

Our classifier exploits several algorithms: the C4.5 algorithm, a decision tree learner, and K-NN showed to be the most accurate ones. As far as the valence classification is concerned, the accuracy of the first algorithm is 83,12 %, very close to the one of the K-NN that is 82,45 %. As far as the arousal is concerned, C4.5 has an accuracy of 79,8 %

while the one of K-NN is 83,63 (validated using a *10 Fold Cross Validation* technique). Other important evaluation metrics to consider include recall, precision, and F-Measure. Tables 1 and 2 report these values for the arousal and valence respectively.

Table 1. F-measure for the arousal

F-measure C4.5	F-measure K-NN	Class
0,89	0,90	high
0,79	0,75	medium
0,77	0,78	low

Table 2. F-measure for the valence

F-measure C4.5	F-measure K-NN	Class
0,81	0,84	positive
0,78	0,82	neutral
0,81	0,78	negative
0,90	0,94	very negative

From the F-measure value, both algorithms perform best in recognizing high arousal and very negative valence. Since the arousal dimension is related to the importance of the goal and the valence dimension is related to the achievement vs. the threatening of the goal, our speech classifier performs well in recognizing negative states like those related to anger and allows distinguishing positive from negative attitudes. However, as expected, some emotions are easier to recognize than others. For example, humans are much better at recognizing anger than happiness; therefore, our result can be considered acceptable under this view.

2.2   Recognizing the Communicative Intention

In order to infer the correct intention from the speech act and emotional content from the presence of cue words, we decided to use a probabilistic model, since understanding human attitude and intention from speech input involves capturing and making sense of imprecise, incomplete and sometimes conflicting data. We learn the inference model using the NPC learning algorithm of Hugin[1] on the labelled database of cases in the corpus. In this way we could learn both the structure and the parameters of the network. We used the 70 % of the dataset for learning leaving a 30 % of randomly selected entries for testing. Variables in this network are related to:

---

[1] www.hugin.com.

- the recognized move category: this information is extracted by the linguistic parser and belongs to one of the categories [12];
- the presence and valence of cue words that may change the linguistic interpretation of the move;
- the valence associated to the move;
- the arousal associated to the move;
- the intention beyond the speech act: this is the variable that we want to monitor using the model and can assume values in the set described in [12].

3   Evaluating the Model

The evaluation of the model was performed on a subset of data extracted randomly from the selected corpora. Then as input evidences of our model we used the following values: first only the linguistic content in terms of speech act of the move and then both the speech act and information about valence and arousal derived from the prosody. Finally, we compared the predicted results with the human annotation in the two cases. The global accuracy of the model, expressed as the total of correct predictions for every category on the total of the moves in the dataset, is of about 60 % in the case of linguistic content only, while it increases to the 75 % when considering both features. Since we use a probabilistic model, we considered as correct a prediction of the value of the intention variable the one of its six states with the highest probability.

Table 3 reports detailed results about the prediction of each feedback category in the two cases. As a general comment, we can say that the model performs a better prediction when using speech + text data, especially in those cases when the linguistic content contrasts the voice tone in identifying the type of feedback intention. This is evident in the case of the "Approve" and "Disapprove" class where the recognition from the linguistic content only performs worse.

Table 3. Confusion Matrix For Intention Recognition - Text Only/Speech + Text

		WantToDo	WantToKnow	KnowAbout	Approve	Disapprove	GetAttention	SocialCue	F1-score
WantToDo	Text	69	37	0	0	0	2	0	0.62
	S+T	74	21	0	5	8	0	0	0.77
WantToKnow	Text	17	30	0	0	0	0	0	0.41
	S+T	9	32	0	0	6	0	0	0.59
KnowAbout	Text	0	0	5	0	0	2	0	0.5
	S+T	0	0	4	1	1	1	0	0.61
Approve	Text	11	12	5	51	0	17	15	0.61
	S+T	0	0	1	85	0	10	15	0.83
Disapprove	Text	17	20	3	0	89	21	19	0.66
	S+T	2	9	1	0	136	9	12	0.81
GetAttention	Text	0	0	0	0	12	38	5	0.52
	S+T	0	0	0	0	9	41	5	0.66
SocialCue	Text	0	0	0	5	0	11	79	0.74
	S+T	0	0	0	3	7	9	76	0.75

4   Conclusions and Future Work

In this paper we proposed a module to interpret correctly the intended meaning of the user's spoken input in interacting with a social assistive robot in the context of Ambient Assisted Living. We focused our research on the analysis of the spoken sentence using two information sources: the linguistic content and the intonation of the voice. First of all, we collected a corpus of data based on the Wizard of Oz approach whose results were annotated and analysed in terms of linguistic communication content, presence of cue words, valence and arousal of the voice tone and intention. Then, this corpus has been used to train a classifier of the valence and arousal in the voice and to learn the structure of the Bayesian network model to be used for recognising the probability for a user to have a particular intention toward the robot and/or the environment situation.

In order to validate our model we used as a testing dataset a subset of moves randomly extracted from the corpus. The performed experiment shows that using both knowledge sources for recognising the user's intention improves the prediction accuracy of the model. The same approach could be used in other domains such as e-learning, e-health, etc. where user's emotion and intention recognition plays a key role for adjusting the behaviour of the system. Understanding user's affective state is a crucial aspect of effective smart environments, therefore monitoring continuously and unobtrusively the user frustration, boredom, enthusiasm, is important for tuning the robot behaviour.

In our future work, we suppose to use this model in the architecture of a social assistive robot aiming at assisting elderly people in their daily life tasks and we plan to improve affect recognition with the analysis of facial expressions and gestures.

References

1. Bierhoff, I., van Berlo, A.: More Intelligent Smart Houses for Better Care and Health, Global Telemedicine and eHealth Updates. Knowl. Resour. **1**, 322–325 (2008)
2. Thrun, S.: Towards a framework for human-robot interaction. Hum. Comput. Interact. **19**(1&2), 9–24 (2004)
3. Cesta, A., Cortellessa, G., Pecora, F., Rasconi, R.: Supporting interaction in the robocare intelligent assistive environment. In: AAAI 2007 Spring Symposium (2007)
4. Pineau, J., Montemerlo, M., Pollack, M., Roy, N., Thrun, S.: Towards robotic assistants in nursing homes: challenges and results. Robot. Auton. Syst. **42**(3–4), 271–281 (2003)
5. Graf, B., Hans, M., Schraft, R.D.: Care-O-bot II – development of a next generation robotics home assistant. Auton. Robots **16**, 193–205 (2004)
6. CompanionAble project (2011). http://www.companionable.net
7. ksera.ieis.tue.nl
8. Drygajlo, A., Prodanov, P.J., Ramel, G., Meisser, M., Siegwart, R.: On developing a voice-enabled interface for interactive tour-guide robots. J. Adv. Robot. **17**(7), 599–616 (2003)
9. Bosma, W.E., André, E.: Exploiting emotions to disambiguate dialogue acts. In: Nunes, N.J., Rich, C. (eds.) Proceedings 2004, Conference on Intelligent User Interfaces, January 13 2004, Funchal, Portugal, pp. 85–92 (2004)
10. Searle, J.R.: Speech Acts: An Essay in the Philosophy of Language. Cambridge University Press, Cambridge (1969)

11. Jensen, F.V.: Bayesian Networks and Decision Graphs, Statistics for Engineering and Information Science. Springer, New York (2001)
12. De Carolis, B., Cozzolongo, G.: Interpretation of User's Feedback in Human-Robot Interaction. J. Phy. Agents **3**(2), 47–58 (2009)
13. www.praat.com

Tracing Shifts in Emotions in Streaming Social Network Data

Troels Andreasen[1], Henning Christiansen[1]($\boxtimes$), and Christian Theil Have[2]

[1] Programming, Logic and Intelligent Systems, Roskilde University,
Roskilde, Denmark
{troels,henning}@ruc.dk
[2] Department of Metabolic Genetics, University of Copenhagen,
Copenhagen, Denmark
christiantheilhave@gmail.com

Abstract. Shifts in emotions towards given topics on social media are often related to momentous real world events, and for the researcher or journalist, such changes may be the first observable sign that something interesting is going on. Further research on why a topic t suddenly has become, say, more or less popular, may involve searching for topics t' whose co-occurrence with t have increased significantly together with the change in emotion. We hypothesize that t' and its increasing relationship to t may relate to a contributing cause why the attitude towards t is changing. A method and tool is presented that monitors a stream of messages, reporting topics with changing emotions and indicating explanations by means of related topics whose increasing occurrence are taken as possible clues of why the change did happen.

1 Introduction

Microblogging through Social media has become a very popular means of communication. Information exchanged is of very diverse nature as is the purposes of posting messages. Messages may report on individuals common day-to-day activities, or be interchanged for chatting. Companies post messages for commercial purposes, while organizations aim to attract attention to their concerns and news media post key headlines to broadcast daily news. It is, however, not always possible to make clear distinctions between types of messages and in many cases there is no need to do so. Patterns of messages and information exchange are, regardless of message types, influenced by events and issues that attract public interest. Often such patterns, as well as changes in these, are good indicators on what's going on and what's considered important issues. Thus analysis of such patterns may in turn lead to expressions of general opinions and trends and may even reveal causes to spotted changes.

In this paper we describe an approach to querying and monitoring trends, events and opinions in streaming messages. In continuation of the querying approach described in [2] we consider sequences of events and search for sequential patterns with emphasis on changes in attitude. We are aiming at

© Springer International Publishing Switzerland 2015
F. Esposito et al. (Eds.): ISMIS 2015, LNAI 9384, pp. 280–289, 2015.
DOI: 10.1007/978-3-319-25252-0_31

identifying significant changes, that relate to shifts in emotions, sentiments and co-ocurrence patterns. Shifts in emotions towards given topics on social media are often related to momentous real world events. For the researcher or data journalist, such changes may be important clues in their research. Changes can be the first observable sign that something interesting is going on, and a possibility to trace whether this is the case or not can be crucial during research. In our approach a pattern describing, for instance, a shift in emotion can be used in search for interesting cases, while possible causes may be studied among topics with simultaneous shifts in co-occurrence. Further research on why a topic t suddenly has become, say, more or less popular, may involve searching for topics t' whose co-occurrence with t have increased significantly together with the change in emotion. We hypothesize that t' and its increasing relationship to t may indicate a cause why the attitude towards t is changing. One example would be a politician P who looses his credibility from one day the other, while the co-occurrence of P and *corruption* raises from none to very high. Here *corruption* is indicated as a possible reason for P becoming unpopular.

In this study we use twitter data for experiments, where topics on messages are identified with hashtags included in these. Thus the features we take into account for individual messages are, apart from the provided hashtags, timestamps as well as sentiments and emotions derived from the message text by sentiment analysis. The sentiment analysis applied in experiments is provided by a tool described in [7].

The present paper is structured as follows. In Sect. 2 we introduce the language EMOEPISODES and in Sect. 3 we describe the special usage of this language covered here: reporting interesting relationships across topics as indication of possible courses to trends discovered. In Sect. 4 the semantics is described and in Sect. 5 we describe experiments and evaluation based on preliminary implementation of the language. In Sect. 6 we discuss related work and finally in Sect. 7 we conclude.

2   An Emotional Episode Language, EMOEPISODES

EMOEPISODES is a general language for formulating hypotheses about streams of time-stamped messages, concerning properties such as emotions. The language was first introduced in [2] and is in this paper further developed for the purpose of indication of cause. The semantics involves measurements of how well such hypotheses are indicative of or apparent for given time intervals. It can be used as a query language concerning the past and for realtime monitoring with known hypothesis or for discovering new interesting hypotheses.

We assume a continuous time and use $\mathcal{TI}$ to refer the set of all possible time intervals, which may be open or closed. The unit of time is unspecified, but we allow standard units such as days, hours, etc., to characterize intervals. A set of *topics* $\mathcal{T}$ is assumed, e.g., $\mathcal{T} = \{\mathsf{Xmas}, \mathsf{beer}, \mathsf{asteroids}, \ldots\}$, and set of *emotions*, e.g., $\mathcal{E} = \{\mathsf{fear}, \mathsf{happiness}, \mathsf{anger}, \mathsf{sadness}, \ldots\}$. We include sentiment as emotions, positive, negative and neutral. The *level* of an emotion is characterized by a finite set of symbols $\mathcal{L}$ ordered by magnitude; in this paper we assume the scale high > medium > low, but more fine-grained scales may also be used. For the present applications, $\mathcal{T}$ corresponds to twitter hash tags, but it may also be meaningfully extended with other important words extracted from the twitter text.

An atomic emotion *statement* about a topic t associates an emotion E and a level ℓ to that topic, written $t : E(\ell)$. For instance "asteroids:fear(high)" measures the degree to which high fear characterizes messages about asteroids. A *data semantics* is given by a *satisfaction degree* function $SD: \mathcal{AS} \times \mathcal{TI} \rightarrow [0; 1]$ where $\mathcal{AS}$ is the set of atomic statements. It measures how well a given statement characterizes a given time interval. For the present application, $SD(x : E(\ell), d)$ reflects the proportion among all messages arriving during time interval d tagged by x that are marked by emotion E; the details are shown in Sect. 4 below.

It is possible for both sentiments positive and negative to be high for a topic t at the same time, if, e.g., 50 % of all messages about t are positive and 50 % negative. This is quite different from all being neutral. The same goes for intuitively opposite emotions, e.g., love and hate.

An additional atomic statement measures co-occurrences of topics. For topics t and t', the following statement measures the amount of messages the *conditional occurrence* of t' relative to the set of all messages containing t.

– $t'|t : (level)$ characterizes the proportion of messages containing t' among those containing t.

As an example: russia|asteroid: (high) measures the degree to which russia characterizes messages about asteroids. Its semantics needs to be defined in a slightly different way than for emotions; details are given in Sect. 4.

EMOEPISODES includes also compound statements that represent an aggregation of atomic statements meant to go for the same time interval; we refer to [2] for details as such statements are not used here.

A *scene* is a statement with an associated *time constraint*; Examples:

asteroids:fear(high)[> 5 days]
asteroids:fear(high)[2013-02-15-15:52:07; 2013-02-16-19:00:00[

The first ones may be applied for different intervals along the time axis, that are longer that 5 days. The detailed language for time constraints is not specified further, and in this paper we need only constraints that fix specific time intervals as shown in the last sample above. we need only constraints that fix specific time intervals as shown in the second sample above.

An *episode* is a sequence of consecutive scenes, separated by semicolon:

asteroids:fear(medium)[2013-02-15;2013-02-17[; doomsday:fear(high)[> 2 days]

The semantics of the full EMOEPISODES language involves first solving the time constraints in order to find a consistent time assignment and then aggregating the SD values for each scene. For the present application, we can do with the following extension of SD for episodes of two scenes with unique and consecutive time intervals d_1 and d_2; here d denotes the concatenation of the two.

$$SD((s_1[d_1]; s_2[d_2]), d) = \min_{i=1,2} SD(s_i[d_i], d_i)$$

Different aggregation operators may be relevant for other applications, as explained in [2], but here, the minimum operator is sufficient.

In fact, EMOEPISODES is a general query language in which variables may stand for unknown constituents, e.g., topics, emotions or degrees, and the query evaluation mechanism may return instantiations of those variables that maximizes the satisfaction degree. An episode S of EMOEPISODES can also be used as a watchdog that signals whenever there is an instance of S and a time assignment ending at the current time, with a satisfaction degree that exceeds a given threshold.

3   Using EMOEPISODES to Report Interesting, Current Relationships

Briefly explained, our mechanism uses a watchdog looking for topics t, for which some emotion is changing in a significant way, and then we search for topics t' whose co-occurrences with t are increasing at the same time. Our rationale is an expectation that t' represents an important aspect related to the change in emotion for t. In case the emotions towards such a t' are unchanged during this period, we hypothesize that t' may reflect a cause of the changed attitude towards t; if emotions towards t' changes in a way similar to that of t, we expect the changes for t and t' to have a common cause, and that t and t' may be strongly connected in a symmetric way, perhaps as synonyms for the same thing.

In the following, the constant *now* refers to a present moment during the monitoring of a data stream. To observe a change in an emotion for some topic x, we need to refer to two time consecutive time intervals ending *now* in which the emotion has different values. The constant k refers to the shortest time period that a journalist (or other user) expects an emotion to hold a fairly stable value in order for a following change in that emotion to be significant. An eager sensation oriented journalists will likely prefer a very small k_e, perhaps an hour or 15 min (if the complete stream of all tweets from the entire world is available realtime, changes within 15 will be significant). A background journalist may prefer k_e to be 7 days or more. Before the interval $[now - k_e; now]$, a period with a different value for the emotion in question must be observed in order to talk about a change of value. We use a constant k for the length of this pre-period, i.e., $[now - k_e - k; now - k_e[$. The magnitude of k should depend on the sort of phenomena of interest (are they normally stable or fluctuating) combined with considerations about the overall arrival frequency of messages.

When a change in emotional value has been observed, we look for topics t' whose co-occurrences with t are increasing. As above, we define a similar constant k_o and reuse k for two similar, successive intervals. The rationale for having two different constants k_e and k_o is an expectation that when a (perhaps drastic) real world event changes the general attitude towards a topic t, it may take some time before more interesting circumstances or speculations about the event becomes known and discussed; this can motivate $k_o < k_e$. Different values of k may be used for measurements of t and t', but we see no good reason for that. We expect identical values for k_e and k_o to be acceptable, although more empirical testing may be needed to find the best choice.

Our system is based on the following abstract algorithm; T_0 is a subset of all topics declared by the user to be of interest. A threshold $\theta \in [0; 1]$ for significant degree of satisfaction is assumed.

---

1. Identify the set of all topics $t \in X \subseteq T_0$ for which there exist emotion E and levels $\ell_0 \neq \ell_1$ such that the satisfaction degree of the following query is $\geq \theta$,

$$t{:}E(\ell_0)[now - k_e - k; now - k_e[\; ; \; t{:}E(\ell_1)[now - k_e; now].$$

2. For each $t \in X$, identify the set of $t' \in Y_t$ for which there exist levels $\ell_2 < \ell_3$ such that the satisfaction degree of the following query is $\geq \theta$,

$$t'|t{:}(\ell_2)[now - k_o - k; now - k_o[\; ; \; t'|t{:}(\ell_3)[now - k_o; now].$$

Notice that the sets Y_t need not be subsets of T_0.

The algorithm may run continuously over time, as *now* inevitably moves forward, although an implementation may need to use a more or less fine-grained discretized time in order to reduce the computational overhead. The results from this algorithm is monitored, showing for each t in the current X set,

- the list of emotions E that may trigger step 1 above,
- the list Y_t with, for each $t' \in Y_t$, whether t' is a *possible cause* of the emotion change (i.e., emotions towards t' stable), or a *related concept* (i.e., emotions change similarly to t); measured analogously to step 1 above.

A user interface to be used in, say, an editorial office may display this rudimentary information as an effective way for the journalists quickly to recognize potentially new hot topics; an proposal for this is shown in Sect. 5 below. Additional information may be called up by click buttons, e.g., about actual satisfaction degrees and detailed information about the t' concepts.

4   A Data Semantics for Mining Trends on Twitter

Different choices for data semantics and aggregation operations are introduced in [2] for a variety of applications and views of the data. Here we give the data semantics used for the present application of measuring changes in streaming twitter messages (which coincides with the so-called elitist semantics of [2].

As indicated above, we include the so-called sentiment as a special sort of emotion. Let us make this precise,

$$\mathcal{E} = \mathcal{E}_1 \cup \mathcal{E}_2, \quad \text{where}$$
$$\mathcal{E}_1 = \{\text{anger, disgust, fear, joy, sadness, surprise}\}$$
$$\mathcal{E}_2 = \{\text{negative, neutral, positive}\}$$

When a twitter message arrives, it is classified by a filter with an emotion $\in \mathcal{E}_1$ and a sentiment $\in \mathcal{E}_2$; however, the filter may fail in identifying emotion or sentiment. The same message cannot be classified with two different proper emotions or two different sentiments. We introduce some notation; we assume topic $t \in \mathcal{T}$, time interval $d \in \mathcal{TI}$, $E \in \mathcal{E}$.

$\delta_\sharp(t, d)$:   The set of messages tagged by topic t during d

$\delta_i(t\!:\!E, d)$: The set of messages tagged by topic t during d and classified by the filter by $E \in \mathcal{E}_i$

$\delta_i(t, d)$:   The set of messages tagged by topic t during d and classified by the filter by some $E' \in \mathcal{E}_i$.

The relative frequency of emotion or sentiment $E \in \mathcal{E}_i$ for a topic t during time interval d is defined as

$$R_i(t\!:\!E, d) = \frac{|\delta_i(t\!:\!E, d)|}{|\delta_i(t, d)|}.$$

For topics t, t', we define the relative frequency of t' given t during d as

$$R_\sharp(t'|t\!:\!, d) = \frac{|\delta_\sharp(t', d) \cap \delta_\sharp(t, d)|}{|\delta_\sharp(t, d)|}.$$

The different levels high > medium > low are treated as simple fuzzy linguistic terms in the definition of satisfaction degrees for atomic statements, using membership functions μ_i^ℓ for $i \in \{1, 2, \sharp\}$ and level $\ell \in \{\text{high}, \text{medium}, \text{low}\}$. The satisfaction degrees defining the data semantic are now given as

$$SD(\phi_i(\ell), d) = \mu_i^\ell(R_i(\phi_i, d))$$

where ϕ_i is one of $\phi_1 = (t\!:\!E_1)$, $\phi_2 = (t\!:\!E_2)$, $\phi_\sharp = (t'|t\!:\!)$. Figure 1(a) shows definitions of relative satisfaction level terms high, medium, low for classifications $\mathcal{E}_1$ and Fig. 1(b) for classifier $\mathcal{E}_2$ (notice that $|\mathcal{E}_1| = 6$ and $|\mathcal{E}_2| = 3$). For occurrences of t' given t, we choose (arbitrarily) the relative frequency $f_{t'}$ of t' among all tweets, measured for a sufficiently large period D in the past, as the midpoint for medium; the remaining membership functions are defined accordingly.

$$f_{t'} = \frac{|\delta_\sharp(t, D)|}{|\delta_\sharp(D)|} \quad \text{where } \delta_\sharp(D) \text{ is the set of tweets arriving during } D.$$

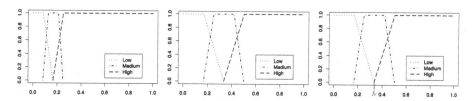

Fig. 1. Membership functions μ_ℓ for fuzzy linguistic terms over relative satisfaction with level $\ell \in \{\text{high}, \text{medium}, \text{low}\}$. Shown in (a) for classifier $\mathcal{E}_1$, in (b) for classifier $\mathcal{E}_2$ (6 and 3 classes respectively), and in (c) for $t'|t$: that depends on t' but not on t.

5   Experiments

Our system is not fully developed with a finished user interface for reporting the results. To illustrate the utility of the mechanism described in the paper we show a manually created report for results generated from a sample run of the system on a large corpus of tweets extracted from the Twitter firehose during the period from December 23 2012 to February 7 2013. The *firehose* is an API which gives access to (a random fraction of) tweets as they occur in real-time. The corpus thus represents a random sample of tweets which can be seen as representative of all tweets in the period.

We use this corpus as if we had continuously been collecting tweets and the current date – *now* – happens to be one of the days up to February 7 2013. We define time constraints in a granularity of days; We use a setting of $k = k_e = k_o = 5\ days$.

After filtering uninteresting nuisance tags such as *teamfollow, followback,* ... which are used in pyramid-schemes to gain followers and retweets, the best scoring result involves the tag *iphone* which co-occurs with the tag *gameinsight*.

In the result shown in the report in Fig. 2 we see that the system has detected an increase in negative sentiment for the topic *iphone* and that it is correlated with an increased co-occurrence of the tag *gameinsight*. This is a tag which is mostly used by certain games when they post automatic updates to Twitter and promote a viral effect with tweets such as,

– "just reached level 19 on Rock the Vegas on my iPhone http://t.co/Z4nLdor1 ♯iphone iphonegames ♯gameinsight"
– "I have completed the quest 'Order 3 Long-Term D...' in the ♯iPhone game The Tribez. http://t.co/nn6tb60a ♯iphonegames, ♯gameinsight"

Of the above tweets, which are announcements of game progress, the second one is classified as having negative sentiment mainly because of the word *quest* which occur in the negative sentiment classifier dictionary. When the word occurs, it often results in negative sentiment classification of the tweet in which it occurs. While in the context of this tweet the classification seems dubious, a sudden surge in popularity in *The Tribez* game, which publishes a lot of tweets about *quests*, gives rise to an increased negative sentiment about *iphone* as can be observed Fig. 2.

6   Related Work

As mentioned above the approach described in this paper is a continuation the querying approach described in [2]. While [2] took a more general approach towards a language for specifying patterns the present paper has a primary focus on identification of cooccurring topics that may candidate as possible causes.

Our terminology is inspired by the seminal work of [8] who suggested a way to define episodes in sequences of discrete events (from a finite alphabet of such) and gave algorithms to search for a sort of association rules among such episodes.

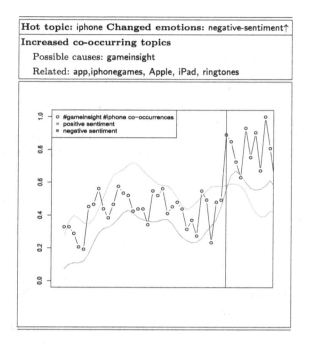

Fig. 2. The figure shows a sample report for the *"hot topic"* result *iphone*. The associated emotion changes are indicated and the most promising candidate is the co-occurring tags, (here only *gameinsight*), are listed. Similarly, *related* highly co-occurring topics which do not correlate sufficiently with the emotion changes are listed. The report also visually plots the change in both emotion (negative sentiment) measured as number of emotion classifications for the topic (*iphone*) scaled to a unit interval and the number co-occurrences also scaled to a unit interval for comparability. For reference, we also provide a similar plot for positive sentiment change (which did not result in an alert).

Before that, [1] described algorithms for mining frequent, sequential patterns in a transaction database. See a recent survey [10] on later work inspired by [1,8]. Our work differs in that we present a logical language for specifying scene and episode sequences that refer to measurements over large sets of timestamped tweets, rather than over a finite alphabet. The language has a well-defined, graduated truth semantics, that by parameterization allows different interpretations of the data.

Analysing trends in blogging corpora based on sentiment is attracting increasing interest by researchers and the micro blogging platform Twitter have shown to be a valuable ressource for this purpose in part due to users widespread use of hashtags that to some extent can be interpreted as topics. In [12] the authors analyse trending topics with emphasis on duration in an attempt to distinguish major public events. In this approach so-called provenance of topics plays a key role. The goal is the same a ours – to search for important topics and trends related to these. However, while we analyse patterns of attention and opinion during consecutive time intervals, the approach in [12] is to apply and additional resource as context, Wikipedia, in an attempt to distinguish important topics and significants trends, such as those related to major public events. Another approach to identification of trends with similarities to ours is described in [9]. They consider real time trend detection over the Twitter stream and based on this provide a monitor system. While we attempt to identify trends that range over changes of attitudes, [9] identifies trends with emerging topics on Twitter, and aims to synthesize accurate descriptions of topics.

An approach that combine sentiment analysis and volume-based measures like ours is described in [4]. They investigate the potential to model political sentiment through mining of social media, and they indicate with their results that their combined approach to analysis may provide prediction for their case, an election campaign, when including examination of sample sizes, time periods as well as methods for qualitatively exploring the underlying content. Further in [3] the authors argues that analysis on the short document length in microblogs provide compact and explicit sentiments. They argue that it is easier to classify the sentiment in short form documents than in longer documents.

Sentiment analysis of Twitter messages over time has previously been demonstrated to correlate with public opinion measured by Gallup polls [11]. However, in this study only positive/negative sentiment is measured and they do not provide a method to search for specific patterns. More detailed emotions are used in a study which correlates Twitter sentiments to socio-economic phenomena [6]. This group have also studied correlation of sentiment of Twitter messages to stock prices [5] and also find sentiment to be highly correlated with stock prices, but their approach does not specifically consider surprising events.

7   Conclusion and Future Work

In this paper we have presented a language for querying development of attitudes towards given topics over time and we have illustrated its use with Twitter data. Expressions in the language may include specification of topics, and emotions and sentiments related to these may be queried within specified time intervals. The general form of a query specifies two consecutive varying length time intervals that can be considered moving along the time line in search for significant changes from the one to the next interval. A query can be open wrt topics as well as wrt emotions. Thus we can look for changed attitude towards specific topics or in general search for topics characterised by such changes. We can also search for specific attitude changes, such as increased surprise, or in general search for significant changes along any emotion or sentiment in any direction. We have put a special emphasis on mining causes. Thus we are not only concerned with significant changes, but also, as part of the framework, aimed at deriving indications of causes by investigating topic co-occurrences.

Being able to answer queries, spot changes and indicate causes, as supported by our language, can be very useful for journalists and researchers looking for interesting new trends and it may have important implications for, e.g., social, socio-economical, political science and for market analytics. Analysis of Social media data and mining for trends is not new, but to our knowledge, a general approach, introducing a query language, and a support for mining, not only for changes, but also for indications of possible causes, has not been seen before.

Our prototype implementation is preliminary and can be improved in a number of ways. Firstly the language is not fully implemented and currently some of the data analytics has to be initiated manually. Secondly the mentioned "watchdog" use of the approach – supporting in principle any number of query expressions introduced as search agents for automated notifications – has not been implemented yet and decisions has to be made concerning its specific functionality.

References

1. Agrawal, R., Srikant, R.: Mining sequential patterns. In: Yu, P.S., Chen, A.L.P. (eds.) ICDE, pp. 3–14. IEEE Computer Society (1995)
2. Andreasen, T., Christiansen, H., Have, C.T.: Querying sentiment development over time. In: Larsen, H.L., Martin-Bautista, M.J., Vila, M.A., Andreasen, T., Christiansen, H. (eds.) FQAS 2013. LNCS, vol. 8132, pp. 613–624. Springer, Heidelberg (2013)
3. Bermingham, A., Smeaton, A.F.: Classifying sentiment in microblogs: Is brevity an advantage? In: Proceedings of the 19th ACM International Conference on Information and Knowledge Management, CIKM 2010, pp. 1833–1836. ACM, New York (2010)
4. Bermingham, A., Smeaton, A.F.: On using Twitter to monitor political sentiment and predict election results. In: Proceedings of the IJCNLP Conference, Chiang Mai, Thailand (2011)
5. Bollen, J., Mao, H., Zeng, X.: Twitter mood predicts the stock market. J. Comput. Sci. 2(1), 1–8 (2011)
6. Bollen, J., Pepe, A., Mao, H.: Modeling public mood and emotion: twitter sentiment and socio-economic phenomena. In: Proceedings of the Fifth International AAAI Conference on Weblogs and Social Media, pp. 450–453 (2011)
7. Jurka, T.P.: Sentiment: Tools for Sentiment Analysis. Version 0.2. http://github.com/timjurka/sentiment
8. Mannila, H., Toivonen, H., Verkamo, A.I.: Discovery of frequent episodes in event sequences. Data Min. Knowl. Discov. 1(3), 259–289 (1997)
9. Mathioudakis, M., Koudas, N.: Twittermonitor: trend detection over the twitter stream. In: Proceedings of the 2010 ACM SIGMOD International Conference on Management of Data, SIGMOD 2010, pp. 1155–1158. ACM, New York (2010)
10. Mooney, C.H., Roddick, J.F.: Sequential pattern mining - approaches and algorithms. ACM Comput. Surv. 45(2), 19:1–19:39 (2013)
11. O'Connor, B., Balasubramanyan, R., Routledge, B.R., Smith, N.A.: From tweets to polls: linking text sentiment to public opinion time series. In: Proceedings of the International AAAI Conference on Weblogs and Social Media, pp. 122–129 (2010)
12. Tran, T., Georgescu, M., Zhu, X., Kanhabua, N.: Analysing the duration of trending topics in twitter using wikipedia. In: Menczer, F., Hendler, J., Dutton, W.H., Strohmaier, M., Cattuto, C., Meyer, E.T. (eds.) ACM Web Science Conference, WebSci 2014, Bloomington, IN, USA, pp. 251–252. ACM (2014)

Machine Intelligence: The Neuroscience of Chordal Semantics and Its Association with Emotion Constructs and Social Demographics

Rory Lewis[✉], Michael Bihn, and Chad Mello

Department of Computer Science, University of Colorado,
Colorado Springs 80918, USA
rlewis5@uccs.edu

Abstract. We present an extension to knowledge discovery in Music Information Retrieval (MIR) databases and the emotional indices associated with (i) various scalar theory, and (ii) correlative behavioral demographics. Certain societal demographics are set in their ways as to how they dress, behave in society, solve problems and deal with anger and other emotional states. It is also well documented that particular musical scales evoke particular states of emotion and personalities of their own. This paper extends the work that Knowledge Discovery in Databases (KDD) and Rough Set Theory has opened in terms of mathematically linking music scalar theory to emotions. We now, extend the paradigm by associating emotions, based from music, to societal demographics and how strong these relationships to music are as to affect, if at all, how one may dress, behave in society, solve problems and deal with anger and other emotional states.

Keywords: Chordal semantics · Computer science · Emotion · KDD · Machine intelligence · Machine learning · MIR · Neuro-endocrinology · Neuroscience · Rough set theory · Social demographics

1 Introduction

The field of neuro-endocrinology has shown that music scales evoke signals in the neurotransmitters of humans called emotions [19] and that these emotions can be mathematically identified [16,23] and mined [8] in the field of Music Information Retrieval (MIR). Research in cognitive neuroscience has shown that perceiving musical and speech rhythms affects oscillatory activity in the brain where Smith *et al* showed that this is because our neural oscillations are hierarchically organized across multiple frequencies which allows a highly flexible process of attentional selection that allows humans to focus and respond emotionally to musical events [7,24] The author's early work created rough set systems that recognized instruments and the notes they played in polyphonic musical wave-forms [13,15].

© Springer International Publishing Switzerland 2015
F. Esposito et al. (Eds.): ISMIS 2015, LNAI 9384, pp. 290–299, 2015.
DOI: 10.1007/978-3-319-25252-0_32

We can also measure, from just the scale of the music, the tension it evokes in humans by synthesizing and measuring the chordal progressions [20]. This paper builds upon this work where details herein can be found, specifically for calculating the fundamental frequency of notes [12], mining music scalar theory [20] in a music database [26] and setting a non-Hornbostel hierarchical manner [11,14]. The interesting work performed by Espinoza *et al* in 2014, formed the ground based hypotheses for this paper: 177 college students rated four, brief musical phrases on continuous happy-sad scales. The research determined that major keys, non-harmonized melodies, and faster tempos were associated with happier responses [4,21]. The idea that one could create machine intelligence using rough sets to associate a human and the emotions their neuro synapses have trained the hypothalamus to evoke – all by just knowing certain information about that human, their age, sex, education, type of degree, race, country of origin, religious and political associations. This research presented in this paper does not, in any way, reach this goal, however, the experimental results presented in this paper are thought provoking and realized the reality of machines calculating what emotions a particular human will feel by just knowing a little about them.

The precise definition of what emotions are in regard to music has been notoriously problematic [9], *'even if there is agreement over their general characteristics and subcomponents'* [3]. Studies associating various feature sets and attributes include psychoacoustic studies that analyze attributes such as loudness and roughness [10]. Structural analysis of scale, mode and harmony attributes and performance [6]. Beat, tempo and articulation studies [1]. Studies of the chordal ontology and semantics in scalar theory [11,14]. Studies in the timbre (pitch, rhythm, harmony, structure) of music [3,17]. Composers think carefully in selecting instruments to play particular section of music on order to bring out emotional nuances in a musical structure [22]. In a study that examined whether frequency ratings of perceived and of felt affect differ significantly, Zentner *et al* ran experiments of Everyday-Life Emotions Versus Music-Induced Emotions for the purposes of measuring the similarity of everyday-life emotions with music-evoked emotions, their findings showed that the frequency ratings of felt musical emotions and everyday emotions differed significantly from each other [27].

2   Data Preparation and Description

In order to associate emotions, music, lifestyle, age, wealth and education, the authors combined multiple datasets from Nielsen, Billboard and the author's tvLander websites were combined with the author's previous datasets to form four datasets described in detail below. We have merged four disparate databases as illustrated in Fig. 1, **Degree, Cable Networks, Age** and **tvLander** into two .csv files *degree* and *merge* at http://www.rorylewis.com/ISMIS2015DATA.zip.

Degree: 19 forms of education comprise the tuples as shown in Table 1 with the first half of the attributes and Table 2 with the second half of the attributes. **Degree Tuples:** A partial set of examples of the 19 degree combinations listed in

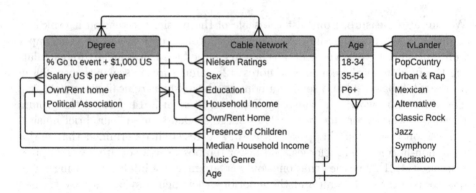

Fig. 1. Four Datasets. Degree is equivalent to degree.csv, while Cable Network (Nielsen), Age (Nielsen / Billboard) and tvLander (tvLander.com) are merged into datasetsmeerge.csv at http://www.rorylewis.com/ISMIS2015Data.pdf.

Table 1 are broken down as follows. *AS - BluCol* - comprises an Associate Degree such as a plumber, electrician or aircraft mechanic. *GED* - comprises a General Education Degree meaning no associate or university degree. *BA* - comprises a Bachelor of Arts Degree such as history. *BS* - comprises a Bachelor of Science Degree such as computer science. *BSN* - comprises a Bachelor of Nursing degree. *MBA* - comprises a Masters of Business. *MS* - comprises a Masters Graduate Degree. *PhD* - comprises a Doctor of Philosophy Degree. *JD* - comprises a Juris Doctor Degree. *MD* - comprises a Medical Doctor Degree. Combinations include: *BS MD* - comprises a Medical Doctor Degree with a Bachelor of Science Degree, *BA JD PhD* - comprises a Juris Doctor Degree and a Doctor of Philosophy Degree a with a Bachelor of Arts degree, *BS JD PhD* - comprises a Juris Doctor Degree and a Doctor of Philosophy Degree a with a Bachelor of Science degree. **Degree Attributes:** First the *Salary US $ per year in K* attribute is given in seven ranges. The ranges are *<30, 30–49, 50–74, 75–99,100–199, 200–499* and *>500*. Then the *Ave* attribute denotes the average salary, followed by the percentages that *own* or *rent* their homes. these attributes are in Table 1. In Table 2, one sees the *Go to event + $1,000 US* attribute which has six distinct events: *Burn* or Burning Man, *Lollapalooza*, *Met* or Metropolitan, *Eag* or an Eagles concert, *Yoga* for those who practice Yoga, and *Hunt* for those that hunt animals. Attributes *Dem* and *Rep* are the percentages of the degree class whose political affiliations are Democrat and Republican respectfully. The *SUM* attribute is the summation of those who are non tvLanders, who do not vote Republican or Democrat, we've called "innertrons". Accordingly, the authors have defined three types of nonTvlanders as (i) innertron Libertarians *iLib*, innertron Independents *iIndep* and pure innertrons *itrn*.

Cable Networks: 63 networks comprise the tuples. The data table presents the distribution of viewers of each network *"NET"* over several age groups attributes, both in number of viewers and as a percentage of the viewers. Cable Networks Tuples The definitions of the 63 networks abbreviations are in Table 1. Cable Networks Attributes The *"Rank for Prime Time (8PM–11PM)"*, the second

Table 1. Degree Part I. Salary & Ave. income associated with education.

| Degree | Salary US $ per year in K | | | | | | | | Home | |
	<30	30-49	50-74	75-99	100-199	200-499	>500	Ave	own	rent
AS BluCol	14	16	32	20	11	5	2	100.84	67	33
GED	72	16	6	3	2	1	0	32.63	19	81
BA	38	46	9	3	2	1	1	52.39	34	66
BS	3	13	26	38	13	4	3	122.58	69	31
BSN	3	4	32	42	16	3	0	90.08	71	29
BA MBA	11	14	19	33	18	3	2	108.19	68	32
BS MBA	7	9	21	20	32	8	3	141.52	72	28
BA MS	36	41	11	6	3	2	1	57.84	47	53
BS MS	1	6	18	26	41	7	2	140.78	71	29
BA PhD	16	24	37	12	6	3	2	88.08	44	56
BS PhD	0	6	17	22	42	9	4	168.38	77	23
BA JD	4	9	23	34	14	11	5	160.24	75	25
BS JD	0	7	18	32	28	11	4	162.1	77	23
BA MD	0	1	2	22	37	26	12	293.68	83	17
BS MD	0	1	2	18	36	29	14	321.6	93	7
BA JD PhD	1	4	19	40	20	11	5	169.38	78	22
BS JD PhD	0	0	15	28	27	19	11	261.36	82	19
BA MD PhD	0	0	0	11	37	34	18	378.67	97	3
BS MD PhD	0	0	0	9	32	40	19	397.13	97	3

attribute, is followed by attributes for the percentage and number of viewers for the average audience *"AA"*, both by household *"HHLD"* and by viewers over the age of two years old *"P2+"* [5,18,25]. Next, the attributes for the time slots, 6AM to 6PM and 7PM to 11PM, are presented. The 6AM to 6PM *"DUR"* attribute indicates each network's broadcast duration in minutes for the measurement period and the day part [2]. The *"Coverage"* attribute is the percentage of homes or persons receiving a particular broadcast signal within a specific geographic area [25]. After the Coverage attribute, one sees four groupings showing audience as a percentage of viewers correlated to age groups, 18–34, 35–54, persons of age 6 and up, and those greater than 55 years of age. For each of these age ranges the percentage of the audience that listens to the specified genre of music is given. The 54th *"male"*, and 55th *"female"*, attributes are given as a percentage of the viewing audience for that network. The *"education"* attribute is followed by three attributes are household income ranges, $75k+, $50k–$74.9k, and $30k–$49.9k. Attributes *"Home Ownership-OWN"* and *"Home Ownership-RENT"*, both are percentage of the viewers. The next attributes are the *"Presence of Children 1+"* in the household, the *"Median Age"* and the *"Median of Household Income"*.

Age: The age attributes of the databases utilized for this study require correlation. The Nielsen Cable Networks Ratings and the Degree data bases have disparate age range definitions.

Table 2. Degree Part II. Education groups that spend more than $1K for events.

Degree	% Go to event + $1,000 US											
	Burn	Lol	Met	Eag	Yoga	Hunt	Dem	Rep	SUM	iLib	iIndep	itrn
AS BluCol	3	60	0	1	3	34	19.33	11.26	68.96	29.57	4.11	35.28
GED	0.38	58.32	0	0.12	0.1	22.32	40.31	6.38	53.05	41.25	0	11.8
BA	1.7	33	1.05	0.36	0.1	3.8	61.31	2.86	35.26	27.72	1.02	6.52
BS	27.6	0.93	8.58	0.78	0.78	0.09	14.49	58.96	26.18	18.28	5.61	2.29
BSN	2.84	1.45	0	0.64	1.12	0.06	12.99	71.57	15.26	11.18	2.05	2.03
BA MBA	3.4	12.8	3.25	1.52	1.08	1.1	28.67	44.27	26.4	17.17	5.19	4.05
BS MBA	23.04	1.4	15.57	1.05	1.92	0.35	14.11	48.3	36.71	27.48	4.99	4.25
BA MS	2.82	15.9	1.16	0.99	0.18	3.6	40.55	27.64	31.66	21.04	5.06	5.56
BS MS	22.01	1.74	16.89	2.16	3.28	0.04	14.62	53.74	31.38	16.44	7.59	7.36
BA PhD	2.2	11.2	3.52	5.18	0.42	1.6	34.5	43.12	21.61	7.84	7.19	6.58
BS PhD	13.86	1.15	25.26	2.72	3.78	0	21.59	69.77	7.65	3.22	4.11	0.32
BA JD	3	2.75	48.07	3.91	0.98	0.44	24.7	69.8	5.26	1.44	2.48	1.33
BS JD	3.85	0.92	17.83	3.24	2.8	0	20.3	66.55	13.07	7.02	3.86	2.19
BA MD	1.66	0	58.74	0.32	2.22	0	39.52	55.28	4.73	2.26	1.95	0.52
BS MD	0.93	0	64.32	0.3	3.24	0	32.94	49.37	17.66	6.06	7.59	4.01
BA JD PhD	2.34	1.1	25.41	2.47	1.2	0.03	31.09	52.99	15.63	6.71	5.64	3.28
BS JD PhD	0	0	49.66	1.65	1.89	0	17.45	69.54	12.63	4.28	4.33	4.02
BA MD PhD	0	0	45.44	0	2.22	0	12.34	69.63	17.2	7.2	5.01	4.99
BS MD PhD	0	0	91.34	0	1.6	0	13.18	66.34	19.83	6.89	6.75	6.19

tvLander: The tvLander Database provides tuples based on age. For each age range, %s of listeners to the population are given for each genre of music. tvLanderTuples The age ranges are *18–34*, *35–54*, and *P6+* for listeners age six years old and above. tvLander Attributes The attributes are the genre of music. These are *PopCountry*, *Urban & Rap*, *Mexican*, *Alternative*, *Classic Rock*, *Jazz*, *Symphony*, and *Meditation*. They are all in percentage of the listening audience.

3   Linking Theoretical Music Attributes with Lifestyles

We first address how artificial intelligence parses scalar patterns in music waveforms and determines what type of music the listener is listening to and what the typical emotions correlate to that particular music scale. The motivation for this paper is that not everybody has the same emotional response to a particular scale. For example, a mid thirty person at Lollapalooza may not react the same way to a Mozart's 40th, that a person at the New York Metropolitan does. Why is this? The author first correlates earlier work in Rough Set Theory and Music Information Retrieval to Nielsen Data bases and with this in place, runs experiments illustrated in the Experiments section of this paper.

Looking at music, the author's previous work procured an artificial intelligence system that enabled a machine to identify the scale of a set of sound

Table 3. Basic Score Classification Scale Table. Jumps between notes in a scale X are represented by $J^I, J^{II}, J^{III}, J^{IV}, J^V$ which correlate to specific scales, regions and genre of music. For details, *see* [11,14]

X	J^I	J^{II}	J^{III}	J^{IV}	J^V	Scale	Region	Genre	Emotion	sma
X_1	2	2	3	2		Pent. Major	Western	Blues	melancholy	s
X_2	3	2	1	1	2	Blues Major	Western	Blues	depressive	s
X_3	3	2	2	3		Pent. Minor	Western	Jazz	melancholy	s
•										
•										
•										
X_{33}	2	1	2	2	3	Minor 11th	neutral	neutral	not happy	a
X_{34}	4	4				Augmented	neutral	neutral	happy	a
X_{35}	3	3	3			Diminished	neutral	neutral	not happy	a

waves in a song. Briefly, the machine first calculates the fundamental frequency and then once established it correlates the scalar theory of that sound. The fundamental frequency, f_1, of a sound wave establishes pitch. The distance of all the other frequencies within an octave, starting from a given note of frequency f_1 and going up in the frequency scale, can be calculated by $f_k = f_1 \cdot 2^{k/12}$, where k is number of semitones separating f_k and f_1 [11,14]. However, as shown in Table 3, we operate in the non-temporal domain the fundamental frequency in each frame i is calculated in the form: $f(i) = \frac{s}{K_i/n_i}$, where s is the sample frequency, n_i is the total number of $r(i,k)$'s local valleys across zero, where $k \in [1, K_i]$ and K_i is the estimated maximum fundamental period. Here we let $r(i,k)$ be the normalized cross correlation of frame i with lag k. To calculate $r(i,k)$, we look at how it reaches its maximum value.

The authors chose to represent basic score classification of music not as a music system but rather as an information system $S = (X, A, V)$, called Scale Table, where $A = \{J^I, J^{II}, J^{III}, J^{IV}, J^V$, Scale, Region, Genre, Emotion, sma$\}$ [11]. To find the dominant key we segment each bar and phrase into notes and then categorize the music based on what scale the most notes have been played. Next, we weight this number by the likelihood value of each note when it is classified to this scale. For example, if all the notes in the music piece are grouped into k bars: $B_1; B_2; ...B_k$, with B_i corresponding to one of the ith scales in X, then we compute a bar-score $\phi(x)$ for each $x \in Note$ as

$$\phi(x) = \Big[\sum_{i=1}^{k} B_i(x)\Big] \Big/ \Big[\sum_{y \in Song} \sum_{i=1}^{k} B_i(y)\Big] \tag{1}$$

and if all the notes in the music piece are grouped into k phrases: $P_1; P_2; ...; P_k$, with P_i corresponding to one of the ith scales in X, then we compute a phrase-score $\psi(x)$ for each $x \in Note$ as

$$\psi(x) = \left[\sum_{i=1}^{k} P_i(x)\right] \Big/ \left[\sum_{y \in Song} \sum_{i=1}^{k} P_i(y)\right] \tag{2}$$

4   Experiments

1,290 users entered their education and then were asked to listen to 20 seconds of four songs and select an emotion they felt with at song with a scale of 1–10 to define how intense that emotion was. The goal is to correlate the attributes $X_1...X_{35}$ in Table 3 to choices humans will most likely make if they associate a (i) a particular emotion, to a (ii) particular X_n. In other words if a person chooses "*melancholy*" with a scale of 10 when listening to *Western & Jazz* song, our machine correlates X_3 3223 *Pentatonic Minor* as shown on line 3 in Table 3. The user has already entered their education.

One the data was collected the user info was run through rapidMiner in accordance to the schema in Fig. 1, resulting in 19 sets of results in terms of each level of education as set forth in the Degree dataset. Resulting in how many people from each degree voluntarily said they would go to one of the four events per the attributes of Table 1. The comparison with what political affiliation was then compared to each event coasting more than $1,000 US comparing that to what the Nielsen ratings expected results. Looking at Fig. 2, we review one of the 19 educational level charts.

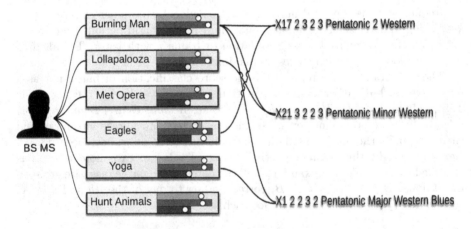

Fig. 2. One of the 19 Resultants: BS MS Those users indicating they have a Bachelor of Science degree and a Masters in Science had at least one person say they would spend $1,000 on each of the selected events. The expected political affiliations are the bars and the resultant affiliations are the circles. The top three musical scales selected with the correct emotion of a level of 80% are shown.

5   Conclusions and Future Work

Referring to Table 2, Figs. 1, and 2 together, for those who indicated that they have a Bachelor of Science and a Masters in Science, we expected, looking at the 9th row of Table 2, that according to the Nielsen Rating that the percentages who would pay \$1,000 US would be, 22.01 % would go to a Burning Man event, but we received a total of 19.96 %, 1.74 % would go to a Lollapalooza event, but we received a total of 1.5 %. 16.89 % would go to a Met Opera Man event, but we received a total of 15.02 %, 2.16 % would go to a Eagles event, but we received a total of 2.3 %, 3.28 % would go to a yoga event, but we received a total of 3.7 %, .004 % would go to a hunt animals event, but we received a total of 0 %. Looking at the bars in Fig. 2 next to each of the aforementioned event results we insert the expected political affiliation of 14.62 % Democrat, 53.74 % Republican and 31.38 % as this should be equal across all events as they were all BS MS in education. The circles in each corresponding bar chart indicates our results.

All resultants for this BS MS category in Fig. 2 were within 86 % of the expected results except for those BS MS innertrons who said they would spend \$1,000 on an Eagles rock concert. The expected percentage was 31.38 % but we received 46.5 % which is the largest inconsistency from what we expected. This could be either a mistake on Nielsen or an anomaly in our code. Nonetheless, the top 90 % of hits picked up 3 scales, the Pentatonic 2, Western, the Pentatonic, Minor Western and the Pentatonic Major, which if one thinks about it, the type of person going to Burning Man, Lollapalooza or meditating with Yoga, that has a BS and an MS, would most probably like rock, that is pentatonic in nature, certainly not Balinese, Japanese or Country Music.

The interesting results confirming the work of this paper is that if one looks at them in the reverse, in other words, play a song of Pentatonic, 2, then a Pentatonic Minor and then a Pentatonic Major and we group those that selected all three domains with high hits, they would fall into a category of going to Burning Man, Lollapalooza and Yoga, and if asked if they went to any of these three shows, we could calculate in future work that they had an MS BS. Certainly not a BSN or GED.

References

1. Baraldi, F.B., Poli, G.D., Rodà, A.: Communicating expressive intentions with a single piano note. J. New Music Res. **35**(3), 197–210 (2006)
2. Cable Network Ranker: FNC No. 1 in Primetime for Week of March 9 (2015). http://adweek.it/1o3as2a. Accessed 05 May 2015
3. Eerola, T., Ferrer, R., Alluri, V.: Timbre and affect dimensions: evidence from affect and similarity ratings and acoustic correlates of isolated instrument sounds. Music Percept. Interdisc. J. **30**(1), 49–70 (2012)
4. Espinoza, F.C., López-Ortega, O., Franco-Árcega, A.: Towards the automatic recommendation of musical parameters based on algorithm for extraction of linguistic rules. Computación y Sistemas **18**(3), 637–647 (2014)

5. Estiri, H.: A structural equation model of energy consumption in the united states: untangling the complexity of per-capita residential energy use. Energy Res. Soc. Sci. **6**, 109–120 (2015)
6. Gabrielsson, A., Lindström, E.: The influence of musical structure on emotional expression (2001)
7. Grahn, J.A., Henry, M.J., McAuley, J.D.: Fmri investigation of cross-modal interactions in beat perception: audition primes vision, but not vice versa. Neuroimage **54**(2), 1231–1243 (2011)
8. Hevner, K.: Experimental studies of the elements of expression in music. Am. J. Psychol. **bf–48**, 246–268 (1936)
9. Lahdelma, I., Eerola, T.: Single chords convey distinct emotional qualities to both naïve and expert listeners. Psychol. Music **39**, 1–18 (2014)
10. Leman, M., Vermeulen, V., De Voogdt, L., Moelants, D., Lesaffre, M.: Prediction of musical affect using a combination of acoustic structural cues. J. New Music Res. **34**(1), 39–67 (2005)
11. Lewis, R., Raś, Z.: Facial recognition. In: Wang, J. (ed.) Encyclopedia of Data Warehousing and Mining, vol. II, 2nd edn, pp. 862–871. Idea Group Inc., Hershey (2008)
12. Lewis, R., Zhang, X., Raś, Z.: Knowledge discovery based identification of musical pitches and instruments in polyphonic sounds. Int. J. Eng. Appl. Artif. Intell. Special Issue on 'Soft Computing Applications' **20**(5), 637–645 (2006)
13. Lewis, R.A., Zhang, X., Raś, Z.W.: Blind signal separation of similar pitches and instruments in a noisy polyphonic domain. In: Esposito, F., Raś, Z.W., Malerba, D., Semeraro, G. (eds.) ISMIS 2006. LNCS (LNAI), vol. 4203, pp. 228–237. Springer, Heidelberg (2006)
14. Lewis, R.A., Ras, Z.W.: Rules for processing and manipulating scalar music theory. In: International Conference on Multimedia and Ubiquitous Engineering, MUE 2007, pp. 819–824. IEEE (2007)
15. Lewis, R.A., Zhang, X., Raś, Z.W.: Knowledge discovery-based identification of musical pitches and instruments in polyphonic sounds. Eng. Appl. Artif. Intell. **20**(5), 637–645 (2007)
16. McClellan, R.: The Healing Forces of Music. Element Inc., Rockport (1966)
17. Menon, V., Levitin, D., Smith, B.K., Lembke, A., Krasnow, B., Glazer, D., Glover, G., McAdams, S.: Neural correlates of timbre change in harmonic sounds. Neuroimage **17**(4), 1742–1754 (2002)
18. Cable news ratings for thursday, March 5, 2015. TVbytheNumbers (2015)
19. Pavel, I.: A hierarchical theory of aesthetic perception - musical scales in theoretical perspectives prometheus. Leonardo **27**(5), 417–421 (1994)
20. Raś, Z.W., Zhang, X., Lewis, R.: MIRAI: multi-hierarchical, FS-tree based music information retrieval system. In: Kryszkiewicz, M., Peters, J.F., Rybiński, H., Skowron, A. (eds.) RSEISP 2007. LNCS (LNAI), vol. 4585, pp. 80–89. Springer, Heidelberg (2007)
21. Schedl, M., Gómez, E., Urbano, J.: Music information retrieval: recent developments and applications. Found. Trends Inf. Retrieval **8**(2–3), 127–261 (2014)
22. Schutz, M., Huron, D., Keeton, K., Loewer, G.: The happy xylophone: acoustics affordances restrict an emotional palate. Empirical Musicol. Rev. **3**, 126–135 (2008)
23. Sevgen, A.: The Science of Musical Sound. ScientificAmerican Books Inc., New York (1983)
24. Smith, R., Rathcke, T., Cummins, F., Overy, K., Scott, S.: Communicative rhythms in brain and behaviour. Philos. Trans. Royal Soc. B: Biol. Sci. **369**(1658), 20130389 (2014)

25. TVBGeneralGlossary (2015). http://www.tvb.org/planning_buying/4757. Accessed 05 May 2015
26. Wieczorkowska, A., Synak, P., Lewis, R., Ras, Z.: Creating reliable database for experiments on extracting emotions from music. In: Kłopotek, M.A., Wierzchon, S.T., Trojanowski, K. (eds.) Intelligent Information Processing and Web Mining, pp. 395–402. Springer, Heidelberg (2005)
27. Zentner, M., Grandjean, D., Scherer, K.R.: Emotions evoked by the sound of music: characterization, classification, and measurement. Emotion **8**(4), 494 (2008)

Network Analysis, Multi-Agent Systems

Communities Identification Using Nodes Features

Sara Ahajjam[✉], Hassan Badir, Rachida Fissoune, and Mohamed El Haddad

Laboratory of Technologies of Information and Communication, National School of Applied
Sciences of Tangier, Tangier, Morocco
ahajjam.sara-etu@uae.ac.ma,
{Hassan.badir,Fissoune.rachida,elhaddad.mohamed}@uae.ma

Abstract. The network sciences have provided significant strides for understanding complex systems. Those systems are represented by graphs. One of the
most relevant features of graphs representing real systems is clustering, or
community structure. The communities are clusters (groups) of nodes, with more
edges connecting to nodes of the same cluster and comparatively fewer edges
connecting to nodes of different clusters. It can be considered as independent
compartments of a graph. There are two possible sources of information we can
use for the community detection: the network structure, and the attributes and
features of nodes. In this paper, we use the features of nodes to detect communities. There are nodes in network that are more able and susceptible to diffuse
information and propagate influence. The main purpose of our approach is to find
leader nodes of networks and to form community around those nodes. Unlike to
most existing researches studies, the proposed algorithm doesn't require a priori
knowledge of k number of communities to be detected.

Keywords: Community detection · Influential node · Complex networks ·
Centrality · Classification

1 Introduction

Graphs become extremely useful as the representation of a wide variety of systems in
different areas (biological, information, and social networks). Graph analysis is
becoming crucial to understand the features of these complex systems.

These networks are complex graphs with high local density and low overall density,
they play a fundamental role in the diffusion of information, ideas and innovation, this
advantage has been the subject of various parts that have moved towards these networks
to achieve advertising goals (ads on Facebook), educational (LinkedIn), or political (Election of USA on Twitter). The key property of a real network is its community structure.
The communities are groups of nodes, with more links connecting to nodes of the same
group and comparatively fewer links connecting to nodes of different groups. Recent
studies have verified that the way in which such nodes are organized plays a fundamental role in spreading processes [1]. Studying the influence of role models can help us
to better understand why some trends or innovations are adopted more quickly than others
and how we can help advertisers and marketers to design more effective campaigns.

© Springer International Publishing Switzerland 2015
F. Esposito et al. (Eds.): ISMIS 2015, LNAI 9384, pp. 303–312, 2015.
DOI: 10.1007/978-3-319-25252-0_33

This fact caused many researchers to look for an efficient method for finding top-k most influential people through social networks.

We are interested to study the problematic of detection of communities and leaders' nodes in complex network. Those nodes have high connectivity with the others nodes, and represent an optimization of the network while maintaining the same characteristics of the network. The major drawback of most of the proposed approaches is that they require knowledge of k leader and communities to detect. In this paper, we introduce a new approach to detect leaders' nodes and communities in the network without a prior knowledge of k nodes to detect. This problem has many applications such as: opinion propagation, studying acceptance of political movements or acceptance of technology in economics.

Actually, identifying influential nodes in networks, also regarded as ranking important nodes has become one of the three main problems in network-based information retrieval and mining [2]. In biological systems, we might like to identify the nodes that are keys to communities and protect them or disrupt them, such as in the case of lung cancer [2]. In epidemic spreading, we would like to find the important nodes to understand the dynamic processes, which could yield an efficient method to immunize modular networks [2]. Such strategies would greatly benefit from a quantitative characterization of the node importance to community structure. For example, suppose that we need to advertise a product in a country or we need to propagate news. For this purpose, we need to choose some people as a starting point and maximize the news or the products influence in the target society. The problem was introduced in [3] for the first time. After that in [4] the authors formalized the problem as follows: given a weighted graph in which nodes are people and edge weights represent influence of the people on each other, it is desired to find K starting nodes that their activation leads to maximum propagation In particular, we will focus our attention in one topological feature: centrality [5, 6]. Since those central nodes can diffuse their influence to the whole network faster than the rest of nodes and they are the most influential spreaders.

2   Overview

The community detection algorithms have been the subject of several research papers. Most studies classify articles and research methods depending on the type of the algorithm. The community detection algorithms are belonging to two main types of approaches namely graph partitioning and classification. The major drawback of methods based on the partitioning of graphs is that they require a prior knowledge of the number and size of groups to determine [9]. Also, the leader detection approaches are divided to two mains types: global and local methods. The global method deals with all the network topology (betweenness centrality) [7], while the local ones treat with local position, i.e. with the node (degree centrality) [8]. Reihaneh Rabbany Khorasgani et al. suggest a new approach to detect leaders nodes that takes into account the nodes that are not associated with no leaders. This algorithm is inspired from k-means, the k nodes to be detected will be randomly selected. Other nodes will be assembled at their closest leaders to form communities, and then find new leaders for each community around which gather followers until no node moves. For each community, the centrality of each member is calculated and the node with the highest degree is chosen as the new leader [10]. Another algorithm of leaders' nodes detection in complex networks proposed by Kernighan and Lin based on partitioning of graphs. This algorithm tries to find a section of the graph minimizing the number of edges between partitions by trading vertices between these partitions. The results of this algorithm are generated by introducing the size of each partition [11]. The results of these two algorithms vary according to the size and number of partitions which are introduced. Other proposed studies use classification. The classification was introduced to analyze the data and partition based on a measure of similarity between partitions. The problem of communities detection can be seen as a problem of data classification for which we need to select an appropriate distance [12]. Indeed, the classification methods are generally appropriate for some networks that have a hierarchical structure. The result obtained by these methods depends on choice of similarity measure that used initially. Blondel et al. have proposed the Louvain method that put each node in a vertex. Other approaches are based on partitioned classification which is like the partitioning of the graph requires prior knowledge of size and number of communities to detect. Another study focuses on the spectral classification. In the Leader-Follower algorithm, we define some internal structure of a community. A community should be a clique and is formed of a leader and at least one "loyal follower" which is a node in the community without neighbors in any other community. The leader is a node whose distance is less than at least one of its neighbors. The nodes will be allocated to the community in which a majority of its neighbors belong by destroying the links arbitrarily. However, parasites communities i.e. leaders without loyal follower assigned will be removed from the network. This can cause a loss of information [13]. Yunlong Zhang et al. propose a greedy algorithm based on user preferences (GAUP) to operate the top-k influential users, based on the model Extended Independent Cascade (EIC said that an active node v is active in t-1, has only one chance to activate all inactive neighbors). During each cycle i, the algorithm adds a record in the selected set such that the vertex S with the current set S maximizes propagation of the influence. This means that the vertex selected in round i is the one that maximizes the incremental propagation influence in this cycle. This algorithm calculates the user's preferences for different subjects, and combines traditional greedy algorithms and preferences calculated by LSI user and calculates an approximate solution of the problem of maximizing the influence of a specific topic. This algorithm provides a good result if k exceeds a certain threshold $k \geq 15$ and it is of complexity $O(n3)$ [14]. More recently, in [14], the authors derive an upper bound for the spread function under the LT model. They propose an efficient UBLF algorithm by incorporating the bound into CELF. Experimental results demonstrate that UBLF, compared with CELF, reduces Monte Carlo simulations and reduces the execution time when the size of seed set is small. Recent research found that the location of the node in the network topology is another important factor when estimating the spreading ability. According to that, [15] propose a new approach to identify the location of node through the k-shell decomposition method, by which the network is divided into several layers. Each node corresponding one layer and the entire network formed the core-periphery structure. K-shell decomposition method indicates that the inner the layer is, the more important the node.

However, in practical applications there are often too many nodes having the same index value by employing these two methods to distinguish which node is more powerful. Generally speaking, DC and k-shell decomposition are suitable to measure the spreading ability of nodes quickly but not very accurate. Another proposed algorithm use both global and local methods of centrality measures to effectively identifying the influential spreaders in large-scale social networks. The main idea, that it reduce the scale of network by eliminating the node located in the peripheral layer (namely relatively small ks value) that will not have much spreading potency comparing with the core node in general, and vice versa. This algorithm uses the k-decomposition centrality to deal only with the nodes in the core of the network. Hence, it reduce the scale of the network by ignoring the nodes whose k_s value is small and the links connected them and retain the nodes in the core layers. At last, the global methods (i.e. betweenness centrality and closeness centrality) are used to rank the most influential spreaders [15]. A novel approach to detect communities and important nodes of the detected communities using the spectrum of the graph defines the importance nodes to community as the relative changes in the c largest eigenvalues of the network adjacency matrix upon their removal. It has two types of nodes, the core nodes who are the central nodes and the most important for the community, and the bridges node who connect the communities to each other's. The main drawback of this approach, it is that to have a better result, they need to know the number of partitions in the network and it cannot identify the important nodes in the small communities when the communities are in very different size has the same size. It cannot identify the important nodes in the small communities when the communities are in very different size [17].

Community and leader nodes detection approaches are diverse. Each proposed algorithm brings a new idea or improvement of existing algorithms. We will propose a new approach to detect communities and leader nodes in complex networks without a priori knowledge of number of communities to detect.

3 Problem Formulation

Social network is represented by a social graph which is an undirected graph G = (V; E) where the nodes are users. There is an undirected edge between users u and v representing a social tie between the users. The tie may be explicit in the form of declared friendship, or it may be derived on the basis of shared interests between users.

There are a number of conflicting ideas and theories about how trends and innovations get adopted and spread. The traditional view assumes that a minority of members in a society possess qualities that make them exceptionally persuasive in spreading ideas to others. These exceptional individuals drive trends on behalf of the majority of ordinary people. They are loosely described as being informed, respected, and well-connected; they are called the leaders, innovators in the diffusion of innovations theory, and hubs, connectors, or mavens in other work [16]. The theory of leaders is intuitive and compelling. By identifying and convincing a small number of influential/leader individuals, a viral campaign can reach a wide audience at a small cost. The theory spread well beyond academia and has been adopted in many marketing businesses, e.g., RoperASW and

Tremor [11]. We need to detect those influential/leader nodes that are responsible for the dissemination of information and form communities around those nodes whose facilitate the spread of influence once we need to.

4 Proposed Algorithm

Identifying social influence in networks is critical to understanding how behaviors spread. In order to detect the catalyst of this influence, we need to detect the central nodes that are responsible for the dissemination of influence. Analysis on social network datasets reveals that in each community, there is usually some member (or leader) who plays a key role in that community. In fact, centrality is an important concept [13] within social network analysis, which measures the relative importance of a vertex within the graph. Different from others methods, our approach detect leaders, and build communities around these leaders without a priori knowledge of k leader to detect.

Given an input dataset, the dataset is modeled as an undirected and unweighted graph $G = (V, E)$. V is the vertex set. Each vertex in V represents an element in the dataset. $|(G)|$ represents the number of vertices in G (or elements in the dataset). E is the edge set. Each edge represents a relationship between a pair of elements. Our approach has three steps as in "Fig. 1":

Nodes centrality: For each node v in the network G, calculate the eigenvector centrality. Eigenvector centrality or Gould's index of accessibility [17] is a measure that describes how well connected an individual is based on direct and indirect relationships (i.e., it takes into account the connections of the individuals the focal individual is connected to [18]. Because eigenvector centrality is proportional to an individual's neighbors' centralities [19], more influential individuals will be more connected with other influential individuals. Lastly, embeddedness quantifies how isolatable an individual is or how involved in the network structure an individual is [20]. If all of an individual's connections with other individuals are severed, the individual would be isolated. Thus, higher embeddedness values mean that it is more difficult to isolate an individual [21].

$$Ax = \lambda x \qquad (1)$$

With: A is the adjacency matrix of the network and λ is the eigenvalue.

Nodes ranking: we rank the nodes by the high centrality score in a list L, and choose the leader V_1 which is the node with the highest centrality.

Form community: we calculate neighborhood function to find the neighbors of the leader node which is the node with the highest centrality score. We assign neighbors to the detected leader node to form a community.

We remove the community i.e. the leader node and its neighbors from the network and we add it to the set of communities detected. After, we deal with the second node with the highest centrality until all the vertices (nodes) will be treated.

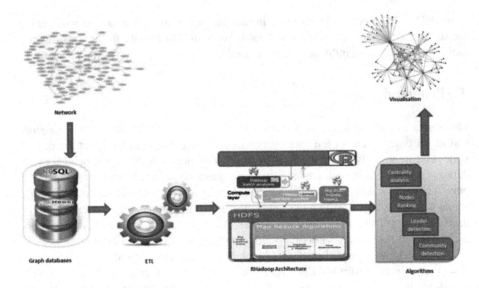

Fig. 1. Architecture of the proposed solution

5   Results and Evaluations

To test our community detection using leader node algorithm, we ran the proposed algorithm on two networks described above:

Zachary's karate club network. This is a well-known benchmark network for testing community detection algorithms. The network is made up of 34 nodes and 78 edges, where every node represents a member of a karate club at an American university. If two members are observed to have social interactions within or away from the karate club, they are connected by an edge. Later, because of a dispute arising between the club's administrator and instructor, the club is eventually split into two factions centered on the administrator and the instructor, respectively [22] (Table 1).

Table 1. Datasets properties

Datasets	Nodes	Edges	Real Communities
Zachary Karaté Club	34	78	2
Word adjacencies	112	425	2

Adjective and noun adjacencies: This is also a famous network widely used as a benchmark to validate community detection algorithms. It's a network of common adjective and noun adjacencies for the novel "David Copperfield" by Charles Dickens, as described by M. Newman. Nodes represent the most commonly occurring adjectives and nouns in the book. Edges connect any pair of words that occur in adjacent position in the text of the book [23].

Figures 3 and 4 show the communities structure in the network for Zachary karate club and Dolphins social network respectively. We compared our community detection algorithm using leader nodes with other community detection algorithm: Label Propagation Algorithm (LPA) [24] and Leading Eigenvalue Algorithm (LEA) [23] using different metrics. For each network we calculate the quality of partition using the modularity Q.

$$Q = \sum_{i=1}^{k}(e_{ii} - a_i^2) \tag{2}$$

Input: undirected, unweighted graph G=(V,E)
Output: Set C=(C$_1$,C$_2$,...,C$_n$)
  1: $i = 0$
  2: While L≠0
  3: Calculate the centrality score of each vertex V ∈ G,
  4: Loop
  5: **Nodes ranking:** Order V via their centrality scores, such that L = (V$_1$, V$_2$, ..., V$_n$) with Cent (V$_1$) ≥ Cent (V$_2$) ≥ ·· ≥Cent (V$_n$).
  6: $i = i + 1$
  7: **Select** where V$_{i_1}$ is the first vertex in the vertex list Q.
  8: Create a new group Ci= {V$_{i_1}$},
  9: New L = L− {V$_{i_1}$}
  10: L = New L
  11: **Community detection:** Calculate the neighborhood function of V$_{i_1}$ to find the **candidate neighbors set "neighbors N(V$_j$)"**.
  12: **insert into** N(V$_j$).
  13: New L=L-N(V$_j$)
  14: End loop

Fig. 2. Pseudo-code of the proposed algorithm

Fig. 3. Community structure in Zachary Karaté Club provided by our algorithm where the leaders are represented by square, by LPA algorithm and LEA algorithm respectively.

where the first term, $\sum_{i=1}^{k} e_{ii}$ is the proportion of edges inside the communities, and the second term $\sum_{i=1}^{k} a_i^2$ represents the expected value of the same quantity in a random network constructed by keeping the same node set and node degree distribution, but connecting the edges between nodes randomly.

Also to evaluate our algorithm, we use the Adjusted Rand Index, the measure penalizes false negatives and false positives. Let a,b,c and d denote the number of pairs of nodes that are respectively in the same community in both G and R, in the same community in G but in different communities in R, in different communities in G but in the same community in R, and in different communities in both G and R. Then the ARI is computed by the following formula:

$$ARI = \frac{\binom{n}{2}(a+d) - [(a+b)(a+c) + (c+d)(b+d)]}{\binom{n}{2}^2 - [(a+b)(a+c) + (c+d)(b+d)]} \tag{3}$$

And we use the Normalized Mutual Information (NMI):

$$NMI(X,Y) := \frac{2I(X,Y)}{H(X) + H(Y)} \tag{4}$$

where I(X,Y) The mutual information corresponds to the quantity of information shared by the variables. Its lower bound is, representing the independence of the variables (they share no information). The upper bound corresponds to a complete redundancy; however this value is not fixed.

The table below presents the result of our algorithm and the Label Propagation Algorithm and Leading Eigenvector Algorithm using the cited metrics.

The results in Table 2 show that for Zachary Karaté Club dataset our algorithm provides the best result for ARI and NMI comparing to LPA and LEA algorithms, while for the modularity that present the quality of founded clusters is quite good compared to LEA which provide the highest one. And for the second dataset, our algorithm provides the best result for the three metrics NMI, ARI and modularity.

Table 2. Comparison results of algorithms.

Network	Algorithm	Communities	Modularity	NMI	ARI
Zachary Karaté club	LPA	2	0.132	0.002	−0.027
	LEA	4	**0.393**	0.006	−0.037
	Proposed algorithm	3	0.318	**0.216**	**0.255**
Word adjacencies	LPA	4	0	0	−1.101
	LEA	5	0.243	0.008	−0.013
	Proposed algorithm	22	**0.584**	**0.109**	**−0.0002**

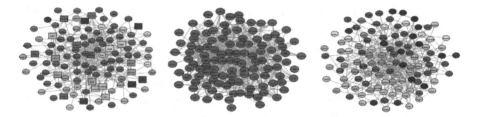

Fig. 4. Community structure in Word adjacencies network provided by our algorithm where the leaders are represented by square, by LPA algorithm and LEA algorithm respectively.

6   Conclusion

This paper presents a study of different detection algorithms communities and especially the leader nodes in complex networks have become increasingly important given the scientific and industrial challenges it represents. The idea is to group objects based on certain criteria. The interest shown by the research in this area is the fact that the dissemination of information i.e. the distribution of influence in complex networks is an element both strategic and particularly sensitive to their use. Thus, we have proposed a new approach for detecting communities using leaders' nodes who unlike the proposed algorithms do not require a priori knowledge of k nodes to detect leaders.

References

1. de Arruda, G.F., Barbieri, A.L., Rodríguez, P.M., Rodrigues, F.A., Moreno, Y., da Fontoura Costa, L.: Role of centrality for the identification of influential spreaders in complex networks. Phys. Rev. E **90**(3), 032812 (2014)
2. Wang, Y., Di, Z., Fan, Y.: Identifying and characterizing nodes important to community structure using the spectrum of the graph. PLoS ONE **6**(11), e27418 (2011)
3. Domingos, P., Richardson, M.: Mining the network value of customers. In: Proceedings of the Seventh ACM SIGKDD International Conference on Knowledge Discovery and Data Mining, New York, NY, USA, pp. 57–66 (2001)
4. Kempe, D., Kleinberg, J., Tardos, É.: Maximizing the spread of influence through a social network. In: Proceedings of the Ninth ACM SIGKDD International Conference on Knowledge Discovery and Data Mining, New York, NY, USA, pp. 137–146 (2003)
5. Shen, H., Cheng, X., Cai, K., Hu, M.-B.: Detect overlapping and hierarchical community structure in networks. Phy. A **388**(8), 1706–1712 (2009)
6. Renoust, B.: Analysis and Visualisation of Edge Entanglement in Multiplex Networks. University of Massachusetts Lowell (2014)
7. Gor, H.R., Dhamecha, M.V.: A survey on community detection in weighted social network. Int. J. **2**(1) (2014)
8. Wu, Q., Qi, X., Fuller, E., Zhang, C.-Q.: Follow the leader: a centrality guided clustering and its application to social network analysis. Sci. World J. **2013**, e368568 (2013)
9. Pons, P.: Detection communities in real networks, Paris 7 (2010). (P. Pons, Détection de communautés dans les grands graphes de terrain, Paris 7, 2010)

10. Khorasgani, R.R., Chen, J., Zaïane, O.R.: Top leaders community detection approach in information networks. In: Proceedings of the 4th Workshop on Social Network Mining and Analysis, 2010, p. 228 (2013). ISSN: 2319-7323

11. Kernighan, B.W., Lin, S.: An efficient heuristic procedure for partitioning graphs. Bell Syst. Tech. J. **49**(2), 291–307 (1970)

12. Fortunato, S.: Community detection in graphs. Phys. Rep. **486**(3–5), 75–174 (2011)

13. Shah, D., Zaman, T.: Community Detection in Networks: The Leader-Follower Algorithm. arXiv:1011.0774, November 2010

14. Zhou, J., Zhang, Y., Cheng, J.: Preference-based mining of top- influential nodes in social networks. Future Gener. Comput. Syst. **31**, 40–47 (2014)

15. Xia, Y., Ren, X., Peng, Z., Zhang, J., She, L.: Effectively identifying the influential spreaders in large-scale social networks. Multimed. Tools Appl., 1–13 (2014)

16. Cha, M., Haddadi, H., Benevenuto, F., Gummadi, P.K.: Measuring user influence in twitter: the million follower fallacy. ICWSM **10**, 10–17 (2010)

17. Wang, Y., Di, Z., Fan, Y.: Detecting important nodes to community structure using the spectrum of the graph. arXiv:1101.1703, January 2011

18. Ruhnau, B.: Eigenvector-centrality—a node centrality? Soc. Netw. **22**, 357–365 (2000)

19. Newman, M.E.J.: Analysis of weighted networks. Phy. Rev. E **70**(5), 056131 (2004)

20. Moody, J., White, D.R.: Structural cohesion and embeddedness: a hierarchicalconcept of social groups. Am. Sociol. Rev. **68**, 103–127 (2003)

21. Fuong, H., Maldonado-Chaparro, A., Blumstein, D.T.: Are social attributes associated with alarm calling propensity? Behav. Ecol. **26**, 587–592 (2015)

22. Zachary, W.W.: An information flow model for conflict and fission in small groups. J. Anthropol. Res. **33**, 452–473 (1977)

23. Newman, M.E.J.: Finding community structure in networks using the eigenvectors of matrices. Phy. Rev. E **74**(3), 036104 (2006)

24. Raghavan, U.N., Albert, R., Kumara, S.: Near linear time algorithm to detect community structures in large-scale networks. Phy. Rev. E **76**(3), 036106 (2007)

Abstract and Local Rule Learning in Attributed Networks

Henry Soldano[1,2]([✉]), Guillaume Santini[1], and Dominique Bouthinon[1]

[1] L.I.P.N UMR-CNRS 7030, Université Paris 13, Sorbonne Paris Cité,
93430 Villetaneuse, France
henry.soldano@lipn.univ-paris13.fr
[2] Atelier de BioInformatique, ISYEB - UMR 7205 CNRS MNHN UPMC EPHE,
Museum d'Histoire Naturelle, 75005 Paris, France

Abstract. We address the problem of finding local patterns and related
local knowledge, represented as implication rules, in an attributed graph.
Our approach consists in extending frequent closed pattern mining to the
case in which the set of objects is the set of vertices of a graph, typically representing a social network. We recall the definition of *abstract
closed patterns*, obtained by restricting the support set of an attribute
pattern to vertices satisfying some connectivity constraint, and propose
a *specificity measure* of abstract closed patterns together with an *informativity measure* of the associated *abstract implication rules*. We define
in the same way *local closed patterns*, i.e. maximal attribute patterns
each associated to a connected component of the subgraph induced by
the support set of some pattern, and also define *specificity* of local closed
patterns together with *informativity* of associated *local implication rules*.
We also show how, by considering a derived graph, we may apply the
same ideas to the discovery of local patterns and local implication rules
in non disjoint parts of a subgraph as k-cliques communities.

1 Introduction

We address here the problem of discovering patterns and associated knowledge
in an attributed graph. Previous work focuses on the topological structure of
the patterns, thus ignoring the vertex properties, or consider only local or semi-
local patterns [4]. In [1] patterns on co-variations between vertex attributes are
investigated in which topological attributes are added to the original vertex
attributes and in [7] the authors investigate the correlation between the support
set of an itemset and the occurrence of dense subgraphs. What we propose in
this article is to consider a graph $G = (O, E)$ whose vertices are labelled by
itemsets and to submit their occurrences in the vertex set O, i.e. their *support
sets*, to connectivity constraints. We consider attribute patterns in the standard
closed itemset mining approach developed in Formal concept Analysis (FCA)[3],
Galois Analysis [2], and Data Mining (see for instance [6]).

In pattern mining, a *support-closed* pattern is a pattern which is maximal,
in size, i.e. in terms of specificity, within the equivalence class of all patterns q

© Springer International Publishing Switzerland 2015
F. Esposito et al. (Eds.): ISMIS 2015, LNAI 9384, pp. 313–323, 2015.
DOI: 10.1007/978-3-319-25252-0_34

sharing the same *support set* $e = \text{ext}(q)$. The corresponding equivalence relation is simply denoted $\equiv$. In standard itemset ming, there is a unique support-closed pattern, i.e. a maximum, in each equivalence class and this support-closed pattern is easily computed using a closure operator f. More precisely, when considering some pattern q its equivalence class is made of all patterns whose support set is $\text{ext}(q)$ and the unique support-closed pattern is obtained as $f(q) = \text{int} \circ \text{ext}(q)$ where int simply intersect the object descriptions of the support set. Support-closed patterns are then simply called *closed patterns*. The set of (support set, closed pattern) pairs is organized within a concept lattice and inclusion of support sets leads to implication rules that hold on the dataset under investigation. The set of frequent closed patterns, i.e. closed elements whose support is greater than or equal to some threshold minsupp, represents then all the equivalence classes corresponding to frequent supports. Such a class has also minimal elements, called *generators*. When the patterns belong to 2^X, the min-max basis of implication rules [6] that represents all the implications $t \to t'$ that hold on O, i.e. such that $\text{ext}(t) \subseteq \text{ext}(t')$, is defined as follows:

$$\{g \to f \mid f \text{ is a closed pattern}, g \text{ is a generator}, f \neq g, \text{ext}(g) = \text{ext}(f)\}$$

2 Abstract Knowledge

In a previous work [9] the attributed graph $G = (O, E)$ was investigated in the following way: each pattern support set $e \subseteq O$, as a set of vertices, induces a subgraph $G(e)$ of G, and this subgraph is then simplified by removing vertices in various ways. The vertices of such an *abstract subgraph* all satisfy some topological constraint, as for instance belonging to a k-clique, and form the *abstract support set* of the pattern. What happens here is that the extensional space is then reduced to a part A of 2^O, called a *graph abstraction*, and that can be generated as the union closure of subsets of O we call *abstract groups*. For instance the k-clique abstraction is made of union of k-cliques and therefore the abstract support set of a pattern is the (maximum) subset of its support set made of k-cliques (Fig. 1).

Example 1. Consider the graph $G = (O, E)$ where $O = \{1, 2, 3, 4, 5, 6, 7, 8\}$ and $E = \{12, 13, 23, 34, 45, 56, 67, 57, 68, 78\}$. Each vertex o is described by $d(o) \in 2^{abc}$, i.e. $d(1) = d(2) = d(3) = ab, d(4) = d(5) = ac, d(6) = d(8) = bc, d(7) = abc$. Consider then the 3-clique abstraction A. The support set of a is $\text{ext}(a) = \{1, 2, 3, 4, 5, 7\}$ and induces the subgraph $G(e)$ whose edges are

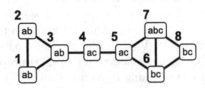

Fig. 1. An attributed graph

$\{12, 23, 13, 34, 45, 57\}$. Its abstract support set is $\text{ext}_A(a) = \{1, 2, 3\}$ as no vertex amongst $4, 5, 7$ belongs to a triangle in $G(e)$.

Abstract support sets are obtained applying an *interior operator* p such that $p[2^O] = A$, i.e. $\text{ext}_A = p \circ \text{ext}$. As an interior operator on 2^O, p has the following properties: for any $e, e' \in 2^O$, *i)* $p(e) \leq e$, *ii)* $p(p(e)) = p(e)$ and *iii)* $e \leq e' \Rightarrow p(e) \leq p(e')$. Abstract implications are then defined by considering inclusion of abstract support sets, i.e. $\square_A q \rightarrow \square_A w$ is valid if and only if $\text{ext}_A(q) \subseteq \text{ext}_A(w)$. Such an abstract rule has the following meaning "whenever the members of some abstract group share pattern q, they also share pattern w". Because of the monotony (condition *iii)*) of the interior operator p, abstraction preserves implication validity:

Lemma 1. *Let A be an abstraction, q and w two patterns, then $q \rightarrow w \Rightarrow \square_A q \rightarrow \square_A w$*

In the case of the k-clique abstraction mentioned above, this means that by restricting the support sets of patterns to be made of k-cliques, we preserve previous valid implications and possibly obtain some new valid abstract implications representing abstract knowledge.

Consider then the equivalence relation $\equiv_A$ defined by $q \equiv_A w$ iff $\text{ext}_A(q) = \text{ext}_A(w)$. Equivalence classes of $\equiv_A$ have a maximum obtained, by applying the closure operator int $\circ$ p $\circ$ ext and called an *abstract closed pattern*, while its minimal elements are called A-generators. We then obtain the *abstract min-max basis* of abstract implications rules where ext_A replaces ext. The abstract min-max basis is made of abstract implications relating A-generators of some equivalence class of $\equiv_A$ to the abstract closed pattern of the same class:

$$\{\square_A g \rightarrow \square_A c \mid c \text{ is an A-closed pattern}, g \text{ is a A-generator}, c \neq g, \text{ext}_A(g) = \text{ext}_A(c)\}$$

Example 2. Consider the data and 3-clique abstraction of Example 1. Intersecting the vertex descriptions of $\text{ext}_A(a) = \{1, 2, 3\}$ we obtain the abstract closed pattern ab. The equivalence class of patterns having abstract support set $\{1, 2, 3\}$ is $\{a, ab\}$ and a is therefore a A-generator. This means that $\square a \rightarrow \square ab$ belongs to the *abstract min-max basis* extracted from G and means "whenever the vertices of a triangle in G share pattern a, they also share pattern ab". Note that $a \rightarrow b$ was not a valide rule, i.e. when considering some vertex o to infer b from a we have to consider some triangle to which o belongs and whose two other vertices also have a.

3   Measuring Abstract Knowledge

When considering frequent abstract closed patterns, we are interested in ordering or selecting them according to to what extent they are related to the graph structure. For that purpose we generalize hereunder the *structural correlation* measure introduced by A. Silva and co-authors [7], originally introduced to compute the ratio of vertices involved in quasi-cliques in the subgraph induced by a pattern, and rename it as *specificity*.

Definition 1. *Let q be a pattern, A an abstraction of some powerset of objects O, the specificity of q with respect to A is defined as:*

$$s_A(q) = \frac{|\,\text{ext}_A(q)\,|}{|\,\text{ext}(q)\,|}$$

Apart from measuring through specificity what is specific to the pattern in its abstract view, we are also interested when considering abstract rules in how informative they are. For that purpose we consider abstract rules whose left and right patterns are equivalent in the abstract space A, i.e. have same abstract support set, as in the min-max abstract rule basis defined above. Whenever these patterns are also equivalent in the original space 2^O intuitively the rule is uninformative. Assume for instance that both $a \to abc$ and $\square_A a \to \square_A abc$ are valid, then the abstract rule did not bring any new information. On the contrary, assume that $\square_A a \to \square_A abc$ is valid while $a \to abc$ has only confidence 0.5, i.e. $\text{ext}(abc) = 0.5 * \text{ext}(a)$, then clearly the abstract rule brings some information. We simply measure here *informativity* as the inverse of confidence.

Definition 2. *Let q be a pattern, A an abstraction of 2^O, the informativity of the valid rule $r : \square_A q \to \square_A w$ is defined as:*

$$I_A(r) = \frac{|\,\text{ext}(q)\,|}{|\,\text{ext}(qw)\,|}$$

An alternative Informativity measure, ranging between 0 and 1, would be the (estimated) probability of not having w whenever we have q i.e. $1 - \frac{|\text{ext}(qw)|}{|\text{ext}(q)|}$. This quantity has value 0 whenever $q \to w$ holds and has limit 1 whenever $|\,\text{ext}(qw)\,|$ approaches 0, i.e. restricting the support set of patterns to elements of A concentrates the support set of q to the very few sharing also w. In the remaining of the article we keep Definition 2 to define informativity.

Considering an implication rule from the abstract min-max basis $\square_A g \to \square_A c$, we are then interested in the specificity $s_A(c)$ of the abstract closed pattern and in the informativity $I_A(r) = \frac{|\text{ext}(g)|}{|\text{ext}(c)|}$ of the rule.

Example 3. Considering the attributed graph and triangle abstraction of Examples 1 and 2, ab has specificity $3 \div 6 = 0.5$ while $\square_A a \to \square_A ab$ has informativity $6 \div 4 = 1.5$.

4   Local Knowledge

Given some attribute pattern, we are now interested in extracting *local support closed patterns*, i.e. maximal attribute patterns each associated to one dense subgraph, so allowing to extract *local implication rules* particular to specific dense groups of objects. Recently the closed pattern mining methodology has been extended to *local closed patterns*: they are obtained by applying a set of *local closure operators* [8]. In the graph case, this means that from the support set of some (closed) pattern c, various dense support sets, called *local support sets* are extracted each associated to a *local closed pattern*, i.e. the most specific pattern l common to the elements of the local support set. Again we obtain a set of *local implication rules* corresponding to inclusion of local support sets, but now such an implication is only valid in the vicinity of some dense group of vertices.

4.1   Direct Local Knowledge

The simplest case appears when the extensional space is reduced to the set F of connected subgraphs induced by vertex subsets belonging to some graph abstraction A. To a pattern q is associated one of its connected component e as a local support set, and $int(e)$ as the corresponding local closed pattern. We may then consider, for instance, as A the 3-clique abstraction and obtain as local support sets connected subgraphs made of 3-cliques. In this simple case, F is a *confluence* of A [8], i.e. a partially ordered set made of several lattices, and that has in general a set $min(F)$ of minimal elements. More precisely, in our connected 3-clique subgraphs case, these minimal elements are the 3-cliques of our graph G. We call such a confluence, whose elements are connected components, a *cc-confluence*. Let q be a pattern, $m \in min(F)$, and $m \subseteq ext_A(q)$, we obtain the connected component containing the 3-clique m as $ext_m^A = p_m \circ ext_A(q)$ where p_m is again an interior operator, and therefore is monotonic. Note that in a cc-confluence, each vertex appears in only one such connected components and we may as well replace m by one of its vertex s in our definitions.

Whenever we have $p_m \circ ext_A(q) \subseteq p_m \circ ext_A(w)$ we rewrite this as the local implication $\Box_m^A q \to \Box_m^A w$ stating that if q has a local support set containing m, then w has a larger than or equal to local support set. Because of monoticity of p_m, again validity of implications is preserved:

Lemma 2. *Let F be a confluence of an abstraction A, q and w two patterns, then $\Box_A q \to \Box_A w \Rightarrow \Box_m^A q \to \Box_m^A w$*

When considering a given abstract closed pattern c which has a local support set e in F that contains m, and whose corresponding local closed pattern is l, we have then that the implication rule $\Box_m^A c \to \Box_m^A l$ holds. The set $\{\Box_A c \to \Box_A l \mid l$ a local closed pattern, c an abstract closed pattern, $c \neq l, ext_m^A(c) = ext_m^A(l)\}$ represents (a basis for) the local knowledge deriving from the reduction of the extensional space from A to the confluence F.

Example 4. Still considering the attributed graph G and triangle abstraction of Examples 1 and 2, we consider the cc-confluence F of vertex subsets inducing connected subgraphs of G made of triangles. We have $ext_A(b) = \{1, 2, 3, 6, 7, 8\}$ that induces a subgraph made of two connected components $\{1, 2, 3\}$ and $\{6, 7, 8\}$. The corresponding local closed patterns are $int(\{1, 2, 3\}) = ab$ and $int(\{6, 7, 8\}) = bc$. As $b = int(\{1, 2, 3, 6, 7, 8\})$, b is an abstract closed pattern and we have the following local implications: $\Box_A^{\{1,2,3\}} b \to \Box_A^{\{1,2,3\}} ab$ and $\Box_A^{\{6,7,8\}} b \to \Box_A^{\{6,7,8\}} bc$ we may rewrite, since A is a cc-confluence, as, for instance: $\Box_A^1 b \to \Box_A^1 ab$ and $\Box_A^6 b \to \Box_A^6 bc$.

4.2 Measuring Direct Local Knowledge

To measure how much a local closed pattern is specific to the associated connected component, and in the same way as in the abstract case where we considered the ratio between the abstract and standard support sets, we are here interested in the ratio between the local and the global (standard or abstract) support set:

Definition 3. *Let q be a pattern, F an extensional confluence of some abstraction A of 2^O, and $m \in F$ such that $m \subseteq \text{ext}_A(q)$, the specificity of q in the vicinity of m is defined as:*

$$s_F(q, m) = \frac{|\text{ext}_m^A(q)|}{|\text{ext}_A(q)|}$$

In the same way as in the abstract implication case, we measure informativity of a local rule with respect to the corresponding global rule. The idea here is that in a valid local implication the patterns left and (left+)right have same local support set while their global support sets are different. Again *informativity* is defined as the inverse of the (abstract) confidence.

Definition 4. *Let q be a pattern, F an extensional confluence of some abstraction A of 2^O, and $m \in F$ such that $m \subseteq \text{ext}_A(q)$, the informativity of the valid local rule $r : \square_m^A q \to \square_m^A w$ is defined as:*

$$I_F(r) = \frac{|\text{ext}_A(q)|}{|\text{ext}_A(qw)|}$$

Intuitively, informativity measures what we have learned when discovering that q and qw had same local support sets with respect to m while they had different abstract support set. Considering a local implication rule $r : \square_m^A c \to \square_m^A l$ we are interested in the specifcity $s_F(l, m)$ of the local closed pattern l and in the informativity $I_F(r) = \frac{|\text{ext}_A(c)|}{|\text{ext}_A(l)|}$ of the rule.

Example 5. Always following Examples 1,2,3 and 4, we obtain bc local specificity w.r.t. triangle $\{6, 7, 8\}$, $s_F(bc, \{6, 7, 8\}) = 3 \div 3 = 1$, i.e. pattern bc is specific of the local support set $\{1, 2, 3\}$. Furthermore, implication $\square_A^{\{6,7,8\}} b \to \square_A^{\{6,7,8\}} bc$ has informativity $6 \div 3 = 2$, i.e. in the abstract extensional space $A \square_A b \to \square_A bc$ has confidence 0.5 while the implication holds at the local level.

4.3 Indirect Local Knowledge and Associated Measures

Local knowledge is related above to a notion of locality in a graph expressed through a confluence structure of the vertex space. This is mainly illustrated on the idea that the subgraph induced by the (abstract) support set of some pattern is made of several connected components, and that there may be specific patterns associated to each connected component. However, we are also interested in locality notions closer to the notion of community in Social Network Analysis.

A well known example of community definition is the k-clique community [5] which is defined as a maximal vertex subset made of adjacent (i.e. sharing $k-1$ vertices) k-cliques. Such a k-clique community may alternatively be defined as a connected component of a graph whose vertices are k-cliques and edges relate two adjacent k-cliques. What we discuss, more generally, in this section is a way to define local knowledge associated to subgraphs which are connected components of a *derived graph* made of particular vertex subsets, as k-cliques in the k-clique community case. This local knowledge, stated as *indirect*, is obtained by using the methodology described in Sect. 4 on the derived graph.

We start from a family T of elements of 2^O, and consider T as the vertex set of a new graph $G_T = (T, E_T)$. We consider then a confluence F of 2^T as the extensional space and search for the corresponding local closed patterns. The corresponding local support sets are afterwards transformed into support sets in 2^O: when considering a (local support set, local closed pattern) pair (e_T, l) we may transform it into the pair (e, l) where e is the union of the elements of e_T. Let $T \subseteq 2^O$, and $u : 2^T \to 2^O$ be such that $u(e_T) = \cup_{t \in e_T} t$. $u(e_T)$ is called the *flattening* of e_T. We consider then two maps ext_T and int_T relating L to 2^T:

- $\text{ext}_T : L \to 2^T$ with $\text{ext}_T(q) = \{t | t \subseteq \text{ext}(q)\}$
- $\text{int}_T : 2^T \to L$ with $\text{int}_T(e_T) = \text{int} \circ u(e_T)$

$\text{ext}_T(q)$ represents the support set of q in 2^T when considering that q occurs in t whenever q occurs in all elements of t. Conversely $\text{int}_T(e_T)$ represents the greatest pattern in L whose support set in T includes e_T, i.e. whose support set in O contains, as subsets, the elements of e_T. We have then the following result when flattening the (local) support sets so found in F:

Proposition 1. *Let F be a confluence of 2^T, u be the flattening operator on O and (e_T, l) be a (local support set, local closed pattern) pair with $e_T \ge m \in \min[F]$, then $u(e_T)$ is the greatest element of $u[F^m]$ among elements e such that $\text{int}(e) = l$.*

This means that the support closed patterns with respect to the confluence F are the same as the support closed patterns with respect to the extensional space $U = u[F]$. Note that as flattened support sets are obtained by joining elements of T, they belong to the abstraction $A = \text{UnionClosure}(T)$.[1]

This will be illustrated by considering T as the set of 3-cliques of G (further called *triangles*) and stating that (t_1, t_2) belongs to G_T whenever t_1 and t_2 share an edge in G. In this case, a flattened local support set of pattern q represents a triangle community in the pattern q subgraph $G(\text{ext}(q))$. An example of both graphs G and G_T is displayed Fig. 2.

It is then natural to extend the definition of specifity to make it relative to the flattened support sets:

---

[1] defined by (i) $T \subseteq \text{UnionClosure}(T)$ and (ii) if q and w belong to $\text{UnionClosure}(T)$ then $q \cup w$ belongs to $\text{UnionClosure}(T)$. By considering any subset $T \subseteq 2^O$ and closing it under union we obtain an abstraction of 2^O [10].

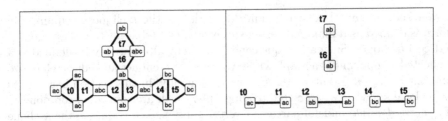

Fig. 2. On the left we have a graph of objects each described as an itemset included in $\{a, b, c\}$. This graph represents the triangle abstraction of some input graph. On the right, the graph G_T whose vertices are the triangles of G. The itemset describing a vertex in G_T is the intersection of the itemsets describing the elements of the corresponding triangle in G.

Definition 5. *Let q be a pattern, F an extensional confluence of 2^T where $T \subseteq 2^O$, A is the abstraction generated from T and $m \in F$ such that $m \subseteq \text{ext}_T(q)$, the flattened specificity of q in the vicinity of m is defined as:*

$$s_F^f(q, m) = \frac{|\, u \circ p_m \circ \text{ext}_T(q)\,|}{|\, u \circ \text{ext}_T(q)\,|} = \frac{|\, u \circ p_m \circ \text{ext}_T(q)\,|}{|\, \text{ext}_A(q)\,|}$$

Coming back to the example of triangles communities, $s_F(q, m)$ states to what extent a pattern q is specific to the community containing a particular triangle m with respect to its abstract support set in O when considering only triangles.

From Sect. 4.1 we know that we may rewrite $p_m \circ \text{ext}_T(q) \subseteq p_m \circ \text{ext}_T(w)$ as a local implication $\square_m q \to \square_m w$. As the flattening operator is monotonic when the rule $\square_m q \to \square_m w$ is valid on the set T, we also have $u \circ p_m \circ \text{ext}_T(q) \subseteq u \circ p_m \circ \text{ext}_T(w)$. We may then define the flattened informativity of $r = \square_m q \to \square_m w$ as

$$I_F^f(r) = \frac{|\, u \circ \text{ext}_T(q)\,|}{|\, u \circ \text{ext}_T(qw)\,|} = \frac{|\, \text{ext}_A(q)\,|}{|\, \text{ext}_A(qw)\,|}$$

Let us consider a (flattened local support set, local closed pattern) pair (e, l), where e is a community containing a given triangle m, l the corresponding local closed pattern, and c an abstract closed pattern whose support set in G induces a subgraph in which e forms a triangle community. This means that $\square_m c \to \square_m l$ is a valid local implication rule stating that when we consider the subgraph induced by the support set of c, all the members of the community containing the triangle m also has pattern l (see Fig. 3). The set of such $\square_m c \to \square_m l$ local implications, with $c \neq l$, represents (a basis for) the local knowledge deriving from the reduction of the extensional space to triangle communities.

Example 6. Let $G = (O, E)$ be the graph displayed on the left part of Fig. 2. Each vertex of G belongs to some triangle in G, therefore G is the same as its triangle abstraction. Each vertex has an itemset included in $\{a, b, c\}$ as a label. The set of triangles is $T = \{t_0, t_1, t_2, t_3, t_4, t_5, t_6, t_7\}$ and forms a triangle graph G_T displayed on the right part of Fig. 2. An edge relates any pair of triangles sharing two vertices in G, as for instance (t_0, t_1). Each triangle in G_T has as its itemset the intersection of the itemsets of its three vertices in G. For instance, the description of t_1 in G_T is $ac = abc \cap ac \cap ac$. The vertex subsets inducing connected subgraphs of G^T form the cc-confluence $F^T = \{\{t_0\}, \{t_1\}, \{t_0, t_1\}, \{t_2\}, \{t_3\}, \{t_2, t_3\}, \{t_4\}, \{t_5\}, \{t_4, t_5\}, \{t_6\}, \{t_7\}, \{t_6, t_7\}\}$.

The support set of the pattern a is $\text{ext}(a) = \{t_0, t_1, t_2, t_3, t_6, t_7\}$. The local support with respect to t_0 is $p_{t_0}(\{t_0, t_1, t_2, t_3, t_6, t_7\}) = \{t_0, t_1\}$, i.e. the connected component containing $\{t_0\}$ of the subgraph induced by $\text{ext}(a)$. The local closed patterns, where $f_i(q)$ denotes a closed pattern which is local w.r.t. triangle t_i, are as follows:

- $f_0(a) = f_1(a) = ac$, $f_2(a) = f_3(a) = ab$, $f_6(a) = f_7(a) = ab$

In the same way, the pattern b whose support set is $\text{ext}(b) = \{t_2, t_3, t_4, t_5, t_6, t_7\}$. leads to the following local closed patterns:

- $f_2(b) = f_3(b) = ab$, $f_4(b) = f_5(b) = bc$, $f_6(b) = f_7(b) = ab$

Note that ab appears both as a local closed pattern resulting from a with respect to f_0, f_1 and to f_6, f_7 and as a local closed pattern resulting from b with respect to f_2, f_3 and again to f_6, f_7. This leads to three different sets of local implications:

- $\square_{t_2} a \to \square_{t_2} ab$, $\square_{t_3} a \to \square_{t_3} ab$, $\square_{t_6} a \to \square_{t_6} ab$, $\square_{t_7} a \to \square_{t_7} ab$,
- $\square_{t_2} b \to \square_{t_2} ab$, $\square_{t_3} b \to \square_{t_3} ab$, $\square_{t_6} b \to \square_{t_6} ab$, $\square_{t_7} b \to \square_{t_7} ab$,

As a whole, a local closed pattern is part of a pair (e_T, l) where l is the local closed pattern and e_T is a local support set corresponding to one of the connected components induced by the support set. Two examples of such pairs are $(\{t_2, t_3\}, ab)$ and $(\{t_6, t_7\}, ab)$. When interested in implication rules, we have to consider triples (c, t_i, l) where c is a pattern whose support set is split in different local support sets one of which, namely e, contains t_i. $\square$

Example 7. The dataset is denoted as *s50-1* and is a standard attributed graph dataset.[2] It represents 148 friendship relations between 50 pupils of a school in the West of Scotland, and labels concern the substance use (tobacco, cannabis and alcohol) and sporting activity (see [9]). We want to answer to the question:"what knowledge can be extracted when considering groups of pupils connected by friendship relationships?". For that purpose, we computed the local abstract closures associated to the cc-confluence representing 3-clique communities in subgraphs of the triangle graph G_T derived from the original graph and the "support$\geq$ 4" constraint on O. In Fig. 3 we represent the flattened local support set e of the local closed pattern l shared in a community (in black lines and dots) of the subgraph induced by the abstract support set of l (in black+grey lines and dots) (w.r.t. the 3-clique abstraction). We also represent (in dashed+ black + grey lines and dots) the abstract support set of the abstract

[2] http://www.stats.ox.ac.uk/~snijders/siena/s50_data.htm.

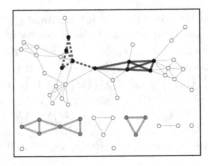

Fig. 3. Representation of a local rule $\square_m c \rightarrow \square_m l$ extracted from a West of Scotland school friendship network. c is the abstract closed pattern shared by the friendship triangles displayed in black+ grey+ dashed lines and dots. l is a local closed pattern specific to the 3-community, to which the triangle m belongs, represented in black lines and dots. This local closed pattern is shared by all the friendship triangles displayed in black+grey lines and dots.

closed pattern c that also induces a subgraph in which e is a connected component. Overall $\square_m c \rightarrow \square_m l$ is a valid local implication rule whose informativity is $I_F^f(r) = \frac{|\text{ext}_A(c)|}{|\text{ext}_A(l)|} = (5 + 9 + 4) \div (5 + 9) = 1.286$. The specificity $s_F^f(r)$ of the local closed pattern l is $5 \div (5 + 9) = 0.357$. Here l means "Never has tried Cannabis, drinks moderately, does not smoke"while c means" Has tried Cannabis at most once, drinks moderately, does not smoke". The specificity of the 3-community, with respect to the whole set of pupils sharing "Have tried Cannabis at most once, drink moderately, does not smoke" is to be composed only of pupils who have never tried Cannabis.

5   Conclusion

We have discussed here a framework extending the closed itemset mining framework to abstract and local information in an attributed network. Our focus in this article was on the abstract and local knowledge to be extracted as abstract and local rules, together with measures about how specific and informative is abstraction or locality.

References

1. Prado, A.B., Plantevit, M., Robardet, C., Boulicaut, J.F.: Mining graph topological patterns: finding co-variations among vertex descriptors. IEEE Trans. Knowl. Data Eng. **25**(9), 2090–2104 (2013)
2. Caspard, N., Monjardet, B.: The lattices of closure systems, closure operators, and implicational systems on a finite set: a survey. Discrete Appl. Math. **127**(2), 241–269 (2003)
3. Ganter, B., Wille, R.: Formal Concept Analysis: Mathematical Foundations. Springer, Heidelberg (1999)

4. Mougel, P.-N., Rigotti, C., Gandrillon, O.: Finding collections of k-clique perco-lated components in attributed graphs. In: Tan, P.-N., Chawla, S., Ho, C.K., Bailey, J. (eds.) PAKDD 2012, Part II. LNCS, vol. 7302, pp. 181–192. Springer, Heidelberg (2012)

5. Palla, G., Derenyi, I., Farkas, I., Vicsek, T.: Uncovering the overlapping community structure of complex networks in nature and society. Nature **435**(7043), 814–818 (2005)

6. Pasquier, N., Taouil, R., Bastide, Y., Stumme, G., Lakhal, L.: Generating a con-densed representation for association rules. J. Intell. Inf. Syst. (JIIS) **24**(1), 29–60 (2005)

7. Silva, A., Meira Jr., W., Zaki, M.J.: Mining attribute-structure correlated patterns in large attributed graphs. Proc. VLDB Endow. **5**(5), 466–477 (2012)

8. Soldano, H.: Extensional confluences and local closure operators. In: Baixeries, J., Sacarea, C., Ojeda-Aciego, M. (eds.) ICFCA 2015. LNCS, vol. 9113, pp. 128–144. Springer, Heidelberg (2015)

9. Soldano, H., Santini, G.: Graph abstraction for closed pattern mining in attributed network. In: Schaub, T., Friedrich, G., O'Sullivan, B. (eds.) European Conference in Artificial Intelligence (ECAI). Frontiers in Artificial Intelligence and Applications, vol. 263, pp. 849–854. IOS Press (2014)

10. Soldano, H., Ventos, V.: Abstract Concept Lattices. In: Jäschke, R. (ed.) ICFCA 2011. LNCS, vol. 6628, pp. 235–250. Springer, Heidelberg (2011)

An Intelligent Agent Architecture for Smart Environments

Stefano Ferilli[1]([⊠]), Berardina De Carolis[1], and Domenico Redavid[2]

[1] Dip. di Informatica, University of Bari, via E. Orabona 4, Bari, Italy
{stefano.ferilli,berardina.carolis}@uniba.it
[2] Artificial Brain S.r.l., Bari, Italy
redavid@abrain.it

Abstract. This paper proposes an architecture for agents that are in charge of handling a given environment in an Ambient Intelligence context, ensuring suitable contextualized and personalized support to the user's actions, adaptivity to the user's peculiarities and to changes over time, and automated management of the environment itself. Functionality involves multi-strategy reasoning and learning, workflow management and service composition. In Multi-Agent Systems, different types of agents may implement different parts of this architecture.

1 Introduction

A situation-aware environment should understand both the status of the environment, and the status and the needs of users in order to proactively support them with the most appropriate configuration of actions of various devices and resources in the smart environment [14]. By 'environment', we mean the set of elements of various kind in which the user and the system operate. By 'context', we mean the specific status of the environment, as determined by data that can be directly sensed in the environment or by derived information that can be inferred by the system starting from those data. Finally, by 'situation', we mean the compound of the context and the specific status of the user, as can be inferred by the system starting from the available data. This requires efficient, flexible and scalable solutions, such as those based on software agents that are able to recognize the user's situational goal, to provide smart (i.e., integrated, interoperable and personalized) services for satisfying this goal and to provide suitable interfaces using the devices and services that are present in the environment. We propose a general architecture for agents in a Multi Agent System (MAS). As a main contribution, it brings to cooperation in one framework several functionalities, including multistrategy reasoning, process handling and service composition. Different kinds of agents may implement different subsets of this architecture and functionalities, depending on their role and objectives in the situation-aware environment. The architecture has been implemented in a prototypical system that works in a Smart Home Environment (SHE), where it is necessary to combine services of the physical environment with net-centric ones according to the recognized situation.

© Springer International Publishing Switzerland 2015
F. Esposito et al. (Eds.): ISMIS 2015, LNAI 9384, pp. 324–330, 2015.
DOI: 10.1007/978-3-319-25252-0_35

The paper is structured as follows. The proposed architecture is described in Sect. 3. Then, Sect. 4 proposes several kinds of roles to be implemented by agents that work in the smart environment. Section 5 concludes the paper.

2   Related Work

A smart environment is able to acquire and apply knowledge about its inhabitants and their surroundings in order to adapt to the situational goals [3]. Developing a smart environment requires software components that may perceive the environment and act on it by providing appropriate responses to the situations. BDI (Belief Desire Intention) agents are endowed with this capability and act on the basis of practical reasoning [12]. Agents can be organized in a MAS that must be modular, flexibile and scalable to handle the complexity of a smart environment [1]. When endowed with appropriate knowledge and reasoning capabilities a MAS represents a way to design and implement a proactive and adaptive environment in relation to individual and changing needs [3].

Early research projects concerned with applying intelligent agents to the realization of a Smart Home focused on the development of a home that programs itself by observing the lifestyle and desires of the inhabitants, and learns to anticipate and accommodate their needs [5]. Applying agent-based approaches on this domain seems to be a promising direction for research [4, 10].

In [9] the problem of service personalization in smart environments is addressed by a Service Oriented Architecture implementation based on MASs. In particular, the mobility features of software agents are leveraged in that work. The proposed hierarchical, agent-based solution is intended to be applicable to different smart space scenarios, ranging from small environments to large smart spaces like cities.

The proactive nature of a smart environment has been investigated by several research works that propose the processing of contextual data for carrying out analysis for erroneous events in order to provide "relevant information and/or services to the user, where relevancy depends on the user's task" [6, 11].

Therefore, developing a smart environment requires designing a complex intelligent system that is able to acquire and apply knowledge about the environment and its residents in order to improve their quality of life in that environment by providing situation-aware services through natural and effective interfaces. To achieve this aim, we recognized the need of developing an architecture of a MAS that combines the practical reasoning capabilities of BDI agents, the learning capabilities of intelligent learning agents and the service oriented approach combined with Semantic Web in order to implement an Ambient Intelligence (AmI) system that is able to unobtrusively and proactively adapt to the individual situational needs.

3   Architecture

The AmICo (Ambient Intelligence Coordinator) architecture is shown in Fig. 1. It consists of three layers. The environment in which the agent operates represents the Sensors, Effectors and Applications Layer (*SeaL*). The Reasoning Layer

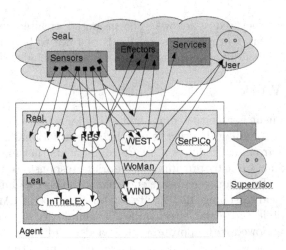

Fig. 1. Agent's architecture.

(*ReaL*) and the Learning Layer (*LeaL*) endow the agent with a multi-strategy reasoning engine useful to exploit and refine relevant domain knowledge. Available types of inference for these functional layers include:

Abstraction simplifies the available information by removing irrelevant details;
Deduction understands what is going on in the environment and determines what actions should be taken consequently;
Abduction hypothesizes useful but sensible missing information;
Induction refines the available knowledge based on the feedback obtained during the system's interaction.

The knowledge base used by ReaL, organized in several cooperating modules referred to specific topics, can be refined and improved automatically as long as the system works by exploiting the functionality of LeaL. Among other kinds of knowledge, it includes user models, context models and process models to be used for adapting and personalizing the interaction and controlling the flow of events. A user model includes knowledge that allows to infer his mental and physical status, his goals and objectives, his preferences and interests, and his needs and requirements. Personalized models may be available for specific users, typically learned by monitoring their behavior. A context model includes knowledge that allows to infer what kind of situation is faced by the system, what is going on in the environment, and how to act in order to properly handle the situations that are occurring. A process model specifies which combinations of activities are allowed to accomplish a given goal. Many models may be available for reaching the same goal. Process models can be combined (e.g., one process may involve a complex activity, for which one or more process models may be available).

RES (Reasoning Expert System) is the component in ReaL that applies the rule-based knowledge using deduction supported by abstraction and abduction. Based on the user profile and the current context, ReaL manages the environment by suitably activating the effectors on which it has direct control, or sending proper notifications or starting dialogues with the users or with external supervisors. When it identifies relevant user's goals, it extracts from its workflow repository the items that specify how to satisfy those goals. Then, using the events coming from the environment, WEST (Workflow Enactment Supervisor and Trainer) checks whether they are compliant with the current process model(s) that have been activated. If several candidate workflow models are considered, WoGue (WOrkflow GUEsser) can hypothesize which models are more likely to be taking place. For each model the system can also try to foresee which will be the next actions of the user, based on the current status and context of the execution. This allows the agent to carry out suitable interventions that support or facilitate the execution of these actions. In some cases, the desired effect is provided by existing services that the agent may call.

While atomic services can be directly activated by the agent, there are complex tasks for which a single service is not available, and there is no known composition of elementary services that solves the problem. In these cases, SerPICo (Service Planner Identifier and Composer) [13] may look for new service compositions that may reach the objective and are compliant with the given constraints, preferences or requirements. It is based on an encoding of OWL-S atomic processes as SWRL rules [8] of the form *inCondition* $\wedge$ *Precondition* $\rightarrow$ *output* $\wedge$ *effect*. The obtained set of SWRL rules is interpreted as a plan that can be described as an OWL-S composite process. According to OWL-S specifications, the service model defines the concept of a *Process* that describes the composition of one or more services in terms of their constituent processes. A *Process* can be *Atomic* (a description of a non-decomposable service that expects one message and returns one message in response), *Composite* (consisting of a set of processes within some control structure that defines a workflow) or *Simple* (used as an element of abstraction, i.e., a simple process may be used either to provide a view of a specialized way of using some atomic process, or a simplified representation of some composite process for purposes of planning and reasoning). A composite process can be considered as an atomic one using the OWL-S Simple process declaration. The compositions obtained using SerPICo can be stored as new OWL-S (Simple) services and reused as atomic services in the future.

Some kinds of feedback from the environment may be interpreted as expressing positive or negative examples for the rule-based reasoning that the system is expected to carry out. Positive examples consist of confirmations of the operations carried out by the system, or of actions that suggest operations that the system was expected to carry out but did not. Negative examples consist of actions that counteract operations carried out by the system, or by confirmations that operations not carried out by the system did not have to be carried out. As regards process execution, when an execution terminates without raising any error, if the system is notified that the execution was correct, this may be considered as a cases of successful process execution to be used to confirm or improve the existing process model. In these cases, LeaL can incrementally refine the knowledge used by ReaL. Again, it includes different operational modules: InTheLEx (Incremental Theory Learner from Examples) is in charge of learning and/or refining the models that make up the rule-based knowledge; WIND (Workflow INDucer) can learn and/or refine the process model based on cases that are provided as successful examples of process execution. WEST, WoGue and WIND are components of the workflow management system WoMan (Workflow Manager) [7]. WoMan and InTheLEx are tightly coupled, in that the latter may also learn and refine models of activity preconditions to be used by ReaL to guide and check process execution, discarding possible paths that are not compliant to the current situation and environment.

4   Types of Agents

The types of agents that implement the AmICo architecture extend those in [2]:

Sensor Agents (SA): provide information about sensor parameters and values (e.g., temperature, light level, humidity, etc.).

Context Agents (CA): determine the current context from sensor events; they are able to reason at a higher level than sensor agents, for instance starting from temperature and humidity data they may determine whether the user is in a comfortable situation [5].

User Profile Agent (UPA): is responsible for determining the preferences profile to be used, and may serve personalization purposes.

Butler Agent (BA): combines intelligent reasoning, machine learning, service-oriented computing and semantic Web technologies for flexibly coordinating and adaptively providing smart services in dynamically changing contexts.

Effector Agents (EA): each appliance and device is controlled by an EA that reasons on the opportunity of performing an action instead of another in the current context.

Interactor Agents (IA): handle interaction with the user. They are responsible for choosing the best interaction metaphor according to the situation and to the user's needs and preferences, and for executing suitable communicative tasks by performing communicative actions.

Housekeeper Agent (HA): acts as a facilitator since it knows all the agents that are active in the house and also the goals they are able to fulfill.

The underlying metaphor is that of a butler in a grandhouse that is in charge of perceiving the situation of the house and of coordinating the housestaff in order to satisfy the needs of the house inhabitants. In particular, it reasons on the user's goals and devises the workflow to satisfy them.

These agents coordinate themselves as follows. Cyclically, or as an answer to a user action, the butler runs its reasoning model about the user. Based on the information provided by the appropriate CAs, it infers the possible goals and needs of the user and ranks them by urgency or certainty by consulting the UPA. Given a specific goal, it selects an appropriate workflow by matching semantically the goal with all the Input, Output, Pre-Condition and Effect (IOPE) descriptions of the workflows stored in a workflow repository. During workflow enactment, semantic matchmaking is also used to select the services/actions to be invoked among those available in the environment. The workflow services are invoked dynamically, matching the user's needs in the most effective way. As regards predicates of Web Services, both simple and complex Web Services will be implemented according to the standard OWL-S.

5 Conclusions

This paper proposed an architecture for agents that are in charge of handling a given environment in an Ambient Intelligence context, ensuring suitable contextualized and personalized support to the user's actions, adaptivity to the user's peculiarities and to changes over time, and automated management of the environment itself. Functionality involves multi-strategy reasoning and learning, workflow management and service composition. In Multi-Agent Systems, different types of agents may implement different parts of this architecture. The architecture has been implemented in a prototypical agent-based system that works in a smart home environment. It is currently undergoing extension and refinement in order to make it able to deal with more varied and complex situations.

Acknowledgements. This work was partially funded by the Italian PON 2007-2013 project PON02_00563_3489339 'Puglia@Service'.

References

1. Ayala, I., Amor, M., Fuentes, L.: Self-configuring agents for ambient assisted living applications. Personal and Ubiquitous Computing, pp. 1–11 (2012)
2. Cavone, D., De Carolis, B., Ferilli, S., Novielli, N.: A multiagent system supporting situation aware interaction with a smart environment. In: 2nd International Conference on Pervasive Embedded Computing and Communication Systems (PECCS-2012), pp. 67–72. SciTePress (2012)
3. Cook, D.: Multi-agent smart environments. J. Ambient Intell. Smart Environ. **1**, 47–51 (2009)
4. Cook, D.J., Augusto, J.C., Jakkula, V.R.: Ambient intelligence: technologies, applications, and opportunities. Pervasive Mobile Comput. **5**, 277–298 (2009)
5. De Carolis, B., Cozzolongo, G., Pizzutilo, S., Plantamura, V.L.: Agent-based home simulation and control. In: Hacid, M.-S., Murray, N.V., Raś, Z.W., Tsumoto, S. (eds.) ISMIS 2005. LNCS (LNAI), vol. 3488, pp. 404–412. Springer, Heidelberg (2005)
6. Dey, A.K.: Understanding and using context. Pers. Ubiquit. Comput. **5**, 4–7 (2001)
7. Ferilli, S.: WoMan: logic-based workflow learning and management. IEEE Trans. Syst., Man Cybern.: Syst. **44**, 744–756 (2014)
8. Horrocks, I., Patel-Schneider, P.F., Bechhofer, S., Tsarkov, D.: OWL rules: A proposal and prototype implementation. J. Web Semant. **3**(1), 23–40 (2005)
9. Marsá-Maestre, I., López-Carmona, M.A., Velasco, J.R., Navarro, A.: Mobile agents for service personalization in smart environments. J. Network. **3**, 30–41 (2008)

10. O'Grady, M.J., Muldoon, C., Dragone, M., Tynan, R., O'Hare, G.M.: Towards evolutionary ambient assisted living systems. J. Ambient Intell. Humaniz. Comput. **1**, 15–29 (2010)

11. Olaru, A., Florea, A., Seghrouchni, A.: A context-aware multi-agent system as a middleware for ambient intelligence. Mobile Network. Appl. **3**, 429–443 (2013)

12. Rao, A.S., Georgeff, M.P.: Modeling rational agents within a bdi-architecture. In: Proceedings of the 2nd International Conference on Principles of Knowledge Representation and Reasoning, pp. 473–484 (1991)

13. Redavid, D., Ferilli, S., Esposito, F.: Towards dynamic orchestration of semantic web services. Trans. Comput. Collective Intel. **10**, 16–30 (2013)

14. Yau, S.S., Liu, J.: Incorporating situation awareness in service specifications. In: 9th IEEE International Symposium on Object and Component-oriented Real-time Distributed Computing (ISORC), pp. 287–294 (2006)

Trust Metrics Exploration in Optimizing Stock Investment Strategies

Zheyuan Su[✉] and Mirsad Hadzikadic[✉]

Complex Systems Institute, College of Computing and Informatics,
University of North Carolina at Charlotte, Charlotte, NC 28223, USA
{zsu2,mirsad}@uncc.edu

Abstract. The decision-making process in stock investment requires not only the rational trading rules practices, but also faith that market information is reliable. A trust metric is an indication of the degree to which one social actor trusts another. In our Agent-Based Model we built a stock-trading model that issues a daily stock trading signal. This paper introduces an agent-based model for finding the optimal degree of trust in stock transactions. The system has been evaluated in the context of Bank of America stock in the period of 1987–2014. The model outperformed both S&P 500 and buy-and-hold Bank of America stock strategy by two to three times.

1 Introduction

The decision-making process in stock investment requires not only the rational trading rules practices, but also faith that market information is reliable. A trust metric is an indication of the degree to which one social actor trusts another. Picking the right time and position is hard, as both endogenous and exogenous factors impact stocks' intrinsic value at any given moment. With the advantage of asymmetric information, which is what happens when one party in a transaction has more or superior information compared to others, some investors will benefit from utilizing it. However, trust metrics will alleviate this impact by simply following the market's trend. Investors with the highest degree of trust in others can adapt their trading rules to the latest market changes, thus making it possible for them to outperform the market and maximize their profit.

The concept of risk control involves the method of identifying potential risk factors and takes actions to reduce or eliminate such threats. It plays a key role in portfolio management, as it can stop the potential loss in the unexpected stock price changes. The use of trust metric can be beneficial in portfolio risk control, allowing investors to see the unanticipated changes in the market and take counter actions to prevent further loss.

Computers and sophisticated analytical techniques have automated many of the previously mentioned approaches by using information technology tools, (Teixeira and DeOliveira 2010) although with limited success. In recent years, complex adaptive systems–inspired methods, primarily using agent-based modeling techniques, have been tried as a way to simulate traders' behavior and capture the intricacies of stock trading (Kodia et al. 2010). This paper introduces an agent-based model for finding the optimal degree of trust in stock transactions. The model is a derivative of multi-sectors trading

F. Esposito et al. (Eds.): ISMIS 2015, LNAI 9384, pp. 331–340, 2015.
DOI: 10.1007/978-3-319-25252-0_36

model (Su and Hadzikadic 2014) published in WCSS 2014. The system has been evaluated in the context of Bank of America stock in the period of 1987–2014. The model outperformed the buy-and-hold strategy on both S&P 500 and Bank of America stock by two to three times.

2   Background

Investment methods are usually classified as passive and active portfolio management (Barnes 2003). Passive portfolio management only involves limited buying and selling actions. Passive investors typically purchase and hold investments, anticipating long-term capital appreciation and limited portfolio maintenance. Consequently, only active portfolio management strategies can bring investors extra profits simply because they bring about the possibility of covering a wide range of stock price movements. An active equity portfolio management (Grinold and Kahn 1995) requires periodic forecasts of economic conditions and portfolio rebalancing based on forecasted conditions. A degree of trust impacts the stock trading activities and risk control strategies (Asgharian et al. 2014). Risk control strategies make it possible for dynamic portfolio management to outperform the market (Browne 2000). A simple momentum and relative-strength strategy could outperform the buy-and-hold strategy 70 % of the time tracing back to 1920 s (Faber 2010). Performance can also be improved by considering a simple trend before taking positions.

However, these methods are not on the agent level. They simply provide a retroactive simulation technique. Since the market consists of trading individuals, taking into consideration interactions among agents provides investors with an important lever for significantly improving portfolio performance. The Complex Adaptive Systems (CAS) framework offers a natural technique for augmenting portfolio management strategies, given that its main advantage lies in its focus on capturing interactions among agents in the market place. A further advantage of CAS stems from its ability to set up different rules for agent interactions and test different trust thresholds among other agents, thus uncovering agent interactions that actually improve portfolio performance.

3   Complex Adaptive System Approach to Portfolio Management

As the Complex Adaptive System tools have the capability to capture the essence of distribution, self-organization, and nonlinear social and natural phenomena, characterized by feedback loops and emergent properties (Holland 1992), they offer a brand new way of modeling inherently complex systems such as stock market.

Interaction patterns and global regularities are important features in the financial markets (Cappiello et al. 2006). It is possible to utilize agent-based modeling (ABM) techniques to model financial markets as a dynamic system of agents. There already have been successful implementations of ABM models in fields as diverse as economics, government, military, sociology, healthcare, architecture, city planning, policy, and biology (Johnson et al. 2013, Hadzikadic et al. 2010). In financial market simulations, a large number of agents engage repeatedly in local interactions, giving rise to global markets (Roberto et al. 2001, Bonabeau 2002).

In this paper we describe an ABM system that looks for an optimized degree of trust among agents, based on the single stock trading model presented at Simultech (Su and Hadzikadic 2015). This model issues a stock trading signal (buy, sell, or hold) for a stock (Bank of America in our example). Agents will trade the Bank of America stock based on the publicly available, adjusted, daily data from January 2, 1987 to December 31, 2014. In addition, agents have the knowledge of the current status of the stock market, either bull or bear, based on the recession data available from the National Bureau of Economic Research (NBER). Here, a bull market indicates a financial market of a group of securities in which prices are rising or expected to rise. Bear market denotes the opposite scenario in financial market terms. Agents use this information to select their trading rules. As the market has been in the bull market in the post-1987 period, the simulation for a single stock will only track the bull market trading decision rules.

3.1   A CAS Stock Trading Model

In our CAS Agent-Based Model, we built a stock-trading model that issues a daily stock trading signal. Agents trade the Bank of America stock based on the current adjusted daily closing price. The interest is distributed based on the agents' cash holdings at the end of the day, using data from the Federal Reserve in the same timeframe. In addition, the transaction cost is set to be $10 each trade to bring in the opportunity cost for agents. This prevents agents to get tiny profits in unlimited trading.

Agents. Acollection of agents constitutes the "trading world" in this ABM simulation. In order to simplify the model as much as possible, we decided to look at the individual investors only. We understand that the institutional investors represent a large component of financial markets. However, we are trying to investigate the best degree of trust among agents to maximize the profits from the proposed trading strategies. Agents are initialized with a certain amount of money. Agents' transactions are triggered by their decision rules, the amount of capital they have, and the current market momentum, based on which agents align their trading strategies with the latest market changes. Based on the knowledge of the latest market status, agents choose the trading rules for the current tick. Table 1 describes the trading rules assigned to individual agents.

Table 1. Trading rules assigned to individual agents

Buy-Threshold	Minimum price change required for taking a long position
Buy-Period	Time window agents observe before evaluating the Buy-Threshold
Sell-Threshold	Minimum price change required for taking a short position
Sell-period	Time window agents observe before evaluating the Sell-Threshold
Self-Confidence	Degree of trust in the trading rules born with agents

Trading Rules. The following formulas describe agents' basic separated decision rules in detail.

> Buy Rule:
>   – Price-Change > Buy-threshold * (1 – (1 – self-confidence) * *momentum of buying*) in past Z days
>   – Agents will take long positions

For example, if the values for *buy-threshold, buy-period, self-confidence and market momentum* for an agent are 0.2, 30, 0.9, and 0.5 respectively, then the buying rule for this agent is:

> Sell Rule:
>   – Price-Change < Sell-threshold * (1 – (1 – self-confidence) * *momentum of selling*) in past Z days
>   – Agents will take short positions

IF the stock price goes up $0.2 * (1 – (1 – 0.9) * 0.5) * 100 \% = 19 \%$ in the past 30 trading days,

THEN take a long position on this stock.

Similarly, if the values for *sell-threshold, sell-period*, self-confidence and market momentum are 0.2, 10, 0.1 and 0.8 respectively, then the selling rule for this agent is:

IF the stock price goes up less than $0.2 * (1 – (1 – 0.1) * 0.8) * 100 \% = 5.6 \%$ in the last 10 trading days,

THEN agent will take a short position.

Also, short selling is allowed at any point. An agent can short sell any amount of stock up to their available cash amount. In financial terms, a long position indicates the purchase of a security such as stocks, commodity or currency, with the expectation that the asset will rise in value over time. A short position denotes the opposite scenario. Short selling indicates the sale of a security that is not owned by the seller. However, in order to close the position in the future, agents have to buy the same amount of securities to cover their short selling positions placed earlier.

Market Momentum

> Momentum ranges in [0, 1]
>   – Count how many people intend to buy/sell
>   – If no one is *buying/selling, momentum of buying/selling* will be 0
>   – If everyone is *buying/selling*, momentum of *buying/selling* will be 1

Not everyone in the market is a rational investor. Investors' greedy and panic-prone behavior impacts the market as well. As a result, market momentum is an important factor that impacts agents' decision rules. Heterogeneous agent models (Hommes 2006) show that most of the behavioral models with bounded rational agents using different strategies may not be perfect, but they perform reasonably well.

The addition of market momentum to the model will potentially increase the performance of investors, as it allows agents to adjust their trading rules to the latest market changes. In the stock trading model, momentum was generated by the overall buy/sell behavior of agents. The more agents buy stocks, the higher the bidding price. Likewise, the more agents sell stocks, the lower the stock prices, as agents are trying to liquidate their inventories.

Degree of Trust. Degree of Trust is a parameter that we are interested in exploring. In the model, a variable called self-confidence is created to control the degree of trust to other agents' trading strategies. The higher the self-confidence, the stronger the agents believe in their own trading strategies, thus decreasing the impact of the trading environment around them.

Agents have access to the latest market information. All agents conduct transactions based on the current stock price. They can track all the past prices, starting with Jan 2, 1987. At the same time, the bandwagon effect, which is characterized by the possibility of individual adoption increasing with respect to agents who have already done so, plays an important role in transactions. As a result, if there are a lot of agents who are buying stocks, then agents will increase their buy-threshold. At the same time, if there are many agents who are shorting stocks, then a substantial number of agents will correspondingly decrease their sell-threshold, as they try to liquidate their assets as soon as possible.

As a result, self-confidence captures the agents' degree of trust among all the other agents. Self-confidence is randomly assigned at the initial stage, making it possible to explore the best degree of confidence one should adopt in order to outperform the market.

The Connection Between Market Momentum and Self-Confidence. The overall trend of buying and selling constitutes the market momentum, according to which agents adjust their trading strategies correspondingly. With the adaptive trading strategies, agents are able to seek more trading opportunities and boost their profits.

However, agents cannot completely trust the market momentum. The degree of trust will affect the participation rate in stock trading (Guiso 2008). Therefore, the variable self-confidence bridges the gap between market momentum and agents' rational trading strategies. Each agent has a randomly initialized degree of trust and the simulation tracks the best performer's self-confidence.

In the agent's buy and sell rules, self-confidence plays a key role in setting the real trading threshold. In some extreme scenarios, agents with the self-confidence of 1 do not listen to the latest market trends, while agents with the self-confidence of 0 follow market trend makers.

Rational investors are risk-averse in transactions. These investors prefer the lower risk choice when facing two investments with similar expected returns. Risks can be categorized into non-systemic risk and systemic risk, which cannot be diversified. As a result, following the market trend alleviates the non-system risk and lets agents keep the systemic risk only.

Genetic Algorithm. In the world of artificial intelligence, genetic algorithms embody a search heuristic that mimics the process of natural selection. This heuristic mainly generates promising descendants that are more adaptive to the changes in the environment.

A hatch-and-die concept of NetLogo was introduced into the simulation regarding the mechanism for regenerating the best performers or eliminating underperforming agents. Agents who lose all their money are eliminated from the environment. At the same time, the hatch mechanism ensures new agents will be initialized and placed into the environment, thus keeping the number of agents constant. This mechanism makes sure that active trading among agents is maintained.

Search Space and Mutation. As agents have a lot of parameters, the search space covering all possibilities counts trillions of states. In order to use a small amount of agents to simulate all possible agents, a mutation mechanism is introduced while all parameters are randomized within small ranges. With the benefits of mutation, some agents' parameters are randomized far beyond the preset small ranges, making it possible to explore the whole space. Monte Carlo simulation (Fishman 1995) is also applied in the model to achieve the best result with the minimal rounds of simulations.

Learning and Interaction. In ABM implementations agents can learn from each other. The process of learning offers agents the possibility to refine their transaction decision rules, thus helping them capture the slight opportunities in the market and turning them into profits. Agents will learn from the neighbors within a certain radius. The neighborhood structure is introduced for efficiency reasons. With learning, the search space can be made smaller so that more agents can be included in the model. Also, the learning mechanism makes it possible to investigate alternative strategies that have not yet been discovered in the market (Outkin 2012). In this implementation, to preserve computational time, there is a radius around agents. Agents move around and seek for the potential learning opportunities from the best performing agents in their radius.

There is a local variable called *aggressiveness* that indicates to agents to what extent they want to adopt their neighbor's behavioral structure. Also, there is a period of time that learning process is prohibited in order to allow agents to prove their trading rules are the best overtime.

Benchmark Agents. Two benchmark agents, using buy-and-hold strategy, are created to evaluate the performance of agents. Their underlying assets are Bank of America and SP500 respectively.

Global Trading Environment. In the CAS stock-trading model, the world is represented in two dimensions, X-axis and Y-axis, ranging from −10 to +10. There is a variable called *radius* defining how far agents can reach out to other agents to learn their trading strategies. The *Radius* has a different value for each agent, making it possible for agents to have a diversified trading and differing learning preferences. Agents have the knowledge of where they are and who else is in their radius.

3.2 Implementation

This stock trading CAS model was implemented using the Netlogo 5.1.0 programmable modeling environment (Wilensky 2009). Netlogo offers a user-defined grid and the possibility of defining agents, normally called turtles in NetLogo.

In this model, the exploration space for all possible trading strategy combinations exceeds trillions of possibilities. As the combination is extreme large, it has a huge impact on the computing speed of the simulation. In order to provide a trade-off between the computing speed and the space exploration, we set the agent number to 1,000. Table 2 shows the setting of parameters in the stock-trading model. All transaction decision rules are randomized within the [-0.4,0.4] range for required returns and within [0,100] range for the trading periods. Aggressiveness is set to 0.3 and 0.001. Self-confidence is randomized from 0 to 1 by 0.01. The small range used in the simulation is to decrease the search space, thus boosting the coverage of each run. However, the full search range will be close to the real world trading. Mutation is introduced to allow a subset of agents to mutate from [-0.4,0.4] to [-1,1] for required returns, and from [1,100] to [1,1000] for trading periods. Agents are assigned with the initial capital in the amount of $50,000. The transaction cost is fixed at $10 per transaction, thus forcing agents to trade off for the opportunity costs. The mutation rate is fixed at 0.1, which allows 10 % of all agents to get buy/sell threshold and buy/sell period generated in [-1,1] and [1,1000] respectively. Also, interest is distributed at the end of each tick, based on the amount cash held on hand.

Learning from other agents is disabled in the first 1,000 days, which leaves enough time for agents to evaluate their initial trading strategies. After that, agents learn throughout the rest of the simulation. This mechanism allows agents sufficient time to optimize their strategies throughout the volatilities of the market, i.e. financial crises or huge price volatility periods.

Table 2. Parameter ranges

Non-Mutation	Buy-Threshold	[- 0.4, 0.4] with step size 0.2
	Buy-Period	[0, 100] with step size 20
	Sell-Threshold	[- 0.4, 0.4] with step size 0.2
	Sell-period	[0, 100] with step size 20
Mutation	Buy-Threshold	[-1, 1] with step size 0.2
	Buy-Period	[0, 1000] with step size 20
	Sell-Threshold	[-1, 1] with step size 0.2
	Sell-period	[0, 1000] with step size 20
Mutation Rate	0.1	
Self-Confidence	[0,1] with step size 0.01	
Aggressiveness	0.001	
Initial Capital	$50,000	
Transaction Cost	$10	

4   Results

In the stock trading model, S&P 500 and Bank of America (BAC) buy-and-hold strategies were used as performance benchmarks. As the timeframe of the data is from 01/02/1987 to 12/31/2014, different settings of training/test experiments were conducted during the simulation.

The S&P 500 index increased from 246.45 to 2059.9 during the period of 01/02/87 to 12/31/13. The Bank of America stock price increased from 2.34 to 17.89 in the same period. If investors use buy-and-hold strategy, they will receive 735.82 % and 664.53 % respectively. Figure 1 shows the cumulative return for different agents in the simulation. The best performer in the simulation achieved 1607.52 % profit, which is 2.18 times better than the SP500 benchmark. Also, the top 10 % performers achieved 678.75 % profit, which is slightly better than the BAC benchmark. The reason for it is that the SP500 is related to banking industries, which were the main cause for the last financial crisis. As a result, BAC stock price is still 60 % below its historical high but SP500 was surpassing its historical highs again and again in the past few years. Table 3 shows the summary of the results.

Table 3. Simulation results

Best Performer	Cumulative Return	1607.52 %
	Self-Confidence	0.42
Top 10 % Performers	Cumulative Return	678.75 %
	Self-Confidence	0.46
Benchmark Agent 1	SP500 Buy-and-hold	735.82 %
Benchmark Agent 2	BAC Buy-and-hold	664.53 %

Fig. 1. Cumulative return in simulation

From the Monte Carlo simulation of the model, the best level of degree of trust is at the level of 0.42. It indicates that the best performer will change its decision rules 58 % according to the market changes, and keep the 42 % of its initial decision rules settings.

> • If the stock price goes down 52% in last 234 trading days, take a long position.
> • If the stock price goes up less than 40% in last 42 trading days, take a short position.

The best decision rule set was described as follows.

However, these rules are subject to 58 % with respect to the degree trust to the market momentum. As market momentum changes everyday, agent will change their decision rules correspondingly. Following shows a particular trading rules set are actually put into practices on Dec. 31 2014.

> • If the stock price goes down 20% in last 234 trading days, take a long position.
> • If the stock price goes up less than 18% in last 42 trading days, take a short position.

The narrative for the trading rule set basically states that an investor should track the past performance for a particular stock and then decide whether to get in. For the exit strategy, if the stock's performance deviates from the expectations, investors should clear their positions immediately.

5   Discussions, Conclusion, and Future Work

Computer simulations allow us to see behind-the-scene actions of agents, and to make forecasts on the possible futures of the markets. Compared with the return of the S&P 500, the CAS model with optimized self-confidence presented here has achieved a much higher return, around 2 times, in the same timeframe as the S&P.

However, the momentum, a measure of the overall market sentiment (Scowcroft and Sefton 2005), plays an important role in the CAS stock stock-trading model. All the rules are adjusted based on the market momentum in a specific time tick. With the benefit of the momentum, the performance of the stock-trading model is far better than a simple buy-and-hold strategy for both S&P 500 and BAC. In the current model, momentum is generated by the agents' desire to conduct transactions. Future refinements in the momentum component will play a key component in improving the performance of the model. Agents should change their rules at a level of 0.42 to the market changes in order to outperform the market.

We are currently working on several strategies for improving the computation of the momentum component. One is to extract the real time tweets from Tweeter and to run a sentiment analysis on those tweets. Then the signals from Twitter will be attached to the current momentum component. Another one is to use the transactions volume to deduce the historical drive in the market and plug it into the current momentum mechanism, leading to a more precise forecast about the upcoming market movements. In return, agents can anticipate the changes in the future investors' actions and adjust their transaction strategies to maximize profits. We are also planning to extend the trading of single stock to multiple stocks, which are selected from different sectors. In that way, the multiple stocks trading model will give out a better sense of different degree of trust across different sectors.

References

Su, Z., Hadzikadic, M.: Application of complex adaptive systems in portfolio management. In: 2014 Proceedings of 5th World Congress on Social Simulation, WCSS (2014)

Su, Z., Hadzikadic, M.: An agent-based system for issuing stock trading signals. In: 2015 Proceedings of Simultech (2015)

Holland, J.: A new era in computation. Daedalus **121**(1), 17–30 (1992)

Johnson, L., Hadzikadic, M., Whitmeyer, J.: The future engaging complexity and policy: Afghanistan citizen allegiance model. Int. J. Humanit. Soc. Sci. **3**(10) (2013)

Wilensky U: NetLogo. Center for Connected Learning and Computer-Based Modeling. Northwestern University, Evanston, IL (1999). http://ccl.northwestern.edu/netlogo

Fishman, G.: Monte Carlo: Concepts, Algorithms, and Applications. Springer, New York (1995)

Faber, M.: Relative Strength Strategies for Investing, Cambria (2010)

Barnes, J.: Active vs. passive investing. CFA Mag. **14**, 28–30 (2003)

Scowcroft, A., Sefton, J.: Understanding momentum. Financ. Anal. J. **61**, 64–82 (2005)

Grinold, R., Kahn, R.: Active Portfolio Management, pp. 367–384. Probus. Publications, Chicago (1995)

Cappiello, L., Engle, R.F., Sheppard, K.: Asymmetric dynamics in the correlations of global equity and bond returns. J. Financ. Econometrics **4**(4), 537–572 (2006)

Raberto, M., Cincotti, S., Focardi, S.M., Marchesi, M.: Agent-based simulation of a financial market. Phys. A Stat. Mech. Appl. **299**(1), 319–327 (2001)

Bonabeau, E.: ABM: methods and techniques for simulating human systems. Proc. Natl. Acad. Sci. U.S.A. **99**(Suppl 3), 7280–7287 (2002)

Hadzikadic, M., Carmichael, T., Curtin, C.: Complex adaptive systems and game theory: an unlikely union. Complexity **16**(1), 34–42 (2010)

Hommes, C.H.: Heterogeneous agent models in economics and finance. Handbook of Computational Economics, vol. 2, pp. 1109–1186 (2006)

Outkin, A.V.: An agent-based model of the NASDAQ stock market: historic validation and future directions. In: CSSSA Annual Conference, Santa Fe, New Mexico, pp. 18–21, September 2012

Browne, S.: Risk-constrained dynamic active portfolio management. Manage. Sci. **46**(9), 1188–1199 (2000)

Asgharian, H., Liu, L., Lundtofte, F.: Institutional Quality, Trust and Stock-Market Participation: Learning to Forget. Trust and Stock-Market Participation: Learning to Forget (2014)

Guiso, L.: Trusting the stock market. J. Financ. **63**(6), 2557–2600 (2008)

Teixeira, L.A., De Oliveira, A.L.I.: A method for automatic stock trading combining technical analysis and nearest neighbor classification. Expert Syst. Appl. **37**(10), 6885–6890 (2010)

Kodia, Z., Said, L.B., Ghedira, K.: A study of stock market trading behavior and social interactions through a multi agent based simulation. In: Jędrzejowicz, P., Nguyen, N.T., Howlet, R.J., Jain, L.C. (eds.) KES-AMSTA 2010, Part II. LNCS, vol. 6071, pp. 302–311. Springer, Heidelberg (2010)

Applications

Audio-Based Hierarchic Vehicle Classification for Intelligent Transportation Systems

Elżbieta Kubera[1]([✉]), Alicja Wieczorkowska[2], and Krzysztof Skrzypiec[3]

[1] University of Life Sciences in Lublin, Akademicka 13, 20-950 Lublin, Poland
elzbieta.kubera@up.lublin.pl
[2] Polish-Japanese Academy of Information Technology, Koszykowa 86,
02-008 Warsaw, Poland
alicja@poljap.edu.pl
[3] Maria Curie-Skłodowska University in Lublin, Pl. Marii Curie-Skłodowskiej 5,
20-031 Lublin, Poland
krzysztof.skrzypiec@poczta.umcs.lublin.pl

Abstract. Nowadays almost everybody spends a lot of time commuting and traveling, so we are all very much interested in smooth use of various roads. Also governing bodies are concerned to assure efficient exploitation of the transportation system. The European Union announced a directive on Intelligent Transport Systems in 2010, to ensure that systems integrating information technology with transport engineering are deployed within the Union. In this paper we address automatic classification of vehicle type, based on audio signals only. Hierarchical classification of vehicles is applied, using decision trees, random forests, artificial neural networks, and support vector machines. A dedicated feature set is proposed, based on spectral ranges best separating the target classes. We show that longer analyzing frames yield better results, and a set of binary classifiers performs better than a single multi-class classifier.

Keywords: Intelligent transport system · Vehicle classification · Audio signal analysis

1 Introduction

Traveling and commuting takes a lot of time in our lives nowadays. Therefore, the efficient use of various roads becomes an important issue, since smart distribution of traffic into available roads can facilitate smooth transportation, save time, and increase our safety. This is why smart use of roads became a topic of interest not only for road users, but also for governing bodies. Intelligent Transport Systems (ITS, Intelligent Transportation Systems) are used in many countries worldwide, to integrate information and telecommunication technologies with road systems. The European Union in its directive defines ITS as follows:

"Intelligent Transport Systems (ITS) are advanced applications which without embodying intelligence as such aim to provide innovative services relating to different modes of transport and traffic management and enable various users to

© Springer International Publishing Switzerland 2015
F. Esposito et al. (Eds.): ISMIS 2015, LNAI 9384, pp. 343–352, 2015.
DOI: 10.1007/978-3-319-25252-0_37

be better informed and make safer, more coordinated and smarter use of transport networks" [3]. In Poland, ITS are implemented and under development in many cities, including Warsaw, Cracow, Łódź, Tricity (Gdańsk, Sopot, and Gdynia), Poznań, Bydgoszcz, and Olsztyn.

ITS can be based on data from video cameras, traffic sensors, Global Positioning System (GPS), license-plate readers, geographic information systems (GIS), wireless networks, traffic signals, computers and software [10]. These data can be manually processed in traffic control centers, but automation of this process is highly desirable. The United States Department of Transportation in the 2015–2019 Strategic Plan indicates advancing automation as one of "the primary technological drivers of current and future ITS work across many sectors" [7].

In this paper we address automatic classification of vehicle type based on audio information. Such data can be received from microphone only recordings, or as audio tracks from video recordings. The resulting information on vehicles passing by can be applied to monitor traffic and noise, redirect vehicles, estimate traffic congestions, adjust speed limits displayed on dynamic signs, and assure smoother and safer traveling.

1.1 Related Work

Vehicle exterior noise comes from the following sources: engine, tires, exhaust system and even air intake system [8]. Automatic classification of vehicles based on audio signal usually starts with sound parametrization based on a downsampled signal. The sampling rate is usually reduced, comparing to standard audio recordings, to reduce computational load. The length of the analyzing window is sometimes set as one of the parameters, but it is often quite short (10–50 ms). Afterwards, various classifiers are applied (possibly with feature selection as a preprocessing step). A literature review of works in this area is presented in [6].

A variety of classifiers has already been used for vehicle classification purposes. For instance, Alexandre et al. in [1] applied artificial neural networks (multi-layer perceptrons) combined with feature selection based on a genetic algorithm, with 3 target classes: car, motorcycle, and truck. Features included mel-frequency cepstral coefficients (MFCC), and zero crossing rate, for 11.025 kHz sampling rate and the analyzing window of 512 samples (46 ms). The authors reported 93 % correctness for 22 features and 75 % for 66 features [1].

In [9], authors used quadratic and linear discriminant analysis, as well as k-nearest neighbor classifier (k-NN) and support vector machines (SVM) to bus, car, motorcycle, and truck recognition, i.e. 4 target classes. They investigated such features as short time energy, average zero cross rate, and pitch frequency of periodic segments of signals, for center clipped signal [12]. According to the authors, better classification accuracy can be achieved by considering high energy feature vectors, due to the correspondence of low energy regions with noises of the background. For 11025 Hz sampling rate and 15 ms analyzing frame, they obtained 80 % correct classification rate for SVM with 12 Mel coefficients [9].

Erb in [6] used audio data downsampled to 8 kHz, for window lengths of 256 and 1024 samples, even 2048 for wavelets (256 ms). He also applied 1 kHz downsampled data for reduced computational effort, mainly to extract features from the motor sound; window lengths of 256 and 512 samples were applied in these analyzes. SVM and feature selection/linear prediction were applied for 3 classes: car, truck, and van. Classification results yielded 87 % overall correctness for vehicles traveling at low speed, 83 % for higher vehicle speeds. For the combined traffic situation without given probabilities, the best result were 80 %, and increased to 83 % if class probabilities were equal to those in the originally recorded traffic scenarios [6].

The authors in [2] approached another problem, namely distinguishing between vehicles (cars, trucks, and vans) and background in audio data, i.e. 2 target classes only. They applied dimensionality reduction methodologies such as random projections instead of traditional signal processing and features extraction. The audio signal was downsampled to 2 kHz, and then PCA was applied, yielding average correctness 84 %.

Other interesting work aims at classifying 4 military vehicle classes: assault amphibian vehicle, main battle tank, high mobility multipurpose wheeled vehicle and dragon wagon [4]. Acoustic, seismic and infrared signals from distributed sensor networks were used as input, and k-NN, SVM, and maximum likelihood estimation were used for classification. The results reached 69 % for acoustic data.

1.2   Our Proposed Methodology

The audio signal recorded today is usually sampled at 44.1 kHz or more, so high audio frequencies are easily available. Also, the noise of the vehicle includes not only the engine sound, but also other noises. We believe that higher frequencies can facilitate distinguishing between vehicles, so we propose to analyze the full spectrum, for 48 kHz sampling rate. Our plan is to analyze data on frame by frame basis, with the goal to classify vehicles from their audio data, so we assume that silence (or low level background noise) is already removed. The investigated data have been manually labeled, based on the accompanying video material. In our first experiments we classify single vehicle sounds. Since vehicles in heavy traffic can be heard simultaneously, we then also investigate classification of simultaneously passing vehicles.

The feature set we propose is based on spectral ranges best separating the target vehicle classes, combined with 5 standard audio features. The classifiers applied in our research include artificial neural networks, SVM, decision trees, and random forests. We start with single classifiers with 3 target classes: car, truck, and tractor. No motorcycles were recorded in the analyzed data. Since the size of vehicles (and thus the noises produced) in car and truck classes can vary, we also performed 2-stage classification, with further division of these classes into subclasses. As a result, we have 5 classes: small truck, big truck (subclasses of truck), van, small car (subclasses of car), and tractor. Additionally, various analyzing frame sizes were investigated. Although the analysis phase is faster for shorter frames, more frames must be analyzed in this case, and the training phase of the classifiers is much longer. Also, better frequency resolution of the sound spectrum is available when a longer analyzing frame is used. Therefore, having a longer analyzing frame can be advantageous.

2   Data Preparation

The investigated audio data originate from 1 h video recording, shot in November 2012 in a suburban area nearby Lublin in Poland (where cars travel with approximately uniform speed), with 48 kHz sampling rate and 24 bit resolution, converted to 16 bits. The recorded data contained cars, trucks, and tractors. Additionally, we decided to further divide cars and trucks into subclasses, depending on the vehicle size. Therefore, we investigated 5 classes altogether: 1: small truck (no separate tractor unit), 2: big truck (heavy goods vehicle - HGV: tractor unit with or without semi-trailer), 3: van (including minibuses), 4: small car, 5: tractor.

Audio segments containing vehicle sounds were manually selected for further investigations and labeled based on video data.

2.1   Feature Set

Audio data were parameterized using the following feature set:

- $A13$, $A31$, $A14$, $A41$, $A15$, $A51$, $A23$, $A32$, $A24$, $A42$, $A25$, $A52$, $A45$, $A54$, $A35$, $A53$ (16 features) - normalized (with respect to the spectrum energy) energies Axy in the spectral ranges determined in such a way that the energy of this frequency range separates classes x and y, namely the class x shows higher energy values than the class y in this range (for these ranges where discerning differences were found; $A12$ $A21$, $A34$ and $A43$ not discovered):
  - A13:   3–867 Hz,   A31:   873–9328 Hz,
  - A14:   114–841 Hz,   A41:   1119–1336 Hz,
  - A15:   683–803 Hz,   A51:   3942–3942 Hz (1 point in the spectrum),
  - A23:   120–938 Hz,   A32:   1119–2555 Hz,
  - A24:   138–938 Hz,   A42:   1336–1342 Hz,
  - A25:   770–829 Hz,   A52:   3228–3258 Hz,
  - A45:   1207–1272 Hz,   A54:   73–82 Hz,
  - A35:   718–2007 Hz,   A53:   79–545 Hz;

  These ranges were selected for averaged spectra calculated using 330 ms frames without overlap for 1-s sounds, 12 s of recordings for each of 5 classes. For the available spectral resolution all frequency ranges were tested and for each pair of classes the range selected gives the biggest margin separating objects from these 2 classes. This was automatically done as follows: for each candidate lower limit ll and upper limit ul (possible candidates resulted from the spectrum resolution) spectral energy in this range was calculated as the sum of the power spectrum points in this range, for each object (1-second sound). For all 12 objects of class x these values were in the range $< min_x, max_x >$, and for y in $< min_y, max_y >$. For all ranges such that $min_x < max_y$, the one of maximal margin was chosen, where margin was calculated as $(min_x - max_y)/(max_x - min_y)$. Such a range ll, ul discerns between x and y class, since spectral energy in this range for x exceeds the energy for y;
- $A13\_A31, A14\_A41, A15\_A51, A23\_A32, A24\_A42, A25\_A52, A45\_A54, A35\_A53$ (8 features): proportion of energies between the indicated spectral ranges;

- *EnAbove4 kHz* - proportion of the spectral energy above 4 kHz to the entire spectrum energy;
- *Energy* - energy of the entire spectrum;
- *Audio Spectrum Centroid* - the power weighted average of the frequency bins in the power spectrum. Coefficients were scaled to an octave scale anchored at 1 kHz [13];
- *Audio Spectrum Spread* - RMS (root mean square) of the deviation of the log frequency power spectrum wrt. *Audio Spectrum Centroid* [13];
- *Zero Crossing Rate* in the time-domain of the sound wave; a zero-crossing is a point where the sign of the function changes;
- *RollOff* - the frequency below which 85 % (experimentally chosen threshold) of the accumulated magnitudes of the spectrum is concentrated.

These features were calculated for 4 analyzing frame sizes: 15, 45, 100, and 330 ms. Spectrum was calculated using fast Fourier transform, Hamming windowed, with about 1/3 frame hop size (i.e. frames overlapped by 2/3).

2.2   Training and Testing Data

Training and testing of the classifiers was mainly performed through 4-fold cross validation. The 1 h recording was divided into four 15 min segments, and in each run sounds from three of these segments were used for training, and the remaining one for testing. This way the training frames were not present in the testing set, even though the analyzed audio frames overlap. The goal of this procedure was to test several classifiers, build for various frame lengths, and then select the best performing ones to become parts of the final, hierarchic classifier. For this purpose 7 separate classes were considered, i.e. car, truck, tractor (higher level, superclasses), small car, van, small truck, and big truck (lower level, subclasses); the feature set was not changed. Small car and van were subclasses of car, and small truck and big truck were subclasses of truck.

The training was performed in 2 versions. Firstly, only single sounds representing one vehicle were used in the training of all classifiers to be tested. Various approaches can be taken to select positive and negative examples in the training phase of a hierarchic classifier, see [11]. We decided to take "siblings" approach, i.e. negative examples represent siblings of a target class at a given level; positive examples represent a target class (and its children). This means that for each of the 3 classes at the higher level, i.e. superclasses (car, truck, tractor), negative examples represented the other 2 superclasses. Positive examples represented the target class, including its children; for instance, truck samples (big truck and small truck) were positive examples for truck class. For the lower level, negative examples represented the other subclass child of the same superclass. For instance, van samples were negative examples for small car, big truck samples were negative examples for small truck, and vice versa.

After the best classifiers were selected for the final hierarchical classifier, the second version of training was performed, where the data representing 2 vehicles recorded simultaneously were added. Additional positive examples contained sounds of a target class, recorded together with a sound from any non-target class; additional negative examples contained sounds from a sibling class, plus a sound from any non-target class. For instance, van samples (single sounds, or together with other, non-small car sounds) were negative examples for small car, big truck samples (single sounds, or together with other, non-small truck sounds) were negative examples for small truck. Positive examples represented single sounds of the target class, plus sounds of the target and any other class. For instance, single sounds of small cars and sounds of a small car accompanied with any other vehicle represented positive examples for small car class.

Testing was performed first on single sound trained classifiers, using single sounds for tests, and then also using simultaneous sounds. The second testing was performed for the classifiers being parts of the hierarchical classifier, trained on both single sounds and simultaneously recorded sounds.

3   Classification

Classification was performed using WEKA [5]. The classifiers applied included the following algorithms:

- decision tree (j4.8),
- random forest with 500 trees (RF),
- artificial neural network (MLP - multi-layer perceptron), and
- SVM (SMO in WEKA); RBF (radial basis function) was chosen, with C=8 and gamma=2; R package was used for SVM parameter tuning – tune.svm function was applied [14].

The following versions of the classifiers were used:

1. single (multi-class) classifier with 5 target classes: small truck, big truck, van, small car, tractor;
2. set of 5 binary classifiers, for comparison with classifier no. 1,
3. hierarchical classifier:
   (a) higher level: 3 binary classifiers to recognize car, truck, and tractor,
   (b) lower level: 2 binary classifiers for subclasses of car (small car, van) and 2 binary classifiers for subclasses of truck (small truck, big truck).

4   Results

In first experiments we compared the performance of multi-class classifiers with sets of binary classifiers, for 5 classes: small truck, big truck, van, small car, and tractor. The classifiers were trained and tested on single sounds only. In order to obtain the results quickly, the training was performed on a limited data set, namely 12 second of recordings were selected for each target class, with one-second objects (manually selected) representing a sound of one target vehicle. Ten-fold cross validation was applied, using randomly selected frames; since analysis and parametrization was performed for overlapping frames, the

Table 1. Percentage of correctly classified instances for vehicle classification for multi-class classifiers; best result in bold

classifier	j4.8	RF	MLP	SVM
frame 330 ms	82.14	92.62	91.19	**92.86**
frame 100 ms	85.38	92.29	88.02	91.37
frame 45 ms	79.54	87.56	82.75	87.28
frame 15 ms	67.63	76.70	71.92	76.37

Table 2. Results (percentage of correctly classified instances) of vehicle classification for 5 binary classifiers, for various frame sizes; best results shown in bold

Small truck	j4.8	RF	MLP	SVM	Big truck	j4.8	RF	MLP	SVM
330 ms	88.10	96.19	**97.62**	97.62	330 ms	91.43	95.95	**97.62**	97.14
100 ms	92.01	96.06	94.01	96.12	100 ms	93.79	96.17	95.79	96.28
45 ms	89.55	93.56	90.82	93.30	45 ms	91.70	94.62	94.02	94.44
15 ms	83.86	87.78	85.01	87.06	15 ms	87.54	90.24	88.75	90.16
Small car	j4.8	RF	MLP	SVM	van	j4.8	RF	MLP	SVM
330 ms	91.43	**94.76**	94.52	94.52	300 ms	92.38	95.00	93.81	95.95
100 ms	93.80	95.84	92.39	95.30	100 ms	92.23	**96.06**	93.20	94.88
45 ms	91.77	94.49	92.63	93.84	45 ms	89.99	93.77	91.54	92.24
15 ms	87.50	90.29	88.88	90.03	15 ms	86.16	89.55	87.27	88.34
Tractor	j4.8	RF	MLP	SVM					
330 ms	98.33	**99.76**	99.76	99.52					
100 ms	97.46	98.60	98.92	99.24					
45 ms	95.83	97.85	97.80	98.42					
15 ms	92.96	95.26	94.35	95.58					

data in training and testing could partially overlap. The results of multi-class classifiers for various frame sizes are shown in Table 1. The shortest frame yields the worst results, and the best results are obtained for 330 ms and 100 ms frames. The results for a set of 5 binary classifiers for the same classes are higher even for short frames, see Table 2 (not surprising, as our features discern pairs of classes); in this case the results do not differ that much for various frame sizes. Still, the best results were obtained for longer analyzing frames. These results are very good, compare favorably with results obtained in any other similar research, and show that the proposed parametrization is very efficient.

Next, a hierarchical classifier was build, based on binary classifiers shown in Table 3, for all sounds from our 1-h recording. These classifiers were trained and tested on single sounds only. For each of 3 classes at the higher level and 4 classes at the lower level, the best performing classifier and frame size were selected, as marked in Table 3. They were next combined to create a hierarchical

Table 3. Results (percentage of correctly classified instances) of vehicle classification for the 3 classes at the higher level and 4 classes at the lower level, for various frame sizes; best results shown in bold

car	j4.8	RF	MLP	SVM
330ms	92.8	95.2	95.3	**96.0**
100ms	92.9	94.7	95.0	95.9
45ms	92.5	94.1	92.9	95.0
15ms	90.2	91.7	91.4	92.6

small car	j4.8	RF	MLP	SVM	van	j4.8	RF	MLP	SVM
330ms	85.9	88.8	85.4	86.5	330ms	85.9	88.75	84.2	86.5
100ms	84.2	88.5	85.9	87.8	100ms	84.2	88.5	85.4	87.8
45ms	85.6	88.4	87.5	88.5	45ms	85.6	88.5	88.7	88.5
15ms	87.5	89.0	87.8	**89.1**	15ms	87.5	89.0	88.4	**89.1**

truck	j4.8	RF	MLP	SVM
330ms	95.2	95.0	94.3	**95.85**
100ms	93.1	94.6	95.1	95.77
45ms	92.7	94.0	94.0	95.1
15ms	90.3	91.8	92.0	92.8

small truck	j4.8	RF	MLP	SVM	big truck	j4.8	RF	MLP	SVM
330ms	64.02	78.00	**78.83**	74.22	330ms	64.02	**79.32**	76.44	74.22
100ms	69.67	75.84	76.27	74.05	100ms	69.67	76.17	75.47	74.05
45ms	67.08	75.67	69.62	73.85	45ms	67.08	75.60	71.14	74.63
15ms	70.89	75.02	76.23	75.80	15ms	70.89	74.96	74.53	75.82

tractor	j4.8	RF	MLP	SVM
330ms	98.13	98.85	98.71	98.15
100ms	**98.86**	98.64	98.10	98.15
45ms	97.56	98.00	97.70	97.90
15ms	96.72	97.31	96.78	97.39

classifier, i.e. performing 2-stage classification. As we can see, the results at the higher classification level (for 3 classes: car, truck, and tractor) are above 95 %.

This classifier was then tested on sounds representing multiple vehicles recorded simultaneously (2 vehicles of different type), yielding the following results:

- higher level: car 86.96 %, truck 45.65 %, tractor 95.57 %,
- lower level:
  - small car 99.58 %, van 68.92 %,
  - small truck 74.07 %, big truck 70.37 %.

As we checked, especially the results for trucks passing simultaneously with cars were lower, because the level of the truck sounds was hardly heard (almost masked by car sounds). Also, when a car was passing just after a truck, the car was hard to recognize by the classifier. Probably these soft sounds should be ignored and the threshold for the minimal sound loudness could improve the classification results; these vehicles could be classified correctly anyway when passing closer to the microphone.

The results indicate that the subclasses of truck are much better classified than the truck at the higher level. This is possible because even though the frames sent to the lower level of the classifier might be incorrectly classified at the higher level, they can still be correctly classified in the lower level. Namely, if they do not belong to the target subclasses, there is still a chance that the dedicated classifiers correctly indicate these frames as not representing the target subclasses. This explains why the results at the lower level of classification can be higher than at the higher level.

Another training of the hierarchical classifier was also performed on both single sounds and sounds representing more than 1 vehicle. In our recordings, no more than 2 types of vehicles were audible simultaneously. For this method of training, the results of hierarchical classification tested on sounds representing 2 vehicles at the same time are as follows:

- car 87.50 %, truck 45.65 %, tractor 95.32 % at the higher classification level,
- at the lower classification level:
  - small car 99.93 %, van 68.79 %,
  - small truck 81.48 %, big truck 74.07 %.

In some cases just slight changes can be observed, but the classification of truck class and subclasses yields better results than in the case when only single sounds were used in training. When looking into the details (not shown here because of space limitations) no false negatives were indicated in the classification of car and small car examples, for both training procedures. It is difficult to compare results with other research because of using different data, but the final average result for 5 classes tested on single or multiple sounds, for both training on single sounds or also on multiple ones, is around 85 % on average for the lowest available classification level (classes small car, van, small truck, big truck, and tractor). This is comparable with the results obtained for 2 or 3 classes in other research, and better than in the case of 4 classes [2,6,9].

5   Summary and Conclusions

Advancement in ITS can improve the use and safety of the road systems, and automation of the road usage monitoring is highly desirable. In this paper we investigated automatic classification of vehicles based on audio signals, including multiple vehicles simultaneously passing by. We proposed a classification-oriented spectrum parametrization, where spectral ranges were selected in such a way that the margin between classes is maximized. The obtained results are satisfactory and they show the efficiency of the proposed parametrization. The experiments were performed for various sizes of analyzing frames. When multi-class classifiers were applied, the longest frames performed best; in the case of binary classifiers, the difference was not that clear. Binary classifiers yielded much higher results, and they also allowed extension of classification to recognition of multiple classes for a single frame, if more than one vehicle was passing by. This task was much more difficult and the results were lower in some cases, but this can be caused by the difference in loudness levels. Also, hierarchical classification allows improving the results at the deeper levels of the hierarchy, if the objects are misclassified at the higher level.

In future research, we plan adding a cleaning stage after the testing phase; if a given class is not recognized in a frame between 2 other neighboring and overlapping frames where it is recognized, then it should probably be recognized also in this frame. Additionally, we plan to perform more recordings, and add automatic silence (or rather background noise) removal as a preprocessing.

Acknowledgments. This work was partially supported by the Research Center of PJAIT, supported by the Ministry of Science and Higher Education in Poland.

References

1. Alexandre, E., Cuadra, L., Salcedo-Sanz, S., Pastor-Sánchez, A., Casanova-Mateo, C.: Hybridizing extreme learning machines and genetic algorithms to select acoustic features in vehicle classification applications. Neurocomputing **152**, 58–68 (2015)
2. Averbuch, A., Rabin, N., Schclar, A., Zheludev, V.: Dimensionality reduction for detection of moving vehicles. Pattern Anal. Applic. **15**(1), 19–27 (2012)
3. Directive 2010/40/Eu of the European Parliament and of the Council of 7 July 2010 on the framework for the deployment of Intelligent Transport Systems in the field of road transport and for interfaces with other modes of transport. http://eur-lex.europa.eu/LexUriServ/LexUriServ.do?uri=OJ:L:2010:207:0001:0013:EN:PDF
4. Duarte, M.F., Hu, Y.H.: Vehicle classification in distributed sensor networks. J. Parallel Distrib. Comput. **64**, 826–838 (2004)
5. Machine Learning Group. http://www.cs.waikato.ac.nz/~ml/
6. Erb, S.: Classification of Vehicles Based on Acoustic Features. Graz University of Technology, Thesis (2007)
7. ITS 2015–2019 Strategic Plan. http://www.its.dot.gov/strategicplan.pdf
8. Iwao, K., Yamazaki, I.: A study on the mechanism of tire/road noise. JSAE Rev. **17**, 139–144 (1996)
9. Mayvan, A.D., Beheshti, S.A., Masoom, M.H.: Classification of vehicles based on audio signals using quadratic discriminant analysis and high energy feature vectors. Int. J. Soft Comput. **6**, 53–64 (2015)
10. Monahan, T.: "War Rooms" of the street: surveillance practices in transportation control centers. Commun. Rev. **10**, 367–389 (2007)
11. Silla Jr., C.N., Freitas, A.A.: A survey of hierarchical classification across different application domains. Data Min. Knowl. Disc. **22**(1–2), 31–72 (2011)
12. Sondhi, M.M.: New methods of pitch extraction. IEEE Trans. Audio Electroacoust. AU **16**(2), 262–266 (1968)
13. The Moving Picture Experts Group. http://mpeg.chiariglione.org/standards/mpeg-7
14. The R Foundation. http://www.r-project.org/

A Novel Information Fusion Approach for Supporting Shadow Detection in Dynamic Environments

Alfredo Cuzzocrea[1](✉), Enzo Mumolo[2], Alessandro Moro[3], Kazunori Umeda[3], and Gianni Vercelli[4]

[1] DIA Department, University of Trieste and ICAR-CNR, Trieste, Italy
`alfredo.cuzzocrea@dia.units.it`
[2] DIA Department, University of Trieste, Trieste, Italy
`mumolo@units.it`
[3] Chuo University, Tokyo, Japan
`moro@sensor.mech.chuo-u.ac.jp, umeda@mech.chuo-u.ac.jp`
[4] DIBRIS Department, University of Genova, Genova, Italy
`gianni.vercelli@unige.it`

Abstract. In this paper we present a system for detecting shadows in dynamic indoor and outdoor environment. The algorithm we propose fuses together color and stereo disparity information. Some considerations on the nature of the shadow improves the algorithm's ability to candidate the pixels as shadow or foreground. The candidate of both color and disparity information are then weighted by analyzing the effectiveness in the scene. The techniques employed allows separate computation and multithreading operations.

1 Introduction

In video surveillance scenario, a fundamental task for automatic systems is to detect moving objects. A commonly used technique for segmentation of moving objects is background subtraction. This technique detects moving regions (i.e. the foreground) in an image by subtracting the background image from the current image. The background is in general a statistical model which is updated over spatial and/or temporal conditions. An hard challenge is represented by the impact of the shadow in foreground segmentation. Shadows can belong to a specific category: *static shadows* or *dynamic shadows* . Static shadows occur due to static objects blocking the illumination from a light source. For example buildings and trees but also parked objects like bikes and cars. Dynamic shadows are due to moving objects (i.e. people, vehicles, etc.). Static shadows can be incorporated to the background model, whereas dynamic shadows are more problematic. The effect of dynamic shadows can be critical for the foreground segmentation, and cause objects to merge, distort their shape and size, create ghost regions.

In a generic environment, dynamic shadows can take any size or shape, can be both *umbra* (dark shadow) and *penumbra* (soft shadow) shadows. These two

F. Esposito et al. (Eds.): ISMIS 2015, LNAI 9384, pp. 353–363, 2015.
DOI: 10.1007/978-3-319-25252-0_38

type of shadows have different properties. Penumbra shadows have low intensity but similar chromacity values w.r.t. the background. Instead umbra shadows can exhibit different properties than the background, and their intensity values can be similar to any object that can appear in the scene. There is a third case. When the chromacity of umbra shadows differs from the chromaticity of the global background illumination, the shadow can be considered *chromatic shadow*. The result of this is that umbra shadows are notably more difficult to detect, and they are often detected as part of the moving objects. The contribution of this paper is to present a method which detect shadow combining the advantage of both color and stereo information. In this work we propose an algebraic method in order to solve the result's ambiguities.

2   Supporting Shadow Detection in Dynamic Scenarios via a Novel Information Fusion Approach

As described in Sect. 1, we initially segment each frame into background, foreground, and shadow. This is performed by combining the results from background subtraction process and shadow detection. The methods are described below. The advantage of these methods is that they do not require a training phase. Nevertheless they give generally good performances as it will be shown in Sect. 3.

2.1   Subtraction of Background

In this work we segment the indoor images using a background subtraction change detection algorithm based on thresholding. Background subtraction methods based on thresholding show that it is difficult to tune thresholds and parameters. In this work, problems of over-segmentation is not a negative aspect, since it is actually a aim effect. We tune the parameter to detect even small changes of the dynamic foreground. While in indoor environment with artificial light, a simple background subtraction give good image segmentation, for an outdoor environment where it is not possible to have an empty background and the light condition changes during the day, we opted for a kernel based algorithm as that described in [9]. For each pixel p, a background model is learned by the nonparametric KDE method in the RGB color space [11], from which the foreground probability can be estimated. Potential moving objects can be extracted by simply thresholding this density distribution, and within the segmentation the cast shadows can be evaluate over both the color and stereo domain.

2.2   Detection of Color Shadow

The method we are going to describe is a modification of the works described in [16] and [14]. The shadow is identified estimating the differences of color and textures between the background image and segmented foreground. Results are fused by exploiting temporal consistency between frames. Shadow, in general conditions, has a direction. Suppose that I(x, y) is the intensity of the pixel located in (x, y), E(x, y) is the direct light and $\rho(x, y)$ the reflected light from a 3D point and projected to the point (x, y). Simplifying the models it is possible to consider that an occlusion as shadow casting reduces the radiance but not the diffuse reflectance. If we consider all the pixels of the image, $\forall_x, y \exists w, \forall_i, j | i \in [x-1, x+1], j \in [y-1, y+1], i \neq x, j \neq x$ the intensity ratio will be:

$$\frac{I_{x,y}}{I_{i,j}} = \frac{I_{x,y}}{I_{i,j}} \cdot sin\left(\frac{(\varphi - \mu)}{2}\right) + \frac{\rho_{x,y}}{\rho_{i,j}} \tag{1}$$

where the angle of the shadow is ρ respect to the Cartesian axes, and μ is the angle of research of the shadow. Unknowing a priori the source light physical properties, and because it may change during time, we look for the intensity ratio which minimizes the variation of intensity. That is, given a pixel (x, y), we have:

$$d_{x,y} = min\left(|\ln \Delta I|\right) \tag{2}$$

The variations in background and segmented image will be used to calculate the error score within a small region, used for discriminating a pixel as shadow. The error score is computed as reported in Eq. (3).

$$\Psi(x, y) = \sum_{c \in R, G, B} \sum_{i,j \in \omega(x,y)} \left| d_c(i, j) - d_c'(i, j) \right| \tag{3}$$

Even if colors cannot be used singularly to extract shadows, they represent an important source of information. The color difference which is used to estimate if a pixel belongs to a cone shadow is calculated comparing the color information between the background and detected foreground image. The color space is modeled as follows:

$$\begin{cases} C_1(x, y) = \arctan\left(\frac{I_r(x,y)}{I_b(x,y)}\right) \\ C_2(x, y) = \arctan\left(\frac{I_g(x,y)}{I_b(x,y)}\right) \end{cases} \tag{4}$$

The color score error due to the variation of color is computed as:

$$\Lambda(x, y) = \left| C_1(x, y) - C_1'(x, y) + C_2(x, y) - C_2'(x, y) \right| \tag{5}$$

The shadow value which is used to estimate if a pixel is shadow or non shadow, is calculated as follows:

$$\Theta_{(t,x,y)} = \begin{cases} \alpha\Psi_{(x,y)} + \beta\Lambda_{(x,y)} + \\ (1 - \alpha - \beta) \cdot \Theta_{(t-1,x,y)} & if \frac{I_{x,y}}{\eta} < I'x, y \\ \infty & otherwise. \end{cases} \tag{6}$$

An high shadow value means that a pixel is probably an object. On the contrary, a low value means shadow. A thresholding is applied to the shadow value in order to estimate whether a pixel is shadow or not. The preference to use global rather than local threshold and parameters is due to the following reason. The use of local threshold and parameters are influenced by the number of people moving in the scene. If the dynamicity in a region is low, the probability to detect shadow become low. In order to be flexible and adaptable, the parameters are adjusted on-line. In the error weight map of cast shadows, if we deal with each pixel independently, the segmentation results may contain many sparse values and generate small pieces. We applied a smoothness cost term inspired by MRF energy function [12], where energy is computed as follows:

$$E(f) = \sum_{p \in P} D_p(f_p) + \sum_{p,q \in N} V_{p,q}(f_p, f_q) \tag{7}$$

where $E(f)$ is the energy of a particular parameter f, p and q are indexes over the pixels, $D_p(f_p)$ is the data cost of assigning the pth pixel to label f_p, and $V_{p,q}(f_p, f_q)$ represents the smoothness cost of assigning pixels p and q in a neighborhood N to respective labels f_p and f_q. In this work, the data cost assigning is set as $-k_1 log(error)$. The smoothness cost term is defined as:

$$V_{p,q} = (f_p - f_q)^2 \, e^{-\beta |I_p - I_q|} \tag{8}$$

where I_p and I_q denote gray-scale intensity of pixels p and q, and β is a constant. The obtained value is associated to the current error parameters. These values are floating during the execution. In order to have a stable value, we perform a correction applying an average calculation as follows:

$$E_t(f) = E_{t-1}(f) \frac{s}{s+1} + \frac{E_t(f)}{size} \frac{1}{s+1} \tag{9}$$

Once the shadow value is estimated by the equation Eq. (6), in order to estimate if a pixel is shadow or no shadow, it is necessary to determine a threshold value. The threshold may vary during the execution. We propose a method which adjusts the threshold value adaptively with time. If μ and σ are the average and variance of all the Θ:

$$T_t = T_{t-1} \frac{s}{s+1} + (\mu - \sigma) \frac{1}{s+1} \tag{10}$$

2.3   Detection of Stereo Shadow

As previously mentioned, in ideal condition, it would be possible to separate an object from the shadow by the estimated distance. Once obtained the distance value of each point of the image from the background, a pixel is considered object if the difference is higher than zero (ideal condition), or ϵ (real condition). Because the calculation of distance generally is problematic in saturation regions, only the points with an intensity less than a threshold are used. Stereo cameras have a range within the error measures calculated, which is lower than the measured distance. The working range changes from camera to camera and it depends on several factors, software (matching algorithms) and hardware (sensors, camera displacement and numbers, focal length). We took in consideration the Bumblebee2 stereo camera and estimated the errors due to the range measuring the real distance and estimated distance. If the measure is taken after the best distance, the error shows a quadratic behaviour. Otherwise it has a quasi-linear behaviour. If the distance information has been calculated both for background and foreground and it is within the best range, then if $d_{x,y} \neq \emptyset \wedge d'_{x,y} \neq \emptyset \wedge d_{x,y} \leq \epsilon$:

$$S_{x,y} = \begin{cases} 1 \text{ if } \left| d_{x,y} - d'_{x,y} \right| \leq m\,(d_{x,y} - \epsilon) + q \\ 0 \text{ otherwise.} \end{cases} \tag{11}$$

If the distance information has been calculated both for background and foreground but it is over the best range, then if $d_{x,y} \neq \emptyset \wedge d'_{x,y} \neq \emptyset \wedge d_{x,y} > \epsilon$:

$$S_{x,y} = \begin{cases} 1 \text{ if } \left| d_{x,y} - d'_{x,y} \right| \leq \frac{1}{m}\,(d_{x,y} - \epsilon)^2 + q \\ 0 \text{ otherwise.} \end{cases} \tag{12}$$

From the hypothesis that the camera is able to estimate a correct distance within a certain range r, we propose a method to automatically adjust m and q. Given the distance difference between background and current images, it is reasonable to suppose that within a certain confidence range r', the pixels have to be labeled as shadow. Instead out of r', pixels are expected to belong to a moving object. A set of points measured around r are collected and labeled based on the previous consideration. If a pixel is recognized as *shadow* and the distance difference is lower than r' then the point is marked as *success* otherwise as *failure*. If a pixel is recognized as *no shadow* and the distance difference is equal or greater than r', then the point is marked as *success* otherwise as *failure*. Once obtained the graph, we modify the parameters only if the percentage of success inside and outside the range r' is lower then an accuracy value. At each iteration, the parameters are adjusted using a random function as follow:

$$m = m + sign \cdot rand \tag{13}$$

$$q = q + sign \cdot \frac{rand}{100} \tag{14}$$

where *sign* is positive if the percentage of failures inside the range is higher or equal to the percentage of failures outside. Negative in opposite case.

2.4   Modeling and Analyzing Colors of Environments

In order to estimate the diffuse chrominance of the scene, we propose a method which analyzes the variation in a sequence of images. If I_c is the intensity of the current frame and I'_c is the intensity of a given background image, the diffuse chrominance is considered the highest variation in a sequence. c can assume red, green, or blue value, and *gray* is equal to gray value. We consider that each channel has 8 bit resolution. Even if it is obvious that colored objects influence the estimation of the color, considering all the points of the image in a sequence will reduce that effect. Given the background image and the current image, the histogram of the differences is computed as:

$$\forall_p, H_{c,|I_p-I'_p|} = H_{c,|I_p-I'_p|} + 1 \tag{15}$$

The color value is then:

$$cv_c = \frac{\sum_{i=0}^{256} H_{c,i} \cdot i}{s} \tag{16}$$

The difference in intensity between the gray-scale image in each channel is computed as reported in Eq. (17):

$$DI_c = \frac{1 - |cv_c - cv_{gray}|}{256} \tag{17}$$

and the proportional variation of each channel respect the grayscale image is computed as:

$$PVI_c = \begin{cases} -\left(1 - min\left(\frac{cv_c}{cv_{gray}}, \frac{cv_{gray}}{cv_c}\right)\right) & \text{if } cv_c > cv_{gray} \\ 1 - min\left(\frac{cv_c}{cv_{gray}}, \frac{cv_{gray}}{cv_c}\right) & otherwise. \end{cases} \tag{18}$$

Differences and proportional variations are gathered for the number of frames necessary to estimate the average and variance of the sequence analyzed. Empirically we estimated that 100 frames are sufficient for that analysis. First the DI are normalize respect the maximum and minimum value of all the color DI:

$$DI_c = \frac{DI_c - min(DI)}{(max(DI) - min(DI))} \tag{19}$$

We compute the average and variance of the differences and proportional variations. The sequence average difference and variation are estimated as shown in Eqs. (20) and (21):

$$SAD_c = \ln\left(\frac{\mu DI_c}{1 - \delta^2 DI_c}\right) \tag{20}$$

$$SAPV_c = \left|\ln\left(\frac{\mu PVI_c}{1 - \delta^2 PVI_c}\right)\right| \tag{21}$$

For each combination of colors the difference of sequence average difference and variation is computed in order to estimate the prevailing color:

$$\begin{aligned} \Delta_{rg} &= SAD_r - SAD_g \\ \Delta_{rb} &= SAD_r - SAD_b \\ \Delta_{gb} &= SAD_g - SAD_b \end{aligned} \tag{22}$$

and finally the color strenght is estimated as described in Eq. (23).

$$CS_r = (\Delta_{rg} >= 0) \cdot |\Delta_{rg}| + (\Delta_{rb} >= 0) \cdot |\Delta_{rb}|$$
$$CS_g = (\Delta_{rg} < 0) \cdot |\Delta_{rg}| + (\Delta_{gb} >= 0) \cdot |\Delta_{gb}| \qquad (23)$$
$$CS_b = (\Delta_{rb} < 0) \cdot |\Delta_{rb}| + (\Delta_{gb} < 0) \cdot |\Delta_{gb}|$$

The suggested best chrominance will be the highest value. Obviously, one color will prevails, unless the image is not completely gray-scale. Because we want to avoid errors due to noisy or particular configurations, we consider colored diffuse light only the best color value which satisfies the following equation:

$$color = \begin{cases} white \ if \ CS_{best} < min \ (SAPV) \\ best \ otherwise. \end{cases} \qquad (24)$$

We consider that a pixel segmented as foreground cannot be a shadowed pixel if its intensity is higher than background. Thus, a pixel is candidate as shadow if:

$$sp_a = \left(I_a^R < \mu^R\right) \wedge \left(I_a^G < \mu^G\right) \wedge \left(I_a^B < \mu^G\right) \qquad (25)$$

Moreover, in the case of bluish effect (similar to greenish and reddish), the changes on the intensity component of the red and blue channels are bigger than the blue channel. To be more flexible, the increment will be proportional. This fact can be used to reduce the shadow region as follows. If bluish:

$$cs_a = (k \left(I_a^R - \mu^R\right) > \left(I_a^B - \mu^B\right) \wedge$$
$$k \left(I_a^G - \mu^G\right) > \left(I_a^B - \mu^B\right)) \wedge sp_a \qquad (26)$$

If reddish:

$$cs_a = (k \left(I_a^G - \mu^G\right) > \left(I_a^R - \mu^R\right) \wedge$$
$$k \left(I_a^B - \mu^B\right) > \left(I_a^R - \mu^R\right)) \wedge sp_a \qquad (27)$$

If greenish:

$$cs_a = (k \left(I_a^R - \mu^R\right) > \left(I_a^G - \mu^G\right) \wedge$$
$$k \left(I_a^B - \mu^B\right) > \left(I_a^G - \mu^G\right)) \wedge sp_a \qquad (28)$$

The mask so obtained is then used to mark the pixels that should be analyzed. If the color of the environment is not white the pixels which are not candidate to be shadow are labeled as moving objects.

2.5   Information Fusion Method

Once the shadow values are obtained with the previously described methods, the difficult task is to find a relationship between result obtained from color information and shadow information. Because the measures are not directly in relation to each other, we estimate the strength of each detection method. The shadow detection method described in Sect. 2.1 returns a value we called shadow parameter. It is possible to combine the shadow parameter with the distance in order to obtain a *confidence* value. Likewise, it is possible to obtain a confidence value from the stereo shadow detection. The confidence values are higher for the stereo information if the distance of a point from the camera, is near the focal point. On the opposite side, confidence value is higher for the points which have a distance lower or higher than the focal point. The curve that define the probability to be shadow is not centered around zero, but differs if a pixel is detected as shadow or no shadow. The equations can be so resumed. If the distance from the camera is less or equal to the focal point:

$$Wc_{x,y} = \begin{cases} pShadow \cdot \left(\frac{d}{m} + q\right) \text{ if } shadow \\ \frac{Fp}{pShadow} \cdot \left(\frac{d}{m} + q\right) \text{ otherwise.} \end{cases} \quad (29)$$

$$Ws_{x,y} = \begin{cases} \frac{1}{\Delta_d} \cdot \frac{1}{\frac{d}{m}+q} \text{ if } shadow \\ 2d \cdot \frac{1}{\frac{d}{m}+q} \text{ otherwise.} \end{cases} \quad (30)$$

However, if the distance from the camera increases:

$$Wc_{x,y} = \begin{cases} pShadow \cdot \left(\frac{1}{m} \cdot (d - Fp)^2 + q\right) \text{ if } shadow \\ \frac{Fp}{pShadow} \cdot \left(\frac{1}{m} \cdot (d - Fp)^2 + q\right) \text{ otherwise.} \end{cases} \quad (31)$$

$$Ws_{x,y} = \begin{cases} \frac{1}{\Delta_d} \cdot \frac{1}{\frac{1}{m}(d-Fp)^2+q} \text{ if } shadow \\ 2\Delta_d \cdot \frac{1}{\frac{1}{m}(d-Fp)^2+q} \text{ otherwise.} \end{cases} \quad (32)$$

Stereo information may be not always available in all the points of the image. This is due to several factors: light conditions, distance of the object from the camera, visibility of an object from both the cameras. If the stereo information is not available, we choose to use only the color information. In the case a pixel is labeled as shadow (or no shadow) with both the methods, the pixel will be labeled with the detected value. If the detection value is different, the pixel will be labeled with the value of the method which have higher weight.

3   Experimental Results and Analysis

The setup of this work is centered around a commercial stereo rig from Point Grey Research,[1]. The Bumblebee2 stereo camera delivers rectified stereo images pairs via FireWire at up to 40 frames per second, depending on resolution. In this work we operate with a 320×240 resolution for the rectified images, obtained from the 640×480 raw images. We tested the performance of the program with a Intel Quad CPU at 2.83 GHz and 4 GB of ram. Since we did not find a suitable video sequence for testing the algorithm, we acquired and hand-labeled several video sequences using our 3D camera. These sequences have been taken in different condition of illumination and environment and camera orientation.

---

[1] http://www.ptgrey.com/products/bumblebee/index.html.

In particular in order to evaluate the shadow detection performance, we analyzed four different light conditions: artificial light, only diffuse light, partially sunlight, and strong sunlight.

The proposed algorithm has been compared with [13] which has a similar approach. For a quantitative evaluation, we calculate the accuracy of the cast shadow detection by using two metrics proposed in [15]. The *shadow detection rate* η measures the percentage of correctly labeled shadow pixels among all detected ones, while the *shadow discrimination rate* ξ measures the discriminative power between foregrounds and shadows:

$$\eta = \frac{TP_S}{TP_S + FN_S} \tag{33}$$

$$\xi = \frac{\overline{TP_F}}{TP_F + FN_F} \tag{34}$$

where: (*i*) TP_F is the foreground pixels correctly detected; (*ii*) $\overline{TP_F}$ is the ground-truth pixels which belongs to the foreground minus the pixels which belong to the shadow; (*iii*) FN_F is the foreground pixels detected as shadow; (*iv*) TP_S is the shadow pixels correctly detected; (*v*) FN_S is the shadow pixels detected as foreground.

Table 1. The table contains the comparison of the proposed method with the method proposed in [13]

Sequence	Outdoor NS		Outdoor LIS		Outdoor SIS	
	η	ξ	η	ξ	η	ξ
Proposed	0.9047	0.896	0.787	0.849	0.884	0.897
[13]	0.916	0.721	0.856	0.704	0.707	0.642

The quantitative comparison with proposed and other approaches are given in Table 1. In this table shows that the shadow discrimination rate is always better than Madsen et al. for every shadow conditions. Shadow detection rate is better than Madsen et al. only for strong intensity shadow. In the other case it is slightly lower. In Table 2 we report the same two measure for indoor environments. However, it is not possible to compare these results with Madsen et al..

Table 2. Performance of the proposed algorithm in indoor environment

Measures	η	ξ
Proposed	0.818	0.886

4   Conclusions and Future Work

In this paper, a new cast shadow detection algorithm that requires a stereo camera is proposed, which exploits the color information, texture and temporal information, and depth information. The results show that the proposed method can work well in both indoor and outdoor scenes. Adaptive parameters gave flexibility of the system. The performances in outdoor environment have been compared with the only comparable work we found, and it shows better result especially in strong shadow condition. In the future, we plan to compare the algorithm performance in indoor environment and to analyze the reason why shadow detection rate is mostly valid for strong shadows. The application of a modified version of this algorithm within medical surgery is also possible (e.g., [10]). Also, our proposed framework by means of several characteristics, as to enhance it significantly, is under study. For instance, some interesting properties to be investigated in future are: *fragmentation* (e.g., [1,3]), *approximation* (e.g., [4,5]), *privacy preservation* (e.g., [6,7]), *big data* (e.g., [2,8]).

References

1. Bonifati, A., Cuzzocrea, A.: Efficient fragmentation of large XML documents. In: Wagner, R., Revell, N., Pernul, G. (eds.) DEXA 2007. LNCS, vol. 4653, pp. 539–550. Springer, Heidelberg (2007)
2. Cuzzocrea, A., Bellatreche, L., Song, I.-Y.: Data warehousing and OLAP over big data: current challenges and future research directions. In: Proceedings of DOLAP 2013, pp. 67–70 (2013)
3. Cuzzocrea, A., Darmont, J., Mahboubi, H.: Fragmenting very large XML data warehouses via k-means clustering algorithm. IJBIDM 4(3/4), 301–328 (2009)
4. Cuzzocrea, A., Furfaro, F., Greco, S., Masciari, E., Mazzeo, G.M., Saccà, D.: A distributed system for answering range queries on sensor network data. In: Proceedings of 3rd IEEE PerCom 2005 Workshops, pp. 369–373 (2005)
5. Cuzzocrea, A., Matrangolo, U.: Analytical synopses for approximate query answering in OLAP environments. In: Galindo, F., Takizawa, M., Traunmüller, R. (eds.) DEXA 2004. LNCS, vol. 3180, pp. 359–370. Springer, Heidelberg (2004)
6. Cuzzocrea, A., Russo, V., Saccà, D.: A robust sampling-based framework for privacy preserving OLAP. In: Song, I.-Y., Eder, J., Nguyen, T.M. (eds.) DaWaK 2008. LNCS, vol. 5182, pp. 97–114. Springer, Heidelberg (2008)
7. Cuzzocrea, A., Saccà, D.: Balancing accuracy and privacy of OLAP aggregations on data cubes. In: Proceedings of DOLAP 2010, pp. 93–98 (2010)
8. Cuzzocrea, A., Saccà, D., Ullman, J.D.: Big data: a research agenda. In: Proceedings of IDEAS 2013, pp. 198–203 (2013)
9. Elgammal, A., Duraiswami, R., Harwood, D., Davis, L.: Background and foreground modeling using nonparametric kernel density estimation for visual surveillance. Proceedings of the IEEE 90(7), 1151–1163 (2002)
10. Gaudina, M., Zappi, V., Bellanti, E., Vercelli, G.: elaparo4d: A step towards a physical training space for virtual video laparoscopic surgery. In: Proceedings of CISIS 2013, pp. 611–616 (2013)
11. Lee, J., Park, M.: An adaptive background subtraction method based on kernel density estimation. In: Sensors, pp. 12279–12300 (2012)

12. Li, S.Z.: Markov random field modeling in image analysis. In: Advances in Computer Vision and Pattern Recognition. Springer Publishing Company (2009)
13. Madsen, C.B., Moeslund, T.B., Pal, A., Balasubramanian, S.: Shadow detection in dynamic scenes using dense stereo information and an outdoor illumination model. In: Kolb, A., Koch, R. (eds.) Dyn3D 2009. LNCS, vol. 5742, pp. 110–125. Springer, Heidelberg (2009)
14. Moro, A., Terabayashi, K., Umeda, K., Mumolo, E.: Auto-adaptive threshold and shadow detection approaches for pedestrian detection. In: Proceedings of Ws. on Sensing and Visualization of CHI, pp. 9–12 (2009)
15. Prati, A., Mikic, I., Trivedi, M., Cucchiara, R.: Detecting moving shadows: algorithms and evaluation. IEEE Trans. on Patt. Anal. Mach. Intell. **25**(7), 918–923 (2003)
16. Yang, M.T., Lo, K.H., Chiang, C.C., Tai, W.K.: Moving cast shadow detection by exploiting multiple cues. Image Process. **2**, 95–104 (2008)

Extending SKOS: A Wikipedia-Based Unified Annotation Model for Creating Interoperable Domain Ontologies

Elshaimaa Ali and Vijay Raghavan[✉]

The Center of Advanced Computer Studies, University of Louisiana at Lafayette, Lafayette, USA
{eea7236,vijay}@cacs.louisiana.edu

Abstract. Interoperability of annotations in different domains is an essential demand to facilitate the interchange of data between semantic applications. Foundational ontologies, such as SKOS (Simple Knowledge Organization System), play an important role in creating an interoperable layer for annotation. We are proposing a multi-layer ontology schema, named SKOS-Wiki, which extends SKOS to create an annotation model and relies on the semantic structure of the Wikipedia. We also inherit the DBpedia definition of named entities. The main goal of our proposed extension is to fill the semantic gaps between these models to create a unified annotation schema.

Keywords: Semantic schema · SKOS · Wikipedia · Annotation modeling

1 Introduction

Semantic annotation is the process of adding formal metadata to web pages. This metadata links instances in a web page to defined concepts (classes) in an ontology. The main challenge in the annotation process is the creation of domain ontologies.

As stated in [1] web annotations have to be: explicit, formal and unambiguous. Hence the challenge of semantic annotation in a nutshell is due to the lack of standardized annotation schema, and the inconsistencies and the differences in the definition of domain concepts and relations. Formal annotation requires an annotation schema that will define the specification of the annotation elements to ensure semantic interoperability. "Interoperability is concerned with ensuring that the precise meaning of the exchanged information is understandable by any other application that was not initially developed for this purpose" [2].

In this paper, we present a semantic annotation schema that is adapted from the Wikipedia semantic network. This schema represents a template that facilitates the learning of lightweight ontologies by extracting corresponding knowledge through the mining of semantics from the Wikipedia, which is main interest of our ongoing research. We represent the Wikipedia annotation schema –named SKOS-Wiki- as an extension of the SKOS – The Simple Knowledge Organization System- to accommodate different entity types and still be generic enough for use in a wide range of domains and support interoperability.

© Springer International Publishing Switzerland 2015
F. Esposito et al. (Eds.): ISMIS 2015, LNAI 9384, pp. 364–370, 2015.
DOI: 10.1007/978-3-319-25252-0_39

The next section briefly discuss the characteristics of modeling lightweight ontologies Then, we briefly present the structure of SKOS as an upper ontology. Then, the proposed annotation schema is presented, followed by a discussion and future work.

2   Dimensions of Semantic Modeling

Major dimensions to consider when modeling information for building a domain ontology, in general, and annotation ontology specifically are expressivity, granularity and interoperability.

Expressivity: Expressivity of a model, is the amount of semantics we decide to embed in an ontology when modeling a given domain. Expressivity controls the level of semantic details to define for a concept, including properties, relations, logical assertions and axioms [3]. Obrst explained the expressivity by the semantic spectrum as in Fig. 1.

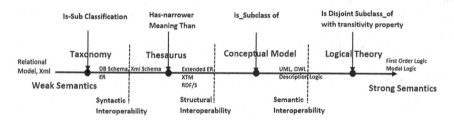

Fig. 1. The Ontology Spectrum [3]

Lightweight ontologies are more expressive than a conceptual model and less than logical theory. Lightweight ontologies would then typically consist of a hierarchy of concepts and a set of relations between those concepts. Conversely, heavyweight ontologies add cardinality constraints, stand-alone axioms, reified statements (a statement about a statement) and more. This will raise the capabilities of the semantic inference and the usage of the ontology, but will definitely influence the computability of the inference and reasoning process negatively.

Granularity: Semantic granularity addresses the different levels of specifications of an entity in the real world [4]. Major granularity levels, as presented in [3] are upper level ontology, mid-level ontology and domain ontology. Upper and mid-level ontologies are sometimes referred to as the foundational ontology. Upper ontology is a generic ontology describing general knowledge with concepts that are shared with most domains, such as time and space. A domain ontology, describes a domain, such as the medical domain or narrower domain, such as the gene ontology. Ontology mapping becomes easier if domain ontologies are derived from the same upper ontology.

Interoperability Semantic Interoperability is the ability of a system to exchange data, with other, new applications, that was not initially developed for use by such applications, while ensuring that the intended meaning is conveyed unambiguously and consistently.

Lack of interoperability is due to heterogeneity conflicts, explained, in detail, in [5,6]. We are concerned with semantic conflicts that should be resolved on the data level and the schema level. The resolution of such conflicts requires the understanding of the domain; this is where ontologies come into the picture. A multi-level ontology standardizes the definition of datatypes, on the upper level, named-entity properties and concept properties on the mid-level, and domain specification on the lower level.

3   Ontology-Based Annotation Scheme

Most ontology-based annotation efforts were domain-specific relying on an expert developed ontology to be used for annotating domain concepts. Some were instances of a generic upper ontology, such as SKOS; but, most were not related to any common structure or upper design. We, therefore, focus in this study on generic ontologies that can be extended and instantiated in various domains.

SKOS – Simple Knowledge Organization System [7].
SKOS allows for web definition of terms, taxonomies and thesauri for referencing and reuse on the semantic web. The history of the development of SKOS from 1990 to 2009 can be found in [8]. The SKOS is a lightweight upper ontology that can be instantiated in various domains. It follows the principle of making minimal ontological commitment as stated by Thomas Gruber [9]. The SKOS data model and the design choices made in modeling those components is explained, in detail, in [8]. The main entities in this schema are, "Concept", "ConceptScheme", "Collection" and "OrderedCollection". Relations and properties are designed to satisfy the specification of taxonomies and thesauri. Relations are either semantic hierarchical relations, such as "broader","narrower" and"related," or mapping relations to define synonyms or equivalent concepts, such as "broadMatch", "narrowMatch" and "closeMatch". Properties are either labeling properties or documentation properties. Concepts can be identified with URIs, labeled with lexical strings assigned notations, documented with various notes, linked to other concept schemes, grouped and labeled into a collection, and mapped to concepts in other schemes. We are proposing a Wikipedia-based semantic annotation schema, as the mid-level ontology that instantiates the SKOS and extends the semantic capabilities far more than thesauri and taxonomies.

DBpedia- DBpedia is a community effort to extract knowledge from the Wikipedia and make it available in the form of RDF triples [10]. DBpedia represents an RDF repository for the infoboxes in Wikipedia pages. Infoboxes are like fact sheets for named-entities. A DBpedia ontology was then constructed to detect inconsistencies in the definitions of similar named-entities [11].

In this paper, we enable the linking of named-entities within the proposed Wikipedia annotation schema to their corresponding definitions in DBpedia.

4   SKOS-Wiki: Wikipedia-Based Semantic Annotation Model

We propose an annotation model, named SKOS-Wiki, which is based on the structure of the Wikipedia knowledge to build a template to be used for extracting lightweight ontologies that can work with most general domains. We chose to extend the SKOS, as it played an important role to enhance ontology engineering interoperability for linked data applications [12]. We extend the upper level ontology SKOS to build a mid-level ontology that can, then, be instantiated to create domain ontologies by using the knowledge in the Wikipedia. By extending SKOS, we follow a top-down approach for ontology development, which it is believed will speed up the learning of the domain ontologies and enhance their quality and interoperability. Basically, we use the structure of the Wikipedia network and the template within the Wikipedia pages to define different types of concepts for building an annotation model.

4.1 Wikipedia-Based Annotation Model

The proposed schema models the concepts that are represented in the Wikipedia categories or articles. A schematic description of the main entities is shown in Fig. 2.

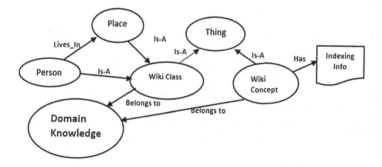

Fig. 2. Wikipedia-based annotation schema

In this model, we divide the concepts that are covered in Wikipedia articles as either a Wiki-Concept or Wiki-Class. A Wiki-Concept is an abstract concept that cannot be instantiated or enumerated but can be used for categorization such as the concept "Engineering", "politics", "computer science". A Wiki-Class is a type that can be instantiated with an Is-A relation e.g. person, place organization and event, or any class that can be listed or enumerated. Introducing this distinction makes it easier to associate appropriate relations and properties to an entity depending on the type of the entity.

We consider the Domain Knowledge as the major category that describes the entities (concepts and classes) of the annotation schema. Both Wiki-concepts and Wiki-classes are linked to the category Domain Knowledge by the property Belongs-to. A Wiki-Concept or a Wiki-Class can belong to multiple domain knowledge depending on the context. This aspect allows for faceted browsing of concepts. Any relation can be defined between two wiki concepts, two wiki classes, or a concept and a class or an instance of a class. In OWL there are primitive built-in relations, such as the owl:subclass relation, and rdf:typeOf relation.

In this schema, we define some of the basic relations that will fit for all domains as Lives-in between a person and place, Has-location between event and a place, etc.

More relations are mentioned in the OWL definition of the schema. The entity Index-ingInfo is an entity that contains indexing information that best defines a Wiki-Concept. TheWiki schema also includes a set of descriptors that are designed to define Wiki-classes and Wiki-concepts, which are Location-descriptor, Person-descriptor, Event-descriptor, Organization-descriptor and Index-descriptor. The Index-descriptor includes important information such as synonyms which are terms that have the same meaning, keywords that represent common search terms for this concept, and ambiguous terms or homonym terms. Index-descriptor also includes highly related terms, which are terms that are frequently mentioned together but the relation cannot be easily extracted such as "Global warming" and "Greenhouse effect". A descriptor acts as an annotation template that includes properties/slots that can define a Wiki-Class, or a Wiki-Concept based on the Wikipedia definition for this entity. Descriptors are inherited from the DBpedia ontology. The DBpedia ontology is a community-provides mappings that help normalize the variations in the properties and classes defined in the Wikipedia as Info-boxes [11].

4.2 Extending SKOS: The Multi-level Wikipedia Semantic Model

In order to build a multi-level semantic model, we extend the SKOS ontology. A multi-level ontology enhances interoperability by resolving heterogeneity. The previously defined Wikipedia schema (SKOS-Wiki) acts as the mediator between SKOS as an upper ontology, Wikipedia definition of concepts and the DBpedia definition of named-entities. Figure 3 illustrates the multi-layered definition of the proposed semantic model.

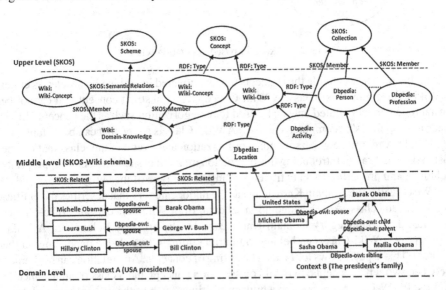

Fig. 3. A general structure of a multi-layered semantic annotation model

As mentioned earlier, SKOS is a standard way to describe thesauri and other sets of terms for the semantic web. We extend the SKOS definition of a "SKOS:Concept" to include "Wiki-Concept" and "Wiki-Class", i.e. Wiki_Class and Wiki_Concepts are subclass of Concept. The multi-layered model enables different levels of abstraction for domain annotations. The middle level SKOS-Wiki schema exploits the definition of concepts and classes according to the Wikipedia structure, to achieve a mapping between SKOS schema and DBpedia definitions of named entities. The domain level layer presents a samples (snippets) of two extracted domain ontologies, where the same entities are shared between both ontologies. For example, the entity Barak Obama mentioned in both ontologies with different contexts, in context A as one of the US presidents and in context B to show the president's family.

There are two levels of semantics here, the semantics of the named-entities or the concepts, -which is defined with the descriptors- and the semantics of the context. The context represents the domain in which certain terms and concepts occur together. Wiki: Domain-knowledge which is extended from SKOS: schema allows for resolving the context conflicts by defining the concepts and entities related to a specific entity in each domain. Different contexts of a concept can be extracted using the Wikipedia link structure, which is one of the key advantages of using Wikipedia for domain ontology constructions.

5   Summary

The goal of an annotation schema is to accelerate the construction and learning of domain ontology, and establish a common vocabulary to ensure interoperability and ease the process of ontology mapping. In view of these goals, we developed a unified annotation schema that will be generic enough to accommodate various domains. This schema represents the backbone information structure of a framework for Wikipedia-based, semi-automatic information extraction and construction of a domain ontology.

In order to accommodate more specific domain, in future work, we plan to model an entity that accommodate relations extracted with text mining and natural language processing tools through the process of ontology construction. For example, in Fig. 3, relations, such as president_of and Ex_president_of, are not defined in any of the three levels. Unfortunately, we do not have any vocabulary that can be instantiated for these relations, other than SKOS:related, or SKOS:Semantic_realtions developed for the hierarchical type of relations.

References

1. Ding, Y., Embley, D.W.: Using Data-Extraction Ontologies to Foster Automating Semantic Annotation. In: ICDE (2006)
2. Chen, D., Doumeingts, G.: European initiatives to develop interoperability of enterprise applications basic concepts, framework and roadmap. Ann. Rev. Control **27**(2), 153–162 (2003)
3. Obrst, L.: Ontological Architectures. In Theory and Applications of Ontology: Computer Applications. Springer, Heidelberg (2010)
4. Fonseca, F., et al.: Semantic granularity in ontology-driven geographic information systems. Ann. Math. Artif. Intell. **36**(1–2), 121–151 (2002)

5. Park, J., Ram, S.: Information systems interoperability: what lies beneath? ACM Trans. Inf. Sys. (TOIS) **22**(4), 595–632 (2004)
6. Sheth, A.P.: Changing focus on interoperability in information systems:from system, syntax, structure to semantics. In: Včkovski, A., Brassel, K.E., Schek, H.-J. (eds.) INTEROP 1999. LNCS, vol. 1580. Springer, Heidelberg (1999)
7. Miles, A., Bechhofer, S.: SKOS simple knowledge organization system reference. In: W3C Recommendation, p. 18 (2009)
8. Baker, T., et al.: Key choices in the design of simple knowledge organization system (SKOS). Web Semant. Sci. Serv. Agents World Wide Web **20**, 35–49 (2013)
9. Gruber, T.R.: Toward principles for the design of ontologies used for knowledge sharing? Int. J. Hum Comput Stud. **43**(5), 907–928 (1995)
10. Auer, S., Bizer, C., Kobilarov, G., Lehmann, J., Cyganiak, R., Ives, Z.G.: DBpedia: A nucleus for a web of open data. In: Aberer, K., et al. (eds.) ASWC 2007 and ISWC 2007. LNCS, vol. 4825, pp. 722–735. Springer, Heidelberg (2007)
11. Töpper, G., Knuth, M., Sack, H.: Dbpedia ontology enrichment for inconsistency detection. In: Proceedings of the 8th International Conference on Semantic Systems, ACM (2012)
12. Corcho, O., Poveda-Villalón, M., Gómez-Pérez, A.: Ontology engineering in the era of linked data. Bull. Am. Soc. Inf. Sci. Technol. **41**(4), 13–17 (2015)

Toward Real-Time Multi-criteria Decision Making for Bus Service Reliability Optimization

Vu The Tran[1(✉)], Peter Eklund[2], and Chris Cook[3]

[1] Danang University of Science and Technology, Da Nang, Vietnam
ttvu@dut.udn.vn
[2] IT University of Copenhagen, Copenhagen, Denmark
petw@itu.dk
[3] University of Wollongong, Wollongong, Australia
ccook@uow.edu.au

Abstract. This paper addresses issues associated with the real-time control of public transit operations to minimize passenger wait time: namely vehicle headway, maintenance of passenger comfort, and reducing the impact of control strategies. The randomness of passenger arrivals at bus stops and external factors (such as traffic congestion and bad weather) in high frequency transit operations often cause irregular headway that can result in decreased service reliability. The approach proposed in this paper, which has the capability of handling the uncertainty of transit operations based on Multi-objective evolutionary algorithm using a dynamic Bayesian network, applies preventive strategies to forestall bus unreliability and, where unreliability is evident, restore reliability using corrective strategies. "Holding", "expressing", "short-turning" and "deadheading" are the corrective strategies considered in this paper.

Keywords: Bayesian network · Multi-objective optimization · public transit · Transit service reliability · Transit modeling and simulation · Control strategies

1 Introduction

Measuring and reducing unreliability in a bus service is the focus of this paper. Unreliability affects passengers because it causes them to wait longer. Particularly on high frequency bus routes headway regularity is important to passengers because of its impact on waiting time and overcrowding. Overcrowding is a key to passengers because it impacts their comfort in a direct way and headway irregularity compounds operations because it slows boarding and alighting [5].

Passenger numbers are also important in transport planning because this measures network efficiency. For transit services with short headways, passengers can be assumed to arrive (more or less) randomly, namely independently of the schedule. Headway variability causes passengers to perceive that a service is unreliable, especially when "bunching" of buses occurs (clustering of the buses within a short distance of one another). The transit industry has (so far) lacked

© Springer International Publishing Switzerland 2015
F. Esposito et al. (Eds.): ISMIS 2015, LNAI 9384, pp. 371–378, 2015.
DOI: 10.1007/978-3-319-25252-0_40

a measure of service reliability in terms of its impact on customers because traditional metrics do not express how much reliability impacts on passengers' perceptions. In this paper, service reliability is measured based on passenger wait time, comfort and bus headway [5].

In order to minimize unreliability, it is important to identify its possible causes in bus operations. Prevention strategies focus on reducing the variability of vehicle running and dwelling times, while corrective strategies focus on reducing negative impacts to passengers. Passenger costs, operation costs and implementation feasibility are used to evaluate corrective strategies. The most common corrective strategies are reviewed in this section: namely "holding", "expressing", "short-turning" and "deadheading" [8].

Corrective strategies, using headway and schedule optimization with bus location tracked in real-time is addressed by Dessouky et al. [4], Chen and Chen [2], Yu et al. [9], Daganzo and Pilachowski [3] and Bartholdi et al. [1]. These approaches develop real-time corrective strategies by coordinating buses along their route.

Among the corrective strategies, "expressing", "short-turning" and "deadheading" all involve station skipping but using varying strategies. "Expressing" involves sending a bus to a stop further downstream and skipping (not servicing) some, or all, intermediate stops. The objective of this strategy may be either to increase the headway between buses (separating bunched buses) or to close a service gap further downstream, both in an attempt to balance headways and improve service past the end of the express segment [8].

Previous studies do not provide methods that have the ability to handle uncertainty in transit operations arising from within the transit environment and via the randomness of passenger arrivals. They also lack any mechanism that supports decision making for bus operations on route and at the bus stop simultaneously. This paper focuses on an approach for real-time multi-criteria decision-making based on dynamic Bayesian networks. These have the ability to handle uncertainty, reason about current states, and predict future states in cooperation with multi-objective optimization at each time slice in order to find appropriate strategies that maintain bus service reliability. The bus service reliability in our work takes into account passenger wait time, headway adherence, in-vehicle time, and passenger comfort, which are combined via Pareto comparisons in the fitness assignment processes.

The remainder of this paper is organized as follows: our proposed methodology for real-time decision making is presented in Sect. 2; simulation and results are reported and discussed in Sect. 3, with conclusions presented in Sect. 4.

2   Proposed Methodology

Figure 1 shows the proposed methodology for controlling bus operations. In the real-time mode, a supervisor will receive evaluations of a current scenario including real-time travel demand, transit demand, bus network and assignment data. This is then used to give optimal proactive adaptation, including guidance for

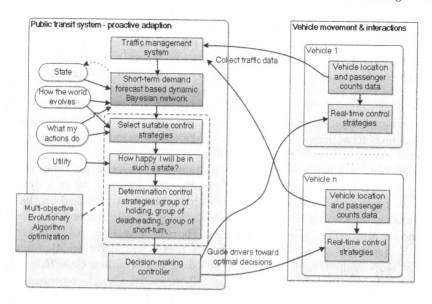

Fig. 1. Contextual Framework for bus operation control.

drivers leading to the goal of optimizing the bus network operations. The intention is to find strategies to guide drivers towards optimizing the overall bus network, not strategies solely for optimizing individual bus usage.

Real-time passenger demand and bus operation data are assumed to be collected from automatic passenger counting (APC) and automatic vehicle location (AVL) systems.

There are two main mechanisms in our control methodology: (1) state reasoning and demand prediction model, (2) multi-criteria decision making.

1. provides a mechanism that allows reasoning about current states and prediction about future states of bus operation based on a dynamic Bayesian network. This provides adequate information for (2) to make in-time and appropriate decision making.
2. provides a mechanism that allows suitable rational decision making for bus drivers on the route, namely preventive strategies, and at the bus stop, namely corrective strategies.

The details of these two mechanisms are described in the following two sub sections.

2.1   State Reasoning and Demand Prediction Model

Rational decision-making in the context of this paper depends upon "both the relative importance of various goals and the likelihood that, and degree to which, they will be achieved" [7]. Probability offers a means of summarizing the uncertainty that originates from "laziness" and "ignorance". "Laziness" here means there is too much work in listing the complete set of antecedents and consequents needed to ensure an exception-less ruleset. The term "ignorance" splits in meaning between theoretical and practical. In theoretical terms "ignorance" here means there maybe no complete theory so the point at which a complete coverage of rules for the problem domain can never be adequately determined. In terms of practical "ignorance", even though all the rules are known, there is uncertainty about specific circumstances because not all the necessary deterministic tests have been (or can be) run [7]. Decision-making Bayesian networks have the ability to handle these types of uncertainty.

In order to monitor the state of the system over a specific period of time, a dynamic Bayesian network model [6,7] is proposed. Figure 2 shows a dynamic Bayesian network model with t time slices for a bus network based on the static network, which includes three types of nodes: chance nodes (ovals), decision nodes and utility nodes (diamonds). The set of variables of chance node is $\mathcal{X}_\Gamma = \{$speed $\mathcal{V}_i$, position $\mathcal{X}_i$, number of passengers alighting $\mathcal{A}_{i,k}$ of vehicle i at stop k, number of passengers boarding $\mathcal{B}_{i,k}$, running time $\mathcal{R}_{i,k}$, dwell time $\mathcal{D}_{i,k}$, in-vehicle load $\mathcal{L}_{i,k}$, headway adherence $\mathcal{H}_{adherence}$, passenger wait time $\mathcal{T}_{wait}$, action impact $\mathcal{T}_{impact}$, passenger comfort $\xi_{comfort}\}$. Action of decision node has state $\mathcal{X}_D = \{$no action, preventive control $px_{i,k}$, holding $hx_{i,k}$, expressing $ex_{i,k}$, short-turning $sx_{i,k}$, deadheading $dx_{i,k}$ $\}$. The utility node represents the expected utility associated with each action.

Each time step models the state of the bus network at a specific point in time; the dashed lines present the separation of the model into time slices. In Fig. 2, smoothing is the process of querying about the state of the bus network at a previous time step from the current time, while filtering is the process of querying and predicting the state of the system from the current time to future steps. The conditional probability distributions $P(\mathcal{V}_i^t|\mathcal{V}_i^{t-1})$, $P(\mathcal{X}_i^t|\mathcal{X}_i^{t-1})$, $P(\mathcal{A}_i^t|\mathcal{A}_i^{t-1})$, $P(\mathcal{B}_i^t|\mathcal{B}_i^{t-1})$ $P(\mathcal{R}_{i,k}^t|\mathcal{R}_{i,k}^{t-1})$, $P(\mathcal{D}_{i,k}^t|\mathcal{D}_{i,k}^{t-1})$ and $P(\mathcal{L}_{i,k}^t|\mathcal{L}_{i,k}^{t-1})$ are the relevant transition probability distributions. The state of the bus network at the current point in time will impact the state of the system in the future and be impacted by the state of the system in the past. The development of the bus network is specified by links between variables in different time-slices. In this paper, the interval between slices is assumed to be fixed. For monitoring bus network states, a practical interval is 5 min.

2.2   Multi-criteria Decision Making

Decision making for bus operations to provide service reliability in this paper is driven by a set f of four objective functions ($f = f_1, f_2, f_3, f_4$), where f_1 is passenger wait time, f_2 is headway adherence, f_3 is passenger comfort, and f_4 is impact of control strategies. These functions are combined via Pareto comparisons in the fitness assignment processes.

$$f_1 = \mathcal{T}_{wait} = \sum_{i=1}^{n}\sum_{k=1}^{m}(\frac{\lambda_k(\widetilde{AD}_{i,k} - AD_{i-1,k})^2}{2} + \mathcal{P}_{i,k}(\widetilde{AD}_{i,k} - AD_{i-1,k})) \quad (1)$$

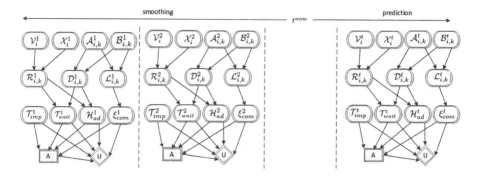

Fig. 2. Bus network time slices.

$$f_2 = \mathcal{H}_{adherence} = \sum_{i=1}^{n}\sum_{k=1}^{m}(\widetilde{\mathcal{H}}_{i,k} - \mathcal{SH})^2 \tag{2}$$

$$f_3 = \xi_{comfort} = \sum_{i=1}^{n}\sum_{k=1}^{m}\frac{\widetilde{\mathcal{L}}_{i,k}}{\mathcal{L}_{max}} \tag{3}$$

$$f_4 = \mathcal{T}_{impact} = \sum_{i=1}^{n}\sum_{k=1}^{m}\widetilde{\mathcal{L}}_{i,k} \times hx_{i,k} + (ex_{i,k} + dx_{i,k} + sx_{i,k}) \times \widetilde{\mathcal{H}}_{i,k} \times \widetilde{\mathcal{PD}}_{i,k} \tag{4}$$

Handling multi-objective problems, namely multi-criteria optimization, can be described as a process of finding the vector of decision variables $x^* = (x_1^*, x_2^*, ..., x_n^*)$, where n is number of buses and $x_i^* \in \{hx, ex, dx, sx, px\}$ is the control strategy applying for bus i at decision time, which minimizes the vector function,

$$minf(x) = (w_1 \times f_1(x) + w_2 \times f_2(x), f_3(x), f_4(x)) \tag{5}$$

where $x = (x_1, x_2, ..., x_n) \in \Omega \in R^n$ is called the decision variable, the set Ω is called the feasible region. Figure 3 depicts control sequence encoding of n buses. Each bus selects a control from the list. The solution will balance the optimization of an individual bus and the whole bus network.

In this paper, objective functions f_1 and f_2 are considered to be optimizeable simultaneously. This can mean that improvement of one can lead to an improvement of the other. f_1 and f_2, hence, use a weighted-sum approach. f_4 is considered to conflict with f_1 and f_2 while f_3 is independent so it does not influence any other objective function.

Station reasoning and demand prediction model in Sect. 2.1 provides information about current and predicted bus service reliability, which are used to decide whether multi-objective optimization process should be run. If service unreliability is predicted, the optimization algorithms will advise the decision making for buses at selected bus stops to restore the reliability.

A multi-Objective Evolutionary Algorithm (MOEA) is proposed for handling multi-objective problems in this paper. Deterministic algorithms are most often

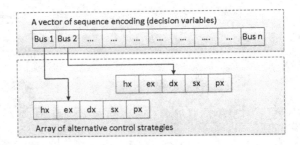

Fig. 3. Control sequence encoding

used if a clear relationship exists between the characteristics of the possible solutions. If the relation between a solution candidate and its "fitness" are not so obvious, as in the case of transit operation, probabilistic algorithms come into play.

3   Simulation and Results

3.1   Simulation

A case study of bus operations on the Gwynneville-Keiraville bus route in the regional city of Wollongong, Australia (population 300,000) is used to demonstrate and test the simulator. The simulator deals with a single time period, namely the peak period from 16:34 to 22:32 on weekdays.

3.2   Results and Analysis

The non-dominated solutions of the three objectives: f_1, f_2, f_3 obtained from 100 iterations are shown in Fig. 4 by blue-colored points. Figure 4 shows the trade-off between passenger wait time and action impact and trade-off between passenger wait time and passenger comfort. Square points represent evolutionary algorithm solutions, circle points represent pseudo-optimal pareto front.

After generating a Pareto optimal configuration with the set of good solutions, several key performance indicators are calculated for each solution. Decision makers can also choose any non-dominated solution from an experimental run based on their preference weight vectors.

Performance parameters used in this paper are used to address transit service reliability from the perspective of passengers. For short-headway services, the variability of headways is the main route-based measure for evaluating transit reliability. An effective control strategy improves service reliability by reducing headway variability, which in turn results in shorter passenger waiting times. Charts 1 and 2 of Fig. 5 presents space-time headway adherence before and after applying control strategies for the peak hour (16:30 – 23:00). There is more bunching before applying control strategies.

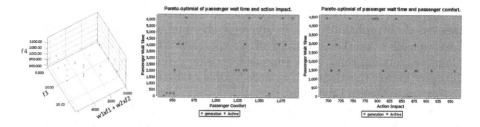

Fig. 4. Pareto-optimial.

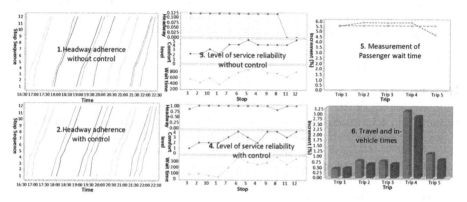

Fig. 5. Measure of Efficiency

Another performance route-based measure is passenger wait time. Charts 3 and 4 of Fig. 5 shows passenger wait time compared with expected wait time and the effect of control strategies on travel and in-vehicle times.

Control strategies may cause delays to on-board passengers and longer travel times that may result in higher fleet costs. However, improved regularity of headways can reduce the in-vehicle time of the passengers at the subsequent stops. In addition, passenger waiting time at bus stops can in practice be considered more important than passenger in-vehicle waiting time.

Bus reliability at the stop level is considered from a passengers point of view, which can be used to enhance reliability from a passengers perspective. Line charts in Fig. 5 used to measure stop-level bus reliability with (and without) employing control strategies.

The results in charts 5 and 6 of Fig. 5 indicate that there was low service reliability for the bus transit network before applying the control strategies. Applying control strategies helps to dramatically improve service reliability.

4   Conclusion

Our Multi-objective Evolutionary algorithm based dynamic Bayesian networks approach provides the ability to reason and predict bus service reliability network as well as to handle multi-criteria decision making to control real-time information. It is able to handle uncertainty which, when presented through variables based on probability and its dynamic choosing action, yields the highest expected utility. Another advantage of our approach is that it considers headway adherence, running time, dwell time and decision-making as continuous values. The effect is that the algorithm is more flexible in decision-making compared to existing transit control methods.

A simulation-based evaluation enables us to verify the efficiency of our approach. The simulation examined performance and level-of-service by capturing the interactions between transit operations and passenger demand. Pareto-optimal analysis is used to measure efficiency of a multi-objective evolutionary algorithm. Route and stop level analysis for transit service reliability improves passenger decision-making processes and enhances daily route service management by the transit agents.

References

1. Bartholdi, J.J., Eisenstein, D.D.: A self-coördinating bus route to resist bus bunching. Transp. Res. Part B: Methodol. **46**(4), 481–491 (2012)
2. Chen, W.Y., Chen, Z.Y.: A simulation model for transit service unreliability prevention based on avl-apc data (pdf) (2009)
3. Daganzo, C.F., Pilachowski, J.: Reducing bunching with bus-to-bus cooperation. Transp. Res. Part B: Methodol. **45**(1), 267–277 (2011)
4. Dessouky, M., Hall, R., Zhang, L., Singh, A.: Real-time control of buses for schedule coordination at a terminal. Transp. Res. Part A: Policy Pract. **37**(2), 145–164 (2003)
5. Furth, P.G.: Using archived AVL-APC data to improve transit performance and management, vol. 113. Transportation Research Board National Research (2006)
6. Kjaerulff, U.B., Madsen, A.L.: Bayesian networks and influence diagrams: a guide to construction and analysis. Springer (2007)
7. Russell, S.J., Norvig, P.: Artificial intelligence: a modern approach. Prentice Hall, Englewood Cliffs (2010)
8. Wilson, N.H.M., Cham, L.C., et al.: Understanding bus service reliability: a practical framework using AVL/APC data. Ph.D. thesis, Massachusetts Institute of Technology (2006)
9. Yu, B., Yang, Z., Sun, X., Yao, B., Zeng, Q., Jeppesen, E.: Parallel genetic algorithm in bus route headway optimization. Appl. Soft Comput. (2011)

Building Thermal Renovation Overview

Combinatorics + Constraints + Support System

Andrés F. Barco[✉], Elise Vareilles, Michel Aldanondo, and Paul Gaborit

Mines d'Albi, Université de Toulouse, Route de Teillet Campus Jarlard,
81013 Albi Cedex 09, France
abarcosa@mines-albi.fr

Abstract. Facade-layout synthesis is a combinatorial problem that arises when insulating buildings with rectangular parameterizable panels. At the core of the problem lies the assignment of size to an unknown number of panels and their arrangement over a rectangular facade surface. The purpose of this communication is to give an overview of the facade-layout synthesis problem and its reasoning by constraint satisfaction problems. Then, we show the combinatorial characteristics of the problem, its modeling by means of constraint satisfaction and a decision support system that solves the problem using several constraint-based algorithms.

1 Problem

In order to achieve a reduction on energy consumption, buildings may be externally renovated by covering each of its facades with an envelope made out of rectangular wood-made panels [6,12]. Given that each facade has its own geometrical and structural properties, panels composing the envelope are parameterized and manufactured once the envelope definition is done. Also, due to manufacturing limitations, size of panels are not fixed but bounded. Further, panels should avoid overlapping, they must cover each window over the facade and must be attached in specific areas strong enough to support their weight. Thus, one of the key problems in this setup is to propose a computational process that allows defining the specific set of non-overlapping panels with respect to each facade to renovate.

Now, although finding a correct allocation of panels over a facade surface can be efficiently made by a human, it is not the same for enumerating several solutions or an optimal one (e.g. with minimum number of panels). This is due to the huge solution space result of the combinatorics within; the problem as well as any other packing or covering problem is NP-hard [4,10]. In consequence, to tackle this problem a non-trivial technique is required. *Artificial Intelligence* and *Operation Research* provide such technique called constraint satisfaction problems (CSPs) [9]. The aim of this paper is then to given an exposition of the combinatorics within the problem, its modeling by CSP and to present a decision support system for architects in charge of the thermal renovation. In Sect. 2 we introduce the underlying combinatorics of the problem. In Sect. 3 we present

© Springer International Publishing Switzerland 2015
F. Esposito et al. (Eds.): ISMIS 2015, LNAI 9384, pp. 379–385, 2015.
DOI: 10.1007/978-3-319-25252-0_41

the constraint model of the renovation. Afterwards we introduce an intelligent support system that implements several constraint-based algorithms to solve the problem. Finally, some conclusions are drawn in Sect. 5.

2 Combinatorics

Combinatorial problems rise in many real-life contexts where the instantiation of discrete structures lead to different results. Typical questions for these kinds of problems are how many solutions exist and how to efficiently enumerate them. The answer depends on the number of discrete structure and possible instantiations for each structure. In the interest of simplicity let us analyze the basic combinatorics that springs from our industrial scenario. In our context, discrete structures are rectangular panels that can be instantiated to different sizes and different positions over a rectangular facade surface:

- Facades. A facade is a rectangular surface with fixed width and fixed height.
- Panels. A panel is a parameterizable rectangular entity whose width and height is not known but that has given lower and upper bounds. The number of panels to allocate over the facade is *not known in advance*.

We will simplify our problem by setting the minimum and maximum panel width and height. The minimum width and height is 10 cm. The maximum width and height is 20 cm. Question: How many possible panel sizes exist in this specification (see Fig. 1)?

As the reader can see, using an interval of [10,20] centimeters for panels size,

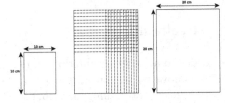

Fig. 1. Panel size possibilities.

the number of possible panel size is already big. How many? 10, 100, 1000? In fact, the answer depends on the discretization of the interval. For instance, if we are counting in centimeters, then there is 10 cm between 10 and 20. In this case the number of configuration would be 100. But, if we are counting on millimeters, then there is 100 mm between 10 and 20. Thus, the possible configuration would be 10000. In our work the discretization is in millimeters.

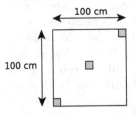

Fig. 2. Panels placement.

Suppose now that we have chosen the size of the panel, say 10 × 10. Now it is the time to allocate it over the facade. But where? Again, we have several possibilities on where to put the panel. To simplify, let us consider a facade with 100 cm as width and 100 cm as height. Figure 2 shows three panels allocated in three possible locations. Again, we must consider that there exists a huge number of possible locations for a single panel. For instance, considering the horizontal axis and counting in centimeters, the panel can be located on 90 possible places. On the vertical axis we have another 90 possibilities. Thus, the panel can be placed on 8100 possible placed over the facade. However, the panel size is not 10×10, but actually, it can be any number between 10 and 20. Therefore, the number of possibilities to configure a single panel over the facade is the product of the number of possibilities of the panel size and the number of possibilities to place the panel, i.e., $100 * 8100 = 810000$ possible configurations[1] for a single panel.

3   Constraint Model

Constraint satisfaction problems are conceived to allow the end-user to state the logic of the computation rather than its flow. This is done by representing the elements as a set of variables $\mathcal{V}$, a set of potential instantiation values $\mathcal{D}$ known as domains, and a set of relations $\mathcal{C}$ over the stated variables referred to as constraints. A CSP solution is a unique assignment of variables in such a way that all constraints in $\mathcal{C}$ are satisfied (see [3,9] for further references). A CSP may evolve into a *constraint optimization problem* (COP) by adding an objective function. We are interested on solving the satisfaction problem as well as the optimization one, with the minimization of used panels as its criterion.

Constraint Variables. We introduce the notation used in the model. Let F denote the set of frames, S the set of supporting areas and fac the facade. Let $o_{e.d}$ and $l_{e.d}$ denote the origin and size, respectively, of a given entity e in the dimension d, with $d \in \{1, 2\}$ (1 for x-axis and 2 for y-axis). For instance, $o_{fr.1}$ denotes the origin in the horizontal axis and $l_{fr.1}$ the width of frame fr. Additionally, min_d and max_d denote the size lower bound and size upper bound, respectively, in dimension d for all panels.

   Intuitively, each panel is described by its origin point w.r.t. the facade origin and its size. For convenience, let us assume that $\mathcal{P}$ is the set of panels composing the layout-plan solution. Then, each $p \in \mathcal{P}$ is defined by $\langle o, l \rangle$ where

- $o_{p.d} \in [0, l_{fac.d}]$ is the origin of panel p in dimension $d \in \{1, 2\}$.
- $l_{p.d} \in [min_d, max_d]$ is the size of panel p in dimension $d \in \{1, 2\}$.

Constraints. The following six constraints express the main relations among panels, and panels and facade that must respect a layout solution.

(a) *Manufacturing and transportation limitations constrain panels size with a given lower bound* min *and upper bound* max *in one or both dimensions.*

$$\forall p \in P, d \in \{1, 2\} \; min_d \leq l_{p.d} \leq max_d$$

(b) *For two given panels p and q there is at least one dimension where their projections do not overlap.*

$$\forall_{p,q} \in P, p \neq q, \exists d \in \{1, 2\} | o_{p.d} \geq o_{q.d} + l_{q.d} \lor o_{q.d} \geq o_{p.d} + l_{p.d}$$

---

[1] The number is an approximation as number of positions for placing a panel decreases when the panel size increases.

(c) *A given panel p must be either at the facade edge or ensure that enough space is left to fix another panel.*

$$\forall_p \in P, d \in \{1,2\}, o_{p.d} + l_{p.d} = l_{fac.d} \vee o_{p.d} + l_{p.d} \leq l_{fac.d} - min_d$$

(d) *Each frame over the facade must be completely overlapped by one and only one panel. Additionally, frames borders and panels borders must be separated by a minimum distance denoted by Δ.*

$$\forall_f \in F, \exists p \in P, d \in \{1,2\} | o_{p.d} + \Delta \leq o_{f.d} \wedge o_{f.d} + l_{f.d} \leq o_{p.d} + l_{p.d} + \Delta$$

(e) *The entire facade surface must be covered with panels.*

$$\sum_{i \in P} \prod_{d \in \{1,2\}} l_{i.d} = \prod_{d \in \{1,2\}} l_{fac.d}$$

(f) *Panels corners must be matched with supporting areas in order to be properly attached onto the facade. Note in the expression that $s1 = s2$ is valid.*

$$\forall_p \in P, d \in \{1,2\}, \exists s \in S | o_{s1.d} \leq o_{p.d} \wedge o_{p.d} + l_{p.d} \leq o_{s2.d} + l_{s2.d}$$

4 Decision Support System

The aforementioned model is used as foundation of a decision support system[2] conceived to support architects decision-making. The input of the system is a numerical model describing the facades (e.g. width and height of facades, position of windows and attaching areas, bounds for panels size, etc.). The output is a numerical model, for each facade to renovate, describing each panel composing the envelope (e.g. width and height of panels, position over the facade, etc.). The support systems throws several solutions (if any) for each facade to renovate. In this Section we briefly describe the internal design of support system and its implemented algorithmic solutions.

Design. Our support system, called Calpinator, has a very basic and modular design that allows to add new algorithms when available. Figure 3 presents the internal design of calpinator at first glance.

Initially, the user inputs its building numerical model (step 1). The system parses the file (step 2) and creates a data base (step 3) that stores all objects of the specification. The system shows a visual representation of the building (step 4) and the user customizes the solving process (step 5).

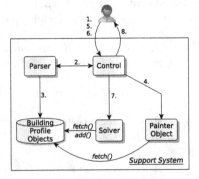

Fig. 3. Internal design of system.

---

[2] The source code can be found in https://bitbucket.org/anfelbar/calpinageprototype/wiki/Home.

The user asks for solutions (step 6) and system calls executes one of its algorithms (step 7). The solutions found are shown and the user may store each one of them in files (step 8).

It is worth mentioning that currently the user is supposed to be an architect, the building owner or a third-party contractor that is in charge of mapping the building data into the appropriated input format. Nevertheless, in a different stage of the renovation, it is expected that drones with pattern and image recognition will be used to obtain most of the facade related information.

Algorithmic Solutions. Up-to-now, we have focused our attention on a subset of variables and subset of constraints from the industrial scenario. We have been developing different algorithms, and their respective implementation, to find solutions and then optimal solutions. Each algorithm tackles the problem by decomposing the constraints in tasks and thus gaining modularity. In what follows we give a description of these algorithmic solutions.

1. *Greedy-recursive algorithm.* A solution that makes local decisions for positioning panels following a constructive approach [7]. In essence, the algorithm tries to put one panel and, if successfully placed, it proceeds by placing the next panel until the definition of the entire layout has been done. Constraint conflicts are solved locally. Details of the solution can be found in [1].
2. *Global constraints algorithm.* A solution model exploiting the capabilities of the constraint solver Choco [8]. It works by placing a panel over each one of the frames over the facade. Then, it expands the size of panels to cover the larger possible surface. If summation of panels areas is less than the facade area then more panels are added iteratively. Constraint conflicts are solved locally. Details of the solution can be found in [2].
3. *Constrain-based solution for symmetrical facades.* A solution using Choco conceived for symmetrical facades, i.e., where windows/doors and supporting areas are uniformly distributed over the facade surface. In essence, it targets the problem of undefined number of panels by creating a matrix of panels whose dimension depends on the size of the facade and bounds for panels size. It generates compliant and often optimal, w.r.t. the number panels, layout-plans solutions for symmetrical facades.
4. *Layout synthesis as two-dimensional packing.* A solution that exploits the constraint solver Choco to solve the problem as a two-dimensional packing problem [5]. As the number of panels is not know in advance, it uses a variant of the diffN constraint to handle *optional* rectangles (open global constraints [11]). A dedicated heuristic exploits the problem structure to provide efficiency when traversing the search tree.

Each of the algorithms provide solutions in reasonable computational time, in seconds or few minutes depending on the facade characteristics. But, according to the industrial case and expert knowledge, no real-time interaction is needed as architects have enough time to select the appropriated panels-made envelope.

5 Conclusions

Controlling energy consumption in buildings is one of the major challenges of the 21st century. Our work is part of a project that investigates the possibility of automated building renovation based on rectangular panels and supported by an intelligent system. In this paper we have shown an overview of the problem and its reasoning under CSP. First, we have presented the combinatorial problem that arises from the industrial scenario. Second, we have modeled the problem under the framework of constraint satisfaction. Finally, we have introduced a decision support system conceived to support architects decision-making in the context of building renovation. To avoid the generation of non-compliant solutions, future work should stress "aesthetic" knowledge and its inclusion to the (constraint) knowledge model. Ultimately, we highlight the flexibility of constraint satisfaction to address and to enumerate different solutions in hard industrial problems.

References

1. Barco, A.F., Vareilles, E., Aldanondo, M., Gaborit, P.: A recursive algorithm for building renovation in smart cities. In: Andreasen, T., Christiansen, H., Cubero, J.-C., Raś, Z.W. (eds.) ISMIS 2014. LNCS, vol. 8502, pp. 144–153. Springer, Heidelberg (2014)
2. Barco, A.F., Vareilles, E., Aldanondo, M., Gaborit, P., Falcon, M.: Constraint-based decision support system: Designing and manufacturing building facades. In: Join Conference on Mechanical, Design Engineering and Advanced Manufacturing. Springer, June 2014 (to appear)
3. Brailsford, S.C., Potts, C.N., Smith, B.M.: Constraint satisfaction problems: Algorithms and applications. Eur. J. Oper. Res. **119**(3), 557–581 (1999)
4. Csirik, J., Woeginger, G.: On-line packing and covering problems. In: Fiat, A., Woeginger, G. (eds.) Online Algorithms. LNCS, vol. 1442, pp. 147–177. Springer, Heidelberg (1998)
5. Imahori, S., Yagiura, M., Nagamochi, H.: Practical algorithms for two-dimensional packing. In: Gonzalez, T.F. (ed) Handbook of Approximation Algorithms and Metaheurististics. Chapman & Hall/CRC computer & information science series, ch. 36, vol. 10, CRC Press (2007)
6. Jelle, B.P.: Traditional, state-of-the-art and future thermal building insulation materials and solutions - properties, requirements and possibilities. Energy Buildings **43**(10), 2549–2563 (2011)
7. Liggett, R.S.: Automated facilities layout: past, present and future. Autom. Constr. **9**(2), 197–215 (2000)
8. Prud'homme, C., Fages, J.G.: An introduction to choco 3.0 an open source java constraint programming library. In: International Workshop on CP Solvers: Modeling, Applications, Integration, and Standardization, Uppsala, Sweden (2013)
9. Rossi, F., Beek, P., Walsh, T.: Handbook of Constraint Programming (Foundations of Artificial Intelligence). Elsevier Inc., NY (2006)
10. Szczygie, T.: Comparing CLP and OR approaches to 2D angle cutting and packing problems (2001). http://citeseerx.ist.psu.edu/viewdoc/download?doi=10.1.1.205.8953&rep=rep1&type=pdf

11. van Hoeve, W.-J., Régin, J.-C.: Open constraints in a closed world. In: Beck, J.C., Smith, B.M. (eds.) CPAIOR 2006. LNCS, vol. 3990, pp. 244–257. Springer, Heidelberg (2006)

12. Vareilles, E., Barco Santa, A.F., Falcon, M., Aldanondo, M., Gaborit, P.: Configuration of high performance apartment buildings renovation: A constraint based approach. In IEEM 2013 International Conference, pp. 684–688, December 2013

Frequency Based Mapping of the *STN* Borders

Konrad A. Ciecierski[1]([✉]), Zbigniew W. Raś[1,2],
and Andrzej W. Przybyszewski[3]

[1] Institute of Computer Science, Warsaw University of Technology,
00-655 Warsaw, Poland
K.Ciecierski@ii.pw.edu.pl
[2] Deptartment of Computer Science, University of North Carolina,
Charlotte, NC 28223, USA
ras@uncc.edu
[3] Department of Neurology, UMass Medical School, Worcester, MA 01655, USA
Andrzej.Przybyszewski@umassmed.edu

Abstract. During deep brain stimulation (DBS) surgery for Parkinson disease, the target is the subthalamic nucleus (*STN*). *STN* is small, (9 × 7 x 4 mm) and typically localized by a series of parallel microelectrodes. As those electrodes are in steps advanced towards and through the *STN*, they record the neurobiological activity of the surrounding tissues. By careful inspection and analysis of such recordings one can obtain a range of depth at which given electrodes passed through the *STN*. Both human made inspection and computer based analysis are performed during surgery in the environment of the operation theatre. There are several methods for the *STN* detection, one of them – developed by the authors – is described in [8]. While the detection of the *STN* interior can be obtained with good certainty its borders can be slightly fuzzy and sometimes it is difficult to classify whether given depth should be regarded as belonging to the *STN* proper or lying outside of it. Mapping of the borders is important as the tip of the final permanent stimulating electrode is often placed near the ventral (bottom) border of the *STN* [12]. In this paper we are showing that analysis focusing on narrow frequency bands can yield better discrimination of the *STN* borders and *STN* itself.

Keywords: STN border · DBS · Wavelet · DWT(Discrete Wavelet Transform) decomposition · Band filtering · Signal power

Introduction

The first line of treatment for Parkinson Disease (PD) is pharmacological one. Still, some patients do not tolerate the anti PD drugs very well, in some others with time the effectiveness of treatment lessens and the dosage has to be gradually increased. This might eventually lead to motor fluctuations between ON and OFF states [4]. For some of those patients, the surgical treatment can be applied as an alternative to ineffective pharmacological one. This treatment is

© Springer International Publishing Switzerland 2015
F. Esposito et al. (Eds.): ISMIS 2015, LNAI 9384, pp. 386–395, 2015.
DOI: 10.1007/978-3-319-25252-0_42

called the Deep Brain Stimulation. During the surgery, in specific part of the brain the permanent stimulating electrodes are placed [4]. The technical problem of this surgery is that the target within which the electrodes are to be placed – Subthalamic Nucleus (*STN*) – is small and difficult to precisely locate using conventional medical imaging techniques i.e. CT[1] and MRI[2].

In the literature and in medical practice there are several approaches to localization of the *STN*. Some of them are purely CT and MRI based while others rely for example on intrasurgical microrecording [10,11]. Microrecording localization is based on the fact that *STN* has a physiology that is distinct from this of adjacent brain territories [3,4].

In microelectrode based methods, during surgery several thin parallel electrodes are inserted into patient's brain and in measured steps advanced towards and through the *STN*. The electrodes, as they advance through the brain, at each step perform typically a 10 seconds long recording of the activity of the brain tissue surrounding their recording tips. *STN* is characterized by high neuronal activity [3] and due to pathological changes that occur in PD, the activity of the *STN* is even more increased [4]. By this, using carefully designed computer analysis one can discriminate between recordings that have been registered within the *STN* and outside of it. Mapping of the interior part of the *STN* is by this achieved with reliable and repeatable results. Of course there are border cases with recordings registered at the dorsal[3] and ventral[4] borders of the *STN*. Those borders may differ from patient to patient, sometimes they are clearly defined in the span of single millimeter while in other cases they might be fuzzed over few millimeters. One might now ask, why the precise localization of the borders is so important as long as one can safely map the interior of the *STN*.

Best results with least risk of complications are achieved when the tip lead of the stimulating electrode is located close to the ventral border of the *STN* [4]. In this way, contacts of the electrode that are above its tip are still within the *STN* and close to its ventral border.

1   Band Power Calculation

In previously published papers [5–9] we computed the signal's power for two distinct frequency ranges, i.e. LFB for frequencies below 500 Hz and HFB for frequencies between 500 Hz and 3 KHz. Both attributes – even alone – provide reasonably good discrimination between *STN* and non *STN* recordings. As shown in Table 1, the ranges of Q1 – Q3 quartiles do not overlap for either of them.

Still, for the 25 % of the *STN* recordings their LFB and HFB are lower than respectively provided Q1 values. Those cases are in some cases related to patients

---

[1] Computer Tomography.
[2] Magnetic Resonance Imaging.
[3] top.
[4] bottom.

Table 1. Electrode rank example

Attribute	Class	Q1	Q2	Q3
LFB	$\neg STN$	0.833	1.034	1.392
	STN	2.253	3.644	5.775
HFB	$\neg STN$	0.922	1.063	1.429
	STN	2.650	4.103	6.272

with unusually quiet STN or simply are border cases recorded at the dorsal or ventral borders of the STN.

For any function that can be calculated for data recorded in course of a single electrode at consecutive depths, there is a need for normalization. This procedure is described in detail in [8]. Here we will only mention, that in this procedure, average from the first five depth is treated as a base value and values of that function taken from all depths for that electrode are divided by this base value.

An example of the normalized LFB (Low Frequency Background power [8,9]) value calculated for a PD patient with unusually quiet STN is presented on Fig. 1.

Only at two depths the value of the LFB attribute was above 3 and never on the whole track of the electrode it exceeded 4. This electrode was also the best one as other two electrodes provided even worse STN localization.

The classifier pinpointed the depths -1000 μm and 0 μm as belonging to the STN. Due to the sheer drop of the LFB at $+1000$ μm it was assumed that at depth 0 μm is the ventral border of the STN. Problematic however was the localization of the dorsal STN border. Does it start at -1000 μm or maybe earlier at -3000 μm. During test stimulations performed in the course of the surgery the STN extent has been mapped to be between -2000 μm and 0 μm.

This means that depth at -2000 μm has not been correctly mapped as STN and this contributed to the lowering of the sensitivity of the classification.

2   Narrow Band Power Calculation

Power calculated summarily for range below $500\,Hz$ (LFB) was clearly not sufficient to fully identify the extent of the STN in the case described above.

What would be the result if one were to cut the power spectrum in much narrower bands.

To obtain power calculation for narrow frequency bands, the recorded signals were DWT fully transformed i.e. both in low and high frequencies [1,2]. Recorded signals are sampled with 24 KHz and have 10 s lengths. For DWT applicability the signals were zero padded to closest length being the power of 2 i.e. 262144 samples. DWT was then performed up to level where single set of coefficients had 64 elements and represented frequency span of about 3 Hz. In this way the frequency range was evenly split into $\mathcal{F}_{DIV}$ set of 4096 consecutive intervals.

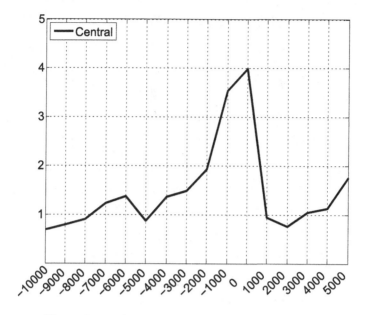

Fig. 1. Low value of LFB obtained from quiet STN

For given frequency range g its coverage $\mathcal{F}_{COV}(g)$ is defined as

$$\mathcal{F}_{COV}(g) \; = \; \{f \; : \; f \in \mathcal{F}_{DIV} \wedge \frac{\overline{\overline{f \cap g}}}{\overline{\overline{f}}} \geq \frac{1}{2}\}$$

If for given depth d and $f \in \mathcal{F}_{DIV}$, computed set of DWT coefficients is denoted as $DWT_c(d, f)$ then power calculated for given frequency range g is

$$PWR(d, g) \; = \; \sum_{f \in \mathcal{F}_{COV}(g)} \sum_{x \in DWT_c(d,f)} x^2$$

When one is now to regard the signal's power in various frequency bands some clear differences might be noticed. In selected frequency bands different parts of the STN are clearly more active then in others. Normalized powers for the same electrode are shown on Fig. 2 as in Fig. 1 but in two selected frequency bands.

Unfortunately this finding cannot be generalized. In some patients given frequency range is characteristic for a given part of STN while in others this range is completely different – to the point that they do not even overlap. There is a saying that there are not two identical cases of Parkinson Disease, just the same there are no two identical brains and subthalamic nucleuses. Even left and right STN in the same patient present different frequency characteristics. Power percentiles calculated from 4138 STN recordings registered during 173 surgeries are shown on Fig. 3 below. Calculations were made for consecutive, 25 Hz wide frequency ranges below 3 KHz.

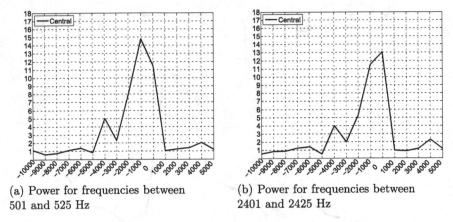

(a) Power for frequencies between 501 and 525 Hz

(b) Power for frequencies between 2401 and 2425 Hz

Fig. 2. Power for frequency bands

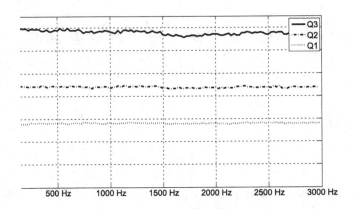

Fig. 3. Power quartile distribution over frequencies for STN recordings

It is immediately evident that no frequency is patients–wide characteristic for the *STN*.

Some general anatomically based frequency findings can still however be found. When above distribution is calculated not for all STN recordings but separately for those from dorsal and ventral borders of the STN differences are clearly present as shown on Fig. 4(a) and (b).

While power is still almost evenly distributed among frequencies, the dorsal parts of the STN gave recordings yielding greater power than ventral ones.

3   Narrow Band Power Based Discriminating Attribute

From the previous section it is evident that various parts of the STN display different frequency characteristics and that those characteristics are highly individual and cannot be easily generalized. All recordings were made with high

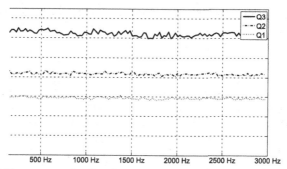

(a) Power quartile distribution over frequencies for dorsal STN recordings

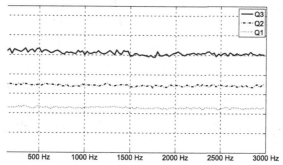

(b) Power quartile distribution over frequencies for ventral STN recordings

Fig. 4. Power quartile distribution for recordings from parts of the STN

impedance electrodes. When using low impedance electrodes one might find frequencies that do carry significantly more power in certain STN areas [12].

In any given patient for any part of the STN we can however find a frequency range at which its normalized power is minimized or maximized. Thus for any given depth we can compute minimal and maximal normalized narrow band power.

$$PWR_{MIN}(d) = \min_{f \in \mathcal{F}_{DIV}} PWR(d, f)$$

$$PWR_{MAX}(d) = \max_{f \in \mathcal{F}_{DIV}} PWR(d, f)$$

Such attributes maximize and minimize the power calculated for given depth regardless at which frequency brain tissue located there is most and least active. This should help discriminate the border cases and help in more finetuned STN border mapping.

For electrode pass shown on Fig. 2 the PWR_{MAX} is shown on Fig. 5. This time value of PWR_{MAX} beside obvious increase at -1000 μm and 0 μm is also more elevated at depth -2000 μm. That is depth which was classified by the neurosurgeon as belonging to the STN and misclassified by the algorithm given in [8,9].

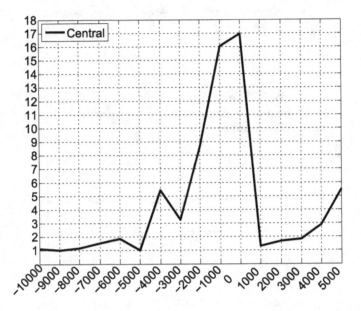

Fig. 5. PWR_{MAX} obtained from quiet STN

4   Evaluation

For evaluation purposes a test set of 14422 recordings was used. Those recordings were registered during 135 neurosurgeries. The attributes used in set were based both on spike occurrence and background neural activity. They are described in detail in [8,9].

The 10–fold cross validation results obtained with Random Forest classifier are shown in Table 2.

$$sensitivity = \frac{3061}{3061 + 189} \approx 0.942 \quad specificity = \frac{11054}{11054 + 118} \approx 0.989$$

When existing attributes were extended by the PWR_{MIN} and PWR_{MAX} attributes the results are similar regarding specificity but clearly better in sensitivity. That means that in this configuration classifier has recognized properly more of the STN recordings. The count of false negatives dropped from 189 to 160 (Table 3).

$$sensitivity = \frac{3090}{3090 + 160} \approx 0.951 \quad specificity = \frac{11039}{11039 + 133} \approx 0.988$$

Table 2. Cross validation results

		\multicolumn{2}{c}{Human classification}		
		STN	MISS	Total
Classifier	STN	3061	118	3179
	MISS	189	11054	11243
	Total	3250	11172	14422

Table 3. Cross validation results

		\multicolumn{2}{c}{Human classification}		
		STN	MISS	Total
Classifier	STN	3090	133	3223
	MISS	160	11039	11199
	Total	3250	11172	14422

Many of those corrected false negatives represented recordings from unusually quiet STNs or described earlier cases of dorsal or ventral STN borders. Among them is also – described in Sect. 1 – corrected misdetection of the STN at $-2000\,\mu m$.

When existing background based attributes [8] were replaced by the set of attributes representing normalized power calculated for $\mathcal{F}_{DIV}$ frequency intervals, the results became slightly worse (Table 4).

Table 4. Cross validation results

		\multicolumn{2}{c}{Human classification}		
		STN	MISS	Total
Classifier	STN	2968	152	3120
	MISS	282	11020	11302
	Total	3250	11172	14422

$$sensitivity = \frac{2968}{2968 + 282} \approx 0.913 \quad specificity = \frac{11020}{11020 + 152} \approx 0.986$$

Both the false positive and false negative counts have increased. Clearly unsegregated power values calculated for $\mathcal{F}_{DIV}$ cannot provide as good discrimination as normalized power calculated for two specific wide frequency ranges i.e. below 500 Hz and between 500 Hz and 3 KHz.

Finally, if one were to build classifier solely on those $\mathcal{F}_{DIV}$ based attributes the results are much worse (Table 5).

$$sensitivity = \frac{2382}{2382 + 868} \approx 0.733 \quad specificity = \frac{10720}{10720 + 452} \approx 0.960$$

Table 5. Cross validation results

		Human classification		
		STN	$MISS$	Total
Classifier	STN	2382	452	2834
	$MISS$	868	10720	11588
	Total	3250	11172	14422

The sensitivity has dropped from 0.989 to 0.733 while the specificity has increased from 0.942 to 0.960. The dramatic decrease of sensitivity proves that while $\mathcal{F}_{DIV}$ based aggregative attributes do improve classification, they themselves alone in their plain form are insufficient for good classification.

5 Summary

Methods described in [8,9] provide good basis for discrimination between STN and non STN recordings. The quality of the classification has been repeatedly confirmed during many neurosurgeries performed at Institute of Psychiatry and Neurology in Warsaw.

In some cases the classifier still fails to correctly identify a recording registered within the STN. This, in most cases comes from one of the two possible causes:

– Patient has STN with unusually low activity
– Wrongly classified location lies on border between STN and another adjacent brain area

While STN may be misdetected due to either one of the above causes or both of them, one might still often find its increased activity at certain frequency ranges. Problem lies in natural individuality of brain functioning as well as in individuality of Parkinson Disease related pathological disturbances in its physiology [3].

Using hight impedance recording microelectrodes no narrow frequency band could have been found such that its elevated power would be STN characteristic for majority of the patients. Increase of signal's power quartiles within STN is evident for all frequency sub bands below 3 KHz.

Still, while it is hard to say which frequency range at which part of the STN should yield increased power, one might hypothesize that at least for some of those frequency bands the power indeed would be increased. That hypothesis led to construction of the PWR_{MIN} and PWR_{MAX} attributes that reflect lowest and highest band based power obtained from given recording.

During classification tests, use of those two additional attributes improved the sensitivity which has increased from 0.942 to 0.951 While the increase by ≈ 0.01 might seem to be small, it is noteless important as it improves classification of the most difficult cases.

References

1. Jensen, A., Ia Cour-Harbo, A.: Ripples in Mathematics. Springer-Verlag, Heidelberg (2001)
2. Smith, S.W.: Digital Signal Processing. Elsevier, Melbourne (2003)
3. Nolte, J.: The Human Brain, An Introduction to Its Functional Anatomy. Mosby year, Maryland Heights (2009)
4. Israel, Z., Burchiel, K.J.: Microelectrode Recording in Movement Disorder Surgery. Thieme Medical Publishers, New York (2004)
5. Ciecierski, K., Raś, Z.W., Przybyszewski, A.W.: Foundations of recommender system for STN localization during dbs surgery in Parkinson's patients. In: Chen, L., Felfernig, A., Liu, J., Raś, Z.W. (eds.) ISMIS 2012. LNCS, vol. 7661, pp. 234–243. Springer, Heidelberg (2012)
6. Ciecierski, K., Raś, Z.W., Przybyszewski, A.W.: Discrimination of the micro electrode recordings for STN localization during DBS surgery in parkinson's patients. In: Larsen, H.L., Martin-Bautista, M.J., Vila, M.A., Andreasen, T., Christiansen, H. (eds.) FQAS 2013. LNCS, vol. 8132, pp. 328–339. Springer, Heidelberg (2013)
7. Ciecierski, K., Raś, Z.W., Przybyszewski, A.W.: Foundations of automatic system for intrasurgical localization of subthalamic nucleus in Parkinson patients. Web Intell. Agent Sys. **12**(1), 63–82 (2014)
8. Ciecierski, K.: Decision Support System for surgical treatment of Parkinsons disease, Ph.D. thesis, Warsaw University of technology Press (2013)
9. Ciecierski, K., Mandat, T., Rola, R., Raś, Z.W., Przybyszewski, A.W.: Computer aided subthalamic nucleus (STN) localization during deep brain stimulation (DBS) surgery in Parkinson's patients. Annal. Acad. Med. Silesiensis **68**(5), 275–283 (2014)
10. Mandat, T., Tykocki, T., Koziara, H., et al.: Subthalamic deep brain stimulation for the treatment of Parkinson disease. Neurol. I Neurochir. Pol. **45**(1), 32–36 (2011)
11. Novak, P., Przybyszewski, A.W., Barborica, A., Ravin, P., Margolin, L., Pilitsis, J.G.: Localization of the subthalamic nucleus in Parkinson disease using multiunit activity. J. Neurol. Sci. **310**(1), 44–49 (2011)
12. Zaidel, A., Spivak, A., Shpigelman, L., Bergman, H.: Israel delimiting subterritories of the human subthalamic nucleus by means of microelectrode recordings and a hidden markov model. Mov. Disord. **24**(12), 1785–1793 (2009)

Planning, Classification

Planning with Sets

Rajdeep Niyogi[1]([✉]) and Alfredo Milani[2,3]

[1] Department of Computer Science and Engineering,
Indian Institute of Technology Roorkee, Roorkee 247667, India
`rajdpfec@iitr.ac.in`

[2] Department of Mathematics and Computer Science,
University of Perugia, 06123 Perugia, Italy
`milani@unipg.it`

[3] Department of Computer Science, Hong Kong Baptist University,
Kowloon Tong, Hong Kong

Abstract. In some real world applications like robotics, manufacturing, the same planning operator with single or multiple effects is instantiated to several objects. This is quite different from performing the same action (or plan) several times. In this paper we give an approach to construct an iterated form of these operators (actions). We call such actions iterated actions that are performed on sets of objects. In order to give a compact and formal specification of such actions, we define a new type of predicate called set predicate. We show that iterated actions on sets behave like classical planning actions. Thus any classical planner can be used to synthesize plans containing iterated actions. We formally prove the correctness of this approach. An implementable description of iterated actions is given in PDDL for an example domain. The implementations were carried out using the state-of-the-art BlackBox planner.

Keywords: Classical planning · Set predicate · Iterated actions · PDDL

1 Introduction and Motivation

In some real world applications like robotics, manufacturing, the same planning operator with single or multiple effects is instantiated to several objects. For instance, a robot may be asked to paint different containers. In this case the same operator, say $paint(x)$, is applied to different objects. Further, the robot may need to check whether the surface of a container is rusty; if yes then paint otherwise no painting is needed. In this case the operator, say $check\_rust(x)$, has two outcomes, the actual outcome being determined only at run time. In this paper we consider planning with these types of repeated actions. We show that richer plan constructs involving branching and iteration, in some contexts, can be handled in the framework of classical planning [4], although it may appear to require the power of NPDDL based planners [3]. NPDDL [2] is an extension of PDDL [8] to model some important features of real world domains (e.g., incomplete state information, actions having multiple possible outcomes), that cannot be modeled using PDDL.

© Springer International Publishing Switzerland 2015
F. Esposito et al. (Eds.): ISMIS 2015, LNAI 9384, pp. 399–409, 2015.
DOI: 10.1007/978-3-319-25252-0_43

We now illustrate our approach with a detailed example. **A conference management domain**: An organizer of a conference would like to send an invitation for an event to *all* the members affiliated to some community. The organizer would usually send the information via email. However in the database, only the phone numbers of some members exist; no email address is recorded for these members. Thus the organizer has to make phone calls to these members and inform them about the event. The goal in this domain is to inform all the members about the event. The initial state consists of a set of members and the invitation message; it is not known which member has email address or phone number. Operators available are *send_mail, phone_call* (we assume that these operators have only one outcome; however, in general, it may also have multiple outcomes) and *check_mail* (has more than one outcome). A general form of plan satisfying the goal would be:

> **for each** $x \in \{m_1, \dots, m_k\}$ //members of the community
>   *check_mail(x)*; // determines whether or not a member has an email
>   **if** *have_mail(x)* **then** *send_mail(x, msg)*;
>   **else if** $\neg have\_mail(x)$ **then** *phone_call(x, msg)*;

It is easy to see that there is no classical plan (a sequence of actions) for this problem. In the above plan we observe that an operator (e.g., *check_mail(x)*, *send_mail(x, msg), phone_call(x, msg)*) is instantiated to distinct objects, i.e., x would be substituted by m_j at each iteration of the 'for-loop'. The branching is due to the sensing operator '*check_mail*'. The above plan, with branching and repetition, can be converted to a sequence of actions if we can construct iterated form of these operators. Suppose that we can construct an iterated form of these operators, where the iteration is defined on a set (for the example: set of members of the community). Then there exists a classical plan corresponding to the above plan. This is represented as:

> *check_mail_iter(S, S_1, S_2)*; *send_mail_iter(S_1, msg)*;
> *phone_call_iter(S_2, msg)*; *dummy(S, S_1, S_2, msg)*;

where $S_1 \cap S_2 = \emptyset, S_1 \cup S_2 = S, S = \{m_1, \dots, m_k\}$ (a detailed description of the actions is given in Sect. 4). We make a brief note that whereas *check_mail(x)* is an operator, an instance of it, say, *check_mail(m_1)* is an action. When an operator does not have any parameter, it is referred to as an action, as in the operator (or action) *chop*, which hits the tree once with an axe. Repetition of the *chop* action k times $(k > 0)$ is different from instantiating the parameter x, as in *send_mail(x, msg)*, with k distinct objects $m_1, \dots, m_k$, since different actions are obtained from each instantiation; i.e., $send\_mail(m_i, msg) \neq send\_mail(m_j, msg)$ for all $i \neq j$. Thus the iterated actions, considered in this paper, are quite different from performing the same action (or plan) several times as in [2,7].

In this paper we show how an iterated form (e.g., *send_mail_iter, check_mail_iter*) of an atomic operator (e.g., *send_mail, check_mail*) can be constructed. We consider two types of atomic actions (operators). The first is the classical planning actions that have a single outcome (e.g., instantiations of *send_mail* and *phone_call*). In non-classical planning, actions may have multiple outcomes. Consider the operator *pickup*(*x*) with preconditions robot hand empty, the object *x* is on the table, and nothing on top of the object. The effects are either the robot hand is holding the object or the object is on the table (when the robot hand fails to lift it). Here the effects are nondeterministic; the state on which it is applied is, however, fully known.

Knowledge producing actions or sensing actions [6, 10] also have nondeterministic outcomes, but such actions are used to obtain information about a proposition that is not known at a state. For instance, a *(fasting) blood glucose test* has possible outcomes like normal (70–100 mg/dL), lower than normal, above normal (100–125), and very high (> 125). Other examples would be of an agent driving a car and *looking at traffic signals*, the outcomes of which are red, yellow, or green, *check_mail*(*x*) that makes *have_mail*(*x*) true when a member has an email, false otherwise. A sensing action has finitely many possible outcomes (effects) but exactly one among these would be determined at run time. The second type of actions considered in this paper is the sensing actions.

In order to construct an iterated action from an atomic action, we introduce the notion of *set predicates*. We show how to implement set predicates in PDDL. Thus a richer type of planning domains can be expressed in PDDL and the power of classical planners can be leveraged as well. To the best of our knowledge, iterated operators and its formalization has not been studied before in the literature. The paper is organized with set predicates being defined in Sect. 2, iterated set operators in Sect. 3, PDDL specifications of iterated operators in Sect. 4, and conclusions in Sect. 5.

2   Set Predicate

Let C_{set} denote a nonempty set of constant symbols denoting sets, a set symbol $S \in C_{set}$ will denote a finite set of objects, $S = \{o_1, \ldots, o_n\}$, where each $o_j, 1 \le j \le n$ is in C_{obj}: a set of object symbols. We take some examples to motivate the meaning and our intuition behind the notion of set predicates. Given that $C_{obj} = \{u_1, u_2, hello, hi\}$, $C_{set} = \{User, Msg\}$, $User = \{u_1, u_2\}$, $Msg = \{hello, hi\}$, and a predicate $sent(u, m)$. This predicate means that a user u has been sent the message m.

Example 1. All users in the set *User* have been sent the message *hello*.

This is expressed as $sent(u_1, hello) \land sent(u_2, hello)$ which is equivalent to $\forall a_1 \in User\ sent(a_1, hello)$. This can be denoted (compactly) by the set predicate $sent(\forall User, hello)$. The scope of the quantifier is only for the set symbol before which it appears. In this case, the scope of $\forall$ is for the set *User*. Thus, $sent(\forall User, hello) \equiv \forall a_1 \in User\ sent(a_1, hello)$.

Example 2. All users in the set *User* have been sent some message from the set *Msg*.

This is expressed as $((sent(u_1, hello) \land sent(u_2, hello)) \lor (sent(u_1, hi) \land sent(u_2, hi)) \lor (sent(u_1, hello) \land sent(u_2, hi)) \lor (sent(u_1, hi) \land sent(u_2, hello)))$ which is equivalent to $\forall a_1 \in User\, \exists e_1 \in Msg\, sent(a_1, e_1)$. This can be denoted (compactly) by the set predicate $sent(\forall User, \exists Msg)$. Thus, $sent(\forall User, \exists Msg) \equiv \forall a_1 \in User\, \exists e_1 \in Msg\, sent(a_1, e_1)$.

Example 3. All users in the set $User$ have been sent all the messages from the set Msg.

This is expressed as $sent(u_1, hello) \wedge sent(u_2, hello) \wedge sent(u_1, h_i) \wedge sent(u_2, h_i)$ which is equivalent to $\forall a_1 \in User\, \forall a_2 \in Msg\, sent(a_1, a_2)$. This is denoted by the set predicate $sent(\forall User, \forall Msg)$. Thus, $sent(\forall User, \forall Msg) \equiv \forall a_1 \in User\, \forall a_2 \in Msg\, sent(a_1, a_2)$.

Note that in a set predicate at least one parameter is a quantified set symbol. This is not the case with a conventional predicate.

Definition 1. Set predicate. Let $\sigma_1, \ldots, \sigma_k, \sigma'_1, \ldots, \sigma'_{k'} \in C_{set}, o_1, \ldots, o_{k''} \in C_{obj}$. Let $b(\forall \sigma_1, \ldots, \forall \sigma_k, \exists \sigma'_1, \ldots, \exists \sigma'_{k'}, o_1, \ldots, o_{k''})$ be a $(k+k'+k'')$-ary ground set predicate such that: $b(\forall \sigma_1, \ldots, \forall \sigma_k, \exists \sigma'_1, \ldots, \exists \sigma'_{k'}, o_1, \ldots, o_{k''}) \equiv \forall a_1 \in \sigma_1, \ldots, \forall a_k \in \sigma_k \exists e_1 \in \sigma'_1, \ldots, \exists e_{k'} \in \sigma'_{k'} b(a_1, \ldots, a_k, e_1, \ldots, e_{k'}, o_1, \ldots, o_{k''})$.

In Example 1, $\sigma_1 = User, o_1 = hello, k = 1, k' = 0, k'' = 1$. In Example 2, $\sigma_1 = User, \sigma'_1 = Msg, k = 1, k' = 1, k'' = 0$. In Example 3, $\sigma_1 = User, \sigma_2 = Msg, k = 2, k' = k'' = 0$.

Set Operators. The definition of set actions are similar to conventional actions, with the difference that preconditions and postconditions (effects) can contain set predicates. By convention we will use lower case for object variables/constant and upper case for set variables/constants. Set actions can be very expressive, for example, the operators $send\_mail$ and $send\_mail\_iter$ are described in PDDL style as:

action: $send\_mail$: parameters: $(?member\ ?m)$ // an atomic action
precond: $(and(have\_mail(?member))(msg(?m)))$
effect: $(and(sent\_mail(?member\ ?m))(got\_msg(?member\ ?m)))$
action:$send\_mail\_iter$: parameters: $(?Member\ ?m)$ // iterated form of the atomic action
precond: $(and(have\_mail(\forall ?Member))(msg(?m)))$
effect: $(and(sent\_mail(\forall ?Member\ ?m))(got\_msg(\forall ?Member\ ?m)))$

In the above description, the argument $(?Member)$ is universally quantified in both the precondition and effect $(\forall ?Member)$; $(?m)$ is not quantified. We discuss the construction of an iterated action from a given atomic action in the next section.

Definition 2. A planning domain with sets is a tuple $D_{Set} = \langle C_{obj}, C_{set}, B, O \rangle$ where C_{obj} is a finite set of object symbols, C_{set} is a nonempty finite set of constant symbols denoting sets, B is a finite set of set predicates, and O is a finite set of operators, with $O = O_{CP} \cup O_{KP} \cup O_{set}$, O_{CP}: classical **planning** operators on objects, O_{KP}: knowledge **producing** operators on objects, O_{set}: classical planning operators on sets.

A planning problem for a domain D_{Set} is to find a (general form of) plan that transforms an initial state I to a final state G. At the end of Sect. 3 we obtain a new domain D_{iter} that contains only iterated classical planning operators on sets. Thus a planning problem for the domain D_{iter} can be solved by any classical planner [1,5].

3   Iterated Set Operators

Object based operators (O_{CP}) or (O_{KP}) can be iterated over sets in order to create set operators that will be called *iterated set operators*.

Given that *send_mail* will send a message to a member (who has an email address), we should obtain *send_mail_iter* that is supposed to send the same message to all the members (who have an email address). The parameters, precondition, and effect of *send_mail_iter* will be the same as that of *send_mail* except that every occurrence of *?member* should be replaced with an appropriate universally quantified set variable (e.g., $\forall?Member$). A PDDL-like description of *send_mail_iter* is given above.

If *send_mail_iter* is supposed to send all the messages to all the members (who have an email address), then both *?member* and *?m* should be replaced with appropriate universally quantified set variables (e.g., $\forall?Member$ and $\forall?Msg$ respectively). Thus given the purpose of an iterated operator, deriving it from the corresponding operator is easy. However we also need to ensure consistency of the effect. This is discussed below.

Definition 3. Conflicting Actions. Let the operators w_1 and w_2 be instantiated with two distinct object assignments $o' = [o'_1, \ldots, o'_m]$ and $o'' = [o''_1, \ldots, o''_n]$ respectively. Let a_1 and a_2 be the corresponding actions for w_1 and w_2 respectively. The actions a_1 and a_2 are said to be conflicting if either $pre(a_1) \wedge post(a_2)$ or $pre(a_2) \wedge post(a_1)$ are inconsistent (i.e., one is the negation of the other), where *pre* and *post* are the precondition and postcondition (effect) respectively.

The above definition can easily be extended for a single operator also. This is useful for iterated actions. Let the operator be w that is instantiated with two distinct object assignments $o' = [o'_1, \ldots, o'_m]$ and $o'' = [o''_1, \ldots, o''_n]$. Let the corresponding actions be a_1 and a_2. These two actions would be conflicting if they satisfy the condition given in Definition 3. We call such assignments o' and o'' as conflicting instantiations. The preconditions and postconditions (effects) are represented as conjunctions of predicates. Thus it can easily be determined whether two actions are conflicting or not.

Definition 4. Iterated classical planning operator on set: Condition. Given a classical planning operator on objects $w \in O_{CP}$, with parameters $(?p_1, \ldots, ?p_n)$, $pre(w), post(w)$, and the parameters $(?p_{j_1}, ?p_{j_2}, .., ?p_{j_m})$ that are to be iterated upon. Then an *iterated classical planning operator on a set, w_iter*, is obtained by replacing each parameter $?p_{j_k}$ with a quantified set variable $\forall?S_{j_k}$ in the precondition and effect, if there does not exist two distinct conflicting instantiations $w(o')$ and $w(o'')$. The precondition and effect of w_iter are respectively $pre(w\_iter) = pre(w)$ and $post(w\_iter) = post(w)$ with the above substitutions.

Example 4. Given the operator $send\_mail$ with parameters $?member$ and $?m$, we obtain $send\_mail\_iter$ with parameters $?Member$ and $?m$; precond: $(and$ $(have\_mail(\forall?Member))(msg(?m)))$, effect: $(and(sent\_mail(\forall?Member, ?m)$ $(got\_msg(\forall?Member, ?m)))$

In the following we establish the conditions that should be satisfied in order to construct an appropriate iterated operator corresponding to a given operator.

Proposition 1 *No common predicate. Let $Pred(x) \subseteq B$ be a function which returns the set of predicates in a formula x. If $Pred(pre(w)) \cap Pred(post(w)) = \emptyset$ then any operator $w \in O_{CP}$ can be iterated over any object variable to obtain w_iter.*

Proof. Consider an operator $w \in O_{CP}$ satisfying $Pred(pre(w)) \cap Pred(post(w)) = \emptyset$ and $p \in pre(w)(q \in post(w))$ denoting a generic precondition (effect) of w. By *reduction ad absurdum* suppose that actions $w(o_i)$ or $w(o_j)$ are conflicting, i.e., there exists at least a precondition p and an effect q such that $p(o_i) \wedge q(o_j)$ or $p(o_j) \wedge q(o_i)$ are inconsistent. We need to show that this is not possible under the given hypotheses of Proposition 1. Note that under the assumption $p(o_i)$ and $q(o_j)$ $(p(o_j)$ and $q(o_i))$ conflicts implies that p and q share the same predicate positively affirmed in one case and negated in the other one. But this contradicts the assumption that $Pred(pre(w)) \cap Pred(post(w)) = \emptyset$. Thus both $p(o_i) \wedge q(o_j)$ and $p(o_j) \wedge q(o_i)$ are consistent. □

Example 5. The operator $w = send\_mail$ satisfies the condition given in Proposition 1, since $Pred(pre(w)) \cap Pred(post(w)) = \{have\_mail\} \cap \{sent\_mail, got\_msg\} = \emptyset$. So we can iterate w to get $w\_iter = send\_mail\_iter$ that satisfies the pairwise conflict free consistency condition.

Proposition 2 *Common predicate. For any operator $w \in O_{CP}$ satisfying $Pred(pre(w)) \cap Pred(post(w)) = \emptyset$ and for some predicate $p \in Pred(pre(w))$ in the precondition, p is also negated in the effect, $\neg p \in Pred(post(w))$ can be iterated over any object variable to obtain w_iter if the following sufficient conditions hold:*

1. *each predicate appears only once either in $pre(w)$ or in $post(w)$ and*
2. *the corresponding parameters in preconditions and effects have the same order.*

Proof. Since all the objects in a set are distinct, if a predicate $p \in Pred(pre(w))$ is applied to the same object in the same order (condition 2) of $\neg p \in Pred(post(w))$ then two distinct instances of w cannot be in conflict because of the affirmed and negated pair, since any other action will have different parameters. The only conflict could arise from another predicate p with different parameters, which is excluded by condition 1. Thus both $\neg p(o_i) \wedge p(o_j)$ and $\neg p(o_j) \wedge p(o_i)$

are consistent. Let the precondition and effect of w_iter be $\neg p(S)$ and $p(S)$ respectively. By Definition 1 , $\neg p(\forall S) = \neg p(o_1) \wedge \ldots \wedge \neg p(o_n)$ and $p(\forall S) = p(o_1) \wedge \ldots \wedge p(o_n)$ are consistent since for any distinct $i, j, \neg p(o_i), p(o_j)$ are consistent. Thus w_iter obtained in this way satisfies the condition given in Definition 4. □

In order to define the iteration of a knowledge producing (sensing) operator we need to extend the domain $D_{Set} = \langle C_{obj}, C_{set}, B, O \rangle$ (see Definition 2) to $D_{Set} = \langle C_{obj}, C_{set}, B \cup \{partition\}, O \cup \{new, add\} \rangle$, where the operator $new(!S)$ returns as output a new empty set $!S$, the operator $add(?o, ?S)$ adds an object $?o$ to an existing set $?S$, and the set predicate $partition(\forall ?S, \exists!S_1, \ldots, \exists!S_n)$ is true iff $S = S_1 \cup \ldots \cup S_n$ and for all $i \neq j$, $S_i \cap S_j = \emptyset$. Whereas S is an input variable, $S_1, \ldots, S_n$ are output variables.

Definition 5. Iterated knowledge producing operator on set. Condition. Given $O \cup new, add$, the special predicate $partition$, and a knowledge producing operator on objects $w \in O_{KP}$ with parameters $(?p_1, \ldots, ?p_n)$, $pre(w)$, the parameters $(?p_{j_1}, \ldots, ?p_{j_m})$ that are to be iterated upon, possible outcomes (effects) $\varepsilon 1_w, \ldots, \varepsilon k_w$, it is possible to iterate w over any set S, if and only if, there exists a plan in the extended domain D_{Set} that for all $o_i \in S$ contains an instantiation $w(o_i)$ of w such that $w(o_i)$'s are not in any order conflict to each other for each possible outcome, and the plan creates a partition $\{S_1, \ldots, S_k\}$ of S, for which each $\varepsilon 1_w(S_j)$ $(j = 1 \ldots k)$ holds after the execution of the plan. The precondition and the effect of the iterated operator w_iter be respectively $pre(w\_iter) = pre(w)$ where each parameter $?p_{j_k}$ is replaced with a quantified set variable $\forall ?S_{j_k}$ and $post(w\_iter) = \varepsilon 1_w(\forall!S_1) \wedge \ldots \wedge \varepsilon k_w(\forall!S_k) \wedge partition(\forall ?S, \exists!S_1, \ldots, \exists!S_k)$.

It is worth noticing that while w is a knowledge producing operator on objects, w_iter is a classical planning operator on sets. Definition 5 considers the general class of knowledge producing operator on objects. Recall that $check\_mail(x)$ determines whether $have\_mail(x)$ or $\neg have\_mail(x)$ holds. For such types of possible effects we have:

Proposition 3 Negated predicate. *Given an operator $w \in O_{KP}$ with two possible effects ε_w and $\neg \varepsilon_w$, if the classical planning operators w_1 and w_2 (obtained by restricting w to each effect) can be iterated on an input parameter $?S$, such that it satisfies the conditions in Proposition 2 , then w can be iterated on the same parameter $?S$ to yield w_iter as per Definition 5.*

Proof. Conditions on w_1 and w_2 guarantee the absence of conflicts in all instantiations of w, while plan existence can be shown constructively as in Fig. 1, where given the iteration on a set S, the plan contains $\forall o_i \in S$ an instance $w(o_i)$, two actions $new(S_1)$ and $new(S_2)$ are executed before all the other actions, and each outcome $\varepsilon_w(\neg \varepsilon_w)$ for an object o_i is followed by an action $add(o_i, S_1)$ (an action $add(o_i, S_2)$) which is executed only in the context of the respective outcome. It is easy to see that after the plan has been executed, $\varepsilon_w(\forall!S_1) \wedge \neg \varepsilon_w(\forall!S_2) \wedge partition(\forall ?S, \exists!S_1, \exists!S_2)$ holds. □

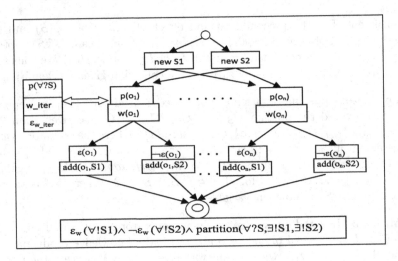

Fig. 1. w iterated as per Proposition 3 with $S = o_1, \dots, o_n$.

Example 6. Consider the sensing action $check\_reg(?x, ?y)$ with precondition $user(?x) \wedge conf(?y)$ and alternative effects $have\_registered(?x, ?y)$ and $\neg have\_registered(?x, ?y)$ that checks if user $?x$ has registered to conference $?y$. Iterating it on the set of users $S = \{u1, u2, u3\}$, $check\_reg\_iter(\forall?S, ABC)$ will produce as output two new sets S_1 and S_2 and $have\_registered(\forall!S_1, ABC) \wedge \neg have\_registered(\forall!S_2, ABC) \wedge partition(\forall?S, \exists!S_1, \exists!S_2)$ holds.

It follows from the conditions of the Propositions 1, 2 and 3 that if we can construct iterated actions corresponding to knowledge producing actions in a plan, then we can use a classical planner on the domain D_{iter} obtained from D'_{set} as given below.

Given a basic domain $D'_{set} = \langle C_{obj}, C_{set}, B \cup \{partition\}, O \cup \{new, add\}\rangle$, where $O = O_{CP} \cup O_{KP} \cup O_{set}$, an extended domain D_{iter} is built by constructing new set operators by iteration of operators on objects. The domain $D_{iter} = \langle C_{obj}, C_{set}, B \cup \{partition\}, O'\rangle$ with $O' = O_{set} \cup O_{iter}$ where operators in O_{iter} are obtained by applying Propositions 1, 2 and 3 to operators in O_{CP} and O_{KP}. Note that O' does not contain any knowledge producing operator. It contains only iterated classical planning operators on sets. Thus a planning problem (to find a plan) for the domain D_{iter} can be solved by any classical planner.

Theorem 1. *If there exists a solution plan P_{iter} for $\langle D_{iter}, I, G\rangle$ then there exists a solution plan P for $\langle D'_{Set}, I, G\rangle$.*

The proof is omitted due to space restrictions.

4   PDDL Specifications and Implementation

We now show how the iterated operators on sets are described in PDDL with respect to the operators of the conference management domain-Section 1.

A set predicate like *have_mail*(∀?*Member*) cannot be directly used in PDDL. So we define a new predicate *set* and the predicate *have_mail*, as given below, to obtain the desired result. The implementable action description in PDDL is given below:

(:predicates (*set* ?*c*)(*partition* ?*c* ?*d* ?*e*)(*member* ?*c*)(*msg* ?*x*)(*get_msg* ?*c* ?*x*)
(*have_mail* ?*c*)(*sent_mail* ?*c* ?*x*)(*called* ?*c*?*x*)(*output* ?*x*)(*not_have_mail* ?*c*))

(:action *send_mail_iter*: parameters (?*c* ?*m*)
:precond (*and*(*set* ?*c*)(*have_mail* ?*c*)(*msg* ?*m*))
:effect (*and*(*sent_mail* ?*c* ?*m*)(*get_msg* ?*c* ?*m*)))
The specification of *phone_call_iter* is similar:

(:action *phone_cal_iter*: parameters (?*c* ?*m*)
:precond (*and*(*set*?*c*)(*not*(*have_mail* ?*c*))(*not_have_mail* ?*c*)(*msg* ?*m*))
:effect (*and*(*called* ?*c* ?*m*)(*get_msg* ?*c* ?*m*)))

A PDDL specification of *check_mail_iter* is given below:

(:action *check_mail_iter*: parameters (?*c* ?*c1* ?*c2*)
:precond (*and*(*set* ?*c*)(*member* ?*c*)(*output* ?*c1*)(*output* ?*c2*))
:effect (*and*(*set* ?*c1*)(*set* ?*c2*)(*partition* ?*c* ?*c1* ?*c2*)(*have_mail* ?*c1*)
(*not*(*have_mail* ?*c2*))(*not_have_mail* ?*c2*)(*not*(*output* ?*c1*))(*not*(*output* ?*c2*))))

The predicate (*partition*?*c*?*c1*?*c2*) is true when the set ?*c* is partitioned into two sets ?*c1* and ?*c2*. For all the objects in the set ?*c1*, *have_mail* holds. For all the objects in the set ?*c2*, *have_mail* does not hold. We need to specify explicitly what holds for the objects in the set ?*c2*. So the predicate *not_have_mail* is included.

Output variables have been implemented according to the technique proposed in [9] by the special predicate *output*(*x*) which denotes the symbol *x* as an output variable. Reasoning on the special predicate partition has been implemented by adding to the domain for each predicate *p* a special dummy action on sets *dummy_union_p*, with precondition (*p*(?*S1*) ∧ *p*(?*S2*) ∧ *partition*(?*S* ?*S1* ?*S2*)) and effect *p*(?*S*). The introduction of *dummy_union_p* makes the BlackBox planner [1] able to do reasoning and draw conclusions such as: if *p* holds for every element of a partition of *S* then *p* also hold for *S*. The specification of the dummy action for *p* = *get_msg* is:

(:action *dummy_union_get_msg* :parameters (?*c* ?*c1* ?*c2* ?*m*)
:precond (*and*(*set* ?*c1*)(*set* ?*c2*)(*set* ?*c*)(*partition* ?*c* ?*c1* ?*c2*)(*get_msg* ?*c1*?*m*)
(*get_msg* ?*c2* ?*m*))
:effect (*get_msg*?*c* ?*m*)))

The above PDDL specifications were executed using the BlackBox planner [1] by taking: (:objects *s1* *s2* *s3* *m1*)

(:init (*set* *s1*)(*member* *s1*)(*msg* *m1*)(*output* *s2*)(*output* *s3*))
(:goal (*get_msg* *s1* *m1*))

The plan obtained is:

$check\_mail\_iter(s1\ s2\ s3); send\_mail\_iter(s2\ m1); phone\_call\_iter(s3\ m1);$
$dummy\_union\_get\_msg(s1\ s2\ s3\ m1);$

5   Conclusions and Related Work

This paper introduced set predicates that render an effective formal specification tool for iterated actions on sets. We showed how such actions can be built from atomic actions and specified using PDDL. The iterated actions on sets have deterministic effect although the knowledge producing actions have nondeterministic effects. Thus plan synthesis of iterated actions can be done using any classical planners.

It is worth noticing that an iterated action on sets, as discussed in our work, is very different from repetition of the same action several times as in [2,7]. In [7] the action *chop*, which hits the tree once with the axe, is repeated. In [2] a conditional plan using the actions *pick-paper, go-right, leave-paper* is repeated. However such actions cannot be instantiated on sets of objects. Thus the ability of operating on sets and fulfilling goals expressed on sets allows to provide a more compact and expressive representation as well as to manage and to solve more complex problem descriptions. The correctness of the proposed approach has been formally proved and experimentally validated. To the best of our knowledge, no other proposal of this kind for iterated actions has been investigated earlier. As part of our ongoing and future work we are studying the applicability of our approach for planning problems with structured data types.

Acknowledgements. The authors thank the anonymous reviewers of ISMIS 2015 for their valuable comments and suggestions for improving the paper.

References

1. Blackbox. http://www.cs.rochester.edu/kautz/satplan/blackbox
2. Bertoli, P., Cimatti, A., Lago, D.U., Pistore, M.: Extending pddl to nondeterminism, limited sensing and iterative conditional plans. In: ICAPS Workshop on PDDL (2003)
3. Bertoli, P., Cimatti, A., Pistore, M., Roveri, M., Traverso, P.: Mbp: a model based planner. In: IJCAI Workshop on Planning under Uncertainty and Incomplete Information (2001)
4. Ghallab, M., Nau, D., Traverso, P.: Automated Planning: Theory and Practice. Elsevier, Melbourne (2004)
5. Hoffmann, J., Nebel, B.: The ff planning system: fast plan generation through heuristic search. J. Artif. Intell. Res. **14**, 253–302 (2001)
6. Levesque, H.J.: What is planning in the presence of sensing? In: AAAI, pp. 1139–1146 (1996)
7. Levesque, H.J.: Planning with loops. In: IJCAI, pp. 509–515 (2005)
8. McDermot, D., Ghallab, M., Howe, A., Knoblock, C., Ram, A., Veloso, M., Weld, D., Wilkins, D.: Pddl. Technical report CVC TR98003/DCS TR1165

9. Milani, A., Rossi, F., Pallottelli, S.: Planning based integration of web services. In: Iinternational Conference on Web Intelligence and Intelligent Agent Technology, pp. 125–128 (2006)
10. Zhang, X.: Reasoning about sensing actions in domains with multi-valued fluents. Stud. Logica **68**, 1–26 (2001)

Qualitative Planning of Object Pushing by a Robot

Domen Šoberl[(✉)], Jure Žabkar, and Ivan Bratko

Faculty of Computer and Information Science, Artificial Intelligence Laboratory,
University of Ljubljana, Večna pot 113, 1000 Ljubljana, Slovenia
{domen.soberl,jure.zabkar,ivan.bratko}@fri.uni-lj.si

Abstract. Pushing is often used by robots as a simple way to manipulate the environment and has in the past been well studied from kinematic and numerical perspective. The paper proposes a qualitative approach to pushing convex polygonal objects by a simple wheeled robot through a single point contact. We show that by using qualitative reasoning, pushing dynamics can be described in concise and intuitive manner, that is still sufficient to control the robot to successfully manipulate objects. Using the QUIN program on numerical data collected by our robot while experimentally pushing objects of various shapes, we induce a model of pushing. This model is then used by our planning algorithm to push objects of previously unused shapes to given goal configurations. The produced trajectories are compared to smooth geometric solutions. Results show the correctness of our qualitative model of pushing and efficiency of the planning algorithm.

1 Introduction

Manipulating objects is usually a fairly straightforward task for many industrial robots. However, it still poses a great challenge to researchers in cognitive robotics, who aim to endow robots with sufficient intelligence to act autonomously within unstructured environments. Advanced robots may learn to move object by grasping. However, another way of moving objects, often used by simple wheeled robots, is by pushing. In many cases it is not practical or even not possible to model the precise robot kinematics, therefore many planning approaches incorporate some form of machine learning [16]. Coupled with qualitative reasoning [5], simple and intuitive models can be learned which are more easily interpreted by humans than models induced by traditional quantitative machine learning techniques [1].

Due to the absence of precise numerics, planning of continuous robotic motion using qualitative models can be achieved in a limited way. Typically, qualitative models are used on an abstract level of planning to produce general execution plans which are not yet directly executable, but still require a sort of numerical fine-tuning. A pure qualitative approach to pushing objects by a robot has been demonstrated in [12]. However, using allocentric approach (changes are observed from the object's point of view) their model was limited to rectangular objects.

© Springer International Publishing Switzerland 2015
F. Esposito et al. (Eds.): ISMIS 2015, LNAI 9384, pp. 410–419, 2015.
DOI: 10.1007/978-3-319-25252-0_44

In this paper we propose a qualitative approach to pushing arbitrary convex polygonal objects by a wheeled robot. Our method covers the induction of the pushing model as well as motion planning with the induced model. Experimentally pushing objects of different shapes we identify common qualitative relations and propose a unified qualitative model of pushing. We introduce a novel on-line qualitative planning algorithm that we use to push objects, and which should easily be adaptable for motion planning in other similar domains. We validate the correctness of our model and evaluate its efficacy through several pushing tasks.

The rest of the paper is organized as follows. In Sect. 2 we give an overview of related work on qualitative induction and planning, followed by a brief description of our robotic domain and experimental setup in Sect. 3. In Sect. 4 we describe the method of inducing the model of pushing using the QUIN program and describe the obtained model. This model is then used by our motion planning algorithm which is described in Sect. 5. Planning results are presented in Sect. 6. Finally, in Sect. 7, we present conclusions and indicate future work.

2   Related Work

Learning qualitative models is a process of inducing qualitative models of a system from a qualitative or quantitative description of the system's measured behavior. Earlier algorithms [6,9,11] used qualitative system behaviors to produce qualitative constraints in the form of *Qualitative Differential Equations* (QDE) which were introduced by De Kleer in the early 1980s [4]. With the emergence of qualitative induction algorithms that induce qualitative models from numerical data, learning qualitative models gained more interest in the field of robotic learning by experimentation. Algorithm QUIN [2] can induce qualitative models in the form of binary decision trees, from the input numerical data. The process is similar to induction of regression trees. Šuc [13] showed how QUIN can be used to induce a qualitative human control strategy for riding a bicycle and operating the crane. Experiments with simulated aircraft flying controllers [14] combine traditionally modeled flying controllers with QUIN generated qualitative rules. In all cases, induced qualitative models provided good and intuitive insight into the human control skill.

Planning with qualitative models is often implemented in combination with some form of trial-and-error learning [10,15]. Qualitative models are used on the abstract level of planning to produce a general plan on how to achieve the goal. Such plans are not executable directly and need additional fine-tuning of numerical parameters. State-of-the-art qualitative planner is embedded in the QLAP (Qualitative Learner of Action and Perception) system [8]. A hierarchical structure of qualitative actions is first experimentally learned by the robot. Given a task, the robot then constructs a plan by chaining higher-level actions towards the lower ones representing immediate motor commands. Random values within qualitative constraints are then sent to the motors while statistics on reliability of plans is kept.

The possibility to produce directly executable plans in an on-line manner was indicated in [12]. Using induced qualitative model of pushing a box the robot was able to push the box to the given goal position. The focus of that paper is on model induction and leaves the problem of planning still to be addressed, while the proposed solution is strictly domain-specific. Our work fills this gap by proposing a more general approach to this problem. We extend the set of possible shapes to convex polygonal and propose a planning algorithm that should easily be ported to similar motion planning problems.

3   The Pushing Domain

To push objects we use a simple two-wheeled robot with a one-pointed frontal bumper. We move the robot by independently sending the desired translational velocity v and rotational speed ω to the controller, which then applies appropriate power to the left and right wheels. The robot tracks the environmental dynamics through a stationary overhead camera. Every item is uniquely colored and marked with asymmetric symbols for computer vision routines to correctly identify its position and orientation. The robot moves an object by making a contact with the tip of its bumper. By the robot's moving forward and rotating, translation and rotation of the pushed object can be achieved. A task is given to the robot to push the object to a goal location and orientation.

The attributes we use are depicted in Fig. 2(a). The coordinate system is always relative to the robot (egocentric approach), therefore its absolute placement is not considered explicitly. However, we do record the robot's rotational differential $\Delta\theta$ when an action is executed. The object's position is denoted by (x, y) and its orientation by β. (x_g, y_g) and β_g are goal position and orientation. Because the coordinate system is bound to the robot, these values would change with every robot's movement, even when no actual pushing is made. The manner of pushing the object through contact with a chosen side e is described by variables τ, φ, v and ω. When a contact is made, $\tau \in [-1, 1]$ denotes the offset from the geometrical center of the contacted edge. If no contact is made, τ is undefined. The angle of pushing φ is the angular offset from the perpendicular to the edge, so that φ increases with CCW rotation of the robot. While pushing, only translational v and rotational ω speeds of the robot are independent variables. When choosing a new contact point, the robot can freely select the edge e and values of τ and φ.

4   Learning a Model of Pushing

4.1   Induction and Generalization

To induce a qualitative model of pushing, we conducted pushing experiments with objects of different shapes. This was done in a real-time robotic simulator. For each object a separate model was induced. Models were then compared and their common properties identified as general. The shapes used were rectangle, equilateral triangle and parallelogram. Such selection of shapes was used to cover the differences in the number of edges and the center of mass. All the objects were of the same weight and had uniformly distributed mass.

We executed a 1000 randomly generated pushing actions on each object, and recorded the observed numerical changes. An action is uniquely determined by the values of independent variables τ, φ, ω, v, and the chosen object's edge e. The observed, dependent variables, here called attributes, were the following: translation of the object's center of mass along the axis of pushing Δy and to the side Δx, the rotation of the object $\Delta\beta$, the rotation of the robot $\Delta\theta$, the change in angle of pushing $\Delta\varphi$ and the sliding of the contact point $\Delta\tau$. The duration of each action was exactly 0.5 s, therefore we interpret all observed attributes as speeds and use the dot notation, e.g. $\dot\beta$ for $\frac{\Delta\beta}{\Delta t}$. Note that the positive rotational speed indicates the rotation in CCW direction and negative in CW direction.

After the data were collected, program QUIN was used to induce a qualitative decision tree for each observed attribute. Every such tree defines qualitative relations between the observed attribute and independent variables, under the given conditions. Since our robot has to initially select the contact point τ, φ, and can freely alter speeds v and ω while pushing, our trees represent functions of the form $f(\tau, \varphi, \omega, v)$. Qualitative rules in the form $z = M^{+-}(x, y)$ are given in the leaves of the tree. This notation means that the value z increases monotonically with parameter x and monotonically decreases with parameter y.

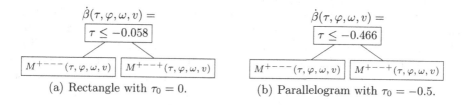

(a) Rectangle with $\tau_0 = 0$.   (b) Parallelogram with $\tau_0 = -0.5$.

Fig. 1. Qualitative classification trees induced for variable $\dot\beta$ (rotational speed of the object). The root condition reflects the object's center of mass.

Figure 1 shows induced trees for attribute $\dot\beta$ in the case of pushing rectangle and parallelogram-shaped objects. In the case Fig. 1(a) the pushed rectangular object had its center of mass projected at $\tau_0 = 0$. With slight error, this fact was captured by the root condition of the induced tree. A similar tree was also obtained by pushing an equilateral triangle. In the case of parallelogram Fig. 1(b) the center of mass was moved to $\tau_0 = -0.5$ which was also clearly captured by QUIN. The difference between the left and the right branch is the opposite qualitative relation between v and $\dot\beta$. Intuitively, higher speed v would increase the rotational speed $\dot\beta$ of the object when the robot is touching it along the right side from its balance point τ_0, and decrease $\dot\beta$ when it is touching it along the left side. Note that negative $\dot\beta$ means rotation of the object in the clock-wise direction.

The presented trees are shape-specific. However, some relations are general in the sense that they hold for all the shapes. In the case of $\dot{\beta}$ the qualitative relation $M^{+--}(\tau, \varphi, \omega)$ holds true if v is kept constant. It turns out that such constraint enforces no real limitations to the pushing task as the robot is always inclined to use its maximum speed to get the job done quickly, while rotations are controlled by altering other parameters.

4.2 Model Interpretation

Using the above approach of identifying commonly valid relations on all induced trees we obtained the following general qualitative model of pushing:

$$
\begin{aligned}
\dot{y} &= M^+(v) & \dot{\varphi} &= M^+(\omega) \\
\dot{x} &= M^{--}(\tau, \omega) & \dot{\tau} &= M^{--}(\omega, \varphi) \\
\dot{\beta} &= M^{+--}(\tau, \varphi, \omega) & \dot{\theta} &= M^+(\omega)
\end{aligned}
\tag{1}
$$

The model describes qualitative planar behavior of an object when pushed by the robot against one of its sides. Note that the observed attributes are differentials and can be interpreted as speeds, and the independent variables are controllable directly by the robot. The first three rules ($\dot{y}$, $\dot{x}$ and $\dot{\beta}$) describe the behavior of the pushed object from the robot's point of view. Increasing the speed v the speed of the object along the y-axis (robot's heading direction) will also increase. The speed of the object along the x-axis can be controlled either by changing the contact point τ or altering the rotational speed ω. Moving the contact point right (increasing τ) or changing the rotational speed ω in favor of CCW direction (increasing ω) will result in decreasing the speed of the object's central point along the x-axis. The inverse also applies. The third rule states three possibilities to rotate the object. The robot can increase the object's rotational speed by decreasing its own rotational speed and vice-versa. Doing so will obviously also change the pushing angle φ, therefore the action of changing ω can be seen as an equivalent to the action of changing φ. The third way to increase or decrease the object's rotational speed is to move the contact point τ to the right or left, respectively.

The remaining three rules ($\dot{\varphi}$, $\dot{\tau}$ and $\dot{\theta}$) describe the behavior of the robot and its relation to the pushed object. They are not directly connected to object's trajectory planning, but can be useful to overcome certain limitations. In our domain there are two constraints which should not be violated while planning. The triangular shape of the bumper allows the pushing angle to be within $\varphi \in (-30°, 30°)$. According to the fourth rule, the robot can adjust this value by changing its rotational speed ω. The second constraint $\tau \in (-1, 1)$ prevents the contact point τ to slide off the edge. This sliding is described by the fifth rule. When the robot increases its pushing angle in the CW direction (decreasing φ), the contact point τ would slide faster towards the right extreme ($\tau = 1$) or loose its speed towards the left extreme ($\tau = -1$). The inverse also applies. Since φ is directly affected by ω, the same rule also holds for ω. If the friction between the robot's bumper and the object is high, this constraint can be omitted. The last rule states that the robot's orientation θ is directly affected by its rotational speed ω.

5 Qualitative Motion Planning

Using qualitative rules (1) the robot can be guided to push arbitrary convex polygonal objects to a given location and orientation. We presume the domain is free of obstacles and the robot is autonomously capable of positioning itself relative to the object, given by (e, τ, φ), where e is the chosen object's edge. The task of our planner is to produce a sequence of qualitative actions to continuously push the object positioned at (x, y, β) to the goal position (x_g, y_g, β_g). Doing so, the planner must avoid violating constraints $\varphi \in (-30°, 30°)$ and $\tau \in (-1, 1)$. Here we define *qualitative action* as a mapping of domain attributes into the set $\{+, 0, -\}$. We will write $a = (A_1^+, A_2^-)$ to denote an action a that maps attributes $A_1 \mapsto +$, $A_2 \mapsto -$ and all others to 0. We also presume all attributes are real-valued, which is certainly true for our domain. Assigning $+$ or $-$ to an attribute means the value of that attribute should respectively increase or decrease, while 0 means the value may change arbitrarily or stay unchanged. The planner outputs only the directions in which the values of attributes should change while the actual rates of change depend on the properties and capabilities of the system itself.

Comparing the current position of the object (x, y, β) to its goal position (x_g, y_g, β_g), we can construct a qualitative action the object should perform, e.g. $a_{\mathrm{obj}} = (\dot{x}^+, \dot{y}^-, \dot{\beta}^+)$ when $x < x_g, y > y_g, \beta < \beta_g$. However, the object cannot move by itself, therefore the desired dynamics must be translated to robot's qualitative actions. The robot directly controls parameters v and ω, so the qualitative action received from the planner is of the form $a_i = (v^{\{+,-,0\}}, \omega^{\{+,-,0\}})$. The robot therefore chooses among 9 possible actions. Generally, using given qualitative rules, every action $a_i = (A_1^{\{+,-,0\}}, A_2^{\{+,-,0\}}, \ldots, A_n^{\{+,-,0\}})$ is evaluated and assigned a score:

$$\text{score}(a_i) = \sum_{k=1}^{n} D(A_k) \cdot W(A_k). \tag{2}$$

The value of the function $D(A_k)$ determines whether under the action a_i the attribute A_k is moving in the desired direction:

$$D(A_k) = \begin{cases} 1 & A_k \text{ is moving towards its goal value,} \\ -1 & A_k \text{ is moving away from its goal value,} \\ 0 & \text{there are no demands for } A_k. \end{cases}$$

Function $W(A_k)$ sets the weight (importance) assigned to the attribute A_k. The weights are determined dynamically, so that attributes are farther from their goal values carry higher weights. However, attribute metrics may differ and are provided to the planner as a part of the background knowledge in the form of attribute norms $N(A_k)$. To every attribute A_k with a goal value A_k^g, the following weight is assigned:

$$W(A_k) = \frac{\|A_k - A_k^g\|}{N(A_k)}. \tag{3}$$

In our case we have three such attributes, namely x, y and β with their goal values x_g, y_g and β_g, respectively. Their norms were set experimentally to obtain good performance. Generally, a lower norm implies a higher attribute importance. Hence a lower norm for β will make the planner rotate the box first and then push it to its goal location.

Attributes with no set goal value are considered if they are bounded, as in our case $\varphi \in (-30°, 30°)$ and $\tau \in (-1, 1)$. The general idea is that the weight of such an attribute should be relatively small until a critical point is reached, when it should increase very rapidly. Any function with such a behavior should suffice. We used the following weight function:

$$W(A_k) = \sec\left(\frac{\pi}{b-a} \cdot \left(A_k - \frac{a+b}{2}\right)\right) - 1, \qquad (4)$$

where $A_k \in (a, b)$. The *secant* function $\sec(x)$ is $\frac{1}{\cos(x)}$. Using this function, the weight of attribute will stay below 1 until its value is at least 1/6-th of its range away from either extreme. As the value approaches an extreme, the weight starts to increase very rapidly.

Additional domain constraints are also considered and fed to the planner. In our case, the robot is unable to pull an object, while the rule $\dot{y} = M^+(v)$ in (1) implies that by negative v a negative $\dot{y}$ should eventually be reached. To prevent the planner from trying to pull the object, constraint $\dot{y}^{\{+,0\}}$ is added. Whenever an action violates such a constraint it is assigned a negative score. In our case only parameter v is qualitatively related to $\dot{y}$ which automatically results in the elimination of all actions containing v^-.

The action with the highest score is considered. A positively scored action will have a beneficial overall effect on reaching the goal configuration. While it might still move certain attributes away from their goal values, the benefit on (currently) more important attributes justifies the decision. If the highest score is not positive, the robot either reached the goal or came to a dead end. In the latter case, configuration of the robot to the object needs to be changed in order to continue planning. The robot first evaluates the extended set of actions $a_i = (v^{\{+,-,0\}}, w^{\{+,-,0\}}, \tau^{\{+,-,0\}}, \varphi^{\{+,-,0\}})$. If a positively scored action is now found, the robot readjusts the contact point (τ, φ) on the spot, without changing the side of pushing. Otherwise, a different side of the object must be selected which is more costly since the robot needs to reposition itself. In that case, possible actions on all edges are evaluated and the configuration, allowing the highest scored action, is selected.

6   Planning Results

We conducted 12 pushing experiments with 4 differently shaped objects as shown in Table 1. All objects had the same surface size, weight and their mass was uniformly distributed. The robot was able to achieve the speed up to 10 cm/s while pushing an object. Each object was tested in three scenarios. In the first setting the object was placed (1 m, 1 m) (as in Fig. 2(b)) from the goal position

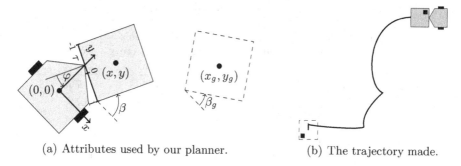

(a) Attributes used by our planner.   (b) The trajectory made.

Fig. 2. A pushing scenario: square object is initially displaced $(1m, 1m)$ from the goal position and rotated by $180°$.

unrotated. This makes the initial distance of $\sqrt{2}\,\mathrm{m}$. The distance for the second setting was the same, but the object was also rotated by $180°$. In the third setting the object was placed on its goal position but rotated by $180°$. The following norms (3) for attributes were used: $N(x) = 1000\,\mathrm{cm}$, $N(y) = 100\,\mathrm{cm}$ and $N(\beta) = 90°$. This means the robot will put the same emphasis on rotation and translation when the object is rotated by $90°$ and displaced by $100\,\mathrm{cm}$. The high norm on x puts more emphasis on translation along the y-axis (the heading direction of the robot).

We captured trajectories made by objects while being pushed, and measured their lengths. It has been shown [7] that a robot, similar to ours, pushing a box, can behave similar to a simple car, known as Dubin's car. Our robot can achieve such behavior by controlling the pushing angle φ. The method for construct-ing shortest possible paths for such systems was proposed in [3]. We compared lengths of trajectories made by our robot with optimal Dubin's trajectories. We also counted the number of times the robot needed to reposition itself, not counting the initial positioning. During each experiment the robot evaluated the accuracy of the given model (1) by comparing the predicted and actual behavior of each action. This was done on the level of attributes, so an action that cor-rectly predicts the dynamics of three attributes out of four is 75 % correct. The errors that occurred were typically caused by noise and occasional unexpected events, such as a sudden uncontrolled slip of the contact point. When objects were displaced from the goal position but not rotated, the robot was inclined to make an "L"-shaped trajectory, avoiding changing already correct orientation. In the case of the equilateral triangle this approach resulted in a zig-zag pat-tern. Due to the shifted point of mass in deltoid and right triangle, these shapes were, from the point of view of a chosen edge, more inclined to rotate in one or the other direction. This forced the robot to put more emphasis into correcting the orientation, hence longer trajectories and more repositionings. Rotating the displaced objects forced the robot to consider its rotations which resulted in a curved path towards the goal, as shown in Fig. 2(b). Such curved trajectories were not much longer from the above straight ones. When an object is already

Table 1. Planning results on four different types of objects.

Object shape	Goal distance (m,deg)	Trajectory length [cm]	Num of. reposi- tionings	Dubin's trajectory [cm]	Model accuracy
Square	$(\sqrt{2}, 0)$	205	2	147	98.7 %
	$(\sqrt{2}, 180°)$	217	3	194	96.7 %
	$(0, 180°)$	61	20	168	91.0 %
Deltoid	$(\sqrt{2}, 0)$	251	3	144	99.8 %
	$(\sqrt{2}, 180°)$	229	5	167	96.2 %
	$(0, 180°)$	79	16	86	98.2 %
Equilateral triangle	$(\sqrt{2}, 0)$	298	8	145	99.8 %
	$(\sqrt{2}, 180°)$	263	8	181	97.6 %
	$(0, 180°)$	76	19	131	97.9 %
Right triangle	$(\sqrt{2}, 0)$	236	8	144	98.7 %
	$(\sqrt{2}, 180°)$	260	7	162	95.7 %
	$(0, 180°)$	72	15	71	86.2 %

placed at its goal position but rotated, the robot tends to rotate it on the spot, avoiding its displacement. This results in comparatively short trajectories and very high number of repositionings. Although the short trajectory may lead us to assume a better solution to Dubin's path, we must also consider the fact that every repositioning can be very time-consuming. It is also interesting to note the smaller number of repositionings for irregular shapes. It turned out that deltoid and right triangle allowed larger rotations due to their longer shape.

7   Conclusion

In this paper we proposed a qualitative approach to planar pushing of objects by a robot, covering the aspects of learning how to push and the method of planning the pushing motion. We considered convex polygonal objects and showed that they exhibit common qualitative behavior when pushed by a single contact point. We introduced a planning algorithm which can use the proposed qualitative model of pushing to plan pushing motions and guide the robot to push the given object to its goal configuration. Results showed high rate of matching between predicted and observed effects of executed actions with small deviations due to sensing and controlling noise or occasional slip of the pushing contact. The robot successfully completed all given tasks with lengths of trajectories being comparable to the lengths of optimal Dubin's trajectories. We leave open the question of using similar methods to pushing concave and curved objects.

Redefinition of touching point τ should be considered for curved objects, especially in the case of cornerless shapes. In the case of concave objects, the frame of the robot could impose additional limitations on pushing configurations.

We believe the presented planning algorithm can be used in other similar continuous robotic domains, given a proper model and domain-specific constraints. Still many improvements are possible. The attribute norms which encode certain domain knowledge should be learned or fine-tuned experimentally by the robot itself. The need to reposition frequently could be tackled by dynamically adding favorable and reachable configurations as intermediate goals. Such intermediate goals could also be used to avoid planar obstacles.

References

1. Bratko, I.: An assessment of machine learning methods for robotic discovery. CIT J. **16**(4), 247–254 (2008)
2. Bratko, I., Šuc, D.: Learning qualitative models. AI Mag. **24**(4), 107–119 (2003)
3. Bui, X.N., Boissonnat, J.D., Soueres, P., Laumond, J.P.: Shortest path synthesis for Dubins non-holonomic robot. In: Proceedings of the 1994 IEEE International Conference on Robotics and Automation 1, pp. 2–7 (1994)
4. De Kleer, J.: A Qualitative Physics Based on Confluences. Artif. Intell. **24**(1–3), 7–83 (1984)
5. Forbus, K.D.: Qualitative process theory. Artif. Intell. **24**(1–3), 85–168 (1984)
6. Hau, D.T., Coiera, E.: Learning qualitative models of dynamic systems. Mach. Learn. **26**, 177–211 (1993)
7. Lynch, K.M., Mason, M.T.: Stable pushing: mechanics, controllability, and planning. Int. J. Robot. Res. **15**(6), 533–556 (1996)
8. Mugan, J., Kuipers, B.: Autonomous learning of high-level states and actions in continuous environments. IEEE Trans. Auton. Mental Dev. **4**(1), 70–86 (2012)
9. Richards, B.L., Kraan, I., Kuipers, B.J.: Automatic abduction of qualitative models. In: Proceedings of the Tenth National Conference on Artificial Intelligence (AAAI 1992). pp. 723–728 (1992)
10. Sammut, C., Yik, T.F.: Multistrategy learning for robot behaviours. In: Koronacki, J., Raś, Z.W., Wierzchoń, S.T., Kacprzyk, J. (eds.) Advances in Machine Learning I. SCI, vol. 262, pp. 457–476. Springer, Heidelberg (2010)
11. Say, A.C.C., Kuru, S.: Qualitative system identification: deriving structure from behavior. Artif. Intell. **83**(1), 75–141 (1996)
12. Troha, M., Bratko, I.: Qualitative learning of object pushing by a robot. In: 25th International Workshop on Qualitative Reasoning, Barcelona, Spain, pp. 175–180 (2011)
13. Šuc, D.: Machine Reconstruction of Human Control Strategies. IOS Press: Ohmsha, cop., Amsterdam (2003)
14. Šuc, D., Bratko, I., Sammut, C.: Learning to fly simple and robust. In: Boulicaut, J.-F., Esposito, F., Giannotti, F., Pedreschi, D. (eds.) ECML 2004. LNCS (LNAI), vol. 3201, pp. 407–418. Springer, Heidelberg (2004)
15. Wiley, T., Sammut, C., Bratko, I.: Qualitative planning with quantitative constraints for online learning of robotic behaviours. In: 28th AAAI Conference on Artificial Intelligence, Quebec City, Canada, pp. 2578–2584 (2014)
16. Zimmerman, T., Kambhampati, S.: Learning-assisted automated planning: looking back, taking stock, going forward. AI Mag. **24**(2), 73–96 (2003)

Musical Instrument Separation Applied to Music Genre Classification

Aldona Rosner[1][✉] and Bozena Kostek[2]

[1] Institute of Informatics, Silesian University of Technology,
Akademicka 16, 44-100 Gliwice, Poland
aldona.rosner@polsl.pl
[2] Audio Acoustics Laboratory, Faculty of Electronics, Telecommunications
and Informatics, Gdansk University of Technology, 80-233 Gdansk, Poland
bokostek@audioakustyka.org

Abstract. This paper outlines first issues related to music genre classification and a short description of algorithms used for musical instrument separation. Also, the paper presents proposed optimization of the feature vectors used for music genre recognition. Then, the ability of decision algorithms to properly recognize music genres is discussed based on two databases. In addition, results are cited for another database with regard to the efficiency of the feature vector.

Keywords: Music Information Retrieval · Musical instrument separation · Music genre classification · Decision systems

1 Introduction

The subject of music genre classification, though visible not only in the Music Information Retrieval (MIR) research but also in commercial applications, still needs attention in terms of effectiveness and quality of classification [6–9, 12, 21]. This is especially important when applied to big music databases available for public [15]. One of the most popular query criteria are: artist, genre, mood, tempo or specific title, when looking for a specific audio track. Lately, genre has become one of the most popular choices, but not always this information is stored in a specific music track. That's why the subject of a more deep content exploring, i.e. taking into consideration sound source separation in the context of music recognition should be addressed, as separation of individual auditory sources, apart from instrument recognition and automatic transcription systems, may be very useful in genre classification. The instrument separation approach seems to be very useful in improving the genre recognition process, since the separation of individual auditory sources and instrument recognition systems get high applicability during the last few years. Therefore, one of the aims of this paper is to propose a set of parameters, which after the audio track separation preprocessing may enable to describe musical content of a piece of music in a more efficient way for the purpose of better distinguishing between selected musical genres. Such an operation may bring the efficiency of genre classification in the databases containing thousands of music tracks similar to earlier experiments carried out on small music databases.

© Springer International Publishing Switzerland 2015
F. Esposito et al. (Eds.): ISMIS 2015, LNAI 9384, pp. 420–430, 2015.
DOI: 10.1007/978-3-319-25252-0_45

This paper overviews related research, then presents databases and parameters used in experiments. In addition main principles of the algorithm used for musical instrument separation are presented. Then, proposed optimization of the feature vectors is shortly discussed. The following Section contains music genre classification results based on two decision systems, separation of music tracks and concerning two databases. A summary is also included providing the most important conclusions.

1.1   Related Research

Because of the extraordinary increase of the available multimedia data, it is necessary to make the recognition process automatic. Although the division of music into genres is subjective and arbitrary, there are perceptual criteria related to the texture, instrumentation and rhythmic structure of music that can be used to characterize a particular genre. Humans can accurately predict a musical genre based on 25 s, which confirms that we can judge the genre using only the musical surface, without constructing any higher level of theoretical descriptions.

The experiments presented by Tzanetakis and Cook (2002) [17] resulted in 61 % of effectiveness for 10 music genres, what is comparable to results aimed by human musical genre classification. The authors proposed three music feature sets for representing timbral texture, rhythmic content and pitch content. Other experiments were made by Kirss (2007) [10] on 250 musical excerpts from five different electronic music genres such as deep house techno, uplifting trance, drum and bass and ambient. By using SVM (Support Vector Machine) 96.4 % of accuracy was reached. It shows that it is possible to obtain a high evaluation efficiency on small datasets and small number of genres, which can also be classified as subgenres. The ISMIS 2011 contest left us with the final results [9] of almost 88 % of accuracy. The tests were concluded in the ISMIS database, which consists of 1310 songs of six genres. It is worth to mention that most misclassification was between Rock and Metal, and Classical and Jazz genres.

The importance of feature extraction in the topic of musical genre classification is confirmed by many researchers. In recent years, an extensive research has been conducted on the subject of audio sound separation, and resulted in interesting ideas and solutions. Uhle *et al.* (2003) [18] designed a system for drum beat separation based on Independent Component Analysis. In contrast, Smaragdis and Brown (2003) [16] applied Non-Negative Matrix Factorization (NMF) to create a system for transcription of polyphonic music that showed remarkable results on piano music. Helen and Virtanen (2005) [3] used NMF, combined with a feature extraction and classification process, and achieved promising results in drum beat separation from popular music. Similar techniques were used by other researchers [1, 13, 19] in percussive-harmonic signal separation or in instrument separation [4].

It is noteworthy that musical social systems, in addition to the area of computer games, have become one of the most profitable financial ventures in recent years. That is why the development of such kind of applications makes it necessary to improve classification of music genre and other most searched criteria, which is still far from being satisfying. Despite major achievements in the field of MIR, there are still some challenges in this area, to name a few: the problem of scalability, large size of the data, different standards (formats) of storing the music information (also media, e.g. a soundtrack), methods of transmitting multimedia data, varying degrees of compression, synchronizing the playback of various media elements, and others.

2   Genre Classification Experiments

2.1   Separation of Music Tracks

Separation of music tracks is based on the NMF method [11]. This method performs well in blind separation of drums and melodic parts of audio recordings. NMF performs a decomposition of the magnitude spectrogram V ($V \approx W \bullet H$) obtained by Short-Time Fourier Transform (STFT), with spectral observations in columns, into two non-negative matrices W and H (where $W \in R_{\geq 0}^{n \times m}, H \in R_{\geq 0}^{n \times m}$ and constant $r \in N$). Matrix W resembles characteristic spectra of the audio events occurring in the signal (such as notes played by an instrument), and matrix H measures their time-varying gains. Columns of W are not required to be orthogonal as is in principal component method. Specifically, an approach based on an iterative algorithm for computing two factors based on the Kullback-Leibler divergence of V given W and H is used in our experiments. This means that the factorization process is achieved by iterative algorithms minimizing cost-functions, which interprets the matrices V and (W, H) as probability distributions [14, 20].

Then, to each NMF component (column of W and corresponding row of H) we apply a pre-trained SVM classifier to distinguish between percussive and non-percussive components. The task of this pre-trained SVM classification which bases on features such as harmonicity of the spectrum and periodicity of the gains is to distinguish between percussive and non-percussive signals bases. By selecting the columns of W that are classified as percussive and multiplying them with their estimated gains in H, we obtain an estimate of the contribution of percussive instruments to each time-frequency bin in V. Thus, we can construct a soft mask that is applied to V to obtain an estimated spectrogram of the drum part, which is transferred back to the time domain through the inverse STFT using the OLA (overlap-add) operation between the short-time sections in the inverting process. It should be reminded that the redundancy within overlapping segments and the averaging of the redundant samples averages out the effect of the window analysis. More details on the drum separation procedure can be found in the introductory paper by Weninger et al. (2011) [19].

2.2   Databases

Music Information Retrieval systems enable to search music basing on metadata: set of parameters which describe the specific track, track title, author, album, year, genre, etc. Another important topic of MIR is related to music databases available for research and experiments. Two music databases were employed in the experiments shown in this paper, namely ISMIS and SYNAT [8] databases.

The ISMIS music database was prepared for a data mining contest associated with the 19th Internat. Symp.on Methodologies for Intelligent Systems [9]. It consists of over 1300 music tracks of 6 music genres: classical, jazz, blues, pop, rock and heavy metal (see Table 1). For each of 60 performers there are 15-20 music tracks, which are then partitioned into 20-s segments and parameterized. Music tracks are prepared as stereo signals (44.1 kHz, 16 bit, .wav format.

Table 1. List of cardinality of the music genres for ISMIS database

Genre	No. of music excerpts
Classical	320
Jazz	362
Blues	216
Rock	124
Heavy Metal	104
Pop	184
Overall	1310

The SYNAT database consists of over 50.000 music tracks of 30-second long excerpts of songs in mp3 format, retrieved from the Internet by a music robot. ID3 tags of music excerpts were automatically assigned to songs by the music robot. The tags were saved in a fully automatic way without human control [5]. SYNAT contains 22 genres: Alternative Rock, Blues, Broadway and Vocalists, Children's Music, Christian and Gospel, Classic Rock, Classical, Country, Dance and DJ, Folk, Hard Rock and Metal, International, Jazz, Latin Music, Miscellaneous, New Age, Opera and Vocal, Pop, Rap and Hip-Hop, Rock, R and B, and Soundtracks. However, for the experiments carried out within this study over 8.000 music excerpts from the SYNAT database representing 13 music genres were used. The cardinality of the music excerpts for the original and separated signals, in relation to specific music genre, is presented in Table 2. From the original audio signal harmonic (H), drum (D), piano (P), trumpet (T) and saxophone (S) signals were retrieved using different options of the Non-Negative Matrix Factorization separation method (cost function; extended KL-divergence method, window sizes: 20, 30, 40 ms, window function (Hann), window overlap (0.5), number of components (5, 10, 20, 30).

Marsyas (GTZAN) [2] is a commonly used database in MIR. It consists of 1000 songs, representing 10 genres (100 songs per each genre). They are as follows: Pop, Rock, Country, Rap and Hip-Hop, Classical, Jazz, Dance and Dj, Blues, Hard Rock and Metal, Reggae. For the purpose of showing the usability of the original feature vector used, preliminary results obtained for this database are recalled here.

2.3 Basic Parametrization

The so-called 'basic' parameters were adapted from previous studies [9] to be able to compare previous results. The list of parameters for ISMIS and SYNAT databases is

Table 2. Cardinality of the classes for original and separated signals, based on 8244 elements received from SYNAT database, representing 13 music genres

Genre	No. of music excerpts	H	%	D	%	P	%	T	%	S	%
Alternative Rock	207	207	100.0	204	98.55	137	66.18	48	23.19	10	4.83
Blues	264	264	100.0	257	97.35	242	91.67	134	50.76	6	2.27
Classical	953	953	100.0	408	42.81	920	96.54	272	28.54	208	21.83
Country	1039	1039	100.0	1034	99.52	837	80.56	261	25.12	51	4.91
Dance &DJ	259	258	99.61	257	99.23	185	71.43	0	0.00	19	7.34
Hard Rock & Metal	602	601	99.83	597	99.17	101	16.78	153	25.42	5	0.83
Jazz	568	567	99.82	527	92.78	537	94.54	211	37.15	33	5.81
Latin Music	444	444	100.0	444	100.00	285	64.19	256	57.66	10	2.25
New Age	567	560	98.77	327	57.67	489	86.24	120	21.16	211	37.21
Pop	796	796	100.0	798	100.25	539	67.71	236	29.65	66	8.29
R&B	610	610	100.0	610	100.00	482	79.02	253	41.48	14	2.30
Rap&Hip-Hop	1012	1011	99.90	1010	99.80	715	70.65	582	57.51	27	2.67
Rock	923	922	99.89	914	99.02	509	55.15	216	23.40	13	1.41
TOTAL	8244	8232	99.85	7387	89.60	5978	72.51	2742	33.26	673	8.16

given in Table 3. Most of the parameters are based on MPEG 7 standard, others are Mel Frequency Cepstral Coefficients (MFCC) as well as some dedicated time-domain-related descriptors.

Before the Feature Vectors (FVs) were used in the classification experiments, two normalization methods were employed for data pre-processing, i.e. Min-Max normalization, Zero-Mean normalization and both methods used jointly. However, the normalization of training and test datasets are performed in that way that the mean and standard deviation values are calculated only for training dataset, and only the current value is retained from training and test datasets (used respectively for normalization of training and test datasets). Also in the classification process separability of data was checked based on Best First, Greedy, Ranker and PCA methods, and in the main experiments reduced FVs were employed.

Table 3. List of parameters for ISMIS and SYNAT databases

#	ID	Audio Feature Description	Comment
1	TC	Temporal Centroid	
2	SC, SC_V	Spectral Centroid – mean and variance	
34	ASE 1-34	Audio Spectrum Envelope (ASE)- average values in 34 frequency bands	for the SYNAT database only 29 subbands occur
1	ASE_M	Mean ASE (for all frequency bands)	

(Continued)

Table 3. (*Continued*)

#	ID	Audio Feature Description	Comment
34	ASEV 1-34	ASE variance in 34 frequency bands	for the SYNAT database only 29 subbands occur
1	ASE_MV	Mean ASE variance (for all frequency bands)	
2	ASC, ASC_V	Audio Spectrum Centroid (ASC) – average and its variance	
2	ASS, ASS_V	Audio Spectrum Spread (ASS) – average and its variance	
24	SFM 1-24	Spectral Flatness Measure (SFM) – average values for 24 frequency bands	for the SYNAT database only 20 subbands occur
1	SFM_M	Mean SFM (for all frequency bands)	
24	SFMV 1-24	SFM variance (for 24 frequency bands)	for the SYNAT database only 20 subbands occur
1	SFM_MV	Mean SFM variance (for all frequency bands)	
20	MFCC 1-20	Mel Function Cepstral Coefficients (MFCC) – first 20 (mean values)	
20	MFCCV 1-20	MFCC Variance – first 20 values	
3	THR_[1,2,3] RMS_TOT	No. of samples higher than single/double/ triple RMS value	Dedicated parameters (24) in time domain based on the analysis of the distribution of the envelope in relation to the RMS value
6	THR_[1,2,3] RMS_10FR_ [MEAN,VAR]	Mean/Variance of THR_ [1,2,3] RMS_TOT for 10 frames	
1	PEAK_RMS_TOT	A ratio of peak to RMS (Root Mean Square)	
2	PEAK_RMS10FR_ [MEAN,VAR]	Mean/variance of PEAK_RMS_TOT for 10 frames	
1	ZCD	No. of transition by the level Zero	
2	ZCD_10FR_ [MEAN,VAR]	Mean/Variance values of ZCD for 10 frames	
3	[1,2,3]RMS_TCD	Number of transitions by single/double/triple levels of RMS	

(*Continued*)

Table 3. (*Continued*)

#	ID	Audio Feature Description	Comment
6	[1,2,3]RMS_TCD_10FR_[MEAN, VAR]	Mean/Variance value of [1,2,3]RMS_TCD for 10 time frames	
	TOTAL	191 (for ISMIS) and 173 (for SYNAT)	

3   Experiments

3.1   Algorithms Used in Music Genre Classification

The most popular methods for music genre classification are: Support Vector Machines (SVMs), Artificial Neural Networks (ANNs), Decision Trees, Rough Sets and Minimum-distance methods, to which a very popular k-Nearest Neighbor (k-NN) method belongs. Since preliminary experiments carried out by the authors showed that SVM (including co-training applied for SVM) classification returns best accuracy of classification, that's why the results obtained while employing this method are to be show [13]. Also, despite computational expensiveness of the kNN algorithm, it is a good method for solving multi-class classification problem and is also commonly used in MIR area, what makes it possible to compare the experimental results. For these reasons results of this method are also presented. The core experiment, as well as normalization methods, were prepared in Java programming language using Eclipse environment. Weka library for Java was used for data managing: selecting attributes by their name, selecting instances (FVs of specific music excerpts) and in classification process. Selecting the best attributes (Best First, Greedy, Ranker and PCA methods) was done in Weka graphical interface.

3.2   Feature Vector Optimization - Adding New Parameters

In the separation process several specific instrument/path (as mentioned earlier, i.e. harmonic, drum, piano, trumpet, saxophone) were retrieved aimed at FV optimization. For this purpose new FVs were created containing the originally extracted parameters and those based on separated music tracks. In that way an original signal (O) with added harmonic (H) components formed OH feature vector, etc. Several mixtures were tried, and a summary of results is shown in Table 4. Table 4 presents the results of overall correctness of classification for 16 mixtures of signals for full set of FVs (p_173 per each signal) for the SYNAT database, as presented above.

The best results were obtained for the OH signal, however the differences between the specific mixed signals are not high. It may also be observed that the mixture of two signals is the most promising resulting in the highest correctness, however a combination of three signals (OHD) was also retained in experiments. In Table 5 results (Precision (Prec) and True Positive (TP)) for the kNN algorithm and optimized FV (using the PCA method only 59 parameters were retained) are shown for SYNAT database.

Table 4. Results of the overall correctness of classification for 16 mixtures of signals for a full set of FVs (p_173 per each signal)

signal mixture/	OH	ORIG	OHS	OD	OP	OHD	OHP	OHDS
correctness	72.37	72.19	71.92	71.57	71.42	71.40	71.35	71.32
signal mixture/	OS	OHDP	OHT	OHDPTS	OHDTS	OHDT	OHDPS	OT
correctness	71.31	71.30	71.29	71.24	70.94	70.91	70.66	70.43

Table 6 presents the results for SVM algorithm. OH signal gave much better results (~ 4 % better) in comparison to Original signal for such genres as alternative rock, blues, jazz, new age, pop (true positive rate) and Latin music (precision). That confirms that separating harmonic path for genres where harmonic plays significant role is useful. On the basis of the results some other conclusions may be derived, such as: for the OH signal an improvement of level of confusion between Pop and New Age, Pop and Latin Music and Pop and Rap and Hip-Hop is also observed and is equal ~ 2.77 % (in case of New Age) and ~ 0.88 % in case of two others. In case of using Co-training for the SVM algorithm results gained approx. 1 % of correctness.

Table 5. kNN-based classification results

kNN	OHD		OH		OD		DH		ORIG	
/genre	TP	Prec	TP	Prec	TP	Prec	TP	Prec	TP	Prec
Alternative Rock	13.04	43.46	14.49	38.68	8.21	24.05	11.11	42.00	14.49	44.76
Blues	7.20	33.01	12.50	48.08	4.55	29.86	4.92	30.35	9.47	39.56
Classical	60.13	82.14	95.80	82.18	39.56	73.87	40.29	69.36	95.17	84.08
Country	90.95	45.85	92.78	47.38	88.64	46.56	87.68	44.13	92.20	50.96
DanceDJ	58.30	95.38	43.25	94.84	59.86	90.28	46.74	89.42	62.17	87.10
Hard Rock/Metal	82.89	88.91	84.39	85.93	82.22	83.19	77.24	87.86	87.87	84.18
Jazz	59.68	65.64	64.96	71.50	54.23	64.82	53.34	61.08	64.79	70.27
Latin Music	38.29	79.84	43.92	79.55	39.64	82.41	30.63	77.74	47.97	81.15
NewAge	44.09	92.40	74.60	84.95	40.74	93.99	36.86	90.16	76.54	87.84
Pop	25.75	40.76	22.10	45.01	25.25	42.58	22.11	40.48	24.87	49.82
Rap HipHop	87.75	64.34	86.66	88.16	86.17	47.11	84.09	45.99	90.71	86.36
RB	61.32	59.63	60.16	64.34	56.73	64.67	57.71	57.03	63.45	69.23
Rock	63.16	51.76	64.90	58.27	62.62	53.56	59.48	52.23	68.26	61.05
Ov. correctn.	60.93		67.52		56.88		54.50		69.84	

Table 7 presents the summary results of the overall CCI (Correctly Classified Instances) for the small (ISMIS) and big (SYNAT) databases conducted with the Co-SVM method, as the one which gives the highest correctness of classification, involving the original FV (191 parameters for ISMIS) and reduced ones (VoP p_52, _59,_60).

Even though the results of CCI for the bigger database (SYNAT) are ~ 10 % lower than for the smaller one (ISMIS), it is still a very good result considering the fact, that

Table 6. Results for the SVM algorithm

SVM	OHD		OH		OD		DH		ORIG	
Genre	TP	Prec	TP	Prec	TP	Prec	TP	Prec	TP	Prec
Alternative Rock	43.48	33.48	41.55	37.30	41.06	33.89	35.27	31.75	33.33	37.14
Blues	41.67	36.10	40.53	38.31	42.80	41.01	37.12	35.08	38.26	45.24
Classical	90.66	87.72	91.08	88.77	89.72	87.82	88.35	85.04	90.87	87.50
Country	77.00	74.01	79.88	73.96	75.55	72.67	73.63	69.42	77.00	72.26
DanceDJ	74.16	67.69	74.18	69.74	72.20	66.08	69.53	63.30	74.13	72.71
Hard Rock/Metal	84.55	83.92	84.22	83.96	84.88	81.86	80.73	80.20	88.03	83.13
Jazz	67.61	68.03	73.24	70.15	68.31	68.24	65.67	66.52	70.43	66.45
LatinMusic	67.79	71.76	69.59	75.27	68.02	71.60	66.22	72.54	69.37	71.96
New Age	77.07	82.57	77.60	81.21	75.13	76.98	71.08	69.56	72.31	68.67
Pop	39.32	39.58	40.83	42.76	40.83	41.65	39.19	40.02	36.68	41.96
Rap HipHop	87.55	90.06	86.56	88.13	88.04	88.35	87.45	88.23	89.92	87.92
RB	60.18	66.83	63.28	69.10	62.14	70.42	57.71	64.24	62.95	69.97
Rock	59.49	63.36	62.41	64.73	60.02	63.91	55.04	59.36	64.46	64.55
Ov. correctn.	70.38		71.79		70.37		67.58		71.05	

Table 7. Comparison of results of overall classification for small and big database obtained for the original and mixed signals involving Co-training method

classifier	Co-SVM			Co-SVM		
type of FV	Original			OD		
database	ISMIS		SYNAT	ISMIS		SYNAT
Vector	FV191	VoP p_60	VoP p_59	FV191	VoP p_52	VoP p_59
no. of classes	4		13	4		13
no. of elements	465		8244	465		8244
CCI [%]	**80.86**	**79.78**	**71.49**	**81.08**	**81.72**	**71.29**

there are as many as 13 genres to be classified instead of only few as in the case of ISMIS database. Moreover some of those 13 genres are similar and very often confused not only by the machine learning methods but also by human listeners, which makes it considerably more difficult to be classified correctly.

Finally, eight music genres common for SYNAT and GTZAN databases were compared in the context of the FV robustness. In the experiments the kNN algorithm (with the Euclidean function) and the 'basic' FV (containing 173 parameters) were used and this experiment resulted in the following classification (approx.): Classical 88 %, Blues 70 %, Country 59 %, DanceDj 52 %, HardRock 78 %, Jazz 68 %, Pop 71 %, Rap 58 %, Rock 51 %. The results obtained for the GZTAN database compared to the SYNAT database are comparable. The reason for the a bit lower classification efficiency for these particular genres may be a greater variety of songs in the GZTAN database. Also, the database contains also recordings with reduced quality, which may have an adverse effect on the effectiveness of the overall parameterization.

4 Summary

In this paper a new strategy to music genre classification was proposed. The main principle is separating music tracks at the pre-processing phase and extending vector of parameters by descriptors related to a given musical instrument components that are characteristic for the specific musical genre. This allows for more efficient automatic musical genre classification. It was also shown that extending the original signal even with one or two parameters (such as OP, OT and OS signals) influences the results for classification for specific genres, what confirms the importance of specific parameters in relation to music genre. It has also to be noted that the overall classification gave better correctness for any type of mixed signals in comparison to the original signal.

Acknowledgments. The study was partially supported by the EU from the European Social Fund (UDA-POKL.04.01.01-00-106/09) and the project PBS1/B3/16/2012 financed by the Polish National Centre for R&D.

References

1. Canadas-Quesada, F.J., Vera-Candeas, P., Ruiz-Reyes, N., Carabias-Orti, J., Cabanas-Molero, P.: Percussive/harmonic sound separation by non-negative matrix factorization with smoothness/sparseness constraints. EURASIP J. Audio, Speech Music Proc. **2014**(26), 1–17 (2014)
2. GTZAN database. http://marsyasweb.appspot.com/download/data_sets/. Accessed May 2015
3. Helen, M., Virtanen, T.: Separation of drums from polyphonic music using non-negative matrix factorization and support vector machine. In: 13th EUSIPCO'2005, Antalaya (2005)
4. Heittola, T., Klapuri, A., Virtanen, T.: Musical instrument recognition in polyphonic audio using source-filter model for sound separation. In: 10th ISMIR'2009, Kobe (2009)
5. Hoffmann, P., Kostek, B.: Music genre recognition in the rough set-based environment. In: Kryszkiewicz, M., Bandyopadhyay, S., Rybinski, H., Pal, S.K. (eds.) PReMI 2015. LNCS, vol. 9124, pp. 377–386. Springer, Heidelberg (2015)
6. ISMIR website [accessed 2015 February]. http://ismir2014.ismir.net
7. Kostek, B.: Music information retrieval in music repositories. In: Suraj, Z., Skowron, A. (eds.) Rough Sets and Intelligent Systems Refer, pp. 463–489. Springer, Berlin (2013)
8. Hoffmann, P., Kostek, B.: Music data processing and mining in large databases for active media. In: Ślęzak, D., Schaefer, G., Vuong, S.T., Kim, Y.-S. (eds.) AMT 2014. LNCS, vol. 8610, pp. 85–95. Springer, Heidelberg (2014)
9. Kostek, B., Kupryjanow, A., Zwan, P., Jiang, W., Raś, Z.W., Wojnarski, M., Swietlicka, J.: Report of the ISMIS 2011 contest: music information retrieval. In: Kryszkiewicz, M., Rybinski, H., Skowron, A., Raś, Z.W. (eds.) ISMIS 2011. LNCS, vol. 6804, pp. 715–724. Springer, Heidelberg (2011)
10. Kirss, P.: Audio based classification of electronic music. University of Jyväskylä (2007). https://jyx.jyu.fi/dspace/bitstream/handle/123456789/13592/urn_nbn_fi_jyu-2007601.pdf?sequence=1. Accessed May 2015
11. Lee, D.D., Seung, H.S.: Learning the parts of objects by non-negative matrix factorization. Nature **401**, 788–791 (1999)

12. Ras, Z., Wieczorkowska, A.: Advances in Music Information Retrieval. Springer Publishing, Berlin, Heidelberg (2010)
13. Rosner, A., Schuller, B., Kostek, B.: Classification of music genres based on music separation into harmonic and drum components. Arch. Acoust. **39**(4), 629–638 (2014)
14. Rosner, A.: Multi-instrumental automatic recognition of musical genres. Ph.D. Dissertation, Silesian University of Technology, Institute of Informatics, Gliwice, Poland (2015)
15. Schnitzer, D., Flexer, A., Widmer, G.: A fast audio similarity retrieval method for millions of music tracks. Multimedia Tools and Applications **58**(1), 23–40 (2012)
16. Smaragdis, P., Brown, J.C.: Non-negative matrix factorization for polyphonic music transcription. In: Proceedings of WASPAA, pp. 177–180 (2003)
17. Tzanetakis, G., Cook, P.: Musical genre classification of audio signals. IEEE Trans. Speech Audio Process. **10**(5), 293–302 (2002)
18. Uhle, C., Dittmar, C., Sporer, T.: Extraction of drum tracks from polyphonic music using independent subspace analysis. In: Proceedings 4th International Symposium on Independent Component Analysis and Blind Signal Separation (ICA), Nara, pp. 843–848 (2003)
19. Weninger, F., Woelmer, M., Schuller, B.: Automatic assessment of singer traits in popular music: gender, age, height and race. In: Proceedings of 12th ISMIR, pp. 37–42 (2011)
20. Weninger, F., Schuller, B.: Optimization and parallelization of monaural source separation algorithms in the openBliSSART toolkit. J. Sig. Process. Syst. **69**(3), 267–277 (2012)
21. Wieczorkowska, A., Kubera, E., Kubik-Komar, A.: Analysis of recognition of a musical instrument in sound mixes using support vector machines. Fundamenta Informaticae **107**(1), 85–104 (2011)

Textual Data Analysis and Mining

Harvesting Comparable Corpora and Mining Them for Equivalent Bilingual Sentences Using Statistical Classification and Analogy-Based Heuristics

Krzysztof Wołk[(✉)], Emilia Rejmund, and Krzysztof Marasek

Department of Multimedia,
Polish - Japanese Academy of Information Technology, Warsaw, Poland
{kwolk,erejmund,kmarasek}@pja.edu.pl

Abstract. Parallel sentences are a relatively scarce but extremely useful resource for many applications including cross-lingual retrieval and statistical machine translation. This research explores our new methodologies for mining such data from previously obtained comparable corpora. The task is highly practical since non-parallel multilingual data exist in far greater quantities than parallel corpora, but parallel sentences are a much more useful resource. Here we propose a web crawling method for building subject-aligned comparable corpora from e.g. Wikipedia dumps and Euronews web page. The improvements in machine translation are shown on Polish-English language pair for various text domains. We also tested another method of building parallel corpora based on comparable corpora data. It lets automatically broad existing corpus of sentences from subject of corpora based on analogies between them.

1 Introduction

In this article we present methodologies that allow us to obtain truly parallel corpora from not sentence-aligned data sources, such as noisy-parallel or comparable corpora [1]. For this purpose we used a set of specialized tools for obtaining, aligning, extracting and filtering text data, combined together into a pipeline that allows us to complete the task. We present the results of our initial experiments based on text samples obtained from the Wikipedia dumps and the Euronews web page. We chose the Wikipedia as a source of the data because of a large number of documents that it provides (1,047,423 articles on PL Wiki and 4,524,017 on EN, at the time of writing this article). Furthermore, Wikipedia contains not only comparable documents, but also some documents that are translations of each other. The quality of our approach is measured by improvements in MT results.

Second method is based on sequential analogy detection. We seek to obtain parallel corpora from unaligned data. Solution proposed by our team is based on sequential analogy detection. Such approach was presented in literature [2, 3], but all applications concern similar languages with similar grammar like English-French, Chinese-Japanese. We try to apply this method for English-Polish corpora. These two languages have different grammar, which makes our approach innovative and let easily

F. Esposito et al. (Eds.): ISMIS 2015, LNAI 9384, pp. 433–441, 2015.
DOI: 10.1007/978-3-319-25252-0_46

broad this method for different languages pairs. In our approach, to enhance quality of identified analogies, sequential analogies clusters are sought.

2 State of the Art

Two main approaches for building comparable corpora can be distinguished. Probably the most common approach is based on the retrieval of the cross-lingual information. In the second approach, source documents need to be translated using any machine translation system. The documents translated in that process are then compared with documents written in the target language in order to find the most similar document pairs.

The authors in [4] suggested obtaining only title and some meta-information, such as publication date and time for each document instead of its full contents in order to reduce the cost of building the comparable corpora (CC). The cosine similarity of titles term frequency vectors were used to match titles and contents of matched pairs.

An interesting idea for mining parallel data from Wikipedia was described in [5]. The authors propose in their word two separate approaches. The first idea is to use an online machine translation (MT) system to translate Dutch pages of the Wikipedia into English and they try to compare original EN pages with translated ones.

The authors of [6] facilitate a BootCat method that was proven to be fast and effective when a corpus building is concerned. The authors try to extend this method by adding support for multilingual data and also present a pivot evaluation.

Interwiki links were facilitated by Tyers and Pienaar in [7]. Based on the Wikipedia link structure a bilingual dictionary is extracted. In their work they measured that depending on the language pair the mismatch between linked Wikipedia pages averages.

What is more, authors of [8] introduce an automatic alignment method of parallel text fragments by using a textual entailment technique and a phrase-base Statistical Machine Translation (SMT) system. Authors state that significant improvement in SMT quality by using obtained data was obtained (increase in BLEU by 1.73).

3 Preparation of the Data

Our procedure starts with a specialized web crawler implemented by us. Because PL Wiki contains less data of which almost all articles have their correspondence on EN Wiki, the program crawls data starting from non-English site first. The crawler can obtain and save bilingual articles of any language supported by the Wikipedia. The tool requires at least two Wikipedia dumps on different languages and information about language links between the articles in the dumps (obtained from the interwiki links). For the Euronews.com a standard web crawler was used. This web crawler was designed to use the Euronews.com archive page. In first phase it generates a database of parallel articles in two selected languages in order to collect comparable data from it.

For the experiments in the statistical machine translation we choose TED lectures domain, to be more specific the PL-EN TED[1] corpora prepared for IWSLT

---

[1] https://www.ted.com/talks.

(International Workshop on Spoken Language Translation) 2014 evaluation campaign by the FBK (Fondazione Bruno Kessler). This domain is very wide and covers many not related subjects and areas. The data contains almost 2,5 M untokenized words [9]. Additionally we choose two more narrow domains: The first parallel corpus is made out of PDF documents from the European Medicines Agency (EMEA) and medicine leaflets [10]. The second was extracted from the proceedings of the European Parliament (EUP) [11]. What is more we also conducted experiments on the Basic Travel Expression Corpus (BTEC), a multilingual speech corpus containing tourism-related sentences similar to those that are usually found in phrasebooks for tourists going abroad [12]. Lastly we used a corpus built from the movie subtitles (OPEN) [10].

In Table 1 we present details about number of unique words (WORDS) and their forms as well as about number of bilingual sentence pairs (PAIRS).

Table 1. Corpora specification

CORPORA	PL WORDS	EN WORDS	PAIRS
BTEC	50,782	24,662	220,730
TED	218,426	104,117	151,288
EMEA	148,230	109,361	1,046,764
EUP	311,654	136,597	632,565
OPEN	1,236,088	749,300	33,570,553

As mentioned, the solution can be divided into three main steps. First the data is collected, then it is aligned at article level, and lastly the results of the alignment are mined for parallel sentences. Sentence alignment must be computationally feasible in order to be of practical use in various applications [13].

With this methodology we were able to obtain 4,498 topic-aligned articles from the Euronews and 492,906 from the Wikipedia.

4 Parallel Data Mining

In order to extract the parallel sentence pairs we decided to try two different strategies. The first one facilitates and extends methods used in Yalign Tool[2] and the second is based on analogy detection. The MT results we present in this article were obtained with the first strategy. The second method is still in development phase, nevertheless the initial results are promising and worth mentioning.

4.1 Improvements to Yalign's Method

In Yalign tool [14] for the sequence alignment A* search algorithm is used [15] to find an optimal alignment between the sentences in two given documents. Unfortunately it

---

[2] https://github.com/machinalis/yalign.

can't handle alignments that cross each other or alignments from two sentences into a single one [15]. To overcome this and other minor problems, in order to improve mining quality, we used the Needleman-Wunch algorithm (originally used for DNA sequences) instead. Because it would require N * M calls to the sentence similarity matrix we implemented its GPU version for accelerated processing [16].

The classifier must be trained in order to determine if a pair of sentences is translation of each other or not. The particular classifier used in this research was a Support Vector Machine [17].

What is more our solution facilitated multithreading and proved to increase the mining time by the factor of 5 in comparison with standard Yalign tool (using Core i7 CPU).

To train the classifier a good quality parallel data was necessary as well as a dictionary with translation probability included. For this purposes we used TED talks [18] corpora enhanced by us during the IWSLT'13 Evaluation Campaign [13]. In order to obtain a dictionary we trained a phrase table and extracted 1-grams from it. We used the MGIZA ++ tool for word and phrase alignment. The lexical reordering was set to use the msd-bidirectional-fe method and the symmetrization method was set to grow-diag-final-and for word alignment processing [20]. For bi-lingual training data we used four corpora previously described. We obtained four different classifiers and repeated mining procedure with each of them. The detailed results for the Wiki are showed in Table 2.

Table 2. Data mined from the Wikipedia for each classifier

Classifier	Value	PL	EN
TED	Size in MB	41,0	41,2
	No. of sentences	357,931	357,931
	No. of words	5,677,504	6,372,017
	No. of unique words	812,370	741,463
BTEC	Size in MB	3,2	3,2
	No. of sentences	41,737	41,737
	No. of words	439,550	473,084
	No. of unique words	139,454	127,820
EMEA	Size in MB	0,15	0,14
	No. of sentences	1,507	1,507
	No. of words	18,301	21,616
	No. of unique words	7,162	5,352
EUP	Size in MB	8,0	8,1
	No. of sentences	74,295	74,295
	No. of words	1,118,167	1,203,307
	No. of unique words	257,338	242,899
OPEN	Size in MB	5,8	5,7
	No. of sentences	25,704	25,704
	No. of words	779,420	854,106
	No. of unique words	219,965	198,599

4.2   Analogy Based Method

This method is based on sequential analogy detection. Based on parallel corpus we detect analogies that exists in both languages. To enhance quality of identified analogies sequential analogies clusters are sought.

However our current research on Wikipedia corpora shows that it is both extremely difficult and machine time consuming to seek out clusters of higher orders. Therefore we restrained ourselves to simple analogies such as A is to B in the same way as C to D.

$$A : B :: C : D$$

Such analogies are found using distance calculation. We seek such sentences that:

$$\mathbf{dist(A, B) = dist(C, D)}$$

and

$$\mathbf{dist(A, C) = dist(B, D)}$$

Additional constrain was added that requires the same relation of occurrences of each character in the sentences. E.g. if number of character "a" in sentence A is equal to x and equal to y in sentence B then the same relation must occur in sentences C and D.

We used Levenshtein metric in our distance calculation. We tried to apply it directly into the characters in the sentence, or considering each word in the sentence, as individual symbol, and calculating Levenshtein distance between symbol coded sentences. The latter approach was employed due to the fact that this method was earlier tested on Chinese and Japanese languages [19] that use symbols to represent entire words.

After clustering, data from clusters are compared to each other to find similarities between them. For each four sentences

$$A : B :: C : D$$

We look for such E and F that:

$$C : D :: E : F \text{ and } E : F :: A : B$$

However none were found in our corpus, therefore we restrained ourselves to small clusters with size of 2 pairs of sentences. In every cluster matching sentences from parallel corpus were identified. It let us generate new sentences similar to the one which are in our corpus and add it to broad resulting data set. For each of sequential analogies that were identified, rewriting model is constructed. This is achieved by string manipulation. Common pre- and suffixes for each of the sentences are calculated using LCS (Longest Common Subsequence) method.

Sample of rewriting model is shown on this example (prefix and suffix are shown in bold)

Poproszę koc i poduszkę. ⇔*A blanket and a pillow, **please***.

Czy mogę poprosić o śmietankę i cukier? ⇔ ***Can I have** cream and sugar?*

Rewriting model consist of prefix, suffix and their translation. It is now possible to construct parallel corpus form non-parallel monolingual source. Each sentence in the corpus is tested for match with the model. If the sentence contains prefix and suffix is considered matching sentence.

Poproszę bilet. ⇔ *A unknown, **please***.

In the matched sentence some of the words remain not translated but general meaning of the sentence is conveyed. Remaining words may be translated word-by-word while translated sentence would remain grammatically correct.

bilet ⇔ ticket

Substituting unknown words with translated ones we are able to create a parallel corpus entry.

Poproszę bilet. ⇔ *A ticket, **please***.

As a result of sequential analogy detection based method we mined **8128** models from of Wikipedia parallel corpus. This enabled generation of **114,000** new pair sentences to extend parallel corpus. Sentences were generated from Wikipedia comparable corpus that is basically an extract of Wikipedia articles. Therefore we have article in Polish and English on the same topic, but sentences are not aligned in any particular way. We use rewriting models to match sentences from Polish article to sentences in English. Whenever model could be successfully applied to a pair of sentences, this pair is considered to be parallel resulting in generation of quasi-parallel corpus (quasi since sentences were aligned artificially using approach described above). Those parallel sentences can be used to extend parallel corpora in order to improve quality of translation.

5   Results and Evaluation

In order to evaluate the corpora we divided each corpus into 200 segments and randomly selected 10 sentences from each segment for testing purposes. This methodology ensured that the test sets covered entire corpus. The selected sentences were removed from the corpora. We trained the baseline system, as well as system with extended training data with the Wikipedia corpora and lastly we used Modified Moore Levis Filtering for the Wikipedia corpora domain adaptation. Additionally we used monolingual part of the corpora as language model and we tried to adapt it for each corpus by using linear interpolation [2]. For scoring purposes we used four well-known metrics that show high correlation with human judgments. Among the commonly used SMT metrics are: Bilingual Evaluation Understudy (BLEU) [11] the U.S. National Institute of Standards & Technology (NIST) metric [20], the Metric for Evaluation of Translation with Explicit Ordering (METEOR) [8], and Translation Error Rate (TER) [20].

The baseline system testing was done using the Moses open source SMT toolkit with its Experiment Management System (EMS) [16] with settings described in [13].

Starting from baseline systems (BASE) tests in PL to EN and EN to PL direction, we raised our score through extending the language model (LM), interpolating it (ILM) and by the corpora extension with additional data (EXT) and by filtering additional data with Modified Moore Levis Filtering (MML) [2]. It must be noted that extension of language models was done on systems with corpora after MML filtration. The LM and ILM experiments already contain extended training data. The results of the experiments are showed in Table 3.

Table 3. Polish to English and English to Polish MT Experiments

Corpus	System	Polish to English				English to Polish			
		BLEU	NIST	TER	METEOR	BLEU	NIST	TER	METEOR
TED	BASE	16,96	5,26	67,10	49,42	10,99	3,95	74,87	33,64
	EXT	16,96	5,29	66,53	49,66	10,86	3,84	75,67	33,80
	MML	16,84	5,25	67,55	49,31	11,01	3,97	74,12	33,77
	LM	17,14	5,27	67,66	49,95	11,54	4,01	73,93	34,12
	ILM	**17,64**	**5,48**	**64,35**	**51,19**	**11,86**	**4,14**	**73,12**	**34,23**
BTEC	BASE	11,20	3,38	77,35	33,20	8,66	2,73	85,27	27,22
	EXT	12,96	3,72	74,58	38,69	8,46	2,71	84,45	27,14
	MML	12,80	3,71	76,12	38,40	8,50	2,74	83,84	27,30
	LM	13,23	3,78	75,68	39,16	8,76	2,78	82,30	27,39
	ILM	**13,60**	**3,88**	**74,96**	**39,94**	**9,13**	**2,86**	**82,65**	**28,29**
EMEA	BASE	62,60	10,19	36,06	77,48	**56,39**	**9,41**	**40,88**	**70,38**
	EXT	62,41	10,18	36,15	77,27	55,61	9,28	42,15	69,47
	MML	62,72	10,24	35,98	77,47	55,52	9,26	42,18	69,23
	LM	62,90	10,24	35,73	77,63	55,38	9,23	42,58	69,10
	ILM	**62,93**	**10,27**	**35,48**	**77,87**	55,62	9,30	42,05	69,61
EUP	**BASE**	**36,73**	**8,38**	**47,10**	**70,94**	**25,74**	**6,54**	**58,08**	**48,46**
	EXT	36,16	8,24	47,89	70,37	24,93	6,38	59,40	47,44
	MML	36,66	8,32	47,25	70,65	24,88	6,38	59,34	47,40
	LM	36,69	8,34	47,13	70,67	24,64	6,33	59,74	47,24
	ILM	36,72	8,34	47,28	70,79	24,94	6,41	59,27	47,64
OPEN	BASE	64,54	9,61	32,38	77,29	**31,55**	**5,46**	**62,24**	**47,47**
	EXT	65,49	9,73	32,49	77,27	31,49	5,46	62,06	47,26
	MML	65,16	9,62	33,79	76,45	31,33	5,46	62,13	47,31
	LM	65,53	9,70	32,94	77,00	31,22	5,46	62,61	47,29
	ILM	**65,87**	**9,74**	**32,89**	**77,08**	31,39	5,46	62,43	47,33

6 Conclusions

The results showed in Tables 4 and 5, to be more specific BLEU, Meteor and TER values in TED corpus were checked whether the differences were relevant. We measured the variance due to the BASE and MML set selection. It was calculated using bootstrap resampling[3] for each test run. The result for BLEU was 0.5 and 0.3 and 0.6 for METEOR and TER respectively. The results over 0 mean that there is significant (to some extent) difference between the test sets and it indicates that a difference of this magnitude is likely to be generated again by some random translation process, which would most likely lead to better translation results in general. [21]

The results of SMT systems based only on mined data were not too surprising. Firstly, they confirm quality and high parallelism level of the corpora that can be concluded from the translation quality especially on the TED data set. Only 2 BLEU points gap can be observed when comparing systems trained on strict in-domain (TED) data and mined data, when it comes to EN – PL translation system. It also seems natural that the best SMT scores were obtained on TED data. It is not only most similar to the Wikipedia articles and overlaps with it in many topics but also the classifier trained on the TED data set recognized most of parallel sentences. In the results it can also be observed that the METEOR metric in some cases rises whereas other metrics decrease. Most likely reason for this is fact that other metrics suffer, in comparison to the METEOR, from the lack of scoring mechanism for synonyms. The Wikipedia is very wide not only when we consider its topics but also vocabulary, which leads to conclusions that mined corpora, is good source for extending sparse text domains. It is also the reason why the test sets originating from wide domains outscore narrow domain ones and also the most likely explanation why training on larger mined data sometimes slightly decreases test sets from very specific domains. Nonetheless it must be noted that after manual analysis we conceded that in many cases translations were good but automatic metric became lower because of the usage of synonyms.

Nowadays, the bi-sentence extraction task is becoming more and more popular in unsupervised learning for numerous specific tasks. The method overcomes disparities between two languages. It is a language independent method that can easily be adjusted to a new environment, and it only requires parallel corpora for initial training. The experiments show that the method performs well. The obtained corpora increased MT quality in wide text domains. From a practical point of view, the method neither requires expensive training nor requires language-specific grammatical resources, while producing satisfying results.

References

1. Wu, D., Fung, P.: Inversion transduction grammar constraints for mining parallel sentences from quasi-comparable corpora. In: Dale, R., Wong, K.-F., Su, J., Kwong, O.Y. (eds.) IJCNLP 2005. LNCS (LNAI), vol. 3651, pp. 257–268. Springer, Heidelberg (2005)
2. Pal, S., Pakray, P., Naskar, S.: Automatic building and using parallel resources for SMT from comparable corpora (2014)
3. Tyer, F., Pienaar J.: Extracting bilingual words pairs from Wikipedia (2008)
4. Clark, J., Dyer, C., Lavie, A., Smith, N.: Better hypothesis testing for statistical machine translation: controlling for optimizer instability. In: Proceedings of the Association for Computational Linguistics, Portland, Oregon, USA (2011)

---

[3] https://github.com/jhclark/multeval.

5. Marasek, K.: TED Polish-to-English translation system for the IWSLT 2012. In: Proceedings of the 9th International Workshop on Spoken Language Translation IWSLT 2012, pp. 126–129, Hong Kong (2012)

6. Smith J., Quirk C., Toutanova K.: Extracting parallel sentences from comparable corpora using document level alignmen (2010)

7. Chu, C., Nakazawa, T., Kurohashi, S.: Chinese-japanese parallel sentence extraction from quasi-comparable corpora. In: Proceedings of ACL 2013, pp 34–42 (2013)

8. Adafree, S., de Rijke, M.: Finding similar sentences across multiple languages in wikipedia (2006)

9. Skadiņa, I., Aker, A.: Collecting and using comparable corpora for statistical machine translation. In: Proceedings of LREC 2012, Instanbul (2012)

10. Koehn, P., Haddow, B.: Towards effective use of training data in statistical machine translation. In: WMT 2012 Proceedings of the Seventh Workshop on Statistical Machine Translation, pp. 317–321, Stroudsburg, PA, USA (2012)

11. Berrotarán, G., Carrascosa, R., Vine, A.: Yalign documentation. http://yalign.readthedocs.org/en/latest/

12. Tiedemann, J.: Parallel data, tools and interfaces in OPUS.: In: Proceedings of the 8th International Conference on Language Resources and Evaluation (LREC 2012), pp. 2214–2218 (2012)

13. Wołk, K., Marasek, K.: Real-Time statistical speech translation. In: Rocha, Á., Correia, A. M., Tan, F., Stroetmann, K. (eds.) New Perspectives in Information Systems and Technologies, Volume 1. AISC, vol. 275, pp. 107–113. Springer, Heidelberg (2014)

14. Kilgarriff, A., Avinesh, P.V.S., Pomikalek, J.: BootCatting comparable corpora. In: Proceedings of 9th International Conference on Terminology and Artificial Intelligence, Paris, France (2011)

15. Strotgen, J., Gertz, M.: Temporal tagging on different domains:challenges, strategies, and gold standards. In: Proceedings of LREC 2012, Instanbul (2012)

16. Cettolo, M., Girardi, C., Federico, M.: WIT3: web inventory of transcribed and translated talks. In: Proceedings of EAMT, pp. 261–268, Trento, Italy (2012)

17. Zeng, W., Church, R.L.: Finding shortest paths on real road networks: the case for A*. Int. J. Geogr. Inf. Sci. **23**(4), 531–543 (2009)

18. Wołk, K., Marasek, K.: Alignment of the polish-english parallel text for a statistical machine translation. Comput. Technol. Appl. **4**, 575–583 (2013). David Publishing, ISSN:1934–7332 (Print), ISSN: 1934-7340 (Online)

19. Yang, W., Lepage, Y.: Inflating a training corpus for SMT by using unrelated unaligned monolingual data. In: Ogrodniczuk, A., Przepiórkowski, M. (eds.) PolTAL 2014. LNCS, vol. 8686, pp. 236–248. Springer, Heidelberg (2014)

20. Musso, G.: Sequence alignment (Needleman-Wunsch, Smith-Waterman). http://www.cs.utoronto.ca/~brudno/bcb410/lec2notes.pdf

21. Joachims, T.: Text categorization with support vector machines: learning with many relevant features. In: Nédellec, C., Rouveirol, C. (eds.) ECML 1998. LNCS, vol. 1398, pp. 137–142. Springer, Heidelberg (1998)

Discovering Types of Spatial Relations
with a Text Mining Approach

Sarah Zenasni[1,3](✉), Eric Kergosien[2], Mathieu Roche[1,3],
and Maguelonne Teisseire[1,3]

[1] UMR TETIS (IRSTEA, CIRAD, AgroeParisTech), Montpellier, France
`sarah.zenasni@teledetection.fr`
[2] GERiiCO, Univ. Lille 3, Villeneuve-d'Ascq, France
[3] LIRMM, CNRS, Univ. Montpellier, Montpellier, France

Abstract. Knowledge discovery from texts, particularly the identification of spatial information is a difficult task due to the complexity of texts written in natural language. Here we propose a method combining two statistical approaches (lexical and contextual analysis) and a text mining approach to automatically identify types of spatial relations. Experiments conducted on an English corpus are presented.

Keywords: Spatial relations · Text mining

1 Introduction

Geographic Information Systems (GIS) are currently used to manage spatial information from maps or geographic databases. With the high amount of information available on the Web (information overload), new media sources can be tapped to extract geographic information from textual data. Spatial information extraction is becoming increasingly important not only to determine(modif vrifier) the spatial entities, but also relationships between spatial entities.

The objective of the work presented in this paper is to discover the type of relationships between spatial entities expressed in texts. Within this scope, we focus specifically on three types of spatial relations: region (e.g. "leading up"), direction (e.g. "going up"), and distance (e.g. "near"). We propose a new approach combining string comparison and contextual methods.

The rest of this paper is organized as follows. Section 2 presents a brief introduction to existing Spatial Information Extraction studies. Then we describe in Sect. 3 the two proposed approaches and their combination. In Sect. 4, we outline the experimental analysis and results. A discussion of our findings is presented in Sect. 5. Finally, Sect. 6 presents the conclusion and prospects of our work.

2 Related Work

Many studies are focused on the identification of Named Entities (NE), and particularly Spatial Entities (SE) from textual data [5]. These methods rely on

© Springer International Publishing Switzerland 2015
F. Esposito et al. (Eds.): ISMIS 2015, LNAI 9384, pp. 442–451, 2015.
DOI: 10.1007/978-3-319-25252-0_47

linguistic methods (e.g. extraction patterns) [6] and/or statistical methods (distribution of terms in a corpus) [8], or the probability of occurrence of a set of terms [19]. Overall, the spatial relations can be identified by linguistic approaches, statistical or hybrid, combining the first two types of approach. Among the linguistic approaches, [12] proposes to identify relations by calculating the similarity between syntactic contexts, and [17] presents an approach that integrates a set of linguistic patterns. Among the statistical approaches, [7] proposes to predict relations using Bayesian networks. Other approaches, such as [10], define the *TF-IDF* (Term Frequency, Inverse Document Frequency) method. This method compares the behavior of a candidate term in the document, and analyzes its behavior in a collection of documents. Among the hybrid methods, spatial relations can be identified via association rules [13], as well as by inferential knowledge using a learning algorithm [3] or by text mining approaches [2,14,22]. Blessing and Schütze [2] proposes an approach to developing and implementing a complete system to extract fine-grained spatial relations by using a supervised machine learning approach without expensive manual labeling. Kirk [22] presented an approach for recognizing spatial containment relations between event mentions. They have developed supervised classifiers for: (1) recognizing the presence of a spatial relation between two event mentions, and (2) classifying spatially related event pairs into one of five spatial containment relations. These methods are interesting, but they do not always identify the semantics of relations. The main contribution of this paper is to automatically identify the type of spatial relations studied. We propose a hybrid method, combining two statistical approaches (lexical and contextual analysis) and a text mining approach to predict the type of spatial relations among the following three types: **region, distance, and direction** [16].

3   Prediction of the Spatial Relation Type

In this Section, we propose an experimental approach involving three steps, as illustrated in Fig. 1.

First, to predict the type of spatial relations, we used string comparison between the "candidate relation" and training set relations through *String Matching (SM)* [1] and *Lin* [4] methods. Specifically, these methods allow for a comparison of strings. We then adapt a data mining algorithm, the *K Nearest Neighbors (KNN)* [15] to return the majority class assigned to each spatial relation.

Secondly, we use another concept concerning the context of a relation as a type indicator. This means that the words around the relation are extracted and weighted with statistical methods, including the *frequency* and *TF-IDF* [10]. As in the previous step, we apply the *KNN* data mining algorithm to select the class, i.e. the spatial relation type, to assign.

Finally, in order to improve the accuracy of the approach, we propose to combine both types of information, i.e. lexical and contextual. To illustrate our approach, the example below shows two sentences with spatial relations in bold. **Sentence 1:** Stairs are **leading up** (region) to the entrance.

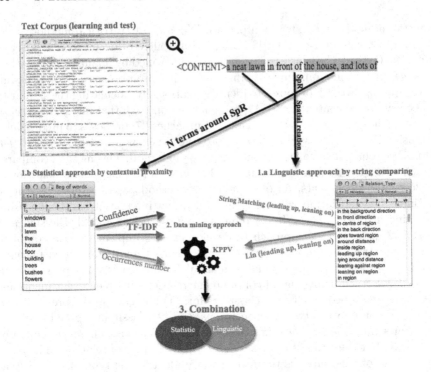

Fig. 1. The proposed process for type prediction.

Sentence 2: Four locals are **sitting on** (region) a bench in a canteen kitchen, **leaning on** (region) a red brick wall.

3.1 String Comparison

Among the many existing similarity measures, we chose *String Matching (SM)* and *Lin* measures that are conventionally used as they produce relevant results [9].

String Matching. *SM* is a lexical measure based on **Levenshtein distance** [11], also called distance Edition *(E)*. *SM* calculates the minimum sum of cost of operations: **delete, insert, and replace,** which are needed to turn a string *Ch1* into a string *Ch2*. From the example shown in Fig. 2, we get E (leading up, leaning on) = 3, as there are three operations for passing the string "leading up" to "leaning on".

One calculating the distance *E*, we apply the formula (1) to calculate the value of *SM*, normalized between 0 and 1.

$$SM(Ch1, Ch2) = max[0; (min(|Ch1|, |Ch2|) - E(Ch1, Ch2))/min(|Ch1|, |Ch2|)] \quad (1)$$

From the example of Sentence 1, *SM* (leading up, leaning on) = max [0, (10-3) / 10] = 0.70.

For each candidate relation, we return the similarities obtained with all relations of a learning set consisting of spatial relations whose type has already been validated (see detailed explanation in Sect. 4). We determine the K closest relations in order to predict the class to associate. When applying the KNN algorithm, the assigned class is the majority class of the K closest relations on the basis of the measurement SM. For special cases, such as relations consisting of two words whose second word is a spatial relation (e.g. **to next, standing in, sitting at** ...), we hypothesize that these relations are the same type as the associated relation, i.e. **to, in, at** ...

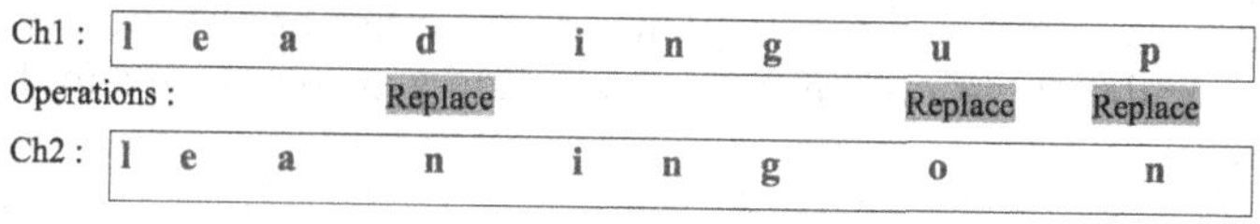

Ch1 :	l	e	a	d	i	n	g	u	p
Operations :				Replace				Replace	Replace
Ch2 :	l	e	a	n	i	n	g	o	n

Fig. 2. Levenshtein distance for relations "leading up" and "leaning on".

Lin. The Lin measure calculates the lexical proximity between two strings. It is based on n-*gram*, indicating the number of consecutive characters. Generally, the value of n ranges fro 2 to 5. By setting $n = 3$ (tri-grams), we get the following result from the example Sentence 1:

- $\mathbf{tr}(Ch2) = \{lea, ead, adi, din, ing, ng , gu, up\} = 8$
- $\mathbf{tr}(Ch1) = \{lea, ean, ani, nin, ing, ng , go, on\} = 8$
- $\mathbf{tr} (Ch1) \cap \mathbf{tr} (Ch1) = 3$

The formula 2 shows the similarity measure Lin based on the calculation of the number of characters (n-gram) in common between the two strings.

$$Lin(Ch1, Ch2) = \frac{1}{[1 + |tr(Ch1)| + |tr(Ch2)| - 2 \times |tr(Ch1) \cap tr(Ch2)|]} \quad (2)$$

From the example of Sentence 1, we get:

$$Lin(leading\ up, leaning\ on) = \frac{1}{[(1+8+8)-(2\times3)]} = 0,09$$

Then we apply the KNN algorithm that returns the majority class of the type for the candidate relation. Note that the lexical information is not always sufficient. Indeed, two expressions can be lexically remote but semantically very close. To solve this problem, in the next section we propose to take the context of relations into account in order to better predict their class.

3.2 Contextual Proximity

In this step, we focus on the context of each relation, i.e. all words in the vicinity of these relations, which we call *"lexical world"*. We hypothesize that it will allow us to improve identification of the spatial relation type. The context is defined in two ways:

– all the sentence,
– N words around the relation.

We then rely on a bag of word approach as follows: we compare different weighting factors (number of occurrences, TF-IDF ...) to construct the lexical world to identify the type of relevant spatial relation.

1. **Number of occurrences (nber_occ)**
   Here we calculate the weight of a term i counting the occurrences of that word in the document j. More specifically, how often does the term appear with the same relation in all sentences? For example the term "wall" is with the relation "leaning on" 3 times. Once the number of occurrences of all the terms is calculated, we obtain for each relation a vector, an example of which is presented Table 1.

Table 1. Extract from the number of occurrences vector by example sentences 1 and 2.

(SpR, nber_occ)	(term 1, nber_occ)	(term 2, nber_occ)	(term n, nber_occ)
(leaning on, 3)	(wall, 3)	(locals, 1)	(term n, nber_occ)

2. **Term Frequency, Inverse Document Frequency (TF-IDF)**
   TF-IDF is a statistical measure. It scores the importance of words (or "terms") in a document based on how frequently they appear across multiple documents.
   We get N vectors representing each relation, whose length is equal to the number of terms of the corpus and the weight of each term equal to $w_{i,j}$. Table 2 is an extract from the *TF-IDF* vector of the relation "leaning on".

Table 2. Extract from the TF-IDF vector by example sentences 1 and 2.

(SpR, nber_occ)	(term 1, $w_{i,j}$)	(term 2, $w_{i,j}$))	(term n, $w_{i,j}$)
(leaning on, 3)	(wall, 5.05)	(locals, 5.74)	(term n, $w_{i,j}$)

3. **The confidence**
   We calculate the confidence (conf) as defined by [20]. It indicates the probability that a transaction contains Y given that it contains X. The confidence levels are given by:

$$conf = \frac{Number\ of\ sentences\ containing\ the\ relation\ X\ and\ the\ term\ Y}{Number\ of\ sentences\ containing\ the\ relation\ X} \qquad (3)$$

For example, if three sentences contain the relation "leaning on" and one sentence contains the relation "leaning on" and the term "locals", we calculate the confidence associated with the word "locals" of the lexical world as follows:

conf = 1/3 = 0.33. As just described with the two previous steps, Table 3 shows an extract of the vector of the relation "leaning on" and the terms contained in the corpus (e.g. Sentences 1 and 2) this time with the weight of each term is equal to *conf* .

Table 3. Extract from the confidence vector by example sentences 1 and 2.

(SpR, nber_occ)	(term 1, conf)	(term 2, conf)	(term n, conf)
(leaning on, 3)	(wall, 1.0)	(locals, 0.33)	(term n, conf)

Based on the three performances and weightings of the words of the lexical world, we measure the proximity between lexical worlds specific to candidate relations and relations of the learning set. This lexical proximity is calculated through the classic Information Retrieval measure, i.e. the *Cosine* measure [21]. To predict the spatial relations type, we apply the *KNN* algorithm and we allocate each candidate relation to the class identified as the closest.

Moreover, in this step, the terms may belong to the lexical world of several relation types. To solve this problem, we choose the type of relation for which the term is the most frequent. To reduce ambiguity and find ways to automatically identify the spatial relations type, we decided to combine the previous two approaches "by chain character comparison" and "contextual proximity" to offer a more robust approach.

3.3   Combination

We propose an experimental approach combining the two previously described approaches (lexical, contextual). As part of our experiments (see Sect. 4), we obtain a list of predicted relations for each approach. Qualitatively, we notice that the approach by comparing strings works best. However, the contextual proximity approach works best when relations consist of more than 4 terms. Looking at this first analysis, we define our combination as follows: if relations consist of more than 4 words, we favor contextual information and Cosine measure, otherwise we choose lexical information based on SM.

In Sect. 4, we outline our results and the analysis of the methods presented in this section.

4   Experiments

In order to assess the ability of our system to automatically predict the type of spatial relations (region, distance, and direction), we chose the annotated corpus SpRL (Spatial Role Labeling) [18] that is a repository from a challenge (SemEval-2012). This corpus consists of 1213 sentences annotated in English. Each sentence contains one or more words corresponding to a spatial relation "spatial indicator". These terms are grouped into 3 types of spatial relations

Region, Direction, and Distance. Our evaluation setting was based on a series of experiments where we change the parameters K for *KNN* and N for the terms around the spatial relations. For the evaluation, we used the following criteria: Precision, Recall, F-measure, and Accuracy.

1. *Accuracy:* percentage of relations correctly classified by the system.
2. *Precision:* percentage of relevant relations correctly retrieved by the system (True Positive "TP") with respect to all relations retrieved by the system (True Positive "TP" + False Positive "FP").
3. *Recall:* percentage of relevant relations correctly retrieved by the system (True Positive "TP") with respect to all relations relevant for humans (True Positive "TP" + False Negative "FN").
4. *F-Measure:* combine in a single measure Precision (P) and Recall (R) giving a global estimation of the performance.

4.1 String Comparison

We apply a cross-validation process. In our case, the corpus is divided into three partitions (P1, P2, P3) and each partition contains 31 relations (18 for region, 10 for direction, and 3 for distance). Table 4 shows in column *String Matching 1* the results obtained in terms of accuracy from the first series of experiments, when applying the *SM* algorithm only. *String Matching 2* shows the results obtained by applying our heuristic presented in Sect. 3.1, indicating that when the relation *r1* consists of two words and the second is a spatial relation *r2*, we attribute for *r1* the same class of *r2*.

The performance of the similarity measure *SM* shows satisfactory results compared to the *Lin* measure, whatever K. *String Matching 2* offers the best accuracy resultand we obtained 0.82 with $K = 1$.

Table 4. Accuracy results of SM and Lin measures.

K	String Matching 1	String Matching 2	Lin
1	0.77	0.82	0.75
3	0.74	0.79	0.73
5	0.73	0.76	0.69

4.2 Contextual Proximity

Here we test the second approach with all lexical worlds (occurrence number, TF-IDF, confidence) while considering the removal of stop words. We test under two different contexts *all the sentence* and *N words around spatial relation*. We evaluate the approach for each lexical world with a K ranging from 1 to 5 and N from 1 to 3. In Table 5, we notice that the context *N words around the relation* gives better results than the context *all sentences*. The lexical world based on TF-IDF with $K \geq 3$ gives significant results.

In conclusion, the best 0.67 accuracy result is obtained with $K = 5$ and $N = 2$.

Table 5. Results of the Cosine method in terms of accuracy without stop words.

k		All the sentence	n words around Rs		
			n = 1	n = 2	n = 3
1	nber_occ	0.61	0.62	0.62	0.60
	TF-IDF	0.56	0.60	0.51	0.53
	conf	0.56	0.62	0.62	0.60
3	nber_occ	0.62	0.60	0.63	0.58
	TF-IDF	0.45	0.63	0.66	0.63
	conf	0.40	0.60	0.63	0.56
5	nber_occ	0.58	0.65	0.67	0.57
	TF-IDF	0.40	0.64	0.67	0.66
	conf	0.41	0.64	0.67	0.56

4.3 Combination

Results obtained from a combination of the previous two steps are given Table 6. We conducted a study to measure the combination using the most appropriate parameters, i.e. $K = 1$ for SM and $K = 5$, $N = 2$ for $Cosine$ using the lexical world based on TF-IDF. The accuracy score is 0.84. That means over 78 to associate to a class, 14 relations are misclassified. For instance, the relations **sailing over** and **walking towards** have to be associated with the class **direction**. But our system automatically classifies these relations as **region**. We can explain this error by the number of relations in each class, where the total number of relations of type **region** is 54, of **direction** 30, and **distance** 9. We note as well that the combination of the two methods (String comparison and consideration the lexical world) performs better than each method individually.

Table 6. Results of combination.

	Precision	Recall	F-measure
Region	0.86	0.94	0.90
Distance	0.75	0.37	0.50
Direction	0.82	0.79	0.80

5 Discussion

Here we presented a hybrid approach to automatically predict the spatial relations type. We note that the SM method with consideration of lexical information performs better than the Lin method. We also note that the consideration of contextual information statically weighted by TF-IDF achieved significant results. Furthermore, we get even better results by combining the two approaches. Since we worked on a corpus in English, we suppose that using a French corpus may influence the results for automatic prediction of the spatial relation type, as the structure of the French language sentences is different. We wonder if the parameter N will change, and if the weighted contextual information statically by *TF-IDF* will always give the best results.

6 Conclusion

In this article, we proposed a comparative analysis of two approaches and their combination for automatic identification of the spatial relation type. Firstly we proposed a heuristic in order to improve the lexical measures. Our results highlight a significant improvement with the best result obtained with an Accuracy score of 0.82. Secondly we defined a lexical world to improve the prediction. Then we proposed a method combining two statistical approaches (lexical and contextual analysis) and a text mining approach. Our results show that the combination improves the quality of identification of the spatial relation type (Accuracy score of 0.84). Further, we plan two future directions. Initially we want to construct a world based on lexical selection morphosyntactic words (verbs, common/proper nouns, adverbs, adjectives, etc.) and/or phrases (noun-adjective, etc.). Secondly, we also plan to create a classification approach for predicting own class: (1) Other types of semantic relations between spatial entities such as the "twinning" relation; (2) Spatial relations between different types of named entities (person, organization, etc.).

Acknowledgments. This work is supported by the Ministry of Higher Education and Scientific Research of Algeria and Cirad. The authors thank David Manley who improved the readability of this paper.

References

1. Maedche, A., Staab, S.: Measuring similarity between ontologies. In: Gómez-Pérez, A., Benjamins, V.R. (eds.) EKAW 2002. LNCS (LNAI), vol. 2473, p. 251. Springer, Heidelberg (2002)
2. Blessing, A., Schütze, H.: Self-Annotation for fine-grained geospatial relation extraction. In: COLING, 23rd International Conference on Computational Linguistics, pp. 80–88 (2010)
3. Giuliano, C., Lavelli, A., Romano, L.: Exploiting shallow linguistic information for relation extraction. In: From Biomedical Literature In Proceedings of EACL (2006)
4. Lin, D.: An Information-theoretic definition of similarity. In: Proceedings of the 15th International Conference on Machine Learning (ICML), pp. 296–304 (1998)
5. Nadeau, D., Sekine, S.: A survey of named entity recognition and classification. Lingvisticae Invest. **30**, 3–26 (2007). John Benjamins Publishing Company
6. Maurel, D., Friburger, N., Antoine, J.-Y., Eshkol-Taravella, I., Nouvel, D.: CasEN: a transducer cascade to recognize French Named Entities. TAL **52**(1), 69–96 (2011)

7. Weissenbacher, D., Nazarenko, A.: Identifier les pronoms anaphoriques et trouver leurs antécédents: l'intérêt de la classification bayésienne. In: Proceedings of TALN, pp. 145–155 (2007)
8. Agirre, E., Ansa, O., Hovy, E.-H., Martìnez, D.: Enriching very large ontologies using the WWW. In: ECAI Workshop on Ontology Learning, Spain (2000)
9. Duchateau, F., Bellahsene, Z., Roche, M.: Improving quality and performance of schema matching in large scale. Ingénierie des Systèmes d'Information **13**(5), 59–82 (2008)
10. Salton, G., Buckley, C.: Term-Weighting Approaches in Automatic Text Retrieval. Inf. Process. Manage **24**(5), 513–523 (1988)
11. Navarro, G.: A guided tour to approximate string matching. ACM Comput. Surv. **33**(1), 31–88 (2001)
12. Grefenstette, G.: Explorations in Automatic Thesaurus Discovery. Kluwer Academic Publishers, Boston (1994)
13. Koperski, K., Han, J.: Discovery of spatial association rules in geographic information databases. In: Egenhofer, M., Herring, J.R. (eds.) SSD 1995. LNCS, vol. 951, pp. 47–66. Springer, Heidelberg (1995)
14. Grčar, M., Klien, E., Novak, B.: Using term-matching algorithms for the annotation of geo-services. In: Knowl. Disc. Enhanced with Semantic and Social, Information, pp. 127–143 (2009)
15. Bhatia, N., Vandana: Survey of Nearest Neighbor Techniques. CoRR, vol. 08, pp. 302–305 (2010)
16. Stock, O.: Spatial and Temporal Reasoning. Kluwer Academic Publishers, Norwell (1997)
17. Kordjamshidi, P., Martijn, V.O., Moens, M.-F.: Spatial role labeling: towards extraction of spatial relations from natural language. ACM Trans. Speech Lang. Process **08**(3), 1–36 (2011)
18. Kordjamshidi, P., Bethard, S., Moens, M.-F.: SemEval-2012 task 3: spatial role labeling.In: Proceedings of International Workshop on Semantic Evaluation, pp. 365–373 (2012)
19. Velardi, P., Fabriani, P., Missikoff, M.: Using text processing techniques to automatically enrich a domain ontology. In: FOIS, pp. 270–284 (2001)
20. Agrawal, R., Imielinski, T., Swami, A.N.: Mining association rules between sets of items in large databases. In: Proceedings of International Conference on Management of Data (SIGMOD), pp. 207–216 (1993)
21. Baeza-Yates, R.A., Ribeiro-Neto, B.A.: Modern Information Retrieval. ACM Press/Addison-Wesley, New York/Reading (1999)
22. Roberts, K., Harabagiu, S.M., Skinner, M.A.: Recognizing spatial containment relations between event mentions. In: Proceedings of International Conference on Computational Semantics (IWCS 2013), pp. 216–227 (2013)

Multi-dimensional Reputation Modeling Using Micro-blog Contents

Jean-Valère Cossu, Eric SanJuan[✉], Juan-Manuel Torres-Moreno,
and Marc El-Bèze

LIA/Université d'Avignon et des Pays de Vaucluse, 339 chemin des Meinajaries,
Agroparc BP 91228, 84911 Avignon cedex 9, France
{jean-valere.cossu,eric.sanjuan,juan-manuel.torres-moreno,
marc.el-beze}@univ-avignon.fr

Abstract. In this paper, we investigate the issue of modeling corporate entities' online reputation. We introduce a bayesian latent probabilistic model approach for e-Reputation analysis based on Dimensions (Reputational Concepts) Categorization and Opinion Mining from textual content. Dimensions to analyze e-Reputation are set up by analyst as latent variables. Machine Learning (ML) Natural Language Processing (NLP) approaches are used to label large sets of text passages. For each Dimension, several estimations of the relationship with each text passage are computed as well as Opinion and Priority. The proposed automatic path modeling algorithm explains Opinion or Priority scores based on selected Dimensions. Model Robustness' is evaluated over RepLab dataset.

1 Introduction

Online Reputation Management is recently being a top-notch subject of studies investigated in several researches from various fields such Marketing, Psychology, Social Media Analysis (SMA), Information Retrieval (IR) and NLP. Although researchers have been processing and managing business information for several decade, ML techniques are still lacking to support the process of providing a relevant Reputation Modeling. An increasing number of SMA services and community managers use Micro-Blog streams from Twitter based on key-word queries to analyze Opinion and Reputation related to some entities. The use of key-word queries instead of analyzing the whole Micro-Blog sphere allows to deal with very large Micro-Blog networks and to avoid complex an expensive pooling strategies, often non-accurate for a micro analysis of specific entities like museums in Barcelona at the time of champion league. The drawback of these streams based on key-word queries is that they only give a partial and incomplete view of the whole Micro-Blog network. Removing or adding some keywords can lead to very different datasets. Moreover, Dimensions (Reputational Concepts) to explain reputation and recommend actions are often vague and depend on individual analyst expertise and background that could be mislead by a peculiar choice of keywords. Given an entity (public figure), a set of pre-defined attributes, we seek to model their impact on two objective measures:

© Springer International Publishing Switzerland 2015
F. Esposito et al. (Eds.): ISMIS 2015, LNAI 9384, pp. 452–457, 2015.
DOI: 10.1007/978-3-319-25252-0_48

1. External public Opinion expressed about the entity,
2. Internal Stimuli Priority [1],

based on a stream of public Micro-Blogs mentioning the entity.

2   Probabilistic Path Modeling

We consider the case where the analyst defines a set of D Dimensions through which he aims to analyze Opinion or to define Priority over a set of Micro-Blogs. To define these Dimensions we suppose that entity provides examples of Micro-Blogs for each Dimension and that for each example it indicates the Opinion (Positive, Neutral, Negative) and the Priority (Alert, Important, Non-Important). The chosen Dimensions were suggested by e-watcher specialists from the Reputation Institute's Reptrak like: Products and services, Citizenship, Governance, Innovation, Leadership, Performance, Workplace.

Therefore, given a trend Ω of Micro-Blogs texts, the analyst provides a family $\mathcal{S} = S_1, ..., S_D$ of subsets of Ω, each representing a Dimension, an Opinion or an *"Alert"* that we shall model as a latent variable. We do not expect S_i to be unbiased pools of Micro-Blogs, but we can still use them to train a sequence of K independent classifiers over each set that will evaluate the possibility for a new Micro-Blog text to be included by the analyst in some S_i. More formally we consider K sets of D scoring normalized functions such as:

$$f_{i,k} : \omega \in \Omega \mapsto f(\omega) \in [0,1] \tag{1}$$

$$\omega \in S_i \Rightarrow (\forall 1 \leq i \leq D) f_{i,k}(\omega) = 1 \tag{2}$$

Each normalized scoring function $f_{i,k}$ defines a discrete smoothed probability function $P_{i,k}$ over Ω defined by:

$$P_{i,k}(\{\omega\}) = (1 - \lambda) f_{i,k}(\omega) + \lambda E(f_{i,k}) \tag{3}$$

where λ is a smoothing parameter in $[0,1]$ and $E(f_{i,k})$ is the expectation of $f_{i,k}$ over a trend of Micro-Blogs and can be simply estimated over large finite pools T of Micro-Blogs as: $\frac{1}{|T|} \sum_{\omega \in T} f_{i,k}(\omega)$. We hypothesize that for each latent variable S_i there exists: $\alpha_i = (\alpha_{i,1}, ..., \alpha_{i,k})$, $\alpha_{i,k} > 0$ such that the real probability P_i for a Micro-Blogs ω to be associated by the analyst with S_i verifies:

$$P_i(\{\omega\}) = \prod_{k=1}^{K} P_{i,k}(\{\omega\})^{\alpha_{i,k}} \tag{4}$$

To analyze Opinion and Priority based on entity's defined Dimensions, we define two path models among latent variables. For Opinion, let us suppose that S_D is associated with "Neutral", S_{D-1} with "Negative" and S_{D-2} with "Neutral" meanwhile $S_1, ..., S_{D-3}$ are the entity's defined Dimensions. Then we consider as path model the acyclic directed graph $G_{op}(\mathcal{S}) = (1, .., D, A_{op}(\mathcal{S}))$ where the set the set of arrows $A_{op}(\mathcal{S}$ is defined as:

$$A_{op}(\mathcal{S} = \{1, ..., D - 3\} \times \{D - 2, D - 1\} \cup \{D - 2, D - 1\} \times \{D\} \tag{5}$$

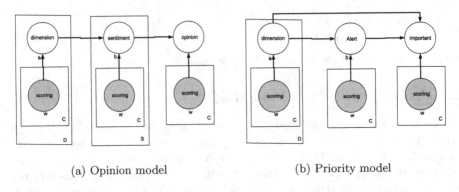

(a) Opinion model   (b) Priority model

Fig. 1. Plate notation for path model

Figure 1a gives the plate representation of this path model. For Priority we define a slightly different path model: we suppose that S_D is associated with *"Important"* and S_{D-1} with *"Alert"* meanwhile $S_1, ..., S_{D-2}$ are the entity's defined Dimensions. Then we consider as path model the acyclic directed graph $G_{al}(S) = (1, .., D, A_{al}(S))$ where the set the set of arrows $A_{al}(S)$ is defined as:

$$A_op(S = \{1, ..., D - 2\} \times \{D - 1, D\} \cup \{(D - 1, D)\} \tag{6}$$

Figure 1b gives the plate representation of this path model. To analyze the impact of Dimensions over Opinion or *"Alert"*, given a smoothed parameter λ, we propose to seek for the probabilistic model with parameter $\alpha = [\alpha_{i,k}]_{1 \leq i \leq D, 1 \leq k \leq K}$ that maximizes the product of conditional probabilities along the Bayesian network induced by $G \in \{G_{op}, G_{al}\}$, i.e.:

$$\alpha(\lambda, S) = argmax_\alpha \{ \prod_{(i,j) \in G} P(S_i/S_j) \} \tag{7}$$

3   Experimental Data and Settings

3.1   Replab Dataset

We use the context of RepLab [2,3] and related tasks [4–11] to evaluate our proposal that is to say: to propose an overview of 61 entity's (drawn in 4 domains: Automotive, Banking, Music and University) e-Reputation regarding experts taxonomies using provided set and pertaining of Micro-Blogs concerning each entity. Data have been annotated by experts from Llorente & Cuenca[1] for either Polarity for Reputation (3 levels), Priority (3 levels) and Reputation Dimensions (only for Automotive and Banking domains). These Dimensions have been identified as relevant interests and are considered as key issues for company' stakeholders.

---

[1] http://www.llorenteycuenca.com/.

3.2   Scoring Protocol

Term Scoring. The proposed supervised corpus-based classification method is entirely based on the tweets contents, i.e., it does not use any external knowledge resource. Besides being the most relevant source of information for classification, the text cannot be directly processed by classifiers and learning algorithms in its unstructured natural form. For that reason, our tweets should be translated into a manageable form: the discriminant bag-of-words representation. In this step, to each term multiple weights (using Frequency-Inverse Document Frequency (TF-IDF-Gini) approach [12]) are assigned, which describe the strength of association of a term with each label, Priority and Polarity levels. We assign a score to each term corresponding to its strength of association with each Dimensions, Priority and Polarity (borrowed from the tweet's label in the training set). Terms mentioned frequently in tweets that have a specific label gets the highest score.

Document Scoring. For an evaluation purpose RepLabs' organizers provided a single annotation for each tweet which correspond to discrete binary score and may remain insufficient to provide a reputation modeling. Dimensions like Innovation or Leadership are vague. Moreover tweets can be misleading and ambiguous this is why we undertake a scoring approach rather than classification using strong learning references. In this step, using statistical NLP approaches [12], we project each document in a #Dimensions × #Opinion or Priority × #Classifiers dimensions spaces. We consider the similarities between each tweet and each label using: Cosine distance, Jaccard index, Linear SVM and k-NN.

Scoring Normalization. As we aim at ranking document in each class, all similarities need to be normalized to the same scale. For each tweet, the degree of similarity of similarity (confidence degree) to each class provided by kNN, Cosinus and Jaccard have been normalized using the sum of all system-provided scores related to this document. For SVM we considered the lowest hypothesis score as a rejected class. Its value was added to all other classes, then it has been normalized just as mentioned before. Cosinus and Jaccard proposals fit the typical IR issue as they rank each document in each class while kNN and SVM approach are nearer human annotator behavior in evaluating the document with a limited number of background neighbors and rejecting hypothesis. The similarity measure operate only on the overlapping document and some documents are assigned with a zero weight which limit the abstraction process and increase the probability of strong errors. It leads us to consider probability re-estimation of a document d in a class c using a smoothing as defined in (3).

4   Dimension Ranking Evaluation

Rather than looking for classification performances we focus on Reputation Modeling. When considering only the best class hypothesis provided by classifiers, the Document Scoring Approach performances have been reported in [12]. We use Replab testing set to evaluate the ability of our probabilistic path model to rank Dimensions by decreasing impact over an Opinion or a Priority. We expect models presented here to be independent from the selected text classifiers used to score Dimensions and Opinions. We seek for models' robustness against incompleteness of data used as training sets. RepLab complete reference are available for two domains: Automotive and Banking. For Banking the test set contains $10,781$ tweets among which $4,524$ are Negative. $5,015$ Positive, the rest being Neutral. Our probabilistic model provides two rankings of these Dimensions:

1. Impact over Negative opinions scored by $P(Negative/Dimension)$
2. Impact over Positive opinions scored by $P(Positive/Dimension)$

We compare each of these rankings against the reference:

$$cor(ref_{Dimension}, ref_{Negative} - ref_{Positive}) \qquad (8)$$

For each Dimension $ref_{Dimension}$ is a binary vector (1 if the tweet is in this Dimension, 0 otherwise) meanwhile $ref_{Negative} - ref_{Positive}$ is ternary (1 if Negative, 0 if Neutral, -1 if Positive). The results of the ranking produced by $P(Negative/Dimension)$ are the following:

– Pearson's product-moment correlation = 0.9759958, p-value = 0.0008574
– Kendall's rank correlation $\tau = 1$, p-value = 0.002778
– Spearman's rank correlation $\rho = 1$, p-value = 0.002778

Therefore, for Banking rankings towards Negative are highly correlated and these correlations are statistically highly significant, meanwhile the same tests over the training corpus are not significant because of data incompleteness. The ranking produced by $P(Positive/Dimension)$ is less correlated because of Neutral Micro-Blogs but correlation remains statistically significant (p-value < 0.1).

For Automotive, there are $12,776$ tweets in the reference with only $1,718$ Negative against $10,995$ Positive and only 63 Neutral. The higher correlation is obtained comparing $P(Negative/Dimension)$ against:

$$cor(ref_{Dimension}, ref_{Negative}) \qquad (9)$$

In this case Spearman's rank correlation $\rho = 0.7714286$ but p-value=0.1028. All other correlations are positive but non significant. In Banking and Automotive cases, the probabilistic model performs better on explaining Negative than Positive. In both cases the Negative class is narrow than the Positive class.

Finally, for *Alerts* in the case of Banks, Dimensions are ranked in the exact same order. However, the Pearson's product-moment correlation between the predicted inner latent variable *Alert* and the reference is 0.49 is significantly high (p-value $< 10^{-3}$) but not significantly higher than the single Cosine estimate. This means that in this case, the use of a single basic classifier would have provided similar results which is not the case for Opinion.

5   Discussion and Conclusions

In this paper we examined an automatic method to order relations between classes and infer more complex latent hierarchies. For that we introduced a probabilistic approach for e-Reputation analysis based on Reputational Concepts Categorization and Opinion Mining from textual content. For each Concepts, several estimations of the relationship with each text passage are computed. We proposed an automatic path modeling algorithm to explains Opinion or Priority scores based on selected Concepts. Robustness of the resulting model has been evaluated over the multilingual CLEF RepLab dataset.

Acknowledgment. This work is funded by the project ImagiWeb ANR-2012-CORD-002-01.

References

1. Mather, M., Sutherland, M.R.: Arousal-biased competition in perception and memory. Perspect. Psychol. Sci. **6**(2), 114–133 (2011)
2. Amigó, E., Carrillo de Albornoz, J., Chugur, I., Corujo, A., Gonzalo, J., Martín, T., Meij, E., de Rijke, M., Spina, D.: Overview of RepLab 2013: evaluating online reputation monitoring systems. In: Forner, P., Müller, H., Paredes, R., Rosso, P., Stein, B. (eds.) CLEF 2013. LNCS, vol. 8138, pp. 333–352. Springer, Heidelberg (2013)
3. Amigó, E., Carrillo-de-Albornoz, J., Chugur, I., Corujo, A., Gonzalo, J., Meij, E., de Rijke, M., Spina, D.: Overview of RepLab 2014: author profiling and reputation dimensions for online reputation management. In: Kanoulas, E., Lupu, M., Clough, P., Sanderson, M., Hall, M., Hanbury, A., Toms, E. (eds.) CLEF 2014. LNCS, vol. 8685, pp. 307–322. Springer, Heidelberg (2014)
4. Villena Román, J., Lana Serrano, S., Martínez Cámara, E., González Cristóbal, J.C.: Tass-workshop on sentiment analysis at sepln (2013)
5. Zhao, W.X., Jiang, J., He, J., Song, Y., Achananuparp, P., Lim, E.P., Li, X.: Topical keyphrase extraction from twitter. In: Proceedings of the 49th Annual Meeting of the ACL: Human Language Technologies (2011)
6. Velcin, J., Kim, Y., Brun, C., Dormagen, J., SanJuan, E., Khouas, L., Peradotto, A., Bonnevay, S., Roux, C., Boyadjian, J., et al.: Investigating the image of entities in social media: Dataset design and first results. In: LREC (2014)
7. Peleja, F., Santos, J., Magalhães, J.: Reputation analysis with a ranked sentiment-lexicon. In: Proceedings of the 37th SIGIR conference (2014)
8. McDonald, G., Deveaud, R., McCreadie, R., Macdonald, C., Ounis, I.: Tweet enrichment for effective dimensions classification in online reputation management. In: Ninth International AAAI Conference on Web and Social Media (2015)
9. Qureshi, M.A., O'Riordan, C., Pasi, G.: Exploiting wikipedia for entity name disambiguation in tweets. In: NLP and Information Systems (2014)
10. Derczynski, L., Maynard, D., Rizzo, G., van Erp, M., Gorrell, G., Troncy, R., Petrak, J., Bontcheva, K.: Analysis of named entity recognition and linking for tweets. Inf. Process. Manag. **51**(2), 32–49 (2015)
11. Damak, F., Pinel-Sauvagnat, K., Boughanem, M., Cabanac, G.: Effectiveness of state-of-the-art features for microblog search. In: The 28th ACM Symposium on Applied Computing (2013)
12. Cossu, J.V., Janod, K., Ferreira, E., Gaillard, J., El-Bèze, M.: Nlp-based classifiers to generalize experts assessments in e-reputation. In: CLEF (2015)

Author Disambiguation

Aleksandra Campar[1], Burcu Kolbay[1], Hector Aguilera[1], Iva Stankovic[1],
Kaiser Co[1], Fabien Rico[2](✉), and Djamel A. Zighed[3]

[1] Erasmus Mundus DMKM, Lyon 2 University, Lyon, France
[2] ERIC Laboratory, Lyon 1 University, Villeurbanne, France
`fabien.rico@univ-lyon1.fr`
[3] ISH, Lyon 2 University, Lyon, France

Abstract. This paper proposes a novel approach in incorporating several
metadata such as citations, co-authorship, titles, and keywords to identify
real authors in author disambiguation task. Classification schemes make
use of these variables to identify authorship. The methodology performed
in this paper is: (1) coarse grouping of article by the use of focus names,
(2) Applying a model using paper metadata to identify same authorship,
and (3) separate the true authors having the same focus name.

Keywords: Author disambiguation · Entity resolution · Focus names ·
Supervised learning · Name matching

1 Introduction

The problem of author disambiguation, is an important concern for digital publication libraries such as DBLP, ... This problem, is an instance of a more general problem: Entity Resolution or Record Linkage. Author disambiguation is
a difficult task due to name variation, identical names, name misspellings, and
pseudonyms. In general, we can find two types of name ambiguities. First, an
author has multiple name labels and, second, multiple authors may share the
same name label. In the literature several proposed supervised machine learning
approaches have been applied in order to tackle the problem of author disambiguation. They are mainly based on similarity measures between papers. Most
of the cases, text-similarity approaches are used in order to compare references.
We have selected some of them in our work. The core of our contribution is
a novel approach that allows us to identify whether two articles belong to the
same real author or do not. This is done by taking into account their components
(authors, title, keywords) and their context (co-author and citation networks).

Our author disambiguation approach is essentially based on three steps.
Firstly, using focus name, we perform a filter to select papers which can possibly belong to the same author. Then, using a fully disambiguated data set,
as a learning data set, we build a classifier using supervised machine learning
techniques. This allows us to be able to predict, if two papers belong to the same
author or do not. Having this result that can be seen as a binary matrix between

© Springer International Publishing Switzerland 2015
F. Esposito et al. (Eds.): ISMIS 2015, LNAI 9384, pp. 458–464, 2015.
DOI: 10.1007/978-3-319-25252-0_49

pairs of articles within the focus names, we can carry out a bi-clustering method in order to separate true authors within a given focus name. The Fig. 1 depicts the workflow.

There are several methods to compute the similarity such as Euclidean, Jaccard and Cosine Similarity. Supervised learning approaches have been proposed [1–3], also some of researches have combined author disambiguation problem with social network analysis [4,5] by considering the distance between two authors in the network. McRae-Spencer and Shadbolt et al. (2006), proposed a graph-based

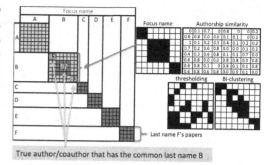

Fig. 1. Workflow of the approach

approach [6] on large-scale citation networks for author disambiguation using self-citation, co-authorship, and document-source analyses.

In an another example [7] that can be considered as an updated version of the previous method for graph-based theory, Fan et al. present a framework named as GHOST (abbreviation for GrapHical FramewOrk for name diSambiguaTion) that is used with a novel similarity metric and utilization of only one type of attribute (co-authorship). The clustering of results is done by affinity propagation clustering algorithm. In [8] is also an example which constructs similarity measures to apply a clustering algorithm. It compares the information presented in citations and context. They construct a context graph based on the co-authors community, topics, and computation of similarity measures between papers by considering the shared neighbor numbers.

The rest of the paper is organized as follows: Sect. 2 shows a general explanation of the method proposed, and explains in a detailed manner the features building scheme (2.1). Section 2.3 introduces the two core approaches (i.e. learning the link between papers and community in entity resolution), Sect. 3 shows the evaluation process and finally Sect. 4 concludes and discusses future work.

2   Our Approach

As presented in [9] it's necessary to process a coarse grouping of article to avoid exhaustive pairewise comparison. This primary step allow to skip the comparison of articles that were certainly written by different author. We use the concept of focus name : this is a string that groups one or more authors with a identical last name in one cluster as shown in Fig. 1. There are various ways to specify a focus name. For example it's possible to group authors that share a surname and have the same first letter of the first name and/or middle name. This lets us select subsets of papers that could belong to the same author. Using focus names allows us to divide the whole dataset into subsets for which we have computed similarity measures. Then similarity measures are computed for each pair of article corresponding to a given focus name. It is worth to mention, that the construction of these features (similarity/distance measures) is the core of our proposal. Lastly, the distance measures are used to feed classification algorithms (i.e. logistic regression, decision tree and random forest) which were trained in order to predict the class variable that states whereas two articles share or do not share the same author.

We have built and used 7 distance measures as features for each pair of papers in one cluster of focus name. They are keyword-based, title-based, communities graph-based and citations graph based. Since each feature has some limitations and advantages, combining them gives a better leverage in classification.

2.1 Features for Titles and Keywords

The similarity between two articles i and j will be considered as a combination of similarities between different topic of the articles. It's first necessary to extract the topics by a topic modelling method and for each article compare obtained topics sets. For the primary part we have decided on using LDA (Latent Dirichlet Allocation)[10] model and the "topics assignments" since LDA give a correspondence between each keyword (or title words) and a topic. Then, we propose to use similarity measures based on "Jaccard" and "Pearson" coefficient.

JACCARD computes the similarities of asymmetric information on binary attributes or set [12]. The measurement is from 0 to 1. For two set A and B it compute three values : $a = \#(A \cap B)$ the number of element in both sets, $b = \#(A \setminus B)$ and $c = \#(B \setminus A)$. Jaccard's basic formula is as follows: $Jacquard(A, B) = \frac{a}{a+b+c}$.

With this measurement we have tried to compute the similarity between keywords and titles. In our case, the set A and B were set of LDA's assigned keyword topics and title topics which give the features $D_{Kw_J}(A, B) = 1 - Jacquard(A, B)$, respectively $D_{t_J} = 1 - Jacquard(A, B)$. Using the same, we have also computed Pearson similarity on both segments [13] : $D_{Kw_P}(A, B) = 1 - Pearson(A, B)$ between two keyword sets and $D_{t_P}(A, B) = 1 - Pearson(A, B)$ between titles. $Pearson(A, B) = \frac{ad - bc}{\sqrt{(a+b)(a+c)(d+b)(d+c)}}$ where $d = \#(A^c \cap B^c)$ is the number of other topics (topics which are not in the two articles).

2.2 Features for Community and Citation Graphs

The logic behind these features is that an author is likely to publish many papers with authors that belong to the same community. So, if two articles share some of the coauthors it is more likely that they were written by the same person. Also, we can observe an indirect connection. For instance, two papers, i and j, do not share directly coauthors, but some coauthors of both paper i and j have participated in the writing of another paper k. There is an indirect link between papers i and j through the paper k that raises the likelihood that papers i and j belong to the same author. Likewise, is two papers share directly or indirectly citations, the likelihood that are by the same author increases.

Dissimilarity Based on Co-authorships (D_a): The article k is neighbor to the couple of articles (i, j) if k shares at least one co-author, different from the focus name, with each of them. The set of articles in the neighborhood of (i, j) will be noted $N(i, j)$.

- *Proportion of Shared Co-authors*: $\pi_{ij} = \frac{n_{ij}}{\max(n_i, n_j)}$. n_{ij} is the number of common authors between i and j; The focus name is excluded. Likewise, n_i and n_j are the numbers of co-authors of i and j without the focus name.
- *The Dissimilarity*: $D_a(i, j) = (1 - \pi_{ij}) \prod_{k \in N(i,j)} (1 - \pi_{ik}\pi_{kj})$.

Dissimilarity Based on Citations (D_c)

- *Proportion of Shared Citations*: $\mu_{ij} = \frac{n_{ij}}{\max(n_i, n_j)}$ where n_{ij} is the number of common citations between i and j; n_i and n_j are the numbers of citations resp. of i and j.
- *Dissimilarity*: $D_c(i, j) = (1 - \mu_{ij}) \prod_{k \in N(i,j)} (1 - \mu_{ik}\mu_{kj})$ where $N(i, j)$ is the set of articles that share citations with (i, j).

Dissimilarity Based on Community: $D_{co}(i, j) = (1 - \frac{|N_i \cap N_j|}{\min(|N_i|, |N_j|)})$ where N_i set of of articles that are neighbor of i (share at least one common co-author) and $|N_i|$ is the number of neighbors of i.

2.3   Classifier

The goal is to build two procedures that respectively (1) identify if two papers, i and j, have been written by the same author and (2) within a given focus name, separate the papers according to the real author.

Step 1: Prediction of the Uniqueness of the Author: We need to discrimin authorship by context using the features introduced earlier. For the learning phase, a set of couples of papers have been disambiguated manually. This gives us a learning data set of 58334 observations (couples of articles) related to many focus names. Each observation is a vector formed by the seven features $(D_{KwJ}, D_{KwP}, D_{tJ}, D_{tP}, D_a, D_c, D_{co})$ explained above and the membership class Y that takes the value 1 if the articles are from the same author and 0 if not. Several supervised machine learning techniques were used to generate the model: logistic regression, decision trees, and random forests.

Step 2: Community in Focus Name (Entity Resolution): After modeling author similarities between papers, discovered matches can be imagined as a set of indirect links. The chosen author will be the most commonly appearing author among the list of papers a particular article is linked to

Table 1. Table 1

Decision Tree		
Reference	0	1
Prediction 0	25794	3471
Prediction 1	1888	27181
Logistic Regression		
Reference	0	1
Prediction 0	23658	3963
Prediction 1	4024	26689
Random Forest		
Reference	0	1
Prediction 0	26085	2456
Prediction 1	1597	28196

(a) Confudion matrices

Method	Accuracy	Confidence Interval	
		Low. CI	Upp. CI
Decision Tree	0.908	0.905	0.910
Logistic Reg.	0.863	0.860	0.865
Random Forest	0.930	0.928	0.932

(b) Accuracy for leave-one-out article

Method	Accuracy	Confidence Interval	
		Low. CI	Upp. CI
Decision Tree	0.794	0.789	0.798
Logistic Reg.	0.792	0.787	0.797
Random Forest	0.930	0.928	0.932

(c) Accuracy for leave-one-out focus name

3   Evaluation

The evaluation procedure of each classifier was the leave-one-out schema. However, two variants were used in order to evaluate the models. In the first, one article was left out, and in the second, one focus name was left out at a time.

All the computations and results presented were done using R software, on a dataset containing 58334 observations. We present the accuracy of the models in terms of correct and incorrect predictions on the testing set. Here, the goal is to predict if two articles are written by the same author or not. Having this, for each focus name, we carry out a bi-clustering that leads us to a set of blocks, each one representing a real author in the considered focus name (Author identification).

The confusion matrices for all three models, when leave-one-out validation method is used, are given in Table 1a. The different models' accuracy and confidence intervals are presented in Table 1b. As we can observe, random forest method yields the best result in accuracy (93 %).

When we observe the accuracy and confidence intervals obtained by the leave-one-out focus name approach (results presented in Table 1c), we can notice a decrease in the accuracy, which was expected. What is interesting to note is that both decision tree and random forest had an accuracy decreased by 11.4 % and 11.6 %, respectively, while logistic regression had an accuracy decrease of only 7.1 %. Random forest still gives the highest value for accuracy (81.4 %). Overall, using leave-one-out method based on articles, results were good in terms of entity resolution. This implies that if the model is used on the existing database, if the focus name of the articles to be identified are already in the database, there is a good chance of correct author identification.

4   Conclusion

In the present research study, we have proposed a method in order to tackle the author disambiguation problem which is based on building similarity measures between scientific articles. Then, by the mean of these features and classification algorithms, we are able to train models which can determine if a couple of articles was written or not by the same individual. We have used the concept of focus names in order to reduce the number of articles to be compared and also with the aim of improving the computational cost of the entire task. In addition, we also have introduced a scheme which is able to determine the real author in a focus name. The proposed similarity measure features are keywords, title and context based.

In summary, our proposal has a more flexible representation for author disambiguation models and described features estimation methods tailored for this novel approach. Further, empirical analysis of these methods on real-world datasets were performed, and the results of the experiments support our claims that error-driven, rank-based training of the new representation can improve accuracy. In future work, we plan to investigate more sophisticated features and parameters estimation that helps to improve the computing time and the accuracy of some algorithms and as well as adding a features selector in order to improving the performance.

References

1. Han, H., Giles, L., Zha, H., Li, C., Tsioutsiouliklis, K.: Two supervised learning approaches for name disambiguation in author citations. In: Proceedings of the 2004 Joint ACM/IEEE Conference on Digital Libraries, IEEE (2004)
2. Huynh, T., Hoang, K., Do, T., Huynh, D.: Vietnamese author name disambiguation for integrating publications from heterogeneous sources. In: Selamat, A., Nguyen, N.T., Haron, H. (eds.) ACIIDS 2013, Part I. LNCS, vol. 7802, pp. 226–235. Springer, Heidelberg (2013)
3. Levin, M., Krawczyk, S., Bethard, S., Jurafsky, D.: Citation-based bootstrapping for large-scale author disambiguation. J. Am. Soc. Inf. Sci. Technol. **63**(5), 1030–1047 (2012)
4. Levin, F.H., Heuser, C.A.: Evaluating the use of social networks in author name disambiguation in digital libraries. J. Inf. Data Manage. **1.2**, 183 (2010)
5. Newman, M.E.J.: Coauthorship networks and patterns of scientific collaboration. In: Proceedings of the National Academy of Sciences, **101**, 1 (2004)
6. McRae-Spencer, D.M., Shadbolt, NR.: Also by the same author: AKTiveAuthor, a citation graph approach to name disambiguation. In: Proceedings of the 6th ACM/IEEE-CS joint conference on Digital libraries, ACM (2006)
7. Fan, X., Wang, J., Pu, X., Zhou, L., Lv, B.: On graph-based name disambiguation. J. Data Inf. Qual. (JDIQ) **2.2**, 10 (2011)
8. Francisco, R.D., Fabien, R., Adrian, T., Djamel, A.Z.: Data Mining based approach for authors disambiguation in large citation networks. In: 60th ISI World Statistics Congress (ISI2015). Rio, 26–31, July 2015
9. Gruenheid, A., Xin, L.D., Srivastava, D.: Incremental record linkage. In: International Conference On Very Large Data Base (VLDB2014)
10. Blei, D.M., Jordan, M.I., Ng, A.Y.: Latent dirichlet allocation. J. Mach. Learn. Res. **3**, 993–1022 (2003)
11. Carpenter, B.: Integrating out multinomial parameters in latent Dirichlet allocation and naive Bayes for collapsed Gibbs sampling. Rapport Technique, 4 (2010)

12. Jaccard, P.: The distribution of the flora in the alpine zone. 1. New Phytol **11**(2), 37–50 (1912)
13. Lerman, I.C.: Classification et analyse ordinale des données. Dunod, ch. Indice de Similarité et préordonance associée (1981)

Author Index

Printed in the United States
By Bookmasters